Introducing Geographic Information Systems with ArcGIS®

About the Cover

The cover shows six images of the same geographic area, demonstrating various GIS data formats, depicting both natural and human-made features. The scene is a river flowing through a canyon. North of the river is a water filtration plant. The top scene is a TIN (Triangulated Irregular Network) indicating the elevation of the surface. Across the bottom of the cover, left to right, the first two images are portions of (a) a DRG (Digital Raster Graphics) file digitized from a US Geological Survey 7.5 minute quadrangle and (b) a DEM (Digital Elevation Model). The last two images, left to right are (d) a raster-based (cell-based, grid-based) depiction of different types of land cover and (e) a DOQ (Digital Ortho Quadrangle), which is an aerial photograph that has been rectified so it can be used as a map. In the center at the bottom is (c) a three-dimensional view in which ArcGIS software was used to "drape" a DOQ over an elevation model. The red dots along the river depict points collected by a GPS (Global Positioning System) receiver on a boat moving along the river.

Introducing Geographic Information Systems with ArcGIS®
Third Edition

A Workbook Approach to Learning GIS

Michael Kennedy

University of Kentucky

Cover image: Courtesy of Michael Kennedy

Cover design: John Wiley & Sons, Inc.

This book is printed on acid-free paper.

Published by John Wiley & Sons, Inc., Hoboken, New Jersey
Published simultaneously in Canada

For general information about our other products and services, please contact our Customer Care
Department within the United States at (800) 762-2974, outside the United States at (317) 572-3993 or
fax (317) 572-4002.

Wiley publishes in a variety of print and electronic formats and by print-on-demand. Some material
included with standard print versions of this book may not be included in e-books or in print-on-demand.
If this book refers to media such as a CD or DVD that is not included in the version you purchased, you may
download this material at http://booksupport.wiley.com. For more information about Wiley products,
visit www.wiley.com.

ISBN 978-1-118-15980-4; ISBN 978-1-118-33034-0 (ebk); ISBN 978-1-118-33103-3 (ebk);
ISBN 978-1-118-33318-1 (ebk); ISBN 978-1-118-51050-6 (ebk); ISBN 978-1-118-51056-8 (ebk)

Printed in the United States of America

10 9 8 7 6 5 4 3 2 1

To the memory of Evan Kennedy,
who had every gift but that of years

Contents

Contents

Contents

Contents

Contents

Contents

Contents

Contents

Contents

Contents

Contents

Contents

Contents

Contents

Contents

Contents

Foreword

by Jack Dangermond

Introducing Geographic Information Systems with ArcGIS offers a unique approach to GIS instruction. In it, Michael Kennedy re-creates his time-tested methods of teaching GIS in the classroom in a step-by-step guidebook to GIS. Students on a journey to learn GIS with Professor Kennedy may feel like he is taking the journey with them, offering them his sage advice each step of the way. Professor Kennedy cares deeply for his students, and the detail of this care and years of teaching GIS come through in this book. In it, he walks students through the multitude of questions that come up daily in the classroom. His goal is to help students understand GIS concepts and learn GIS skills. It takes a master teacher to map GIS knowledge, making it clear to students and enabling them to gain confidence in their growing skills.

Once GIS students have learned the basics, the next step is to learn how to analyze spatial data and identify problems and create solutions. Learning to analyze spatial data moves students beyond exploration, beyond locating places on maps, and helps them create maps that guide better decisions.

All of us learn GIS skills in different ways. Some people are visual learners, some are auditory learners, and some need a hands-on approach. As the learning styles of students in general vary, so do the learning needs of students of GIS. Some students will need classroom study, with conversations and time to process information about GIS concepts, spatial data, geodatabases, map projections, attribute tables, feature classes, datasets, and building maps, while others need only a guidebook with clear graphic illustrations. So, a variety of approaches to teaching GIS will help ensure that the increasing number of students worldwide have opportunities to gain GIS skills in ways that best suit their needs.

GIS is becoming part and parcel of the daily work lives of most people in many fields, from architects to zoologists, from academia to the business world, from city planning to national and international spatial data portals. Teachers are now taking on the essential task of opening the door for students to learn GIS. In *Introducing Geographic Information Systems with ArcGIS*, Professor Kennedy opens such a doorway for students to learn the skills basic to understanding GIS and to prepare students to make our communities better places.

Preface[1]

It turns out to be hard, for me anyway, to write the preface for a third edition. As I tried to compose this I put a lot of electrons in the recycle bin. Most of what I have to say was said in the prefaces to previous editions. And who wants to want to wade through eleven pages of those in addition to this one!

So what I will do is just to tell you about the new material in the text and then just abstract and reference earlier information and ideas. I'm eliminating the Preface to the Second Edition. If you haven't used the book before, you probably should read the Preface to the First edition, included after this one.

First, of course, is that the material is oriented to ArcGIS Desktop versions 10.0 and 10.1. Where there are differences between these two, and there are several, I have usually pointed them out. However, those using 10.0 will occasionally have to adapt the Step-by-Step instructions, which favor version 10.1. I recommend using 10.1 if it is available and you are familiar with it. (To indicate the extent of the changes, we can start with the fact that the functionality level names have changed from those in version 10.0 and before. In 10.1 ArcView is Basic, ArcEditor is Standard, and ArcInfo is Advanced.)

The CD-ROM used in earlier editions has been replaced by a DVD, because the data sets are more extensive and all the figures in the book are available.

Since the first edition, sections and exercises have been added on the topics of:

❏ Publishing maps on the Internet, using ArcGIS.com.

❏ Using the Esri online data service to add basemaps to the student's map.

❏ The terrain data structure, made possible by the emergence of LIDAR as a remarkable method of very dense data collection, is covered both in theory and by exercise.

❏ Layer packages – a welcome invention which facilitates the transfer of feature classes of all formats from one computer to another, without worries like relative path names and separate data transfers.

❏ Since this book is primarily aimed at preparing professionals for using GIS to do analysis and synthesis (topics separate from display and mapmaking, which, for completeness, is covered in considerable detail in Chapter 3), topology plays an important role. A number of exercises, therefore,

---

[1] If this text is used in a classroom/laboratory setting, this preface is for the instructor and may be skipped by students. If you are using the book to learn GIS on your own you should probably read it.

emphasize the use of the topology capabilities of geodatabases, which is considerably different from those of coverages and completely absent from the shapefile format.

Changes from the Previous Editions

❏ Use of and information about coverages has been demoted to an optional exercise on converting an Esri coverage to a geodatabase.[2] All references to ArcInfo Workstation have been removed, since its functions have been taken over by ArcToolbox, and Workstation has been "depreciated" (although many of us "appreciated" it a lot in times past!).

❏ A couple of the more arduous exercises (making feature classes by key entry and digitizing) have been improved so that the student or reader understands the concepts without having to experience the all-too-real tedium of data entry. Other exercises have input data provided for them on the DVD to cut down on digitizing and typing.

❏ More flexibility has been built into the text. I suggest exercises that might be omitted in the interest of compressing the learning of essential GIS material into a shorter time span.[3]

❏ All the figures in the book are reproduced, many in color, on the DVD that accompanies the text. At the beginning of each Step-by-Step section, I encourage students to open both the Color Figures file and their Fast Facts File (both to access reference information and to add new material). The Fast Facts File, into which the students write the information they consider relevant, thus making their own reference guides to ArcGIS Desktop, is emphasized. The past several years have convinced me that the Fast Facts File is an important tool for long-term learning of the material – as the software grows in facility and complexity.

❏ In previous editions, students were asked to write, in their textbook, the names of some menus and tabs. The third edition has no blanks for this. Instead students are encouraged to record the name of tabs and menu items in their Fast Facts Files – and to think about what each item might mean. Since there is no way to cover all of ArcGIS (except perhaps in an intensive, year-long, full time course), having a list that at least hints at the capabilities of the software that are not covered in the text (represented by these tabs and menus) is beneficial. Further, the lists in the student's Fast Facts File can be updated as ArcGIS evolves in the years to come.

The purpose and structure of the book remains essentially the same. (Please see the Preface to the First edition). Chapters are divided into (a) Overviews (a top-down look at GIS theory and other relevant information) and (b) Step-by-Step (sequential) exercises. All the data needed for the exercises is provided on the included DVD. (The DVD does not include ArcGIS software. I assume that the several mechanisms that Esri provides (e.g., site licenses, student one-year licenses, and so on) for access to ArcGIS will be in place.)

This third edition is meant to educate a wider group than the first edition. The subtitle—A Workbook Approach to Learning GIS—is intended to convey that the book has been specifically revamped for

---

[2] Some coverages are still used as data as part of exercises, but primarily to let the student know that such objects still exist and that a lot of data still resides in them.

[3] I want to emphasize, however, that everyone should take the time to read the Afterword by Dr. Michael Goodchild on GIScience at the end of the book, which will be critical for the effective use of GIS in the coming years.

community college and technical institute courses, where almost all students can become proficient with many ArcGIS software abilities in a single semester.

The combination theory-workbook approach is designed to bring the reader from GIS neophyte to well-informed GIS user—from both a general knowledge and practical viewpoint—in a single semester or, used by an individual outside of class, in about 60 hours of self-study.

It is appropriate to repeat some ideas and warnings from the First Edition Preface:

❏ Do not use any of the sample databases on the DVD for anything other than tutorial purposes. Many of the data sets are not current. Many have been modified for instructional purposes. Some of it is totally bogus.

❏ Exercise 5–8 is a cooperative exercise for eight to twenty-four students. Preparation and management on the part of the instructor is a really good idea. Information on the book's companion website at www.wiley.com/go/kennedygis can help.

❏ If you, as an instructor, are quite sure that your students will not need more than the most basic knowledge about coverages and shapefiles, you can have them skip considerable portions of Chapter 4. You should perhaps read those sections and, if needed, supplement the student's knowledge of the concepts that apply to geodatabases.

❏ If you serve ArcGIS, or even just its license manager, over a network, you should thoroughly test the process. Also, in Chapter 8, the unsupported CellTool is used. Students may not be able to install it, so someone from network services may have to be involved.

❏ Students learn the software at their own pace, pretty much regardless of what the instructor does. They learn by doing, and paying attention to and recording what they are doing. As the text proceeds, the sophistication required to operate the software increases. For students who aren't paying attention, the exercises will get harder and harder because it is expected that they will learn (or be able to quickly find in their Fast Facts Files (see next paragraph) how to perform operations that they have performed before. Careful explanation of basic procedures (e.g., finding the properties of something), which is extensive at the beginning of the book, is reduced gradually but considerably as the text unfolds. Warn your students about this: The handholding diminishes as the chapter numbers increase.

❏ Students are asked to develop a Fast Facts File in which they record what it is they have learned about the software. This is a computer file that they keep open during their work sessions, both for adding new material and ascertaining how to do a particular procedure that they have used previously but cannot remember. They periodically revise and augment this file. Then, at the end of the course, they have their own reference manual for the software. I have used this technique for some years now, and it pays dividends. Some students who have graduated and now work in the GIS field tell me they take their Fast Facts File with them and maintain it in their new positions. One failure of other workbooks and web-based courses is that, while students can go through the exercises and even pass a test at the end, they simply cannot operate the software when handed a new exercise. Now with twelve-plus years of teaching GIS with the Overview-Step method behind me, insisting that students make a Fast Facts File to provide themselves a guide through the very complex GIS software, I'm convinced that the not-always-popular-with-the-students Fast Facts File is more than worth the trouble.

❏ One way this book has been used is in a two-semester course sequence for advanced students with an intensive theory text (e.g., Longley, Goodchild, Maguire, Rhind) using *Introducing Geographic Information Systems with ArcGIS—A Workbook Approach* providing the needed practical experience.

I don't know if it's me, the students of today, the multi-media culture, or something else, but I find the traditional lecture to be less and less useful. Lately I have confined my lectures, which I keep short, to those topics that seem to give some students trouble conceptually. My teaching environment has not been an easy one; it has usually involved a mixture of civil engineering graduate students, geography sophomores, and students from other departments (29 such departments as of this writing.) Given the varied computer experience and maturity of students in such a diverse group, I find that an environment in which students work from the text at their own pace, with reasonable deadlines and the opportunity to ask questions individually, seems to work best.

Instructors who want answers to exercises: please write to me on school, college, or university letterhead and just ask.

Michael Kennedy
Department of Geography
Patterson Office Tower 817
University of Kentucky
Lexington, KY 40506-0027

Or obtain the answers from the Instructor Companion website at www.wiley.com/go/kennedygis

For those who want to provide comments, criticisms, corrections (many thanks), or complaints: email me at Michael.Kennedy@uky.edu.

Acknowledgments for the Third Edition

For both the second and third editions, I must foremost thank my son, Alexander Kennedy, who edited the manuscript and worked all the exercises twice, using ArcGIS Desktop 10.0. No less a contribution was made by my daughter, Heather Kennedy, who also contributed to the editing and who developed the images on the cover from figures in the book itself.

I want again to thank

Jack Dangermond, President of Esri, for his encouragement and for writing the Foreword.

Michael F. Goodchild, Professor Emeritus, Department of Geography, University of California, Santa Barbara, for contributing the Afterword on GIScience.

Clint Brown, Director of Software Products for Esri, for his quick decision regarding an administrative problem that occurred during the development of the text, and Ashley Pengelly and AjmalYourish for their on-the-spot help for solving said problem.

Ken Bates, extension specialist with Kentucky State University and unquestioned GIS expert, for his willingness to be the answer man for complex ArcGIS problems and as the source of the Internet map publishing section.

Demetrio Zourarakis of the Division of Geographic Information, Kentucky Commonwealth Office of Technology for yet another dataset Kentucky-wide land cover data.

Joseph Kerski, Esri Education Manager, for his continuing support and embarrassingly complementary tweet regarding the book,

Damian and Meena Spangrud, and Mike Hogan, who helped with respect to the 10.0 and 10.1 Esri Beta programs.

Folks with the Esri Support and Customer Service teams:

Allan R, Archana G , Barbara S, Boro O, Cassandra L, Charles F, Don G, Harshal S, Hashad D, Joy S, Kailai , Michelle B, Prasanta B, Radaha K, Stacey M, Sunil P, Tarun J, Timothy H, Vijay P, and a couple of others whose names have escaped me.

Gretchen Gallegos, with the Lawrence Livermore National Laboratory, who fought her way through an early version of the third edition, both to help me and to become more proficient with GIS.

The Lexington Herald Leader for the photograph of the water filtration facility on the Kentucky River.

And finally, Bob Argentieri and Dan Magers—my editors at John Wiley and Sons—who had to put up with a number issues beyond the normal problems in dealing with authors, and were assisted therein by Bob Hilbert, who managed production, and David Riedy who put the cover together with images from the book, after several iterations and challenges created by yours truly.

Acknowledgments for the Second Edition

Great thanks are due to Mr. Mike Richie, Owner and President of Photo Science (which is among the most comprehensive aerial remote sensing firms in the United States, www.photoscience.com) for the special effort in providing the LIDAR data for the second edition of the book.

The author is indebted (for help with the second edition) to Ms. Ryan Bowe, who read the text and worked all the exercises twice. She is a remarkably good editor, and her detailed knowledge of ArcGIS was invaluable.

Much appreciation is owed to staff and teaching assistants at the University of Kentucky:

Chris Blackden, Sarah McCormack, Amanda Corder, Priyanka Ghosh, and Tim Guenther

Thanks also go to several people who taught with previous editions of the text in their classes and provided feedback: Brad Baldwin, Lee De Cola, James W. Craine, Charla Gaskins, Richard A. Lent, Mark MacKenzie, Jack Mills, Emmanuel U. Nzewi, Thomas Orf, Jim Pimpernell, Brian Scully, Anne Stearns, Fred Sunderman, Raymond Tubby, and Christopher Urban.

Preface

to the First Edition[1]

The purpose of Introducing Geographic Information Systems with ArcGIS is threefold.

1. To acquaint the reader with the central concepts of GIS and with those topics that are required to understand spatial information analysis.

2. To provide the person who works the exercises either (a) a considerable ability to operate important tools in the ArcGIS software or (b) a demonstration of other capabilities of the software.

3. To lay a basis for the reader to go on to the advanced study of GIS or to the study of the newly emerging field of GIScience, which might be described as the scientific examination of the technology of GIS and the fundamental questions raised by GIS.

Introducing Geographic Information Systems with ArcGIS is meant to serve as a text book for a standard one-semester course. It is suitable for a university, college, technical school, or advanced high school course, meeting for three hours per week. Between two and five additional hours per week are required for laboratory work, depending on the capabilities and computer experience of the students. The text may also be used for self-study.

The book, and any course taught from it, depend on having ESRI's ArcGIS Desktop and Workstation software, version 9.0, 9.1, or higher, available. The assumption is that the students will have access to full the ArcInfo package offered to colleges and universities under the generous site license agreement that ESRI offers to educational institutions. For more information about this program, point your browser at: http://www.esri.com/industries/university/education/faqs.html. However, if ArcInfo is not available, many of the exercises can be done with the ArcView level of ArcGIS, available to students with a free, one-year license.

While the author is impressed with the ArcGIS software (and with the aims of ESRI of being a force for conservation, preservation, and sustainable development worldwide), this book is not meant as a promotional text for ESRI. Like all large software packages, ArcGIS has its shortcomings, limitations, and

---

[1] If this text is used in a classroom/laboratory setting, this preface is for the instructor and may be skipped by students. If you are using the book to learn GIS on your own you should probably read it.

bugs. When these arise in the process of working through the exercises, they are candidly pointed out to the reader. All bugs have been reported to ESRI, and most have been repaired or are scheduled for repair. By the way, the ESRI support staff is excellent responsive and friendly.

The function of GIS software is to make a computer think it's a map—a map with characteristics that let the user analyze it, display its elements in a variety of ways, and use it for decision making. This text is oriented more toward preparing the student for doing analysis with GIS, rather than display, mapping, or standard data processing.

Contents of Introducing Geographic Information Systems with ArcGIS

- ❏ Part I: Basic Concepts of GIS

    - ❏ Chapter 1: Some Concepts that Underpin GIS (and introduction to ArcCatalog)

    - ❏ Chapter 2: Characteristics and Examples of Spatial Data (and introduction to ArcMap)

    - ❏ Chapter 3: Products of a GIS: Maps and Other Information

    - ❏ Chapter 4: Structures for Storing Geographic Data (and introduction to ArcToolbox and Workstation)

    - ❏ Chapter 5: Geographic and Attribute Data: Selection, Input, and Editing (and introduction to ArcScene and ArcGlobe)

- ❏ Part II: Spatial Analysis and Synthesis with GIS

    - ❏ Chapter 6: Analysis of GIS Data by Simple Examination

    - ❏ Chapter 7: Creating Spatial Data Sets Based on Proximity, Overlay, and Attributes

    - ❏ Chapter 8: Spatial Analysis Based on Raster Data Processing (and introduction to Spatial Analyst)

    - ❏ Chapter 9: Other Dimensions, Other Tools, Other Solutions (and introductions to 3-D Analyst, Historical Data, Address Geocoding, Network Analyst, and Linear Referencing)

In my view, the pedagogical theme of a first course should be breadth, with depth in vital areas. The text covers virtually all the general GIS capability that ArcGIS has to offer. Vector and raster storage, analysis, and synthesis are, of course, discussed extensively, with many examples and exercises for the student. Other areas receive less attention, such as 3-D GIS, time and GIS, network analysis (path finding and allocation), surface creation, spatial analysis, statistical and numerical analysis, model builder, GIS & GPS, and so on. In some later instances, the exercises are primarily demonstrations of the capabilities of the ESRI software, but, in my opinion, a student in a first course needs to get at least a glimpse of almost all of what GIS can do. Omitted from the text is most of customization, programming, and the more esoteric capabilities of geodatabases, which I believe belong in a second course. Also not included is GIS on the Internet and the issues related to large, enterprise implementations of GIS. To mention it again the thrust of the text is to lay a foundation from which the reader can move toward doing analysis and synthesis with GIS.

The emphasis in terms of data structure is on geodatabases. However, extensive use is made of shapefiles and coverages, since most existing GIS data sets are in these formats. The student will become comfortable with switching and converting among the various formats. Another reason for using all three formats is that, at this stage of ArcGIS development, there are operations that can be done with coverages that cannot be performed with geodatabases.

ArcMap, ArcCatalog, ArcToolbox, ArcScene, and ArcGlobe are all explored in considerable detail. ArcInfo Workstation is introduced. Enough of command-line ArcInfo Workstation is used to make the student aware of its existence and its capability to perform operations that are cumbersome or impossible with the point and click software. This is a book that creates knowledge for the student that is realistic and at least touches on virtually all the ArcGIS capabilities and products.

In the four years the text has been under development, most of the exercises in the book have been performed by scores of students. All of the exercises have been tested and they work, both from a technical and pedagogical standpoint.

In terms of time required to do the exercises, most students will require:

Chapter 1—3:00 to 5:00 hours

Chapter 2—3:45 to 5:45 hours

Chapter 3—4:30 to 6:30 hours

Chapter 4—3:30 to 5:30 hours

Chapter 5—4:45 to 6:45 hours

Chapter 6—4:00 to 6:00 hours

Chapter 7—5:15 to 7:15 hours

Chapter 8—5:00 to 7:00 hours

Chapter 9—4:45 to 6:45 hours

Theory and Practice

Of the myriad of GIS textbooks available, some are long on theory but don't train the student, while the rest are pretty much manuals on how to use software, but don't promote an understanding of what lies behind the mechanics. So frequently GIS is taught either with texts that teach only theory and leave it to the instructor to select software and data to illustrate points or taught with manuals and demonstrations.

The book is unusual, if not unique, in that it serves both as a general introduction to GIS (serving an education function) and a manual on ArcGIS software (serving a training function). This is accomplished by dividing each chapter into an

❑   Overview section, and a

❑   Step-by-Step section

The Overview section is descriptive. It is a top-down discussion of theory and ideas relating to GIS.

The Step-by-Step section is prescriptive. It operates in a sequential fashion—do this, then this, then this. Here the student learns about and practices ArcGIS. There are more than 60 exercises in the book, not counting the 9 review exercises. Almost 60 percent of the book consists of step-by-step instructions on how to use ArcGIS software.

All the data sets for the exercises are on the CD-ROM that accompanies the book.

Teaching with This Book

The contents of the following folders on the CD must be available for downloading by students:

❏ IGIS-Arc—the primary source of data sets for the exercises

❏ IGIS-Arc_AUX—a source for datasets occasionally needed for exercises

❏ IGIS_with_ArcGIS_FastFactsFile_Checklists—a combination chapter summary and set of Fast Facts File prompts

❏ IGIS_with_ArcGIS_Selected_Figures—full-color versions of some figures in the text that suffer from black-and-white reproduction

If you are an instructor, you should consider copying the four folders above from the CD-ROM to a location on a network where the students can access their contents.

Exercises are roughly put into categories of length or difficulty, with such notes as "Warm-up" (least effort), "Project" (greater effort), and "Major Projects" (most effort).

Some warnings:

Do not use any of the sample databases on the CD-ROM for anything other than tutorial purposes. Most of the data is old. Much of it has been modified for instructional purposes.

For students who aren't paying attention, the exercises will get harder and harder because it is expected that they will learn (or be able to quickly find) how to perform operations that they have performed before. The "hand holding" diminishes as the chapter numbers increase.

Exercise 5–8 is a cooperative exercise for eight to twenty-four students. Preparation and management on the part of the instructor is a really good idea.

If you, as an instructor, are quite sure that your students will not need more than the most basic knowledge about coverages and shapefiles, you can have them skip considerable portions of Chapter 4. You should read the sections on coverages yourself and, perhaps in lecture sessions, supplement the student's knowledge of the coverage concepts that apply to geodatabases.

If you serve ArcGIS, or even just its license manager, over a network, you should thoroughly test the process. Also, in Chapter 8, the unsupported CellTool is used. Students may not be able to install it, so someone from network services will have to be involved.

More Resources for the Instructor

If you are an instructor who is using this text, you are encouraged to register on the website www.wiley
.com/college/kennedy. There you will find advice on how to use the book to its fullest potential. Included
there are answers to the questions posed in the text, sample assignments with blanks for the students
to complete, test data for some assignments, and suggestions of how to use the text—avoiding some
pitfalls that lurk, especially when the datasets are served across a network. The Instructor's Guide there
can be a valuable resource for those teaching with this text. Also look at the folder IGIS_with_ArcGIS_
Instructor's_Guide on the CD-ROM.

Concepts, Devices, and Techniques that Underlie
the Philosophy of the Book

How can one textbook touch on almost all of GIS when it takes thousands of pages of manuals to do this?
Two ways:

❏   There are few figures, and, compared to the standard computer manual, there are few screen shots.
    When a student follows the instructions, he or she sees the proper screens. When a figure can be
    better understood by the use of color, the figure is available on the CD-ROM in the folder IGIS_with_
    ArcGIS_Selected_Figures. Such figures are designated in the text reference with three asterisks. For
    example, "See Figure 8-4***."

❏   As the student progresses through the later chapters, the exercises do not contain detailed instruc-
    tions. The students are expected to be able to do steps that were explained in detail earlier. For exam-
    ple, in early chapters, detailed instructions are given for finding or changing a property of a data set
    or data frame. In later chapters, the students will simply be told to take that action. When students
    can't either remember or find out how to perform an action that has been previously detailed, teach-
    ers should take it as a clue that the students are simply going through the motions of executing the
    software tools and that learning is not really taking place.

I believe that students learn best by doing—while observing and recording what it is they are doing.
Students are asked to develop a Fast Facts File in which they record what it is they have learned about
the software. This is a computer file that they keep open during their work sessions, both for adding new
material and ascertaining how to do a particular procedure that they have used previously but cannot
remember. They periodically revise and augment this file. Then, at the end of the course, they have
their own reference manual for the software. I have used this technique for some years now, and it pays
dividends. Some students who have graduated and now work in the GIS field tell me they take their Fast
Facts File with them and maintain it in their new positions. One failure of other workbooks and web-
based courses is that, while students can go through the exercises and even pass a test at the end, they
simply cannot operate the software when handed a new exercise. Now with ten-plus years of teaching
GIS with the Overview-Step method behind me, insisting that students make a Fast Facts File to provide
themselves a guide through the very complex GIS software, I'm convinced that the not-always-popular-
with-the-students Fast Facts File is more than worth the trouble.

The exercise material is project oriented; students learn the software as needed for the particular project
at hand. So rather than learning everything about labeling features at one time or everything about
selecting, the students learn as they complete projects and record what has been learned in their Fast

Facts Files, which are later reorganized. At the risk of losing adoptions and sales, please let me candidly point out that this textbook does not serve very well as reference material. The idea behind the book is to make things click in the students' brains, to promote comprehension of concepts, not to serve as a reference guide to the software. However, the diligent students—indeed even those who follow the instructions—will emerge from the course with their own reference guides, done in a style suitable for each student the Fast Facts File. Some students resist creating the file, so I make it count for 5 percent of their grade. Further, the Fast Facts File will be a reference document that the learners can maintain and upgrade in future months and years. A student of mine of a decade ago came to my class to give a guest lecture. He was in charge of the GIS program of a state unit. He brought his Fast Facts File with him to show to the class. Over the years, he had updated it many times.

The text is workbook-like in that there are blanks in the text which the students are asked to complete, showing that they have performed and comprehended a particular section. This also serves as a mechanism for letting the instructors know how students are progressing. The Web site www.wiley.com /college/kennedy contains forms with these blanks, in context, so an instructor can copy and paste the material into assignment sheets. Student progress can be monitored using these assignments.

The text is set up so students can work at their own pace, respecting different learning styles and speeds of the students. For example, some students create entries in their Fast Facts Files with each step. Others make two passes through the material.

In a few places in the text, students are asked to record the names of menu choices or tabs in windows. Of course this information could have been printed for them, but having them write it in reinforces the words and the concepts behind them in the students' minds. It is also a modest protection against software changes (e.g., addition of menu or tab items).

The last exercise in each chapter is a checklist that can serve the students in development of the Fast Facts File. The students are given prompts that they fill in. The prompts appear in the text and also on the CD-ROM in the folder IGIS_with_ArcGIS_FastFactsFile_Checklists so they are available in machine readable form to the students. This allows students to copy the prompts into their Fast Facts Files and complete them.

The book simulates a teacher sometimes a lecturer, standing in front of students, imparting information or giving directions. More often the instructor is a colleague, sitting beside the student, making suggestions, prompting, and, occasionally, making mistakes that he or she and the student rectify. One might describe the tone of the book as conversational. I believe the most important thing, after correctness, is engaging the student. I believe economy in writing is important. But sometimes additional words can set a tone. I actually use several tones in the text to provide variety. I change pace. I change style. I change attitude. I change the level of formality.

The writing style, for the most part, is informal—to convey the idea that the author is closely involved with the student, guiding her or him. Humor is used, but sparingly. Irony is used, but sparingly.

The ArcGIS software is so complex that there is no way to explore it "depth first." We must look at an overview. The book attempts to teach, or at least demonstrate, the major capabilities of ArcGIS. As you can tell from the weight of ESRI manuals, compared with the size of this text (which also serves as a general GIS theory text), I could hardly cover even a large portion in detail. However, the student will come away with considerable facility with the software and will know how to find and use additional capabilities.

Finally, I believe it is important to emphasize that computer is not a black box. An educated GIS professional should have some understanding of what makes a computer tick. So there is some general material on computers and representation of information, especially as they impact answers from a GIS.

Acknowledgments

This book was something of a family affair. My daughter Heather Kennedy provided help with the 3-D material[2]. My son Alex Kennedy carefully worked all the exercises in all chapters, making corrections and providing insightful observations. He also helped collect some of the GPS data.

Jack Dangermond, for his encouragement in general and writing the Foreword in particular, and ESRI for allowing use of numerous datasets and figures. In fact, this text was originally conceived of as a new edition of *Understanding GIS—The Arc/Info Method (UGIS-tiam)* reworked for the ArcGIS point-and-click version of the software, beginning with ArcGIS 8.x. It has clearly grown way beyond that first idea, with the addition of textual matter dealing with GIS itself, discussion of other capabilities of the software besides vector-based GIS site selection (such as a discussion of Spatial Analyst, the addition of material on several of the other important extensions to the software, and the major emphasis on geodatabases). Teachers who in the past used *UGIS-tiam* (last published almost a decade ago) will, however, recognize the site selection problem of that text (a laboratory to do research in "Aquaculture") as the Wildcat Boat Testing Facility of this text, albeit highly modified.

I am indebted to Dr. Michael Goodchild for writing the Afterword. As mentioned before, the aim of this textbook is to provide a general introduction to GIS and prepare students to use ArcGIS primarily for analysis. Some of those students will want to go on to study GIScience, and I commend Dr. Goodchild's Afterword to them.

Many colleagues and friends contributed to bringing this book to fruition. In particular, I want to thank:

Christian Harder, founding publisher, *ESRI Press*, for suggesting and encouraging the development of the text.

Gary Amdahl, an early, helpful editor with *ESRI Press*.

Randy Worch of ESRI, for support and encouragement over the years.

Damian Spangrud, ArcGIS product manager for ESRI, for being the person I could always count on when I had difficult questions about the software.

Ken Bates, extension specialist with Kentucky State University, for working through many of the exercises with the ArcView level of ArcGIS.

Dan Carey, Ph.D., of the Kentucky Geological Survey and the University of Kentucky, for his thorough reading.

---

[2]She is author of *Introduction to 3D Data: Integrating and Modeling with ArcGIS 3D Analyst, Virtual Earth, and Google Earth* (John Wiley & Sons, 2009, ISBN 978-0-470-38124-3), *Data in Three Dimensions: A Guide to ArcGIS 3D Analyst* (Onward Press, 2004, ISBN 1-4018-4886-9) and editor of *The ESRI Press Dictionary of GIS Terminology* (2001, ISBN 1-879102-78-1).

Preface

I greatly appreciate the help of Richard Greissman, who facilitated my finding the time to write.

Demetrio Zourarakis of the Division of Geographic Information, Commonwealth Office of Technology for Kentucky-wide data.

The staff of ESRI technical support—friendly and helpful people too numerous to mention.

Jim Harper and Amy Zarkos of John Wiley and Sons, who facilitated the editing process in spite of the author's frequent lack of organization and tardiness.

Richard Peal of Publishers' Design and Production Services, Inc. in Sagamore Beach, MA, who looked after the figures, again against complications created by the author.

Teaching assistants Chris Blackden and Amber Ruyter worked through the exercises and helped both students and me over rough spots.

Scores of students at UK had early drafts of the text inflicted on them. In particular Rebecca McClung, who reviewed the text exhaustively and helped prepare the index and Travis Searcy, who combed through the later chapters.

I also wish to thank:

Taylor and Francis (and CRC Press) for allowing me to use portions of my textbook *The Global Positioning System and GIS* (ISBN 0-415-28608-5).

Chris Kimball of Digital Data Services, 10920 West Alameda Avenue, #206, Lakewook, CO (www.usgsquads.com, 303-986-6740) for the data behind Figure 2-3: Frankfort, MI and Crystal Lake DOQQs.

The Lexington Herald Leader for the photograph of the water facility on the Kentucky River.

Sue McCowan, account manager, GIS Markets of Tele Atlas, for San Francisco Street data for the Network Analyst exercise.

I'm appreciative of the University of Kentucky and its College of Arts and Sciences and Department of Geography, for providing the opportunity to develop the text.

Introduction

A geographic information system (GIS) software package is basically a computer program designed to make a computer think that it's a map. This new sort of map is a dynamic entity, designed to assist people in making decisions. Such decisions might be as simple and short range as determining an efficient way to get from place A to place B. Or as complex as designing a light rail transportation system for a city or delineating flood planes. The difference between a paper map and a GIS map is that the latter exhibits "intelligence." You can ask it a question and get an answer.

Geographic information systems are transforming all the activities and disciplines that formerly used maps as the basis for decision making. It's about time. Most fields of human endeavor have long since been heavily impacted by the digital computer; in fact it's hard to think of one that hasn't. Fifty years have gone by since computers began changing accounting, census taking, physical sciences, and communication, to name a few. Even the field of music has been altered. Most of these "nonspatial" fields already couched their problems in terms of discrete symbols (such as A, r, 5, and $) that are easily converted to the binary language (using only the symbols 0 and 1) that the computer understands. The spatial fields such as geography, planning, and land use management, had to stick with maps because, while maps also use symbols, they are not so neat and tidy as to fit on the keys of a keyboard. A symbol for a road might be three feet long! Determining how to efficiently represent the real-world environment in the memory of a computer turned out to be quite a challenge. So until a decade or so ago, those who relied on maps usually could not use computers effectively as the primary source of data from which to work.

Why has computer-based GIS come to influence how decisions are made about land use planning, navigation, resource allocation, and so on? First, the shortcomings of maps for decision making are many. Second, computers have become greatly faster, bigger (in terms of memory size), and cheaper. And, third, we have developed sophisticated data structures and learned how to efficiently program computers to represent the huge, almost infinitely detailed environment that we live in. So those of you who are just now beginning to learn about GIS are not pioneers, but if you enter the field now, I bet that in a decade you will feel like a pioneer because the field is growing so rapidly. You are off on a great adventure!

Some Concepts That Underpin GIS

OVERVIEW

IN WHICH you begin to understand the rather large and complex body of ideas and techniques that allow people to use computers to comprehend and design the physical environment. And in which you use Esri's ArcCatalog to explore geographic data.[1]

You Ask: "What Is GIS About?"

A poem "The Blind Men and the Elephant" tells the story of six sightless men who approach an elephant, one by one, to satisfy their curiosity.

> *It was six men of Indostan*
> *To learning much inclined*
> *Who went to see the Elephant*
> *(Though all of them were blind)*
> *That each by observation*
> *Might satisfy his mind.*
>
> *The First approached the Elephant,*
> *And happening to fall*
> *Against his broad and study side,*
> *At once began to bawl,*
> *"God bless em! But the elephant*
> *Is very like a WALL!"*
>
> *The Second, feeling of the tusk*
> *Cried: "Ho! what have we here*
> *So very round and smooth and sharp?*
> *This wonder of an Elephant*
> *Is very like a SPEAR!"*
>
> *The Third approached the animal,*
> *And, happening to take*
> *The squirming trunk within his hands,*
> *Thus boldly up and spake:*
> *"I see," quoth he, "the Elephant,*
> *Is very like a SNAKE!"*

---

[1]Esri is the Environmental Systems Research Institute. Esri makes the software, ArcGIS, which you will use in this text to understand the concepts of geographic information systems.

The Fourth reached out an eager hand,
And felt about the knee
"What most this wondrous beast is like
Is mighty plain," quoth he:
"'Tis clear enough the Elephant
Is very like a TREE!"

The Fifth, who chanced to touch the ear,
Said: "E'en the blindest man
Can tell what this resembles most;
Deny the fact who can,
This marvel of an Elephant
Is very like a FAN!"

The Sixth no sooner had begun
About the beast to grope,
Than seizing on the swinging tail
That fell within his scope,
"I see," quoth he, "the Elephant
Is very like a ROPE!"

And so these men of Indostan
Disputed loud and long,
Each in his own opinion
Exceeding stiff and strong
Though each was partly in the right
And all were in the wrong!

[. . .][2]

Excerpted from "The Blind Men and the Elephant"
(based on a famous Indian legend)
John Godfrey Saxe
American Poet (1816–1887)

And So You Ask *Again*: "What Is GIS About?"

Poet Saxe's lines could apply to geographic information systems (GIS) in that relating to the subject may well depend on your point of view. Asking what GIS[3] is about is sort of like asking "What is a computer about?" The capabilities of GIS are so broad and its uses so pervasive in society, geography, urban and regional planning, and the technical world in general that a short, meaningful description is impossible. But for starters, here is a generic definition of GIS that you might find in a dictionary:

A geographic information system is an organized collection of computer hardware and software, people,
money, and organizational infrastructure that makes possible the acquisition and storage of geographic

---

[2][. . .] indicates an omission.
[3]In this text "GIS" stands for "a geographic information system," or for the plural "geographic information systems," depending on the context.

and related attribute data, for purposes of retrieval, analysis, synthesis, and display to promote understanding and assist decision making.

To better understand one facet of GIS, consider how you might use the technology for a particular application. Solve the following site selection problem:

Exercise 1-1 (Project)

Finding a Geographic Site by Manual Means

Wildcat Boat Company is planning to construct a small office building and testing facility to evaluate new designs. They've narrowed the proposed site to a farming area near a large lake and several small towns. The company now needs to select a specific site that meets the following requirements:

❏ The site should not have trees (to reduce costs of clearing land and prevent the unnecessary destruction of trees). A regional agricultural preservation plan prohibits conversion of farmland. The other categories (urban, barren, and wetlands) are also out. So, the land cover must be "brush land."

❏ The building must reside on soils suitable for construction.

❏ A local ordinance designed to prevent rampant development allows new construction only within 300 meters of existing sewer lines.

❏ Water-quality legislation requires that no construction occur within 20 meters of streams.

❏ The site must be at least 4000 square meters to provide space for building and grounds.

Figure 1-1 is a key to the following maps. It shows the symbols for land cover, soil suitability, streams, and sewers.[4]

Figure 1-2 is a map showing landcover in the area from which the site will be chosen. Different crosshatch symbols indicate different types of land cover; the white area in the northern part of the map is water. The land cover codes (LC Codes) and categories (LC Type) are as follows:

LC CODE	LC TYPE
100	Urban
200	Agriculture
300	Brush land
400	Forest
500	Water
600	Wetlands
700	Barren

---

[4]These maps are also available on the DVD that accompanies the text. They are the image files: Key_to_maps.jpg, Landcover.jpg, Soil_suitability.jpg, and Streams_&_Sewers.jpg located in the folder IGIS-Arc_AUX.

Keys to Maps

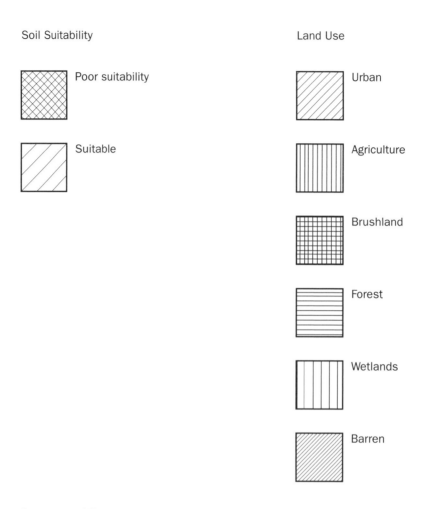

Soil Suitability

Poor suitability

Suitable

Land Use

Urban

Agriculture

Brushland

Forest

Wetlands

Barren

Streams and Sewers

Light solid lines: Stream
Heavy solid lines: Sewers
Light dashed lines: Coastline (and border)

FIGURE 1-1   Key to maps of the Wildcat Boat Facility area

Figure 1-3 is a soil suitability map. Lines separate soils of different types. The different soils are categorized as suitable or unsuitable for building. Therefore, you will see the same symbol on both sides of a dividing line, indicating that, while such soil types may be different, their suitability is the same.

Figure 1-4 is a map that shows the streams (narrow lines) and sewers (broader lines).

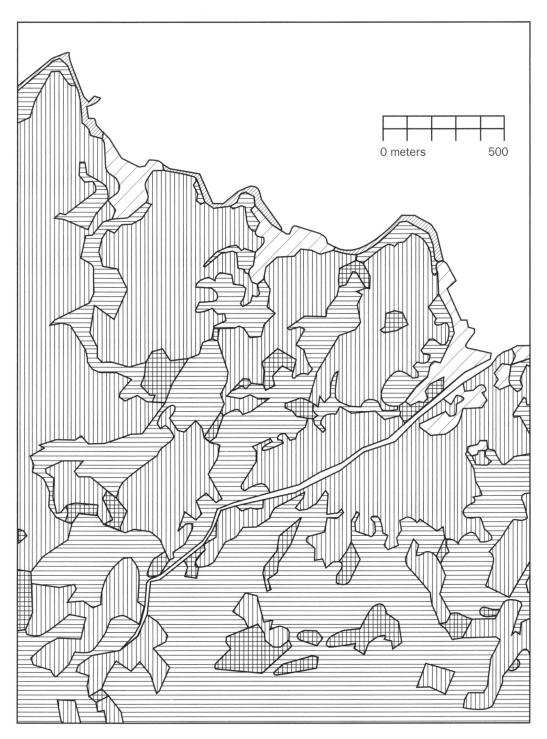

0 meters   500

FIGURE 1-2   Land Cover

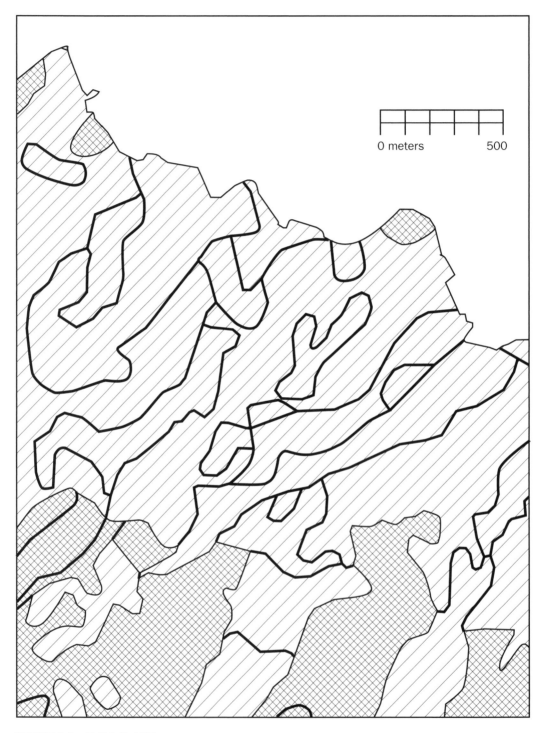

0 meters   500

FIGURE 1-3   Soil Suitability

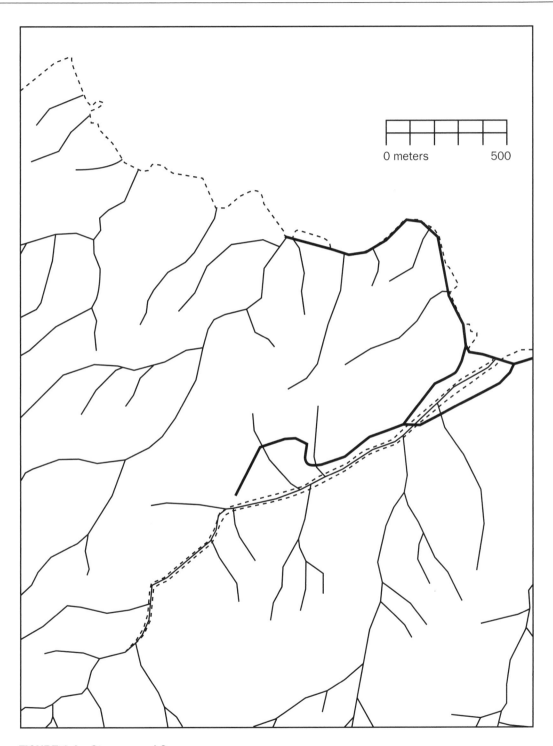

0 meters   500

FIGURE 1-4   Streams and Sewers

You may use scissors, xerography, a computer-based drawing program, a light table, and any other tools to solve the problem.

You are asked to present a map that shows all, repeat ALL, the areas where the company could build, while meeting the requirements stated previously. Make your map the same scale and size as those maps provided on the DVD. Outline in red all the areas that meet the requirements. You don't need to produce a high-quality cartographic product. The main objective here—indeed the object of this textbook—is to analyze geographic data. While making maps is important, it is not the primary focus of this book.

Write a brief description (100 to 200 words) of the procedure you used to make the map.

The problem is much easier than it might otherwise be because the maps provided cover exactly the same area, have the same underlying assumptions regarding the shape and size of Earth, are at the same scale, and use the same projection of Earth's sphere onto the flat plane of the map. These benefits are often not available in the real world, where you frequently need a considerable amount of data preparation to solve such a problem. ArcGIS has many tools to aide in "lining up" geographic data. Despite these advantages, the process can be somewhat daunting.

More of What GIS Is About

Completing Exercise 1-1 showed how GIS can help you solve one kind of problem. There are many others. Computer-based GIS not only serves the purpose of traditional maps but also helps you perform activities that involve spatial analysis, even without maps. Understanding conditions that occur in the vicinity of Earth's surface are important in building structures, growing crops, preserving wildlife habitat, protecting ourselves from natural disasters, navigating from one point to the next, and a myriad of other activities.

Among the many uses of GIS are:

Land use—Helps determine land uses, zoning, environmental impact analysis, locational analysis, and site analysis.

Natural environment—Identifies, delineates, and manages areas of environmental concern, analyzes land-carrying capacity, and assists in developing environmental impact statements.

Energy—Examines costs of moving energy, determines remaining available energy reserves, investigates the efficiency of different allocation schemes, reduces waste, reduces heat pollution, identifies areas of danger to humans and animals, assesses environmental impacts, sites new distribution lines and facilities, and develops resource allocation schemes.

Human resources—Plans for mass transit, recreation areas, police unit allocation, and pupil assignment; analyzes migration patterns, population growth, crime patterns, and welfare needs. It also manages public and government services.

Areas of environmental concern—Facilitates identification of unique resources, manages designated areas, and determines the relative importance of various resources.

Water—Determines floodplains, availability of clean water, irrigation schemes, and potential and existing pollution.

Natural resources—Facilitates timber management, preservation of agricultural land, conservation of energy resources, wildlife management, market analysis, resource allocation, resource extraction, resource policy, recycling, and resource use.

Agriculture—Aids in crop management, protection of agricultural lands, conservation practices, and prime agricultural land policy and management.

Crime prevention, law enforcement, criminal justice—Facilitates selection of sites or premises for target-hardening attention, establishment of risk-rating procedures for particular locations, tactical patrol allocation, location selection for crime prevention analysis, crime pattern recognition, and selection of areas or schools for delinquency prevention attention.

Homeland security and civil defense—Assess alternative disaster relief plans, needs for stockpiling of foods and medical supplies, evacuation plans, and the proper designation of disaster relief areas.

Communications—Facilitates siting of transmission lines, location of cellular equipment, and education.

Transportation—Facilitates alternate transportation plans, locational analysis, mass transit, and energy conservation.

Next Steps: Seemingly Independent Things You Need to Know

Before we launch into the theory and application of GIS, let's look ahead at the remaining text in the Overview of this chapter. You may know some or all of this material already, depending on your background. To use GIS effectively, you should know something about several topics that may seem unrelated at first glance. The next few sections briefly review the relevant aspects of the following:

❑   Determining where something is: coordinate systems

❑   Determining where something is: latitude and longitude

❑   Geodesy, coordinate systems, geographic projections, and scale

❑   Projected coordinate systems

❑   Geographic vs. projected coordinates: which should you use?

❑   Two projected coordinate systems: UTM and state plane

❑   Physical dimensionality

❑   Global positioning systems

❑   Remote sensing

❑   Relational databases

❑   Another definition of GIS

❑   Computer software: in general

❑   Computer software: ArcGIS in particular

Determining Where Something Is: Coordinate Systems

Cartesian Coordinate Systems

A coordinate system is a way of determining where points lie in space. We are interested in two-dimensional (2-D) space and three-dimensional (3-D) space. In general, it takes two numbers to assign a position to a 2-D space and three numbers in 3-D space.

Coordinates may be thought of as providing an index to the locations of points in space, and hence to the features that these points define.

To make a 2-D Cartesian[5] coordinate system, draw two axes (lines) that cross at right angles on a piece of paper. The point at which they cross is called the "origin." The sheet of paper is the x–y plane. Arrange the page on a horizontal table in front of you so that one line points left–right and the other toward and away from you. The part of the line from the origin to your right is called the positive x-axis. The line from the origin away from you is called the positive y-axis. Mark each axis in equal linear units (centimeters, say) starting at the origin, as shown in Figure 1-5. Now, a pair of numbers serves as a reference to any point on the plane. The position x = 5 and y = 3 [shorthand: (5,3)] is shown.

You can create a 3-D Cartesian coordinate system from the 2-D version: Imagine a vertical line passing through the origin; call it the z-axis; the positive direction is up. Now you can reference any point in three-dimensional space. The point x = 5, y = 3, and z = 4 [written more concisely as (5, 3, 4)] is shown in Figure 1-6.

This is called a right-hand coordinate system. The thumb, forefinger, and middle finger represent the positive axes x, y, and z, respectively. With your right hand outstretched, arrange those three digits so that they are roughly mutually orthogonal—that is, with 90° angles between each pair. Point your thumb to the right and your forefinger away from you. Now your middle finger will be pointing up.[6]

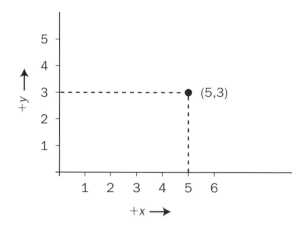

FIGURE 1-5   2-D Cartesian coordinate system

---

[5]Descartes, who lived from 1596 to 1650, made major contributions to both mathematics and philosophy. Descartes is credited with integrating algebra and geometry, by inventing the coordinate system that (almost) bears his name.
[6]It's best not to practice this exercise where other people can see you.

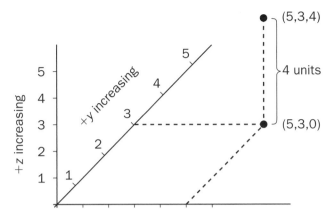

FIGURE 1-6   3-D Cartesian coordinate system

Spherical Coordinate Systems

A spherical coordinate system is another way to reference a point in 3-D space. It also requires three numbers. Two are angles, and the third is a distance. Consider a ray (a line) emanating from the origin. The angles determine the direction of the ray. See Figure 1-7.

The latitude–longitude graticule (a gridded reference network of lines encompassing the globe) is based on a spherical coordinate system. As often happens, different fields of endeavor use different descriptive approaches. Here, referencing navigation and Earth location issues requires a different system from

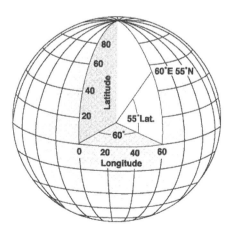

FIGURE 1-7   Spherical coordinate system showing latitude 55° North and longitude 60° East

the one mathematicians use in more abstract systems.[7] The origin is considered to be the center of the Earth. The equator serves as intersection of the x-y plane and the hypothetical sphere of the Earth. To determine the coordinates of a point, one angle (latitude) is measured from the x-y plane. The other angle (longitude) is contained in the x-y plane and is measured from the meridian that passes through Greenwich, England. The third number in a mathematical spherical coordinate system is the distance along the ray from the origin to the point. When added to the latitude-longitude system, altitude is usually defined instead to be the distance to the point along the ray from mean (average) sea level (MSL) or from a gravity-defined pseudo-ellipsoid used with the NAVSTAR Global Positioning System to be discussed shortly.

By using three numbers, you can determine and communicate the position of any point on Earth. Of course, an externally defined set of parameters must qualify these numbers. Any given point on the surface has probably been addressed by dozens of different sets of numbers, based on the parameters (e.g., units) of the coordinate system chosen. These parameters must match when you combine GIS data.

Determining Where Something Is: Latitude and Longitude

A fundamental principle underlies all geography and GIS: Most things on Earth don't move (or move very slowly) with respect to each other. Therefore, we can talk about the position of something embedded in or attached to the ground and know that its position won't change (much). It seems like a straightforward idea, but position confuses a lot of people when it is described as a set of numbers.[8]

Let's suppose that in 1955 somewhere in the United States you (or your parents, or their parents) drove a substantial metal stake or pin vertically into solid ground. Now consider that the object, unless disturbed by human beings or natural forces such as erosion or an earthquake, would not have moved with respect to the planet since then.[9] In other words, it is where it was, and it will stay there. Three numbers—latitude, longitude, and altitude—could identify the location of the object in1955. However, over the last half century, teams of mathematicians and scientists (skilled in geodesy) developed other sets of numbers to describe exactly the same spot where your object now resides. The actual position of the object didn't change, but additional descriptions of the where of the object have been created.

Ignoring the matter of altitude for the moment, suppose that the object was driven into the ground at latitude 38.0000000° (North) and longitude 84.5000000° (West), according to calculations done before 1955 that indicated the location of the center of the Earth, its shape, and the location if its poles. Most people and organizations in the United States in 1955 used the North American Datum of NAD27 (NAD27) to estimate the latitude-longitude graticules, based on parameters of the earth-approximating ellipsoid determined by Clarke in 1866.[10]

---

[7]For example, in the two dimensional Cartesian plane, a mathematician will measure angles starting from the positive x-axis (east) as zero and increasing counterclockwise, to 360 degrees (which is, again, zero). The navigator (think: compass) or cartographer will also use 360 degrees to represent a full circle, but measures clockwise from the positive y-axis (north).
[8]Descriptions of points aren't always just numbers. A possibly apocryphal "metes and bounds" description of a point in Kentucky a couple of centuries ago was "Two tomahawk throws from the double-oak in a northerly direction."
[9]Well, hasn't moved much. If it was on the island of Hawaii, it has moved northwest at about 4 inches per year. Also, if you are unfortunate enough to be in a place where there was an earthquake, it might well have moved and not returned to its original position.
[10]Based on a monument on Meades Ranch, in Kansas; the Clarke 1866 ellipsoid was meant to minimize the error between itself and the geoid in the United States.

The datum described as the World Geodetic System of 1984 (WGS84) offers the most recent, widely accepted view of where the center of the Earth is, its shape, and the location of its poles. The ellipsoid of WGS84 is virtually identical to the GRS80 ellipsoid.[11] In the coterminous states of the United States, this datum is virtually identical (within millimeters) to the North American Datum of 1983 (NAD83), although they result from different approaches and calculations.

According to the WGS84 latitude-longitude graticule, the object previously described would be at latitude 38.00007792° and longitude 84.49993831°. The difference might seem insignificant, but it amounts to about 10 meters on the Earth's surface. Or consider it this way: According to NAD83, a second object placed in the ground at 38° N and 84.5° W would be 10 meters away from the first one. Does that sound like a lot? People have exchanged gunfire in land disputes over smaller distances. Given a latitude and longitude, a GIS must know the datum that is the basis for the numbers. Hundreds of datums exist, and many countries have their own.

Geodesy, Coordinate Systems, Geographic Projections, and Scale

First, a disclaimer: This text does not pretend to cover in detail such issues as geodetic datums, projections, coordinate systems, and other terms from the fields of geodesy and surveying. Nor will the text rigorously define most of these terms. Simply knowing the definitions would mean little without a lot of study. Many textbooks and Web pages are available for your perusal. These fields, concepts, and principles may or may not be important in your use of GIS, depending on your projects. However, the datum, projection, coordinate system designations, and measurement units must be identical when you combine GIS or map information. If not, your GIS project may well produce inaccurate results.

How we apply the mathematically perfect latitude-longitude graticule to points on the ground depends partly on human's understanding of the shape of the Earth. This understanding changes the more we learn. Geodesy is the study of the shape of the Earth and the validity of the measurements human beings make on it. It deals with such issues as spheroid and datum. You don't have to know much about geodesy to use a GIS effectively, provided your data are all based on the same spheroid and datum (and projection and units, as you will see later). It is the application of geodesic knowledge that caused the differences in the coordinates of that hypothetical object I discussed earlier that was put into the ground six or seven decades ago. The object hasn't moved. We simply have a better idea of the location of the object relative to the latitude-longitude graticule.

Projected Coordinate Systems

For several reasons, it's often not convenient to use latitude and longitude to describe a set of points (perhaps connected by straight lines to make up a coastline or country's boundaries) on the Earth's surface. One is that doing calculations using latitude and longitude—for example, determining the distance between two points—can involve complex operations such as products involving sines and cosines. For a similar distance calculation, if the points are represented on the Cartesian x-y plane, the worst arithmetic hurdle is a square root.

---

[11]GRS80 is a global geocentric system based on the ellipsoid adopted by the International Union of Geodesy and Geophysics (IUGG) in 1979. GRS80 is the acronym for the Geodetic Reference System of 1980.

Latitude and longitude measures for many geographic applications do not work well for several aspects of mapmaking. Suppose you plot many points on the Earth's surface—say, along the coastline of a small island that is a considerable distance from the equator—on a piece of ordinary graph paper, using the longitude numbers for x-coordinates and latitude numbers for y-coordinates. The shape of the island would look strange on the map (it would appear horizontally stretched) compared to how it would appear from an airplane. You would not get useful numbers if you measured distances or angles or areas on the plot. This is due to a characteristic of the spherical coordinate system: The length of an arc of a degree of longitude does not equal the length of an arc of a degree of latitude. Those lengths are almost equal near the equator, but the difference grows as you go further north or south from the equator. At the equator, a degree of longitude translates to about 69.17 miles. Very near the North Pole a degree of longitude might be 69.7 inches. (A degree of latitude, in contrast, varies only between about 68.71 miles near the equator and 69.40 miles near the poles.)

For relatively small areas, mathematically projecting the spherically defined locations onto a plane provides a good solution to problems associated with calculations and plotting. Geographic projection might be thought of as imagining a process that places a light source inside a transparent globe that has features of the Earth inscribed on it. The light then falls on a piece of paper (or one that is curved in only one direction and may be unrolled to become flat).[12] The shadows of the features (say, lines, or areas) will appear on the paper. Applying a Cartesian coordinate system to the paper offers the advantages of easy calculation and more realistic plotting. However, distortions are inherent in any projection process; most of the points on the map will not correspond exactly to their counterparts on the ground. The degree of distortion is greater on maps that display more area. Accuracy suffers when you flatten a curved surface and thus convert a three-dimensional coordinate system to a two-dimensional one.

After constructing geographic data sets according to latitude and longitude, based on some agreed-upon spheroid (such as GRS80) and datum (such as WGS84), you decide how to represent them graphically for viewing. At one time, cartographers went directly from the latitude-longitude description of an area or feature to a graphical portrayal on paper. This usually involved "projecting" the data from a three-dimensional (3-D) spherical coordinate system to a two-dimensional (2-D) Cartesian one, and setting a scale: A certain number of units on a map represented that number of units on the ground. Using GIS changes this. As I commented earlier, latitude-longitude is the most fundamental and accurate way to represent spatial data. So large areas—those that are more subject to distortion by being projected—may best be left in latitude-longitude coordinates, and subareas projected to other coordinate systems when needed. Computers are really good at doing the complex computations necessary to project data.

The larger the area projected, the greater the tendency for things not to be where the projected coordinates say they are.

Four considerations for viewing and analysis are size, shape, distance, and direction. Myriad projections have been invented. Many distort size, shape, distance, or direction, and some preserve one or two of them.

Regarding linear units, GIS differs from cartography in that matters of scale may be left until the very end. The position defining numbers in the database should be real-world coordinates—not scaled coordinates. That is, GIS "maps" are stored in ground units rather than map units. Besides being a more

---

[12]This description is a sort of a cartoon to describe a map projection. A map projection is actually a mathematical transformation that "maps" points on the globe to points on a plane; the process may be quite complex depending on the projection; a single light source at the center of the globe does not suffice to explain it.

fundamental way to store data, this makes it possible to easily make maps of any desired scale on the computer monitor or on paper. Scale is only a consideration when you measure on a physical map. Computers take the worry out of determining scale accuracy. Because of the vast computational power of a computer, there is no difficulty in scale conversion. The days of: "Let's see, this distance is 5.3 inches on the map, and 1 inch is 12 miles, so the distance is about 64 miles" are over. Scale is a minor concept in GIS—one used only on final output.

Geographic vs. Projected Coordinates: A Comparison

- ❑ *Advantages of the spherical coordinate system*—You can represent any point on the Earth's surface as accurately as your measurement techniques allow. The system itself does not introduce errors.

- ❑ *Disadvantages of a spherical coordinate system*—You will encounter complex and time-consuming arithmetic calculations in determining the distance between two points or the area surrounded by a polygon determined by a set of points. Latitude-longitude numbers plotted directly on paper in a Cartesian coordinate system result in distorted—sometimes greatly distorted—figures.

- ❑ *Advantages of a projected coordinate system on the Cartesian plane*—Calculations of distances between points are trivial. Calculations of areas are relatively easy. Graphic representations are realistic, provided the area covered is not too large.

- ❑ *Disadvantages of a projected coordinate system on the Cartesian plane*—Almost every point is in the wrong place, although maybe not by much. All projections introduce errors. Depending on the projection, these errors are in distances, sizes, shapes, or directions.

Whether you use geographic or projected coordinates, ensuring that the parameters of geographic data match is of paramount importance in combining GIS data sets if you want the right answers![13]

Two Projected Coordinate Systems: UTM and State Plane

A coordinate system called Universal Transverse Mercator[14] was developed based on a series of 60 projections onto semi-cylinders that contact the Earth along meridians. (To consider, for example, one of these projections, imagine a sheet of paper curved so that it becomes a half cylinder whose radius is that of the Earth's. Then, with the axis of the cylinder oriented in an east–west direction—hence the term transverse—the paper is brought into contact with a globe along the meridian designating 3° longitude. Then, the surface of the Earth between 0° and 6° is projected onto the paper). This process is repeated for central meridians of 9°, 15°, 21°, and so forth up to 357°. The term "zone" is ambiguously used for this swath of territory. However, UTM projections are further subdivided into areas, also called zones, covering 6° of longitude and, for most zones, 8° of latitude. Further, ArcGIS divides a total zone into a

---

[13]The parameters of data sets may not always match but may be close enough so that any error introduced is trivial. In many instances and places, for example, NAD83 and WGS84 match within centimeters. However, you must research carefully to know when you can use data sets that don't match exactly. The Esri manual *Understanding Map Projections* can help.

[14]The idea of the Mercator projection was developed in 1568 by Gerhardus Mercator, a Flemish geographer, mathematician, and cartographer.

northern and a southern part. In any event, a coordinate system is imposed on the resulting projection such that the numbers in any given zone:

❏   Are always positive

❏   Always increase from left to right (west to east)

❏   Always increase from bottom to top (south to north)

The representation of our previously discussed object (at 38° N and 84.5° W) in the UTM coordinate system, when that system is based on WGS84, is a "northing" of 4,208764.4636 meters and an "easting" of 719,510.3358 meters. The northing is the distance to the point, in meters,[15] from the equator measured along the surface of an "Earth" that has no bumps. The easting is somewhat more complicated to explain because it depends on the zone and a coordinate system that excludes negative numbers. Consult a textbook on geodesy or cartography, or review the thousands of Web pages that come up when you type the words

```
UTM     "coordinate system"    "Transverse Mercator"
```

into an Internet search engine (e.g., www.google.com).

One version of the UTM coordinate system is based on NAD27. In this case, our object would have different coordinates: northing 4,208,550.0688 and easting 719,510.6393. This produces a difference of about 214 meters. If you combined WGS84 UTM data with NAD27 UTM data, the locations they depict might be in error by just enough not to be obvious at some scales, but great enough to cause trouble.[16]

Each state of the United States has one or more State Plane Coordinate Systems (SPCSs). They were developed, originally in the 1930s, by the U.S. Coast and Geodetic Survey. These systems are based on different projections (usually Transverse Mercator or Lambert Conformal Conic), depending on whether the state is mostly north-south (like California) or mostly east-west (Tennessee). The units of an SPCS may be international feet, survey feet, or meters, depending on decisions made by the state itself.[17] Zone boundaries frequently follow county boundaries. The coordinate system(s) used in one state are not applicable in neighboring states. Nor can you apply the SPCS of one zone to areas in the state only a short distance away in another zone. Furthermore, the difference between NAD27 and NAD83 (WGS84) can be startlingly large. In Kentucky, for example, 38.0000000° (North) and 84.5000000° (West) would translate into a northing of 1,568,376.1900 feet and an easting of 182,178.3166 feet when based on WGS84. However, when the basis in NAD27, the coordinates are 1,927.939.8692 and 182,145.9821, which makes a difference of some 68 miles!

Why the large differences in projected coordinate systems based on NAD27 and those based on WGS84? Because those responsible for the accuracy of the other coordinate systems took advantage of the

---

[15]In fact, the meter was originally defined as one tenth-millionth of the distance from the equator to the North Pole, along a meridian that passed through Paris.

[16]Comparing these coordinates with the WGS84 UTM coordinates, you see that virtually all of the difference is in the north-south direction. While true for this particular position, it is not true in general.

[17]The meter is the standard unit of length in most places in the world. Two different lengths of "foot" are defined in terms of the meter. A "U.S. Survey Foot" is one in which a meter is considered to be 39.37 inches, exactly; the other sort of foot is the international foot, where an inch is 0.0254 meters, exactly. The survey foot and the international foot are almost, but not exactly, the same length.

development of WGS84—a worldwide, Earth-centered, latitude-longitude system—to correct or improve or change the origin of those projected coordinate systems.

State plane coordinate systems are generally designed to have a scale error maximum of about 1 unit in 10,000. Suppose you calculated the Cartesian distance (using the Pythagorean theorem) between two points represented in a state plane coordinate system to be exactly 10,000 meters. Then, with a perfect tape measure, pulled tightly across an idealized planet, you would be assured that the measured result would differ by no more than 1 meter from the calculated one. The possible error with the UTM coordinate system may be larger: 1 in 2500.

Coordinate Transformations

Coordinate transformation of a geographic data set is simply taking each coordinate pair of numbers in that data set and changing it to another pair of numbers that indicates exactly the same spot on the Earth's surface, but using a different system of assigning coordinates.

Let's make up a coordinate system for a garden delineated by lines between stakes. Suppose the origin (0,0) is at the southwest corner. Now drive several stakes in the ground so that if you passed a string around them it would outline a polygon. Suppose you use a surveyor's tape calibrated in survey feet to measure the Cartesian coordinate for each stake. Stake 0 is at the origin (0.0). Stake 1 is 33 feet east and 0 feet north (i.e., 33,0). Stake 2 is at (33,10), and so on. See Figure 1-8.

Suppose the garden grows and then becomes overgrown. Next year someone locates the origin and wants to find the original stakes. He also has a surveyors tape, but it is calibrated in meters instead of feet. He asks you to provide the data on where the stakes are in meters. In this case, the transformation of coordinate systems is easy: You simply multiply each number by a conversion factor (the number of meters per survey foot, which is 0.3048). So Stake 1's coordinates are 10.06, 0). Those of Stake 2 are (10.06, 3.05), and so on. See Figure 1-9.

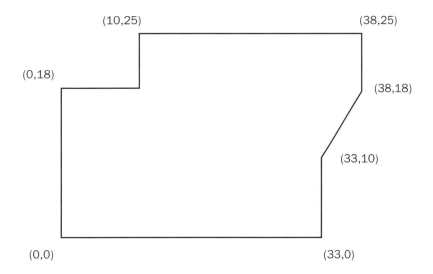

FIGURE 1-8   Locations of garden stakes

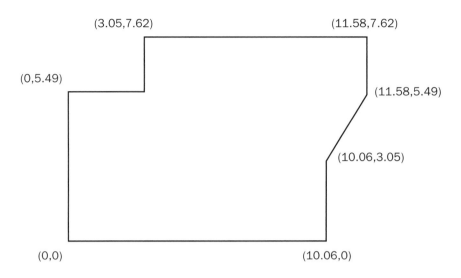

FIGURE 1-9   Locations of garden stakes

The stakes are in the same place that they were before. Their positions are simply referred to by a different set of numbers. Geographic coordinate or datum transformation is nothing more or less than this, except that the mathematical operations applied to each number are usually more involved.

Physical Dimensionality

All matter exists in four dimensions—roughly characterized as left-right, toward-away, up-down, and time. The first three are conceptually clumped together and called "spatial dimensions." Time is treated as a dimension here because we want to talk about "position" in space and time.[18]

Any physical object occupies space and persists in time. This is true of the largest object in the universe and the smallest atom. However, when the measure of one or more dimensions of an object is small, or insignificant, with respect to the measure of others or is tiny with respect to its environment, it is useful to describe and represent objects by pretending that they occupy fewer than three spatial dimensions. For example, a sheet of paper can be considered a pseudo-two-dimensional object; it has thickness (up-down), but that dimension is miniscule compared with left-right and toward-away. A parking meter in a city could be considered a pseudo zero-dimensional object, because the measures of all its dimensions are insignificant with respect to its environment. (The parking meter would certainly not be considered a spatially zero-dimensional object by its designer, manufacturer, or the driver who uses it. Therefore, the pseudo-dimensionality of an object depends on its context.)

---

[18]In physics, time is treated as the fourth dimension because it is inexorably bound up with the other three according to Einstein's theories (special in 1905 and general in 1916) of relativity. This connection need not concern us except for some esoteric technical matters related to time on global positioning system satellites (for one, time on a satellite runs faster because it is farther from Earth, where there is less gravity).

It is important to consider the pseudo-dimensionality of both an object and the field (the space) in which it resides. A point on a line is a zero-dimensional object in one-dimensional space. A line segment on a plane is a one-dimensional object in two-dimensional space. A polygon is a two-dimensional object in a two-dimensional space. A point in a volume is a zero-dimensional object in three-dimensional space. The dimensionality of the object must be less than, or equal to, the dimensionality of the space in which it resides.

This issue of dimensionality comes into play here because great economies of computer storage are achievable if an object is considered to have fewer dimensions than it actually does. For example, describing a fire hydrant as a complex three-dimensional spatial object would be an arduous task and involve many numbers and much text. If it is considered a zero-dimensional object, its location can be described (very precisely, but still inexactly) with three numbers, perhaps representing the latitude, longitude, and altitude of some point on the hydrant or, perhaps, its center of mass.

The spatial dimensions of a spatial pseudo-zero-dimensional object are insignificant with respect to the context or environment in which it resides. A theoretical geometric point is the prototypical example. Examples of such zero-dimensional objects on maps or in a GIS could be streetlights, parking meters, oil wells, census tracts, or cities—depending on the spatial extent of other objects or features in the database.

A spatial pseudo-one-dimensional object or feature has no more than one significant dimension or is made up of essentially one-dimensional objects. A straight-line segment is the prototypical example. In terms of making a one-dimensional object up from component parts, consider the example of a telephone wire strung from one pole to the next to the next and so on. It is considered a one-dimensional object, even though the poles may not be in a straight line, so that the phone line zigzags over a two-dimensional field. (The line itself sags and persists in time—and therefore exists in a four dimensional context—but these facts notwithstanding, it can be simplified or generalized into a one-dimensional feature—saving us lots of computer storage, processing time, and conceptual complication.) Roads, school district boundaries, pipes for fluid transportation, and contour lines on a topographic map are all examples of pseudo one-dimensional entities.

You can disregard the up-down dimension in a spatial-pseudo-two dimensional object. Plain areas such as voting districts and soybean fields are examples. Such pseudo-areas lack vertical components—variability in altitude—that can be important. The true surface area of hilly terrain may be considerably underrepresented by the plane area within the borders of the plane figure that represents it.

You must consider all dimensions of a spatial three-dimensional object in order to represent it. Three-dimensional features are volumes—say, a coal seam or a building.

Relative to each other, positions of features that we record with a GIS do not move, or do not move quickly, with respect to each other. Sometimes we want to know how conditions change or object move. The representation of the objects may be zero-, one-, two-, or three-dimensional. At its most complex, GIS would involve all four dimensions. Most GIS operations and data sets are two-dimensional; the data are assumed to pertain to moments in time, or stable phenomena or conditions over a period of time. Three-dimensional GIS could involve either three spatial dimensions (such as representing the volume of a limestone quarry) or two spatial dimensions and varying time (showing historical changes in a landscape). Rarely will you use four-dimensional GIS.

Results from a GIS are always only an approximation of reality. One of the reasons, among others to be discussed later, is that we simplify objects by reducing their dimensionality.

Global Positioning Systems

A global positioning system (GPS)[19] is a satellite-based system that provides users with accurate and precise location and time information. Using NAVSTAR GPS, you can determine locations on Earth easily within a few meters, and, with more difficulty and expense, within a few centimeters or better. Timing within 40 billionths of a second (40 nanoseconds) is easily obtained. Timing within 10 nanoseconds is possible.

The U.S. Department of Defense operates NAVSTAR GPS in cooperation with the U.S. Department of Transportation. The acronym NAVSTAR stands for NAVigation System Timing And Ranging Global Positioning System." Informally, it is the Navigation Star.

The Russian GLONASS (Global Navigation Satellite System) operates similarly. Concerns about U.S. control over NAVSTAR led Europe to begin development of its own independent Galileo system in 2002, but it is just now becoming operational. China is developing the BeiDou (Compass) GNSS.

A GPS receiver, which "remembers where it has been," is becoming a primary method of providing data for GIS. For example, if you drive a van with a GPS antenna on its roof along a highway, recording data every, say, 50 feet, you will develop an accurate and precise map of the location of the highway. The NAVSTAR GPS is discussed in Chapter 5.[20]

Remote Sensing

Remote sensing probably stated with photographs taken from balloons in the 1840s. The first automated system (not requiring human beings to be with the sensors) may have been in the 1890s in Europe when cameras programmed to take pictures at timed intervals were strapped to pigeons!

Evelyn Pruitt probably introduced the modern use of the term "remote sensing" in the mid-1950s when she worked as a geographer/oceanographer for the U.S. Office of Naval Research (ONR). Remote sensing uses instruments or sensors to capture the spectral characteristics and spatial relations of objects and materials observable at a distance, typically from above them. Using that definition, everything we observe is remotely sensed. More practically, something is sensed remotely when it is not possible or convenient to get closer.

We can categorize remote sensing for GIS many ways. Data can be taken from aircraft or satellite "platforms." The energy that the sensor "sees" can come from the objects or areas being examined as a result of radiation emanating from them (caused by the sun or other heat or light sources) or from radiation bounced off them by an energy source associated with the sensor (e.g., radar or lidar). The images produced may be developed on film or produced by digital sensors. Satellites in geosynchronous orbits can hang in a single spot over the equator. Those in near polar orbits can see different areas of the Earth as it turns. Chapter 2 offers examples of remotely sensed data.

---

[19]Also referred to as a positioning, navigation, and timing (PNT) system, or a global navigation satellite system (GNSS).
[20]See also Prof. Kennedy's textbook *The Global Positioning System and ArcGIS*, Third Edition, 2010, CRC Press (Taylor & Francis Group), 301 pages.

Relational Databases

GIS relies heavily on databases of text and numbers. The relationship between such information and geography is discussed later in the chapter. For now, you simply need to know how text and number data sets are stored.

For this discussion, a database is a collection of discrete symbols (numbers, letters, and special characters) located on some physical medium with at least one principal underlying organization or structure. An old-fashioned library card catalog is an example of a database with a single underlying structure: an alphabetical list of authors; the medium is 3 x 5 index cards, and the data are the symbols on the cards describing books and their locations in the library. Most libraries have substituted computer-based catalogs with the advantage that a user can search and find not only authors but titles and subjects as well, so a number of organizing themes may underpin a database.

Another example is a "hard disk" that has recorded the most common type of soil found in a specific acreage in a county. The disk is the physical medium, the codes assigned to a soil type constitute that data, and the location of each acre—as understood from the position of each datum on the disk—could be the underlying structure.

Existing general purpose databases usually:

❏ Result from some sort of project; some individual or team constructs it—frequently going to considerable effort.

❏ Need to be *updated* (modified and corrected as time progresses) if they are to continue to be of value.

❏ Contain errors *regardless* of size, care of construction, simplicity of data, or quality or physical medium used.

❏ Serve a function when allied with some process. The function may be as simple as supplying a telephone number from a physical phone directory (structure—alphabetical by name; medium—cheap bound paper; data—phone numbers, in very small type; process—looking up a name, finding the adjacent number). The function served by the database might also be quite sophisticated—supporting far-reaching land use decisions, for example.

Numerous databases are used to solve problems at all levels of government. Access to these databases is achieved through the use of referencing schemes. Following are some examples:

Referencing Scheme	*Examples of Data Contained*
Names of people	Salary, Social Security number, medical history, criminal record
Auto license plate number	Car color, owner, serial number
Street address	House value, lot size
Job title	Person employed, duties, salary
Transaction number	Money received, paid, transferred, invested
Events	Schedules, orders, crimes, accidents

Many schemes exist for presenting or storing data. Suppose parents want to find a name for a newborn child. A list of potential names in random order might be provided. It would probably be more useful if the list were divided into girl's names and boy's names. Another improvement could be to alphabetize the list, or present it in terms of current popularity of names.

As a second example, consider how you might store telephone numbers associated with names and addresses. You could order the list alphabetically, so you could easily find a phone number, knowing the name. Or you could order the list numerically by telephone number, so you could find a name, given a number.

A different sort of data structure is "hierarchical." You could use this approach to store the names and positions of people in a corporation or a military organization based on who reports to whom. Likewise you could store voting districts within counties within states. In the computer world, this "tree-structured directory" approach is used to organize the folders and files contained on a hard disk drive on a computer. Any folder on the disk may store other folders and files.

Many schemes exist to store information in the memory of a computer or on its secondary devices, such as disk drives or tapes. The primary method used to store large amounts of information is called the "relational database," or RDB, developed by E. F. Codd.[21] The software is described as a RDBMS (relational database management system). The idea is simple: use a set of two-dimensional tables; for a given table there is a prescribed format in which the rows relate to entities (objects, people, things in general), while the columns relate to attributes (characteristics, properties) of entities. The intersection of a given row and a given column is a cell, containing a value, which defines the particular attribute of the particular entity. See Figure 1-10.

You can use a database table to store names, occupations, and pay schedules of employees, as in Figure 1-11.

Another example shows part of a database of automobiles registered in a state. Each row would represent one car; each column would represent one property of cars. See Figure 1-12.

Here is some terminology: The structure that contains the entity, the row, is also called a *tuple* or *record*. The structure that contains the attributes, the column, is also called a *field* or an *item*.

Relational Database Nomenclature

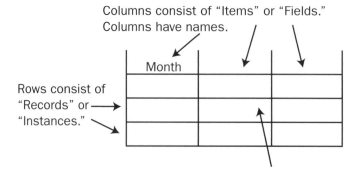

FIGURE 1-10   Components of a relational database table

---

[21]Edgar F. Codd was the originator of the relational approach to database management, which he introduced in 1970. For this work, he was the recipient of the Association for Computer Machinery (ACM) Turing award in 1981.

Garden-Variety Database

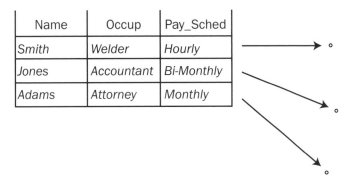

Name	Occup	Pay_Sched
Smith	Welder	Hourly
Jones	Accountant	Bi-Monthly
Adams	Attorney	Monthly

FIGURE 1-11 Trivial example of a relational database table

Mfgr	Num_Doors	VIN	Color	Weight
Porsche	2	123XXX	Silver	2300
Porsche	2	887ABC	White	2100
Toyota	4	9880123	Grey	2350
Honda	4	456789	Blue	2999

FIGURE 1-12 An RDB of automobiles

All the cells of a given column must contain the same sort of value. Some common ones are as follows:

❏ *Character*—Any valid ASCII character in a string of almost any length.

❏ *Short integer*—Can range from slightly less than −32,000 to slightly more than +32,000.

❏ *Long integer*—Can range from slightly less than negative two billion to slightly more than positive two billion.

❏ *Floating-point number*—Can have an exponent as small as 10^{-38} and as large as 10^{38}, and you can be assured of six significant digits of precision.

❏ *Double-precision floating-point number*—Can have an exponent as small as 10^{-308} and as large as 10^{308}, and you can count on 16 significant digits.

❏ *Boolean*[22]—A value that is either true or false.

---

[22]Named after George Boole, mathematician (which is why it is not spelled Boolian).

Getting Information from a Relational Database: Queries

Relational databases are designed to give you information. You can obtain the information by selecting a subset of records from the total set by writing an expression that is a mixture of attribute names, arithmetic and logical operators, and values.[23] For a trivial example, suppose that research has found that gray cars that weigh less than 2500 pounds put their occupants in greater risk than average. You want to select those records from the statewide automobile database. You might first get all the records of the cars that are gray.

```
SELECT: COLOR = 'Gray'
```

Given that subset, you might then write

```
SELECT: WEIGHT < 2500
```

Given this sub-subset of records, you could perhaps write letters to the owners of those cars, making them aware of the danger they face.

Languages to get subsets of records can provide flexibility and efficiency. For example, to do both of the preceding operations with one expression, you might write

```
SELECT: COLOR = 'Gray' AND WEIGHT < 2500
```

Suppose also that the research study showed increased danger to those occupying cars that were built in 1985 or before, regardless of color or weight. You might add to the preceding selection by saying

```
ADDSELECT: YEAR <= 1985
```

(ADDSELECT means add to the current set of selected records.)

Or you may use a single query to select all the records you want at once:

```
SELECT: (COLOR = 'Gray' AND WEIGHT < 2500) OR YEAR <=1985
```

Note the use of parentheses to indicate the order in which operations are done.

Economies in Relational Databases

In theory, the information in a relational database system could reside in a single table. This isn't the best policy and in reality may not even be possible. A relational database usually consists of a set of tables that relate to one another—thus, the word "relational."

Each relational database table must contain a column that has a unique identifier for each record in the table. This is known as the *key field*. If the relational database references people, the key might be Social

---

[23]There are many relational database software products and several different languages used to query them. What is shown in the following is not a specific query language but rather a generic representation of such a language. ArcGIS may be used with several relational database products, each with a somewhat different query language.

Security number, employee number, or student number. In the case of automobiles, the key code might be the vehicle identification number (VIN).

To illustrate why multiple tables might be used, suppose that, in our relational database, it is possible for one person to own more than one automobile. If the entire database is all in one table, then the names and addresses of each multicar owner must be repeated, which increases the amount of storage required. If a multicar owner gets a new street address, several records in the database table would have to be changed. Further, some of the data about owners might be located in another database and might be private. The elegant answer to these problems is to have several tables that contain the information, divided to minimize the repetitions. In the automobile example, it might be that only the owner's identification number is stored in the record of the car itself. This number could be the key field in the database table containing information about owners. If some of the owner information is private, it could be stored in a separate table as well, with the same key field. The database could be set up so that this table could not be accessed by the automobile table.

Relational database tables that meet certain requirements of efficiency are said to be in first normal form (1NF), second normal form (2NF), and so on. The higher the number, the more efficient the database.

In summary, it is useful to partition the information in a relational database into a number of tables. Such partitioning can

❏   Reduce the amount of redundant data stored

❏   Aid in updating the database

❏   Reduce the chances of inconsistency and instability in the database

❏   Aid in protecting private or sensitive data

Languages have been developed for retrieving information from a relational database. One is the Structured Query Language (SQL), developed by the American National Standards Institute (ANSI). You use it to execute queries of a database. Vendors of relational database management systems also have developed proprietary languages.[24] Esri products interface with a number of RDBMS from various sources.

Databases—What's Meant by "Relational"

What you have seen so far is one aspect of the RDB: a two-dimensional table in which you store entities as rows and attributes as columns. What is it that makes the relational database such a powerful approach to storing information? One answer lies in the fact that a RDB can be much more than a single table. Usually it is a number of tables that are related to one another, as previously discussed, that provide for efficiency, flexibility, and ease of updating.

Table 1-1 illustrates a database that might be created by a motor vehicles licensing department. Parts of the database connect to other databases formed by other government departments. Shown in Table 1-1 are nine relational database tables; the RDB table name is on the top line; the attribute names are on the second line. None of the tens of thousands of cells is shown.

---

[24]Some names of past and present relational database management system software products are Access, dBASE, Informix, Ingres, Oracle, SQL Server, and Sybase.

Here, mainly for purposes of illustration, the MASTER table is miniscule, consisting only of a license plate number and an owner identification. The key field, shown in bold font, consists of a unique character string (no two license plates are the same) and an owner identification number. The contents of this second column may not be unique, since one person my own several vehicles.

TABLE 1-1

MASTER		
	Plate_#	Owner_ID

OWNER			
	Owner_ID	Name	Address	SSN

VEHICLE						
	Plate_#	Color	Model	Year	Type_code	VIN

VEH_TYPE						
	Type_Code	Weight	Length	Width	H_Power	Fuel_Cnsup

VOTERS				
	SSN	Name	Precinct_Code	Regis_Party

PRECINCT				
	Prec_Name	Location	Supervisor	**Precinct_Code**

ARRESTED						
	Name	Alias	SSN	Offense	Convicted	**Court_Doc_#**

ACCOUNTING			
	Plate_#	Date_Paid	Insurance_Co_Code

INSURERS				
	Insurance_Co_Code	Co_Name	Location	Phone

Both fields in MASTER refer to—link to—the other eight tables, either directly or indirectly. Plate_# allows the user access to both VEHICLE and ACCOUNTING.

The VEHICLE table describes some of the attributes of the car or truck in question. Since some aspects of all vehicles of a certain year, model, and manufacturer are identical, it would be a waste of space and an updating nightmare to place this information in a record describing a particular vehicle. So each record carries a Type_Code that refers to a key field in the table VEH_TYPE. There you will find the vehicle weight, length, width, horsepower, and fuel consumption data.

The ACCOUNTING table indicates whether taxes and registration have been paid. Also, here is a reference to the INSURERS table, which carries information about the companies that insure vehicles in the state.

The other column in the MASTER table contains an Owner_ID code that matches up with the key in the table OWNER. That table could contain a host of information about the owner of the vehicle. Shown in the table are names and addresses. Also there is the owner's Social Security number (SSN). Perhaps there is a "motor-voter" effort to register all vehicle owners. With the SSN, which is the key field in the VOTERS table, a user could determine which owners were already registered to vote. Further, with the Precinct_Code, information about the location of the voting precinct could be obtained.

Finally, the SSN also allows a link to the table ARRESTED so that drivers with moving violations or driving while intoxicated could be identified. Both the VOTERS and ARRESTED tables probably would reside in some other department, so maintenance of those tables would not fall to the Department of Motor Vehicles, yet the DMV would have access to the information.

Most governments and organizations have extensive and perhaps sophisticated techniques or systems for storing and manipulating data that can be referenced by these and other schemes. Here, we are interested in the quite-useful referencing basis that has not been developed as extensively, however. It may be known as several names: geographic, land locational, geodetic, or spatial position.

Relational Databases and Spreadsheets

A relational database table may look a lot like an electronic spreadsheet, such as Microsoft Excel. There are some important differences:

❏ A relational database is structured, with rows and columns strictly defined. In a spreadsheet, you can put anything anywhere.

❏ In a relational database, the headings of columns are not stored in cells of the relational database. The column headings (attribute names) are known to the database software and are displayed, but are not part of the data. In a spreadsheet, column headings occupy cells.

❏ A relational database is a logical object about which conjectures can be made and theorems proved. An entire branch of computer science is devoted to work with relational databases.

Searching (and Indexing) in General

In the exercises for this chapter, you will be addressing the subject of searching for data. A considerable part of the discussion will relate to indexing names of data files. You may wonder why. The discussion below will explain the rudiments of indexing (which is basically sorting) and why it is important.

A computer can easily search a list of items for a particular item. It simply checks each item in the list against the "search string," and when it finds the two are equal, it declares success. For a trivial example, suppose we had the list of strings, each consisting of three characters and each associated with a number:

```
GHU    2343
UCK    7765
OPR    8828
PRO    1234
ZYX    7876
QWA    9500
ASD    3456
```

If we wanted to find the number associated with QWA, we could program the computer to compare the query (text string) QWA with the first string in the list: GHU. No match, so we go on to the next string (UCK). Again, no match. Finally, we would come to QWA in the list and get our answer: 9500. As you can probably guess, on average we would have to look through half the list to find a given string. In this case, it took six comparisons before the string was matched.

However, if we sort the list alphabetically (carrying along the associated number) we would have:

```
ASD 3456
GHU 2343
OPR 8828
PRO 1234
QWA 9500
UCK 7765
ZYX 7876
```

Now we can apply a clever strategy. We look first at the middle of the list: PRO. It doesn't match QUA, but we know that QUA is alphabetically "greater than" PRO, so we only need to be concerned with the bottom half of the list. We pick the middle of that for the next comparison: UCK. Now we know we've gone too far, so we go halfway back and find QWA. This process, called a binary search, took only three comparisons.

This example, which we've kept very small, only hints at the power we have. If our list to be searched consisted of two million strings, then the first searching method (sometimes called brute force) would, on average have to do a million comparisons. In the worst case, it would have to do two million comparisons. How many would a binary search have to do? In the worst case, about 22.

You may say that you realize the tremendous advantage of a binary search (in which you essentially "throw out half the remaining list each time and, thus, reap the rewards of reducing the list in an exponential fashion) but you point out that the list had to be alphabetized to begin with. Four considerations come in to play here. First, computers are very good at sorting—lots of very fast algorithms exist for the process. Second, the sorting can be done "in the background"—during times when you aren't sitting in front of the machine waiting for a search to succeed. Third, once the alphabetization has been done, innumerable sorts can be run—each highly efficient. Last, if a new element is to be added to the list, it can be inserted in its proper place in the list, without resorting the entire list.

Another Definition of GIS

Each record in a relational database references something: a person, object, idea, equation, feature, subject—some unique entity. In a GIS, the entity is usually a spatial three-dimensional feature such as a parking meter, lake, railroad, and so on.[25] These 3-D features are almost always reduced to abstract objects of fewer dimensions. The parking meter is a point, the railroad is a set of lines, and the lake is an area bounded by a sequence of lines.

Usually, a record in a relational database references a person or object without respect to its current location. A subject of a record in a relational database frequently moves around—like cars or people. However, when the subjects of a relational database are fixed in space, the position, or positions, of the feature may be included in the description of the feature along with the attributes. In one sense, the location of the object becomes one of its attributes.

---

[25]In raster-based GIS, the entity is a set of (usually) square areas that share common values. This approach will be discussed extensively in Chapter 8.

Points

Here, for example, is a map and part of a relational database that together describe fire hydrants in a town. (See Figure 1-13.) Each hydrant has a unique number and is, thus, the key field. Fire hydrants (each of which has a latitude and longitude specification) correspond one-to-one with the records in the database.

Let's consider another definition of GIS: From a computer software point of view, a GIS is the marriage of a (geo)graphical database and an attribute database (frequently a relational database).

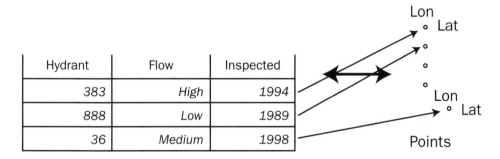

Hydrant	Flow	Inspected
383	*High*	1994
888	*Low*	1989
36	*Medium*	1998

FIGURE 1-13   Marriage of a point geographic database and a relational database table

Lines

As mentioned earlier, a GIS can store and analyze pseudo-one-dimensional entities such as roads. Here, each RDB table record would refer to segments of a roadway between intersections. The attributes might be number of lanes, highway number, street name, pavement type, length, and so on. See Figure 1-14.

Again: A GIS is the marriage of a (geo)graphical database and an attribute database.

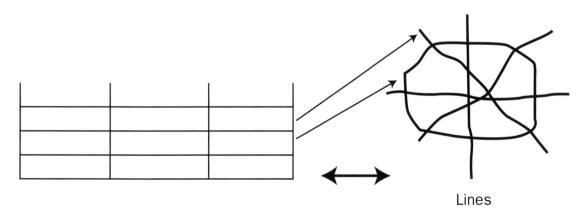

FIGURE 1-14   Marriage of a line geographic database and a relational database table

Areas

A GIS needs to be able to store information about areas, as well as points and lines. For example, such areas might be ownership parcels. Each parcel would be delineated by the lines determined by a land surveyor. Those lines define an area. A record would exist in the relational database that would correspond to the area. The attributes in the record might be owner's name, tax identification number, area of the parcel, and perimeter of the parcel. See Figure 1-15.

Did I mention? *A GIS is the marriage of a (geo)graphical database and an attribute database.*

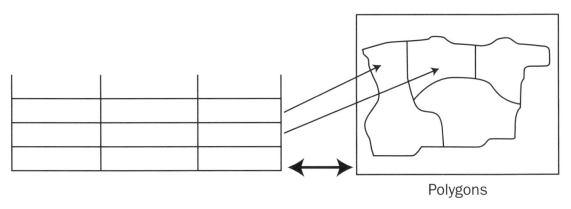

Polygons

FIGURE 1-15   Marriage of a polygon geographic database and a relational database table

Computer Software: In General

Computers, considered at the most fundamental level, do only three things:[26] get input, manipulate data, and produce output; that is, read bits, stir bits, and write bits. What tells the computer what to do? In the early days of computers, more than half a century ago, each individual instruction to a computer came from outside, one at a time. This process was soon automated so that external media, such as perforated paper tape or magnetic tape, contained the instructions. Then, several scientists[27] got the idea of placing the instructions in the store of the computer itself. This major breakthrough in computer development allowed the computer to execute one group of instructions and then, on the basis of testing bits in its memory (e.g., Is number "a" larger than number "b"?), execute another group of instruction in a different place in that memory. This let the computer simulate "reasoning" and "decision making." This concept of a *stored program* revolutionized computer use.

The modern computer may hold an immense number of data sets and store a large number of programs. If a given program or data set not being used usually resides in "slow" disk (usually electromechanical) memory. Dealing with large numbers of diverse elements requires a management scheme. With computers this takes the form of an *operating system* that allows the user to execute specific programs

---

[26]Fundamentals that we'll discuss further in Chapter 6.
[27]John von Neumann, J. Presper Eckert, John Mauchly, Authur Burks, Maurice Wilkes, and others.

when desired, connects the computer to other computers (e.g., by way of the Internet or other network), makes copies of files, and performs many housekeeping activities.

Writing sets of instructions to be stored in the computer's memory is known as *programming*. Developing and marketing *programs* (software packages) is an immense business. The operating system that your computer uses is a software package. The word processor used to write this text is a software package. And the GIS you are about to learn, known as ArcGIS from Esri, is a software package, or rather a suite of software packages.

Computer Software: ArcGIS in Particular

ArcGIS is an integrated GIS that consists of different principal parts. We are working with *ArcGIS Desktop* software, which is an integrated suite of advanced GIS applications. Esri also has interfaces for managing geodatabases in a database management system (DBMS). Also there is an Internet-based GIS for distributing spatial data and services. Esri also has software extensions such as ArcPad, which provides GIS capability for palmtop or handheld computers.

ArcGIS is called a *scalable product* because it allows for deployment of GIS of many different "sizes." A user can choose to have only a system running on a single computer—a *personal GIS*—that allows viewing and simple selection of spatial data without capabilities of editing or extensively analyzing that data. At the other end of the scale, a *multiuser GIS* can serve an entire company or governmental agency. This *enterprise GIS* allows all the capabilities of the software. Options also include systems of intermediate functionality known as *workgroup* GIS and *department GIS*.

For many products, the price charged depends on the features and utilities that the product provides. When one manufacturer produces many different products, users must learn how to combine and package them to provide the correct degree of utility required (you do not need a school bus if a sedan will do). This is also true of Esri products.

Software Focus of This Textbook

In this introductory text, we confine ourselves largely to ArcGIS Desktop. The other products pertain primarily to managing GIS data for large applications or organizations and to distribute GIS capability and data to users over networks. Our emphasis is rather on the analysis, synthesis, and display of spatial and attribute information. You will see an exercise on publishing map data on the Internet.

ArcGIS Desktop itself is divided in a number of different ways. It has three main software packages:

❏ ArcMap

❏ ArcCatalog

❏ ArcToolbox

ArcMap lets you put together graphic and geographic elements to make sophisticated maps and interact with them. It allows you to obtain information from those maps and the attributes associated with their components, by using a variety of processing methods.

ArcCatalog is mainly an operating system for geographic data. ArcCatalog allows you to set up shortcuts to reach particular files of data quickly and easily and examines those files visually and textually. You can use ArcCatalog to rename and copy spatial data sets. You can use a part of ArcCatalog, called the Catalog Tree, to place data into ArcMap.

You use ArcCatalog or ArcMap to access ArcToolbox, which provides immense geo-processing and analysis capability. It has tools to do 3-D analysis, spatial analysis, and spatial statistical analysis, to convert from one spatial data paradigm to another (vector to raster, and vice versa), and to convert spatial data into myriad geographic projections.

These three packages are fundamental to ArcGIS, but you have probably noticed many other Arc*Xxxxxxx* terms floating around. As a scalable product, ArcGIS desktop allows you to purchase what you need. The different levels of capability are bundled into four different packages. In terms of increasing utility (and cost) they are as follows:

ArcReader (free)

ArcView (as of version 10.1 of the software it is called "Basic")

ArcEditor ("Standard")

ArcInfo ("Advanced")

Knowing the capabilities of each is not important for our immediate purposes. You should know that some of these terms are recycled from earlier Esri product names. In particular, ArcView 10.x is a different animal from ArcView 3.x. ArcView 10.x is a level of capability of ArcGIS Desktop; ArcView 3.x is an older, but still supported and useful, independent software package.

CHAPTER **1**   **STEP-BY-STEP**[1]

Some Concepts That Underpin GIS

Exercise 1-2 (Project)

Developing a Fast Facts File for the Information You Learn

To make this textbook work well for you, I strongly recommend that you create and maintain a Fast Facts File—a computer text file in which you can record what you learn so that it is at your fingertips. It will serve you whenever you need to know a particular bit of information or perform an operation. Much of what you do you will put into your own fast memory (contained in your cranium), but there will be a number of facts that your fast memory may not contain when you need them. Here's where the Fast Facts File comes in. It is a computer-based equivalent of a loose-leaf notebook that you continually revise and update. There's where you should put procedures and concepts you might forget after a couple of weeks of doing work other than GIS.

For example, you might use the file to note techniques for changing symbology (colors and symbols that represent features on a map), which are addressed at various points in the text and the software. You can note the techniques down as you work with them and reorganize them later. The computer-searchable file helps you find what you need when you need it, even if you fall behind in organizing.

Periodically reorganize your notes. Occasionally print out the file and put it in a notebook. Periodically back up the file onto a flash drive, CD-ROM, or e-mail it to yourself. As you progress through this text, you will develop your own little book that will aid you in this course and thereafter. Also, because you write it, it will be organized in a way that meets your needs—both as a tutorial and a reference document.

——— **1.** Use a text editor (e.g., WordPad) or word processor (e.g., Microsoft® Word) to create the Fast Facts File. Put your name and other contact information in it. Initially include your computer account name (not your password—put that somewhere else), how to start various software

---

[1]The length of time most students will require to complete the exercises in the Step-by-Step sections can be found in this textbook in the Preface to the First Edition.

programs, and so on. The file will evolve as you study. For the moment, start the file and keep it open. It should at least contain the following:

Name: _____

User or Logon Identifier: _____

Password hint or secure location: _____

Do not save the file yet.

Understanding the File Structure for the Exercises

You will be working primarily with two major folders or directories. The first of these folders will be IGIS-Arc. It will contain the data for the exercises. It will start off as, and will remain, a direct copy of the IGIS-Arc folder that is on the DVD in the back of the book. If you are in class, operating off a network, your instructor will probably have put this folder somewhere in the network file structure. If you are using this book on your own, you should load the IGIS-Arc folder that is on the DVD directly onto your local hard drive. In either case, the IGIS-Arc folder should be protected so that changes cannot be made to it.

___ **2.** Locate or create the IGIS-Arc folder. It may be in the root folder of a hard drive on your computer, or it may be several levels down in a hierarchy. Whenever it is referred to in this textbook, it will be represented as:

[___] IGIS-Arc

The symbols [___] might represent something as simple as C:\ or it might be a long path such as:

U:\ABCNet\GIS_Students\GIS401

Write the path you will associate with the symbol [___] below:

[___] means _____.

You will use a second folder to store the work that you do. It will be specifically yours; your initials will be made part of the folder name. For example, if your name were John W. Stephenson, the folder would be called IGIS-Arc_JWS. In the next step, you will create this folder.

___ **3.** Decide where on the computer or network you want to store your work. Use the operating system of the computer to make a new folder by navigating to and selecting the appropriate drive or path, clicking the New Folder button in Windows 7 or, in Windows XP, opening the File Menu, and clicking Folder when it shows up under New. Once the folder is created, change its name to

IGIS-Arc_YourInitials

(For example, IGIS-Arc_JWS).

In this textbook the simple blank, ___ will designate the disk drive and/or path to the folder where you keep your work.

Write the path associated with the symbol ___ below:

___ means _____.

_____ **4.** Save your Fast Facts File in the folder you have just created, giving it the name FastFactsFile. In other words, the full path and filename for your new Fast Facts File will be as follows:

`___IGIS-Arc_YourInitials\FastFactFile.someextension`

To recap: when I use the designation

`[___]IGIS-Arc`

(note the square brackets), I mean the place where the data sets on the DVD have been placed on your computer system. You should not change any of these folders or files.

When I use the designation

`___IGIS-Arc_`**`YourInitials`**

(note the presence of **YourInitials**), I am referring to the location that contains your work.

Exercise 1-3 (Minor Project)

Getting Set Up with ArcGIS

A click or two of the mouse should get you to the ArcGIS software package. This should provide easy access to two major components: ArcCatalog and ArcMap. They, in turn, provide access to a third component; ArcToolbox.

Depending on how your computer is set up, you may reach a component by clicking an icon on the desktop, or by navigating to it by clicking Start > All Programs > ArcGIS > the component name—or via some other way prescribed by your instructor.

_____ **1.**[2] Find the name or icon for ArcCatalog, right-click it and choose Properties. In the ArcCatalog Properties window, the Target should be something like:

`C:\Program Files\ArcGIS\Bin\ArcCatalog.exe`

Is it? If not, write the target here:

_____.

Find and write the target for ArcMap.[3]

_____.

---

[2]The blank to the left of each step number provides a space where you can put a check mark when you have completed the numbered step.
[3]As you work through the text, you will find blanks in which to enter requested information. Doing this keeps you on the right track. Rather than writing in the book, you might want to use notebook paper or type into a document file. Ask your instructor how best to proceed.

—— **2.** Start ArcCatalog by double-clicking the icon you found in Step 1.

ArcCatalog serves as sort of an operating system for GIS. You manage data with it. The Step-by-Step part of this chapter is largely an introduction to ArcCatalog. (In ArcGIS 10, a part of ArcCatalog can also be accessed from within ArcMap as a sidebar item as you will see shortly.)

—— **3.** Dismiss ArcCatalog by selecting Exit from its File menu.

—— **4.** Start ArcMap. Look under New Maps, then My Templates and agree to start using ArcMap with a Blank Map.

ArcMap is the software package that performs primary and major GIS operations, such as making maps and analyzing spatial data. The Step-by-Step part of Chapter 2 serves as the introduction to ArcMap. Both ArcMap and ArcCatalog allow you to access ArcToolbox, which is used for many advanced GIS operations, such as data conversion.

—— **5.** Dismiss ArcMap by clicking the "X" in the red box in the upper-right corner of the window.[4]

—— **6.** If available, start Arc, which is one of the modules within ArcInfo Workstation.

The programs Arc, Arcedit, Arcplot, Grid, and some others are instructed through text commands that are typed by the user after a prompt (such as Arc:). Contrast his with most familiar programs in which you "point and click." Typed input from a command line directed Esri software prior to ArcGIS version 8. The Graphic User Interfaces (GUIs) of ArcCatalog and ArcMap are relatively efficient and elegant ways to invoke Arc and do serious GIS work. Unfortunately, ArcInfo Workstation is not going to be around much longer; Esri is phasing it out with ArcGIS Desktop 10.1. If you have it available, learning ArcInfo Workstation provides a realization as to how much GUIs can slow things down while at the same time making life easier.

—— **7.** If you started Arc Workstation, leave the Arc program by typing Quit at the Arc: prompt.

Exercise 1-4 (Project)

Looking at the ArcCatalog Program

Esri has developed ArcCatalog as an entire product that is basically designed to help you find, select, understand, and manage geographic data files. You'll begin by starting this software and exploring it. Then, considering the definition of GIS—that it is an information system (IS) whose database is a marriage of geographic database (GDB) and relational database (RDB)—you'll use ArcCatalog to look at a ridiculously simple GIS dataset. A village has developed a system to help with town planning and maintenance. One part of the GIS is a set of features consisting of fire hydrants. You will find the Village fire hydrants theme, named HYRDANTS and explore that dataset.

---

[4]You dismissed ArcCatalog in a different way. When there is more than one way to accomplish something (almost always the case), I will usually show you alternatives, just by having you use them. Note them in your Fast Facts File.

Anatomy of the ArcCatalog Window

Assuming your computer is on and some version of Microsoft Windows® (or UNIX or other operating system) is running:

___ **1.** Start the ArcCatalog component of ArcGIS. Make it occupy the full screen by double-clicking the ArcCatalog title bar. Open the File menu and select Connect to Folder. In the Connect to Folder window navigate to Computer, then to the Local Disk (C:), which you will find under Computer Click that. Click OK.

___ **2.** Click the word Folder Connections at the top of the left-hand subwindow (pane). If you see any negative (minus) signs in the left pane, click them so they become positive (plus) signs. Now click the plus sign to the left of Folder Connection. You should see C:\ in the ArcCatalog window under the Contents pane, similar to Figure 1-16. Double-click on it and the names of the folders on the C:\ drive of your computer will appear.

___ **3.** Click Folder Connections again. Examine the ArcCatalog window.

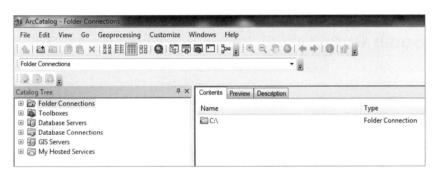

FIGURE 1-16

That left pane is called the Catalog Tree. At the top level, it provides a view of the slower-speed storage devices (hard drives, network drives, CD-ROMs, and so on[5]) of the computer, plus some other entries to be explained later. At the very top left, in the title bar of the overall window, will be some text. What does it say? (Fill in the blank below, if there is any text between ArcCatalog and Folder Connections.)

ArcCatalog - _____ - Folder Connections

In ArcGIS Desktop version 10.0, examining the upper-left corner of an ArcGIS software window is a way to know (1) which software component you are using (ArcCatalog, ArcMap, and others), and (2) the license your computer is operating under: ArcView (the least expensive with the least capability), ArcEditor, or ArcInfo.

With ArcGIS Desktop version 10.1, examining the upper-left corner of an ArcGIS software window is still a way to know which software component you are using (ArcCatalog, ArcMap, and others). But the licensing level is no longer revealed. There are still three levels (now called Basic, Standard, and Advanced) but to

---

[5]If you don't have any other folder connections, you are most likely starting ArcCatalog for the first time. These drives may or may not show up, depending on how your system is configured. Even if they don't appear here, ArcCatalog has a way to make them accessible to you, in the same way that you added the "C" drive.

find out which you are using you can click on Help in the Main menu and then About ArcCatalog. What License Type are you using?

_____

(Several of the exercises in this text require ArcInfo (Advanced) to run through completely. However, for those of you who have only ArcView (Basic) or ArcEditor (Standard), the files that ArcInfo (Advanced) would have developed have been included on the DVD so that you can continue past the spots that require this highest level of functionality.)

The title bar of the ArcCatalog window also shows you the path (disk drive and folder) that is currently being referenced by the software (in this case, it is the root folder of the C drive, which is C:\). This information also appears in the Location text box, which, unless it has been moved or removed, may be found near the top of the window.

Many of the menus, icons, and buttons of the ArcCatalog window may be familiar to you from your work with other software. Assuming that your ArcCatalog window has not been customized, you will find the menu headings File, Edit, View, Go, Geoprocessing, Customize, Window, and Help.

Setting Some Options[6]

___ **4.** Be sure that the "C" drive icon is selected in the Catalog Tree. Under File, click Properties, then click the General tab. (From now on I may abbreviate an operation like this by saying File > Properties > General.) Since C:\ is selected, you will see some of the properties of the C drive, including the amount of disk space—both used and free. See Figure 1-17. Dismiss the Properties window by clicking the "X" in its upper-right corner or the Cancel button.

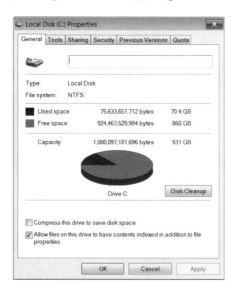

FIGURE 1-17

_____

[6]In general, even though you set options, you cannot count on those remaining set if you terminate ArcCatalog and restart it later. You may have to return to this procedure to reset options.

5. On the Main menu click Customize > ArcCatalog Options > General tab. Several check boxes will show up in the ArcCatalog Options window. These relate to the types of top-level entries that the catalog will display. Since you are at an early-learning stage, you want to have ArcCatalog display everything possible. By clicking the boxes next to the options in the pane, you can toggle a check mark on or off. Make sure each option has a check mark in the box to its left. At the bottom of the window, make sure both of the options (especially Hide file extensions) are unchecked. Click Apply. (If you made no changes in the pane, Apply will be "grayed out"—a standard feature of ArcGIS software if an action is not possible, not needed, or not appropriate.) The resultant window should look approximately like Figure 1-18.

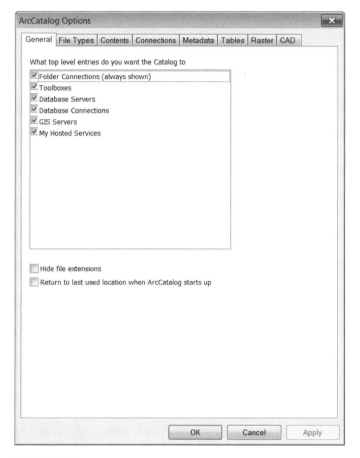

FIGURE 1-18

6. Click the Contents tab in the Options window. Here, you may specify the information you want to see about each spatial data set in the catalog when you ask for Details. Read the title over each pane. In the upper pane, put check marks by Size and by Modified. In the lower pane, put a check by Projected. The window should look like Figure 1-19. Click Apply. Click OK.

FIGURE 1-19

The Catalog Tree

As you recall, the main portion of the default ArcCatalog window is usually divided into two panes. On the left you find the Catalog Tree.[7] The Catalog Tree, when expanded, gives an overall view of the names of your data sets. You can also look with varying levels of detail at all the data on all the disk drives of the computer by expanding the entries (items)[8] in the tree. In this way, ArcCatalog's presentation is like Windows Explorer. A plus sign (+) indicates that (usually) a given disk drive or folder contains additional folders or files that are hidden from view. A minus sign (−) preceding an entry indicates that any additional folders or files that are directly contained within an entry are displayed below that entry. A click on a plus sign changes it to a minus and expands the entry; a click on a minus sign collapses the list and shows a plus again.

—— **7.** With ArcCatalog running, press the Contents tab in the right pane to make it active. In the Catalog Tree, click the hard drive designator of the path [____].[9] The folders on that hard drive will appear in the right pane. Click the plus sign next to the entry for the hard drive. The folders

---

[7]For future reference, you can (1) "Auto-Hide" the Catalog Tree by clicking the pushpin (experiment with this). If you close the Catalog Tree (either accidently or on purpose to provide more visual space for the right-hand pane) you can make it reappear using the Windows menu.

[8]If you are a user of ArcInfo Workstation, or earlier versions of ArcInfo, you know that the term "item" refers to a column heading indicating a coverage's attributes. Although ArcGIS Esri manuals refer to a line of text in the Catalog Tree as an "item," I will refer to it as an "entry."

[9]This could be as simple as C:\. Recall that you recorded the drive and path (if any) of the [____] designator in Step 2 of Exercise 1-2 above. In the event the path to IGIS-Arc does not appear, click File > Connect Folder and navigate to the IGIS-Arc Folder. Select it with a single click. Then say OK to the Connect to Folder window.

that are within that path are then displayed, approximately mirroring the contents of the right pane. From the left pane, navigate (if necessary) to IGIS-Arc and highlight it. Click the plus sign in front of the name. In the right pane you should see sixteen entries: Address_Geocoding, Elevation_Data, and so on.

You may select an entry name in the Catalog Tree, such as River, by clicking the name. When you do so, the list of contents of the selected entry shows up in the right-hand pane (provided the Contents tab is selected). (That is, the contents of the right-hand pane is the same as what you get if you expand the selected entry in the left-hand pane.) This conveniently lets you see your data names at two levels in different areas of the screen. You can select a disk drive, folder, or file from the left pane, and the selection is reflected in the right pane.

In the *left-hand* pane (with the Contents tab of the right-hand pane active):

❏   If you click a folder or file, it becomes selected. If it is a folder, its contents are revealed in the right-hand pane. If it is a file, it may be described in the right-hand pane, or its contents may be shown.

❏   If you double-click a folder, it is selected. It is also expanded or collapsed in the left-hand pane, depending on its previous state.

❏   If you double-click a file, the operating system does something with it—executes it, opens it with an appropriate program, or opens a window that displays its properties. For example, if the file is a text file, it may be displayed in the Notepad text editor.

❏   If you right-click an entry, you have access to a menu of commands that let you do things like copy, paste, delete, rename, make something new (such as a folder, geodatabase, shapefile, coverage,[10] and so on), search, or reveal the properties of the entry.

In the *right-hand* pane, with the Contents tab active:

❏   If you click a folder or file, it becomes selected. (The parent folder name is semi-highlighted, perhaps in light gray, in the left-hand pane, just for reference.)

❏   If you double-click a folder, its contents are revealed. To return to the parent folder, click the bent up-arrow icon at the extreme left of the Standard toolbar.[11]

❏   If you double-click a file, the operating system does something with it—it executes it, opens it with an appropriate program, or opens a window that displays its properties.

❏   If you right-click an entry, you receive a menu of commands that may let you do things like copy, paste, delete, rename, make a new something (folder, shapefile, and so on), search, or reveal the properties of the entry.

—— **8.**   Experiment with selecting entries in each pane. Try out the auto-hide feature, which looks like a push pin, on the Catalog Tree. Click the push pin symbol. Then "mouse over"[12] the

---

[10]Geodatabases, shapefiles, and coverages are types of GIS data sets, which will be discussed shortly.

[11]In the default ArcCatalog window (Figure 1-16), this icon will appear on the line below the Main menu, just under the File keyword, but it may have been moved in your ArcCatalog window.

[12]Mouse over means to use the mouse to make the cursor hover over the item you are interested in.

words "Catalog Tree" and observe the results. Click the push pin again to re-establish the Catalog Tree, thus turning off Auto Hide. Another feature to experiment with: drag the dividing line separating the panes to the left, then to the right, to change the space devoted to each. Notice that, if there is not sufficient room for the length of the longest entry's text string, a scroll bar appears at the bottom of the window, which lets you look at different parts of the window. Further, if you pause the cursor over a text string that is not completely visible, the entire contents of the string will show up in a box that overlays the incomplete string.

When starting ArcCatalog for the very first time after installing it, the pane on the left side will not have any Folder Connections. No local drives, no network drives. Step 1 of this chapter gave one method of connecting to the C:\ drive so that at least that will be visible. You may (and should!) generate additional references to data sets that you are particularly interested in or want to work with so they will be easily accessible. This is called "connecting to a folder" and is a special feature of ArcCatalog that provides a shortcut to data. In the next steps, you will make a connection to a folder containing the HYDRANTS data for the village you will explore.

____ **9.** Collapse all the entries in the Catalog Tree completely by clicking each "-". Click Folder Connections in the Catalog Tree.

Connecting to a Folder

____ **10.** From the File menu, select Connect To Folder. The Connect to Folder window appears. Using that window, expanding the entries in its tree as necessary, starting with Computer, navigate to a folder named[13]

```
[____]IGIS-Arc\Village_Data\
```

Click that folder so that its name is highlighted. See Figure 1-20. Press the OK button. Observe that the Catalog Tree (in the left pane) now contains an entry (currently selected) that gives you access directly to the desired folder. Also, the path to that folder is shown in the Location text box and in the title bar of the ArcCatalog window.) Now whenever you want to access that folder, you merely need to find this direct shortcut in the Catalog Tree, rather than having to navigate to it through levels of hierarchy.

Basically, connecting to a folder provides a shortcut that you should always set up when you anticipate needing a data set more than once or twice.

____ **11.** Using File > Disconnect Folder, remove the entry you just made. Then, restore it with the Connect To Folder icon, which you will find on the Standard toolbar that is located under the Main menu. The icon resembles the one associated with File > Connect Folder. Finally, make a folder connection with the [____]IGIS-Arc folder itself.

---

[13]Recall that [____] is the path designation of the data for this text. You determined it in Step 2 of Exercise 1-2.

FIGURE 1-20

The Toolbars and the Status Bar

___ 12. Under View, turn off the Status Bar. The information bar at the bottom of the window disappears.

___ 13. Turn the Status Bar back on so that messages are displayed at the bottom of the window. To see how the Status Bar shows information, find a red icon that looks like a tool chest on the Standard toolbar. Move the mouse cursor pointer over it. What happens now depends on what version of ArcGIS Desktop you are running. In Version 10.0 of the software the Status Bar should read Opens the ArcToolbox window so you can access geoprocessing tools and toolboxes. A label (called a ToolTip) identifying the tool as the ArcToolbox Window will appear next to the cursor.[14] In version 10.1 and later, the status bar does not indicate ArcToolbox, but a more extensive explanation appears in the ToolTip. In version 10.1 you can press F1 for more explanation. Also experiment by clicking some files and folders in the Catalog Tree and observing the Status Bar. In particular, look at the Status Bar as you select [___]IGIS-Arc. Expand that entry. Select Village_Data. Expand. Select Water_Resources.gdb. Expand. Select

---

[14]Provided ToolTips is set to on. To control this: Choose Customize > Toolbars > Customize... > Options (tab) > Show ToolTips On Toolbars (make sure box is checked).

Hydrants. Expand. Select Fire Hydrants. Press the Preview tab at the top of the right-hand pane. In the pane, you will see the points that represent the hydrants in Village Data.

_____ **14.** Under Customize > Toolbars, make sure the Standard, Geography, Location, and Metadata toolbars are all turned on. Active toolbars have a check to the left of their names.

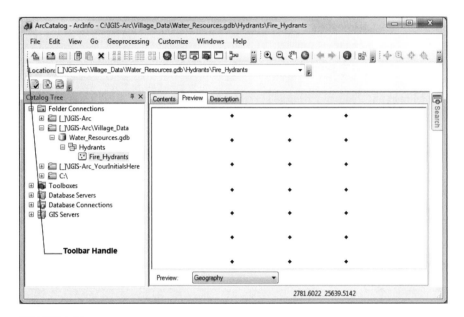

FIGURE 1-21

_____ **15.** Locate a subtle column of three or four dots at the extreme left of the Standard Menu bar. Consult Figure 1-21 to see where this "toolbar handle" is. Using that handle, drag the toolbar to the upper left of the right pane. Also use this technique to drag each of the other toolbars so that the window resembles Figure 1-22. By "undocking" the toolbars in this manner, you can see the name of each. (When the toolbars are docked, the names are hidden to save space.) Also move the Catalog Tree by dragging it by its top. Notice the various icons that appear as you drag the Catalog Tree. Experiment with the options offered by dragging the mouse pointer to an icon as you drag the Catalog Tree. Another experiment: turn the tree off by clicking the "X" in its upper-right corner. Turn it on again by clicking the Catalog Tree icon in the Standard Toolbar. Re-dock it in its original position by dragging its title bar into the furthest left-hand arrow.

_____ **16.** Run the mouse pointer over each of the buttons on the Standard toolbar. Notice that, if you pause for a second or so over an option its name appears—possibly with a description, depending on the version of the software you are using. In version 10.0 of the software you can see an instantaneous and more complete description of the tool by looking at the Status Bar at the bottom of the window.

The Standard Toolbar provides icons that let you perform frequently used actions. Of course, as with many software packages, there are often several ways to perform an action. Some of these alternative procedures are also given.

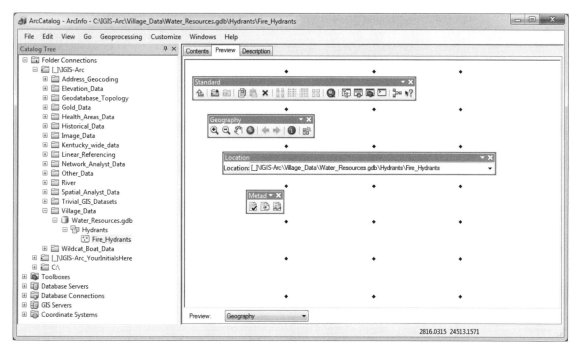

FIGURE 1-22

	Go up a level from whatever selection you have made in a pane. (You can access the top of the Catalog Tree by repeatedly clicking this Up One Level icon on the toolbar.)
	Make a folder connection (substitutes for File > Connect To Folder).
	Remove a folder connection (instead of File > Disconnect Folder).
	Copy, paste, or delete a selection (or use Ctrl-C, Ctrl-V, or Delete keys).
	View folder contents in a variety of ways.
	Launch the ArcMap software package.
	Show the Catalog Tree
	Search for data on the computer, a network, or the Internet, using a variety of criteria.
	Show the ArcToolbox window.
	Show the Python window.

Start the ArcGIS Model Builder Program

 Access specific help on a tool or command.(This icon appears only in Version 10.0. Version 10.1 provides help by mouse-ing over a particular tool or command.)

17. Explore the Geography toolbar. This toolbar lets you look at (geo)graphic[15] data in a number of ways.

 See features in more or less detail by zooming in and out.

 Pan: shift your view of the area of interest

 Look at the entire geographic region depicted by the data set

 Go to Previous or Next extent.

 Identify, with text, various features indicated by the pointing cursor.

 Make thumbnail sketches of the geographic file of interest so you can recognize it when searching for particular data sets.

18. Explore the Location Toolbar. (What's being referred to here is a location in a computer system, not a location on Earth.) The Location Toolbar displays a text string showing the current selection from the Catalog Tree and the complete path to that selection. Also, you can type in a different path and selection. If you type in a path and selection (or simply select the text string that is shown) and press Enter, the string will be placed in the drop-down text box menu below the Location Toolbar, making it easy for you to select this location next time. This provides another way to quickly access data you are using. ArcGIS is full of shortcuts!

19. Explore the Metadata Toolbar. Metadata is "data about data." It is usually in text form. There are different standards and styles.

If you click the Description Tab (next to the Contents and Preview tabs), the toolbar will become active.

This toolbar lets you check the metadata against a standard (validate), export the metadata, and examine the metadata properties:

 Validate metadata.

 Export metadata.

 Metadata properties.

You may have to check with your instructor or network supervisor to be sure that you have the metadata service packs and patches installed.

---

[15]What you are looking at is, of course, graphics. But since the image is tied to positions in the real world, I use the nonstandard term (geo)graphic.

(You can add one more Metadata button, Item Description, which will open a separate window. It displays the item description (metadata) and a preview. This is helpful for editing metadata because you can see the changes you're making and the data without exiting the metadata editing session. You will learn how to add buttons to a toolbar later.)

——— **20.** *Restore the toolbars to their "home" positions:*[16] When you are finished exploring, you will want to put the toolbars back into the positions shown in Figure 1-16. This can be done almost automatically. If you double-click the title of a toolbar, it will align itself horizontally, in the position it was in before you moved it; its title will be hidden.

An Optional Step

By default, ArcGIS rewrites the metadata for a feature class every time you look at it. I recommend that you not let it do that. Not using the default puts the metadata under your control, so you determine when it is to be rewritten. The disadvantage is that you might forget to update metadata when something changes. So, the following step is optional but recommended.

——— **21.** From the ArcCatalog menu, select Customize > ArcCatalog Options. Pick the Metadata tab. See Figure 1-23. Read the information provided on the tab. Change the Metadata Style to FGDC CSDGM Metadata. (If you do not see this option, check with your system manager as to whether the latest service pack is installed.) Then remove any check in the box that relates to automatically updating metadata when it is viewed. This prevents ArcGIS from automatically updating the Metadata each time you access it. However, it now puts the responsibility on you to update the Metadata when it is appropriate to do so. The box asking that you be shown a prompt for metadata upgrades should be checked. Click Apply, then OK.

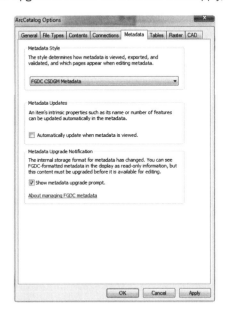

FIGURE 1-23

----

[16]A note on using this text: When ***bold italic*** follows the check box (e.g., __20), it constitutes a general direction. You should read what follows the bold italics to see what specific actions you are to take.

Exploring Basic GIS Data Storage Models

As you know, large amounts of computer data are usually stored on devices called disk drives. They are usually mechanical, with a motor spinning a disk with a magnetic coating, but they can also be completely electronic – as in flash or thumb drives. Disk drives constitute the "slow memory" of a computer. It takes a computer thousands of times longer to retrieve data from slow memory than from fast memory. An analogy with human processes might be that, in your brain—where you remember your friend's names, your home telephone number, or the route you drive to work—is located your "fast memory" or your "electronic memory." On the other hand, the name of your uncle's second cousin, the phone number of the dry cleaner, or how to navigate to Punxsutawney, Pennsylvania, are things that you probably have to look up; they reside in slow memory—family documents, telephone book, and road map (or a GIS).

Most of the data that exists in the world on computers is, of course, not in use (not being printed, analyzed, or otherwise processed) at any given moment and, therefore, is not contained in the "fast" electronic memory of a computer. Data sets in the slow memory of a computer are, in their most basic form, just sequences of 0s and 1s. However, beyond that, well beyond that, such data sets are organized by storage paradigms. You are familiar with the idea of folders on a disk drive. Each folder can contain files and other folders. The operating system of the computer is responsible for keeping files and folders straight. Files may consist of binary sequences that result in typed text, music, photographs, and other products when processed by a computer. For a number of reasons, GIS data sets are composed of fairly complex combinations of folders and files. Also, techniques for storing spatial data sets have evolved over a number of years, so that, even within the single company Esri, you will find several storage paradigms or formats. Some of these formats have been devised because of the types of data being represented.

Other formats come from different inventions based on progress in hardware and software development. In what follows you use the Status Bar to look at the terminology associated with several of them: geodatabases, shapefiles, coverages, rasters, terrains, and TINs.

___ **22.** Check that the options you set in steps 5 and 6 in this exercise are still as you left them. If not, reset them. Make sure the Contents tab is active. Collapse the entries in the Catalog Tree as much as possible; you should see nothing in the leftmost boxes but little plus signs. Expand Folder Connections. Click on the text [___]IGIS-Arc designation. Write here what the status bar indicates: _____. Expand the entry. Find the entries indicated below, expanding them when possible. Write the Status Bar text string you see when you click on the icon or text for each entry. (Those who are artistically inclined may draw the icon they see next to the entry.) These are many of the various structures and data formats you will be encountering as you work with ArcGIS.

Village Data _____

   Water_Resources.gdb _____

      Hydrants _____

         Fire_Hydrants _____

River _____

    Boat_SP83.shp _____

    COLE_DRG.tif _____

    COLE_DOQ64.jpg _____

    COLE_TIN _____

    Terrain_from_LiDAR.gdb _____

    KY_SP_North_NAD83 _____

    LIDAR Points _____

    River_Bend_and_Water_Plant_Terrain _____

Dismiss ArcCatalog.

Exercise 1-5 (Major Project)

Exploring Data with ArcCatalog—Fire Hydrants in a Village

Copying Data over to Your Personal Folder

As mentioned earlier, ArcCatalog serves as a sort of operating system for GIS data. One function of an operating system is to make copies of folders and files. Important for avoiding catastrophes: When working with GIS data in Esri formats you should *always* copy data sets using ArcCatalog—*never* use Windows, UNIX, Linux, or any other primary operating system to copy data sets. You can, however, use the computer's operating system to copy entire folders, provided that they encompass all related spatial data.

In using this text, you will use your working folder

    ___IGIS-Arc_*YourInitials*

to do most of the exercises. The idea is to leave the [___]IGIS-Arc folder contents in pristine condition. So, when you are going to work with the sample data from this textbook, you will usually copy over the data sets to your personal folder: ___IGIS-Arc_*YourInitials*. We'll start with the Village Data.

___ **1.** Start ArcCatalog. Expand [___]. Expand IGIS-Arc. Select Village_Data. Right-click on the selection and pick Copy. Select

    ___IGIS-Arc_*YourInitials*

Right-click on the selection and pick Paste.

___ **2.** Make a folder connection with

    ___IGIS-Arc_*YourInitials**Village_Data*

Recall that you previously made a folder connection to [___]IGIS-Arc\Villiage_Data. Remove that folder connection (so you won't get confused and work on the wrong data set).

3. Collapse the Catalog Tree as much as possible by clicking on all the "minus boxes." Expand Folder Connections, then select the folder connection you just made (with Village_Data) and expand it as much as possible. Select Fire Hydrants. The Catalog Tree should look something like Figure 1-24. If you see serious discrepancies, delete the folder Village_Data from ___IGIS-Arc_*YourInitials* and try again. (Be careful not to remove the entire folder ___IGIS-Arc_*YourInitials* because, if you do, your FastFactsFile will go away.)

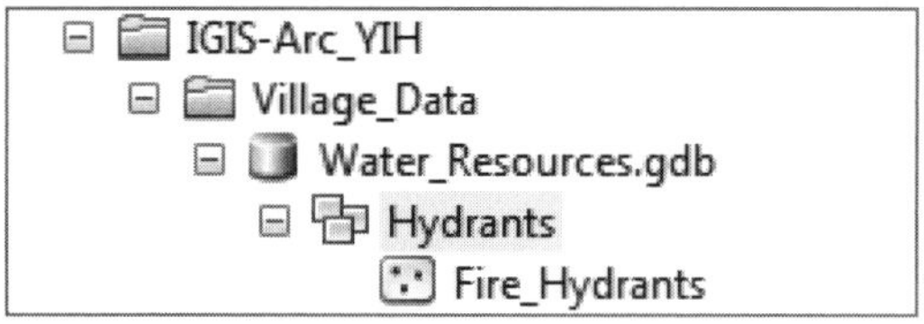

> ⊟ 📁 IGIS-Arc_YIH
>   ⊟ 📁 Village_Data
>   ⊟ 🗄 Water_Resources.gdb
>   ⊟ 🔁 Hydrants
>   ⊡ Fire_Hydrants

FIGURE 1-24

4. Select Fire_Hydrants. Click the Contents tab in the right pane. You now see references to Fire_Hydrants in both panes.

As you determined earlier, Village_Data is a file folder. Within it is a File Geodatabase named Water_Resources.gdb. Within Water_Resources.gdb is a Geodatabase Feature Dataset named Hydrants. Within Hydrants is a File Geodatabase Feature Class named Fire_Hydrants consisting of 21 points that represent the locations of the fire hydrants in the village. Explore Fire_Hydrants in the following steps.

5. Select Fire_Hydrants in the left pane. In the right pane, you see the name and type of that component (feature class). Because of settings you made earlier,[17] you also see its size (in kilobytes) and the date it was last modified, (and you would see its geographic projection if it had one). You also see a blank rectangular image. This will later become a miniature image of the coverage called a thumbnail; you will create it shortly. See Figure 1-25. If you now select the Preview tab at the top of the right pane, you will see an image showing the positions of these hydrants. See Figure 1-26. This is hardly an exciting picture but there is a lesson here so

FIGURE 1-25

---

[17]Repeat Step 6 in Exercise 1-4 if ArcCatalog Options have become unglued.

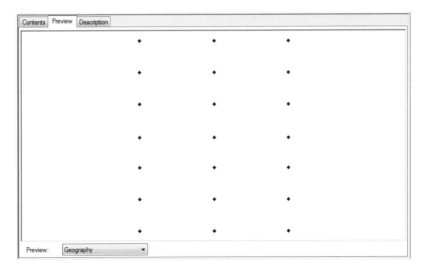

FIGURE 1-26

OBJECTID *	Shape *	X_COORD	Y_COORD	COLOR	TYPE	FLOW_RATE	CONDITION
1	Point	3300	25733	Red	Fire	250	OK
2	Point	3900	25733	Red	Fire	250	OK
3	Point	4500	25733	Red	Fire	250	OK
4	Point	3300	25483	Red	Fire	250	OK
5	Point	3900	25483	White	Fire	365	OK
6	Point	4500	25483	White	Fire	365	DESTROYED - Mr. Watkins ran over it with his Buick
7	Point	3300	25233	White	Fire	365	OK
8	Point	3900	25233	White	Fire	365	OK
9	Point	4500	25233	White	Fire	365	OK
10	Point	3300	24733	White	Fire	365	OK
11	Point	3900	24733	White	Fire	365	OK
12	Point	4500	24733	White	Fire	365	OK
13	Point	3300	24483	White	Fire	365	OK
14	Point	3900	24483	Yellow	Fire	400	OK
15	Point	4500	24483	Yellow	Fire	400	OK
16	Point	3300	24233	Yellow	Fire	400	OK
17	Point	3900	24233	Yellow	Fire	400	OK
18	Point	4500	24233	Yellow	Fire	400	OK
19	Point	3300	24970	Yellow	Fire	400	OK
20	Point	3900	24970	Yellow	Fire	400	OK
21	Point	4500	24970	Yellow	Fire	400	OK

Preview: Table

FIGURE 1-27

please be patient. If you now select Table from the Preview drop-down menu at the **bottom** of the right pane, you will see a table of 21 rows and several columns. See Figure 1-27.

Each *row* (record) of the table represents one hydrant. Each *column* (field) of the table is an attribute (that is, a property or characteristic) of the feature class Fire_Hydrants. Each *cell* in the table (where a given row and given column intersect) indicates the particular attribute value for the particular hydrant.

6. Locate the record of the hydrant that has an OBJECTID of 4. Click on the 4, and the little gray box to its left will show a small triangle. Scroll the table horizontally (if necessary), using the scroll bar at the bottom of the window. You will see that the hydrant COLOR is Red.

Illustrated here is the essence of GIS: the marriage of a geographic database and an attribute database. Each feature in the spatial field has associated with it a row in a relational database table that provides information about the feature. This idea, with variations, is the fundamental underpinning of most of what you will learn about GIS.

_____ **7.** Reread the preceding paragraph.

Examining the Table

_____ **8.** The table has a "current record" that, when you first view the table, is the first record. You can change that. Since you clicked on a cell in the fourth row, the current record is now 4. It is marked by a triangle in the box on the left of the record. The number of the current record is also shown **in** the Record text box at the bottom of the window. You also see the total number of records. You can change the current record by doing any of the following:

❏ Clicking in a box to the left of the OBJECTID field

❏ Clicking a cell of a record

❏ Typing the desired record number in the Record text box (and pressing Enter)

❏ Clicking on the buttons on either side of the Record text box.

Experiment with the procedures described above, making various records the current record.

You can scroll the viewable area of the table with the horizontal scroll bar and, if the table has a lot of records, the vertical scroll bar. You can also change the widths of the columns for better viewing by dragging the dividing lines between the column headings. (To see more of the table, you can shrink, hide, or dismiss the Catalog Tree, but the Tree is probably something you will be using frequently, so that should come into consideration.)

_____ **9.** Both the X_COORD and the Y_COORD columns are excessively wide. Place the mouse pointer on the column heading text line and drag the column separators to the left to reduce the widths of these columns. If you double-click a column separator, the column to the left immediately tailors itself to the size that will show the entire column heading with little space wasted.[18] However the column may not be wide enough to completely show the values in the cells of that column.

_____ **10.** Make record 12 the current record. With the keyboard arrow keys, locate the COLOR value of the record. What is the FLOW_RATE of the Hydrant? _____ What are the geographic coordinates of the hydrant? _____, _____.

(Should you wish to do so, you can change the cosmetic appearance of a table through Customize > ArcCatalog Options > Tables.)

_____

[18]If an entire column becomes selected while you are doing this don't worry about it. Double-click on the separator.

Deriving Information from the Table

Having attribute information in table form allows you to obtain statistics, to search for specific text strings, and to sort the information.

___ **11.** You can find out the average of the values of a column along with the count of the number of values, minimum, maximum, sum, and standard deviation.[19] To do so, you right-click the column heading to bring up a menu of choices from which you choose Statistics. For example, select the column FLOW_RATE by placing the mouse pointer over the heading of the column (the pointer becomes a down-pointing arrow) and right-clicking. (You may have to use the scroll bar at the bottom of the table to find the FLOW_RATE column.) Choose Statistics from the drop-down menu. You see information about the values in the column from the Statistics of Fire_Hydrants window. There are 21 values. The minimum flow rate is 250. What is the maximum flow rate? _____. You also receive a frequency diagram that shows the *number* of hydrants that have each given value (e.g. there are four hydrants with a low flow rate, nine with a moderately high rate, and eight with a high rate). If all the hydrants were turned on at once, how much water would flow? _____. The mean (i.e., average) and standard deviation, although they probably don't mean much in this situation, are about 356 and 54, respectively.

___ **12.** Once you have the Statistics window up, you can determine the statistics of other columns by selecting the column name in the window's Field drop-down menu. What is the average y-coordinate (to two decimal places)? _____. Dismiss the Statistics window.

Sorting the Records

___ **13.** Suppose now that you wanted to have the list ordered so that all the westernmost hydrants (those with the smallest x-coordinates) appeared at the top of the table. Click the X_COORD column heading to make it active. The column becomes highlighted in blue/cyan/turquoise—take your pick. Now right-click the column heading, then sort the values in the column by picking Sort Ascending from the drop-down menu. Notice that the values of the column are now in order from smallest to largest.

Note that when a value in a given record is moved, say during a sort, the entire record moves with it. That is, all of the values in a given row stay together (including OBJECTID), *regardless*. (If you are familiar with spreadsheets (e.g., Excel), you should note that this is different behavior from results you may receive with that software.)

If you wanted the Y_COORD column values to be a secondary sort (sorted "within" the X_COORD values), you could select both columns by holding down the CTRL key while clicking). (An example to clarify the terminology: in a telephone directory the first names are sorted "within" the last names.) When two or more columns are selected, the leftmost one will contain the primary sort, the next selected on the right will contain the secondary sort, and so on.

---

[19]You are probably familiar with these concepts. A discussion of basic statistical measures may be found in the Overview of Chapter 6.

___ **14.** Sort ascending the X_COORD and, within it, COLOR. Examine the results.

If having the leftmost column as the primary sort order doesn't suit you, you must rearrange the columns, as demonstrated next.

___ **15.** Select the COLOR column with a click. Hold down the left mouse button. The cursor changes—a little box is attached to it. You can now drag the column. The column will be placed to the left of whatever column the cursor is in when you release the mouse button. Drag the COLOR column to the left of the X_COORD column. Select both columns. Select Sort Descending (from the drop-down menu as before) and again note the results.

___ **16.** Make the COLOR column active by itself. Locate and press the Table Options icon at the lower left of the window. Click Find. Move the Find window if necessary to see the entire table. Pressing Find Next, find all instances of the text string "Red".[20] Each instance, when found, will have a darker border around it. How many records are there with the text string "Red"? _____

___ **17.** Notice that the Find window gives you considerable flexibility:

❏ You can search all the values in a table, or just those in the selected columns (fields).

❏ The text you type may match only a part of the value or may be required to match the whole string.

❏ The "Find what" string may be required to match the uppercase and lowercase of each letter or not, as you choose.

❏ You may search the entire table or restrict the search to records at or below the current record, or search those records at or above the current record.

___ **18.** In the Search drop-down menu in the Find window, pick ALL. With Match Case checked, search for "yellow." No records will be found, because of the lowercase "y." Now click Match Case off. How many instances do you find? _____. Dismiss the Find window.

Finding Values in a Table

___ **19.** Move the COLOR column to the right of the CONDITION column. If you have only a limited view of the table, you may not be able to tell which hydrant number (OBJECTID) you are looking at when a given cell in the COLOR column is selected. You can set things up so you can always see a given column or columns. To "freeze" the OBJECTID column, select it, right-click, and pick Freeze/Unfreeze Columns. It will now appear always as the leftmost column of the table regardless of the displayed width of the table. Scroll the table horizontally to see the effect (if necessary, reduce the width of the ArcCatalog window to see the effect). (You can thaw a frozen column in much the same way.)

When you terminate the ArcCatalog program, the results of some actions you took are retained, while others are discarded. The folder connections you have added to the Catalog Tree are remembered by the

---

[20]Do not type the quotes in the Find What text box.

software. So are the positions of toolbars and some other settings, but if you have sorted or rearranged a table using ArcCatalog, these changes will not be retained. Operations in ArcCatalog such as sorting do not change the attribute table that is stored with the data set. You are merely changing the table's *appearance* during the time ArcCatalog is being used to look at a particular geographic data set. In fact, to restore the view of a table to its original form, you need only select some other folder or feature class and then return to the table.

___ **20.** Click Folder Connections at the top of the Catalog Tree. Click Contents. Now go back and again preview the Fire_Hydrants table (in IGIS-Arc_*YourInitials*. Notice the table is back to its former configuration—no frozen or rearranged columns and no sorted records.

___ **21.** Click the Preview tab at the top right-hand pane. Select Geography from the Preview drop-down menu at the bottom of the pane.

Identifying Geographic Features and Coordinates

Coordinates are a big deal, and frequently a big headache, for those doing GIS work. HYDRANTS is referenced by a local (and, in this case, fictional) coordinate system devised by the village or perhaps by the county. The units are survey feet. Suppose the village enters into an agreement of some sort with a regional water system agency to maintain the pipes and hydrants. If the agency uses a different coordinate system, the question might arise as to what the latitude and longitude values, rather than the local coordinate system values, were for these hydrants. As mentioned previously, the latitude and longitude coordinate system is the primary basis for the most accurate geographic information.

___ **22.** Select the Identify tool from the Geography toolbar. Write here the description of the tool that you see on the ToolTip. (In version 10.0 of ArcGIS this will be found on the Status Bar.) _____ _____. Notice the appearance of the cursor as you move it around the geographic area. Click in a blank area of the window that shows the hydrants. An Identify Results window appears. (You can move the window around by dragging its title bar and resize it by dragging a side or corner to better see the geographic image – but don't make it too small, since it will shortly display more information.)

___ **23.** Click one of the points representing a hydrant. As you click the feature, it flashes momentarily. The relational database information for that hydrant appears in the Identify Results window. Notice that the Identify Results window, in an information box labeled Location, gives you a precise value of the Cartesian (x and y) coordinates of the location of the tip of the cursor pointer. These coordinates should agree closely, but probably not exactly, with the X_COORD and Y_COORD values shown as the attribute values.

It is not always the case with point data that the coordinates are part of the attribute database; you usually have to take action to put in X_COORD and Y_COORD. How this is done will be described later. However, the information in the Location box is always calculated. *It comes from the geographic location specified by the cursor, not from the relational database.*

___ **24.** With the Identify tool, click again in a place where no fire hydrants exist. Attribute information disappears, but coordinate values of the tip of the pointer are still revealed.

___ **25.** Roughly (to the nearest tenth of a foot) what are the coordinates of the lower left-hand corner of the window? _____ , _____ How about the upper-right corner? _____ , _____ .

___ **26.** Dismiss the Identify Results window. If you right-click on the string Fire_Hydrants in the left pane, you receive a menu of options. Of interest to us now is Properties. Click that. A Feature Class Properties window appears. Under the General tab, you see that the feature type is Point and that the data set is stored in high precision (of which more later). Click the Fields tab. Here you see the names of all the attributes and the types of fields that are used to contain the data. For example, click on the row that says COLOR to select it. Notice that COLOR is a Text field of length 9, meaning that it uses 9 bytes of memory. (You will also sometimes find "length" of a string referred to as "width.") Now look at Y_COORD. It is listed as Float, meaning it is a floating decimal point number. Since you previously saw that the field is stored in high precision it will occupy 8 bytes of storage. (A Float number consists of two parts—a mantissa and an exponent—which allows both high precision (lots of significant digits) and large range (very large and very small numbers). (If this doesn't mean anything to you now, don't worry about it.) Look at FLOW-RATE. It is also a numeric field, but it is an integer (a Long Integer can store an integer somewhat larger than two billion) and may, therefore, be stored in a simple binary format. It uses 4 bytes of storage for the flow rate of each fire hydrant. Dismiss the window.

Looking at GeoGraphics

___ **27.** On the Geography toolbar, click the Zoom In icon (a magnifying glass with a "+" sign in it). Copy the text in the ToolTip or Status Bar here: _____. Click on the upper-left hydrant. Note that it moves the center of the window and the distance between it and its neighbors has increased. You have zoomed in on the layer—not that you will see any more detail in this particular image, but you get the idea.

___ **28.** Notice that although you have zoomed up on the image, the symbols representing the features did not get any bigger. So, this zooming action is somewhat different from looking at a paper map with a magnifying glass, which would increase both the distance between the points and the symbol size.

___ **29.** Click again on the northwest hydrant and observe the results. Click between the hydrant and its neighbor to the south. Click the Full Extent icon (on the Geography toolbar) to restore the view of the entire layer and bring all hydrants back into view. With the Zoom In tool active, drag a box around the middle three hydrants in the middle column and observe. The lessons: You can zoom in by clicking on a point and also by dragging a box. The image is always re-centered, either at the point clicked or the center of the box.

___ **30.** Click on the "hand" icon on the Geography toolbar—this is the Pan tool. Check out its function with the ToolTip. Move the cursor over the middle point, and drag that point to the left side of the pane. When you release the mouse button, you will see that the focus on the image has been shifted ("panned") to the right. The window center is now in between the two columns of hydrants. Click the Full Extent icon. Now click the left-pointing arrow on the Geography toolbar. Note the results. Continue to click it until it "grays out." Now click the right-pointing arrow several times. By using these arrows you can return to previous or later levels of zoom or pan.

A "thumbnail" is a small sketch that can aid you in recognizing the contents of a data set. You make a thumbnail sketch in the step that follows.

_____ **31.** ***Create a thumbnail of the hydrants layer:***[21] On the Geography tool bar, find the Create Thumbnail icon. (Remember: You can determine what an icon does by looking at the ToolTip that appears when you mouse over the icon.) Click it, then click the Contents tab and observe. Where there was before a blank rectangle, there is now an image of the layer. See Figure 1-28. Suppose that you want only a portion of the image to be represented in the thumbnail. Go back to the Preview tab. Use the zoom tool to focus on the six hydrants in the northeast corner. Make a thumbnail sketch of these. Check that it worked by clicking the Contents tab. Now make the thumbnail a third time with all the points shown.

FIGURE 1-28

A First Look at Metadata

Meta comes from Greek, where it means "after" or "beyond." You can generally assume that it means "about." So "metadata" is information about data, or data about data. There are so many things to know about geographic data sets that those experts who work with them have established standards for describing such data sets. You will look at metadata in more detail later, using some real data sets. The lesson from what follows is just that metadata exists, that it consists of many elements, and that it can be presented in different formats. A warning: The ways in which metadata is handled changed considerably between ArcGIS versions 10.0 and 10.1. To really explore metadata you should use the help files, or perhaps better, a search engine, to make effective use of the form of metadata you choose. There are too many options and intricacies to go into here.

_____ **32.** Click Fire_Hydrants again in the Catalog Tree. Click the Description tab. (If an Upgrade Metadata window appears click No.) Scroll down to the bottom of the Fire_Hydrants window. There you should see the two "sections" ArcGIS Metadata and FGDC Metadata.[22] The little triangles should be pointing down, indicating that there is hidden information below them, as in Figure 1-29. If not, click on the text; the arrows will change direction. If you do not see both sections, first check to make sure the options you set in Exercise 1-4 are still the same (Customize > ArcCatalog Options and the Metadata tab). If there is still a problem check with your instructor or the manager of your computer system or network.

Under Description you see the thumbnail of the data set that you created and textual information that the creator of the data set provided about the general nature of the data. Some of the metadata comes about because ArcCatalog has inspected the data set and written it into the description.

---

[21]As mentioned earlier: When ***bold italic*** follows the check box (e.g., __30), it constitutes a general direction. You should read what follows the bold italics to see what specific actions you are to take.

[22]FGDC stands for the Federal Geographic Data Committee of the United States government. The Web site Internet address is www.fgdc.gov. The Committee coordinates the development of the National Spatial Data Infrastructure (NSDI).

FIGURE 1-29

To expand or contract metadata click the text heading. It works somewhat like the Catalog Tree expansion and contraction. A down-pointing triangle is like a plus sign; it means that there is hidden material below. A right pointing triangle is like a minus sign; it means that the entry has been expended.

____ **33.** Expand the ArcGIS Metadata section (scroll down to find it) and expand any sections that aren't already expanded. Explore. Look specifically for Extents, which are coordinates that Esri derives automatically. Also automatically generated is information about attributes such as data type, precision and so on.

____ **34.** Now look again at the ArcGIS Metadata and the Resource Identification information. Spend a bit of time exploring. This metadata is incomplete. Because of that you can get an idea of what is desired for complete metadata.

____ **35.** Continue to explore the metadata by expanding all the triangles under ArcGIS Metadata as well as FGDC Metadata. When they are expanded, they point to the right instead of pointing down. Also continue to compare metadata as best as you can and decide for yourself which set is more useful.

____ **36.** Check out the description by expanding the various sections. Compare the two types of metadata. What is the size of the data set? _____MB. (Hint: It's under Distribution Information.)

___ 37. Explore the Spatial aspect of metadata by reading further down in the Resource Identification section of ArcGIS Metadata. What is the westernmost (leftmost) bounding local coordinate? _____.

___ 38. Expand the Esri Fields and Subtypes headings to explore the Attributes given by the metadata. What is the data type of the FLOW_RATE? _____. What is the Output width of X_COORD? _____.

Other stylesheets are available for different metadata compliances. With the exception of ArcGIS's Item Description (which is the default when you install ArcGIS 10), each has its own compliance standard. Along with this, each will have a varying degree of usefulness and user friendliness. Ultimately, it is up to the organization you are working with as to which standard it uses.

After exploring the metadata, you should have a good idea of what is useful information to have within a metadata file. Were there pieces of information you could not find within the ArcGIS metadata? If the FGDC metadata were completely populated, would it have the information? Consider what information you would have to gather to generate FGDC complaint metadata for this data set.

Using ArcCatalog to Place Data in ArcMap

You have examined ArcCatalog and have used ArcCatalog to find and connect to data sets you want and to explore them. As previously described, ArcMap is the other major component of ArcGIS Desktop. ArcMap lets you see, create, examine, query, edit, and develop geographic data and maps. Your first step in using ArcMap is to place the data you have found with ArcCatalog into ArcMap. There are several ways to do this. Four are described in the following discussion. First, you need to perform a couple of setup steps.

___ 39. From the ArcCatalog Standard toolbar, launch ArcMap. You want to start with a new, empty, blank map. One way to do this is just to press Cancel. Make ArcMap occupy the full area of the monitor screen.

___ 40. Now do either of the next two steps to put ArcCatalog and ArcMap on the screen at the same time.

___ 41. Right-click an empty area of the Windows taskbar to bring up a menu. (The Windows taskbar will be located on one of the four edges of your monitor screen; it contains the Start button.) From this menu, select "Show windows side by side" (In Windows XP: " Tile Windows Vertically"). If necessary, minimize any windows besides ArcCatalog or ArcMap that appear, and "re-tile."[23]

___ 42. If the preceding procedure didn't work for you, try this: Click the middle icon (named Restore Down) of the three in the far upper right of the ArcMap window. That has the effect of reducing the size of the window. By dragging sides and corners of the ArcMap window, make it occupy one half of the monitor screen. Make ArcCatalog occupy the other half. (If you are lucky enough to have two monitors on the same computer that serve as a single monitor (a dual-monitor display), put ArcMap on one screen and ArcCatalog on the other.)

---

[23]Windows 7 users have another interesting option: drag the title bar of ArcMap off to one side of the screen so that Windows 7 automatically fills in half the screen with the ArcMap window. Repeat the procedure for ArcCatalog on the other side of the screen. An advantage to this procedure is that you get to instruct Windows as to which window you want on the left and which window you want on the right.

___ **43.** ***Method 1 of inserting data from ArcCatalog into ArcMap:*** Drag and drop Fire_Hydrants from the left pane of ArcCatalog to the pane in ArcMap that says Layers. The "map will show up in the right pane of ArcMap. (Disregard any warning message about missing spatial reference information. Dismiss any other windows that show up, such as ArcToolBox, Search, and Catalog, if necessary.)

___ **44.** Dismiss ArcCatalog. Maximize the ArcMap window. Right-click the text string "Fire_Hydrants". From the menu that appears select Open Attribute Table. You can see that the feature class Fire_Hydrants—the combination of the geographical database and the attribute database—is now in ArcMap. (Of course, you have been writing information into your FastFactsFile all along. Note down particularly how to open an attribute table in ArcMap!)

___ **45.** Right-click again the reference to the Fire_Hydrants feature class. Select Remove to take the feature class out of ArcMap.

___ **46.** ***Method 2:*** In the ArcMap File menu, click Add Data, then click on Add Data. In the Add Data window that appears, use the Up One Level icon (repeatedly if necessary) so that the Look In box reads Home – Documents\ArcGIS (or something similar, depending on your settings). Notice that the window looks remarkably similar to ArcCatalog before you made your first folder connection in Step 2 of Exercise 1-4. Double-click Folder Connections. Now you see that the connections you made within the Catalog Tree of ArcCatalog are now accessible to you in ArcMap. Contained here is the reference to Village data. Navigate (you may have to scroll to do it) to the Fire_Hydrants feature class by doing the following:

❏ Double-click_IGIS-Arc_*YourInitials*\Village_Data,

❏ Double-click Water_Resources.gdb file geodatabase.

❏ Double-click on the Hydrants file geodatabase feature data set,

❏ Click the Fire_Hydrants file geodatabase feature class, and

❏ Press Add.

This, then, is a second way that ArcCatalog facilitates getting data into ArcMap.

___ **47.** Right-click the reference to the Fire_Hydrants data. Remove the feature class from ArcMap. Hide the ArcCatalog window if it is showing.

___ **48.** ***Method 3:*** To see a third, efficient way of getting data from ArcCatalog into ArcMap, start ArcCatalog from the Windows Start menu.

___ **49.** Make the ArcCatalog window occupy the entire monitor screen. Now make ArcMap occupy the full monitor screen, covering up the ArcCatalog window. Note that both ArcCatalog and ArcMap are represented by buttons on the Windows taskbar. (Recall, the Windows taskbar will be located on one of the four edges of your monitor screen. The button for ArcMap will depict a globe with a magnifying glass. In Windows 7, when you "mouse over" the button a miniature screen, also indicating Untitled, will appear. By clicking on one button or the other, you can bring up the associated window. (You can also bring up applications by holding down the Alt key and pressing Tab to select the application you want. Flip back and forth between the two applications a couple of times.) End this step by making the ArcCatalog window active.

___ **50.** After navigating to the Fire_Hydrants feature class, in the Catalog Tree (you know the drill by now), position the cursor over the it, press and hold down the left mouse button, move the cursor to the ArcMap button on the task bar, wait until the ArcMap window has replaced the ArcCatalog one, move the cursor to the left pane (labeled Layers) of the ArcMap window, and drop (by releasing the mouse button) the Fire_Hydrants feature class. Bingo!

___ **51.** For the last time in this project, remove the Fire_Hydrants data from ArcMap. Close ArcCatalog, leaving ArcMap open.

___ **52.** *Method 4:* This is probably the best way if you don't need to use the full ArcCatalog program or your screen is crowded. Open the Catalog window within ArcMap by clicking the Catalog window icon on the standard toolbar. The button looks like ArcCatalog's icon with a window behind it. The Catalog window will pop out and display the familiar Catalog Tree portion of ArcCatalog. Expand, by clicking the + sign next to each item and thereby changing it to a – sign, the following: Folder Connections; your connection to your ___IGIS-Arc_*YourInitials* folder; the Village_Data folder; Water_Resources.gdb; and Hydrants (Feature Dataset). You should now see the Fire_Hydrants Feature Class. Drag the Fire_Hydrants Feature Class into either the blank map display area or the left pane labeled Layers.

___ **53.** Close ArcMap without saving changes.

Exercise 1-6 (Project)

A Look at Some Spatial Data for Finding a Site for the Wildcat Boat Facility

You usually don't get the luxury of working on one GIS project from beginning to end. Life in the GIS world frequently isn't like that. In this textbook we'll be working on one rather large project (the search for sites for the Wildcat Boat facility that you did manually in Exercise 1-1), but along the way, we'll do several other projects.

This is the beginning of the GIS part of the Wildcat Boat project. Using ArcCatalog, you will find data for the site. Then, using the techniques you learned in Projects 1-4 and 1-5, you will use ArcCatalog to explore data sets associated with this project.

Most of the Wildcat Boat data is contained in a Personal Geodatabase.[24] Both personal and file Geodatabases have a number of advantages over the older data structures you will encounter: the coverage data structure and the shapefile data structure. If you have been in the GIS profession for a while, most of the data you saw was probably in the coverage or shapefile format. Now, the move to file

---

[24]The Personal Geodatabase is quite similar to the File Geodatabase that you met with the Village Data. The differences that affect the user are in the details of how searches are done, how large the data sets can be, and some others. Details of the differences can be found by pointing your browser at www.esri.com and typing "file personal" (without quotes) into the Search box. Also the HELP files, to be discussed later, contain a detailed list of differences. For now, don't concern yourself with the differences, except to know that the File Geodatabase is the most current standard.

and personal geodatabases has continued to proceed at a fast pace. So, most of the work you do in this text will be with geodatabases, and you will probably be converting data to geodatabases in your GIS work after completing this textbook. I will illustrate that conversion shortly.

Using the Area on the Disk for Your Own Work

Again, you want to keep the [___]IGIS-Arc data sets intact and unchanged, so that you can go back to them at any time. But I want you to be able to modify the Wildcat Boat data. To do that, you need to use your own folder, which you made earlier and in which you are keeping your Fast Facts File and the Village_Data. The purpose of

___IGIS-Arc_YourInitials

is to let you store files in a personalized folder, without compromising the information in IGIS-Arc.

Copying Data over to Your Personal IGIS Folder

____ **1.** Start ArcCatalog: make it occupy the full monitor screen. Expand [___]. Expand IGIS-Arc. Click Wildcat_Boat_Data. Right-click the selection and pick Copy. Select

___IGIS-Arc_YourInitials

Right-click the selection and pick Paste.

____ **2.** Using almost the same technique as in Step 1, copy and paste

[___]IGIS-Arc\Other_Data

into your folder, except this time use Ctrl-C and Ctrl-V for copy and paste.

____ **3.** Collapse the Catalog Tree as much as possible, then expand the entries (except for Village_Data) in

___IGIS-Arc_YourInitials

as much as possible. The Catalog Tree should look something like Figure 1-30. If it doesn't, remove the folder Wildcat_Boat_Data from

___IGIS-Arc_YourInitials

and try again.

(Again, be careful not to remove the folder

___IGIS-Arc_YourInitials

because, if you do, your Fast Facts File will depart with it.)

FIGURE 1-30

Come to think of it, now would be a good time to back up your Fast Facts File. Put it on a thumb drive, a flash drive, a network drive, or e-mail it to yourself. Maybe you should make a couple of backups, using different methods. *It is rare that a person has too many backups; it is quite common to not have enough.*

Searching for GIS Data

In the 1990s and 2000s, the major issue related to spatial data and GIS was how to create the data sets you wanted—either directly from the environment or by converting maps. Now the initial emphasis has turned, in many instances, to finding already-existing data—on the Internet and elsewhere. Tremendous stores of spatial data exist, and more comes in every day from satellites and ongoing projects. But you may discover that the data set you want is buried with a lot of other data; it may have an obscure name; the data sets you find may not meet your standards, even if they cover the correct subject and correct geographical area. A major feature of ArcCatalog is the capability to help you discover spatial data sets and, through inspection of their metadata, determine if they meet your needs. Assume that a client or your supervisor asks you to find all the geographic data that could apply to the Wildcat Boat project. You begin by setting up ArcGIS to search your personal folder.

———— **4.** In ArcCatalog, use the Up One Level arrow to get to the top of the Catalog Tree—that's where it says Folder Connections. Collapse any folder designations that are expanded. The catalog can provide the basis for a search for geographic data, which you will now illustrate to yourself. Click the Contents tab.

———— **5.** There are several ways to initiate a search in ArcCatalog: the Windows menu item has a drop-down menu with Search in it; there is a Search icon on the Standard toolbar, or you can just type CTRL-F in some circumstances. Use one of these methods to bring up a Search window. If necessary, click the pushpin in the upper-right corner of the Search window so that it is pointing down. This will ensure that the Search window stays open all the time, not just when you mouse over the window. If you are unfamiliar with the docking and auto-hiding of windows, experiment with it until you are comfortable. Close the Search window (with the "X") and reopen it by pressing CTRL-F. This will allow you to start fresh, while also reinforcing what you have learned about auto-hiding. Although the pushpin to dock the window should default to pointing down, make sure it is.

———— **6.** Familiarize yourself with the layout of the search window, while referring to Figure 1-31. You have buttons to take you to previous searches, or, if you are in a previous search, to take you back to a later search. There is a Home button that will return to the default screen. What is the fourth button used for? (Hint: use the ToolTip.) _____. The fifth button allows you to set up the index and search options—vital to finding data. The last portion of the uppermost bar is a drop-down menu, which allows you to choose where ArcCatalog will search for data: (1) on your local computer, (2) enterprise (throughout your organization's network), or (3) ArcGIS online. Dropping down to the next line, you can see the different types of searches allowed: Maps, Data, Tools, and ALL, which is the default. Click on the different choices to acquaint yourself with the different possibilities. Click Data > Feature Data. The software knows that Feature Data is Vector data, so it automatically provides the phrase [typekeyword "Vector"]. If there are Vector feature classes in the folders that have been indexed, they will be shown to you. If nothing comes up, you can assume that no Vector feature class exist in the indexed folders. If the window doesn't seem to cooperate with you, close it and bring back a fresh one with CTRL-F.

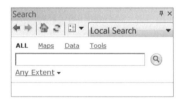

FIGURE 1-31

You can partially reset the Search window by clicking the house icon; this avoids closing and reopening the window. (However, you still have to choose from ALL, Maps, Data, and Tools, whereas if you just click the Search icon you get a completely fresh window.) Click the Index/Search Options icon (in 10.0) or Search Options (in 10.1). By using the window that presents itself (see Figure 1-32), you are going to make sure that Search indexes your personal folder: ___IGIS-Arc_*YourInitials*. Click the Index tab. Remove any folders that are registered. Click Add, navigate to the folder ___IGIS-Arc_*YourInitials*, and select it Click the button that says "Re-Index From Scratch". If asked if you want to delete the existing index, say "Yes". This ensures that your choices will be indexed and gets rid of any previous index. The line Indexing Status will now display the word Indexing, followed by repeating dots. ArcGIS is doing the "alphabetizing" I discussed earlier[25], which may take up to a minute. While you are waiting, note the Indexing Options panel. Accept the defaults for now, but note that you could change the time indexing is done—if, for example, you typically are working at midnight, you might want to change the time automatic indexing is done to 3:00 A.M.

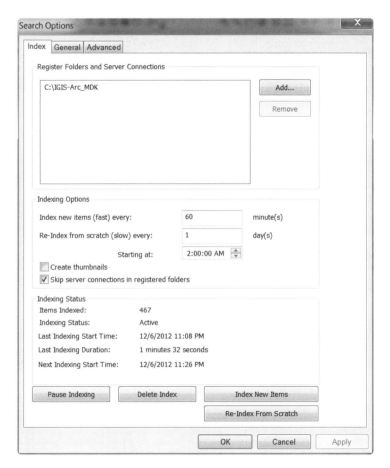

FIGURE 1-32

When the indexing process is finished, Indexing Status will say Active. Click Apply and, if necessary, wait a few more seconds until Active reappears. Click OK.

Once the indexing process is done, ArcGIS will be able to efficiently look in the folders in

_____IGIS-Arc_*YourInitials*

for data. **You should be aware: it won't look for data anywhere else!**

---

[25]If you have questions about what indexing means refer back to the section on "Searching (and Indexing) in General" in the Overview of this chapter.

7. **Search for the name Wildcat:** Restart the Search window to make sure that the location in which the search will be performed is a Local Search and that the text box for entering a search string is empty. Then, with ALL selected, type Wildcat in the text box, and click the magnifying glass icon to the right (or press Enter). Your search will return a geodatabase:

_____IGIS-Arc_*YourInitials*\Wildcat_Boat_Data\Wildcat_Boat.mdb

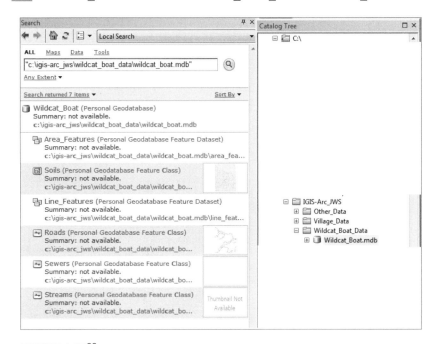

FIGURE 1-33[26]

8. Look at the first line in the body of the search window: **Wildcat**_Boat. Note, in the following parentheses, that it is a Personal Geodatabase. Click it (or the path to it)[27] to take a look the contents of the geodatabase. Two things happen: (1) In the bottom pane of the Search window you see the entries in the folder Wildcat_Boat_Data (see Figure 1-33), and (2) the Catalog Tree expands so that every entry[28] that has the search string "Wildcat" above it is expanded. For each entry in the Search window, there are two links: blue on the name and feature type and green on the path name (which, if you hover over it, shows the full path in case the green path is truncated). Clicking on either link produces the same result. Click the blue link for Soils. Click the Description tab. Under Extent, determine, to two decimal places, the latitudes of the Soils feature class?

North _____

South _____

---

[26]Search-result paths appear in lower case. Don't let this concern you. In general, text strings in Esri products are not case sensitive. This text will continue to show names in the form you are used to.

[27]Different versions of ArcGIS Desktop take different actions. Sooner or later you will see the results described below.

[28]Every entry, that is, if folders that have been indexed. Search doesn't work on unindexed folders.

Use the ArcGIS Metadata to look at the ESRI[29] feature class information for Soils. How many features are there? _____. Look at the ESRI Thumbnail for a picture of the feature class. What are the names of the attributes (the Field names) of the feature class Soils, as shown under ArcGIS Metadata—ESRI Fields and Subtypes?

_____

_____

9. The Wildcat_Boat.mdb, where mdb is the extension for a personal geographic database,[30] is a geodatabase with two feature data sets (Area_Features and Line_Features). The Area_Features feature data set contains one feature class, Soils, and the Line_Features feature data set contains three feature classes: Roads, Sewers, and Streams. You know you are in the ___IGIS-Arc_*YourInitials* folder because you can verify the Wildcat_Boat.mdb's location by glancing at the Catalog Tree, the location toolbar at the top of the ArcCatalog window, or the title bar for the ArcCatalog window itself.

So, you can see that you have quite a hierarchy going on here: Your folder contains a folder (Wildcat_Boat_Data) that contains a personal database; the personal database contains two feature data sets; each feature data set contains one or more feature classes. Visually the hierarchy looks like Figure 1-34. So, you have the Soils, Roads, Sewers, and Streams that you need for the Wildcat Boat facility problem. Let's look at some of the data.

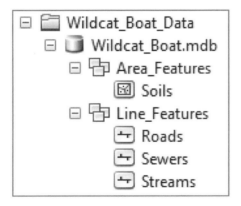

FIGURE 1-34

---

[29]It is worth noting, for those who may be confused by the fact that I sometimes say ESRI and sometimes Esri, that the company changed its name from Environmental Systems Research Institute (acronym ESRI) to simply Esri, pronounced variously "ezri," "esri," or with the letters sounded out "E-S-R-I".

[30]As you have seen you also use File Geodatabases. The filename extension for those is gdb, which stands for Geographic DataBase. File Databases use a variety of types of database software; Personal Geodatabases use only the Microsoft Access database (mdb).

Exploring Soils

___ **10.** Click the Preview tab. Select Soils in the Catalog Tree. You see a map of the Soils you worked with in Assignment 1-1. In the pull-down menu in the Preview text box (bottom of the window), select Table. How many Soils polygons are there? _____.

___ **11.** Examine the table. Notice the fields Shape_Length and Shape_Area. These attributes are the perimeters and areas of polygons. This is information you get "for free" by using a GIS. Sort the Shape_Area column into ascending order. To the nearest tenth of a meter, what is the perimeter of the polygon with the smallest area? _____ meters.

___ **12.** Click the Description tab. Note the information there, some of which is incomplete. Expand the ArcGIS Metadata (click the link) and then see if you can read through the data and find the following values:[31]

Geographic Coordinate System Name: _____

Projected Coordinate System Name: _____

Lat-Lon West Longitude: _____

Lat-Lon East Longitude:_____

Lat-Lon North Latitude: _____

Lat-Lon South Latitude: _____

Now look at the extents in the Item's Coordinate System

West Longitude: _____

East Longitude: _____

North Latitude: _____

South Latitude: _____

Expand the FGDC Metadata > Identification and record the following information:[32]

West Bounding Coordinate: _____

East Bounding Coordinate: _____

North Bounding Coordinate: _____

South Bounding Coordinate:_____

---

[31]I really can't give you more explicit information about where to find those values because the metadata formats change frequently, with new software versions and updates. For example, version 10.0 is different from version 10.1. However, the relevant information is bound to be in there somewhere. You just have to find it.

[32]If necessary, change the stylesheet (Customize > ArcCatalog Options > Metadata) to add FGDC CSDGM Metadata, so you can also access this form of metadata. Again, metadata is handled differently between version 10.0 and version 10.1.

___ **13.** Someone suggests to you that these bounding projected coordinates look very small, given the usual size of UTM coordinates. That is, since the northing is supposed to be the number of meters from the equator (that would be in the millions for 40 degrees latitude),[33] a number like 4500 (that's only forty-five hundred meters or four and a half kilometers) obviously isn't right. So, it is necessary to look more closely at the metadata. Somewhere (pretty much hidden in the ArcGIS metadata) you should be able to find the following values:

False Easting: _____

False Northing: _____[34]

The question of the small number in the northing is now explained. Each UTM northing coordinate has been adjusted by subtracting 4,540,000 meters from the true northing coordinate to produce the local coordinates, which are around 4000 to 6000. Why did the developers of the information decide to do this? To minimize the size of the numbers that are relevant to the problem. Perhaps, at the time the data sets were developed, the software operated in single precision, using only 4 bytes of storage for each coordinate. Numbers as large as 4½ million would contain too many digits to provide sufficient precision.

But Something Is Missing

Assume at this point that you show these results to your client or supervisor, who insists that there is some land use or land cover data somewhere for this region.[35] Your client or supervisor also is sure the data set is in the coverage format. Perhaps the name is simply Landcover, so you will try searching for a data set with that name.

___ **14.** *Initiate another search:* Press the "Go to desktop home search page" icon, which looks like a house, to start a new search. You could also delete the information in the search bar, but pressing the button will ensure you clear everything out. You still want a local search. Since you are looking for a coverage, you will search for Data instead of ALL. With Data active (black instead of blue), select Feature Data, the second option. This populates the Search text box with

typekeywords:"Vector"

which means it will search for nearly all sorts of data except rasters. Type a space, then "Landcover" (without the quotes) at the end of the text box, and press enter. Five entries should

---

[33]The distance from the equator to the North Pole is about 10 million meters. (By the way, this is the way the meter was originally defined: one-ten-millionth of that distance.)

[34]If you can't find the values you can press Ctrl-F while the Metadata is showing a Find window will come up. Type False in the text box and you will be led to something called "Well-Known Text". (NOW he tells us!)

[35]The concepts of land use and land cover have a lot in common—and also significant differences. Both describe "what's going on" with a particular piece of real estate. A land cover classification basically tells what sort of surface component is present, such as forest, water, urban, wetland, and so on. In a land use classification, the surface component is considered but so also is the human use of the landscape, so one might find more detailed information such as residential, commercial, and agricultural. Land use might also include information related to prescribed zoning, such as single-family residential or industrial.

be displayed (see Figure 1-35), all with different symbols beside them.[36] Widen the Search pane by dragging its leftmost border to the left. Look through the names and locations. Click on the entry that indicates the Polygon Feature Class stored at

___IGIS-Arc_*YourInitials*\other_data\landcover\polygon

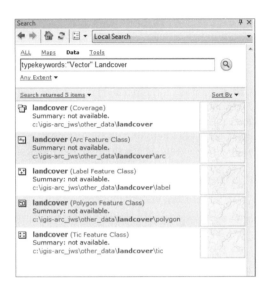

FIGURE 1-35

___ **16.** Click the Description tab. Look at the metadata for the polygon component of Landcover as you did for the Soils geodatabase feature class before. Using the ArcGIS Metadata and the information under the ESRI Spatial Information section (version 10.0) or Resource Information section (version 10.1), record all of the following that you can. *The information that isn't available is shown by Xs in the following blanks:*

Projected Coordinate System: _____XXX_____

Geographic Coordinate System: _____XXX_____

West Longitude (projected): _____

East Longitude (projected): _____

North Latitude (projected): _____

South Latitude (projected): _____

---

[36]Coverages are very interesting data sets. They are made up of points, arcs, polygons, labels, and tics. For those interested in the details of coverages: search the web for: coverage Esri.

___ **17.** Expand the FGDC Metadata > Identification and record the following information (if you can):

West Bounding Coordinate (geographic): _____

East Bounding Coordinate (geographic): _____

North Bounding Coordinate (geographic): _____

South Bounding Coordinate (geographic): _____

So there seems to be a little problem here. We don't really know the coordinate system. No geographic coordinates are shown. The bounding coordinates are presented in "projected or local coordinates" but not in decimal degrees. Does this data set relate to the Wildcat Boat problem?

Is the Newly Found Data Applicable?

Although the Landcover coverage and the Soils personal geodatabase feature class probably occupy the same piece of real estate (note that the projected coordinates you wrote down previously are approximately the same for both), you have to make sure. Suppose that, by looking at some paper documentation, you are able to determine that the projected coordinates shown are in fact in meters and the projection of the coverage is WGS84 UTM Zone 18, just like your other data. So you should fix up the coverage so that it has the same spatial metadata as the other data sets. You can do this with ArcCatalog. When you were reviewing the Landcover coverage (in Other_Data), you may have noticed a SOILS coverage also. Maybe, since they're in the same folder, they are related and SOILS has the needed metadata.

___ **18.** In the ___IGIS-Arc_*YourInitials*\Other_Data folder, check the spatial metadata for the SOILS coverage against the spatial metadata for the Soils personal geodatabase feature class, located in

___IGIS-Arc_*YourInitials*\Wildcat_Boat_Data

\Wildcat_Boat_Data.mdb\Area_Features

Also check the coordinates for the SOILS coverage against the coordinates for both the Soils personal geodatabase feature class and Landcover coverage.

❏   Do they have the same coordinate systems (including the false easting and false northing)?
_____

❏   Are the projected coordinates similar? _____

So at this point you should feel comfortable simply assigning the Soils datum and projection to the land cover data.

Making a Personal Geodatabase Feature Class from a Coverage

What you want to do is to convert the *polygon* component of the Landcover coverage to a Personal Geodatabase Feature Class, so you will have all the data in the same format when you apply analysis tools to it in the future. Also, you need to fix the problem of not having the proper coordinate system associated with the land cover data. It turns out that you can do both these things at the same time, by lodging the converted feature class in a personal geodatabase data set that has the proper datum and projection. You will put it into the Area_Features feature data set that is within Wildcat_Boat_Data.mdb.

_____ **19.** In the ArcCatalog Tree, navigate to

```
___IGIS-Arc_YourInitials\Other_Data\Landcover\polygon
```

and right-click the icon. Choose Export > To Geodatabase (single). In the Feature Class To Feature Class window that appears read the Help panel on the right-hand side.[37] Click in the Input Features field and read the Help panel. You will see that the text box filled in for you. Click the Browse button next to the Output Location text box, and navigate and fill in the blanks so that the window looks like Figure 1-36, making sure that Area_Features appears in the Name text box. You want the full location path to be

```
___IGIS-Arc_YourInitials\Wildcat_Boat_Data
```

```
\Wildcat_Boat.mdb\Area_Features
```

Click Add. Name the Output Feature Class: Landcover, but don't OK the window.

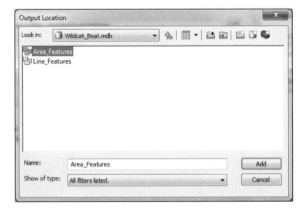

FIGURE 1-36

---

[37]If this is the first time you've run a tool, you will probably have to expand the help by clicking the Show Help >> button at the bottom-right. Also, if the help isn't as helpful as you'd like, click the Tool Help button to see a more detailed explanation and get all the information you're about the tool in one place.

Coverages have a number of fields that will not be useful in a personal geodatabase feature class. You can use the Field Map to eliminate them. You want to lose the fields AREA, PERIMETER, Landcover_, and Landcover-ID so they don't show up in the personal geodatabase feature class table, so click on the field name in the field map area to highlight it, and then click the delete button (which looks like a small x). Do *not* delete the COST_HA and LC_CODE. The resulting window should resemble Figure 1-37. Assuming that it does, press OK. When the conversion has completed a pop-up window will appear for a brief period of time. (If you miss it, open a Results window by navigating to the Geoprocessing drop-down menu and choosing Results. There you can review all the settings you used in the recent process. Dismiss the results window when you are finished.)

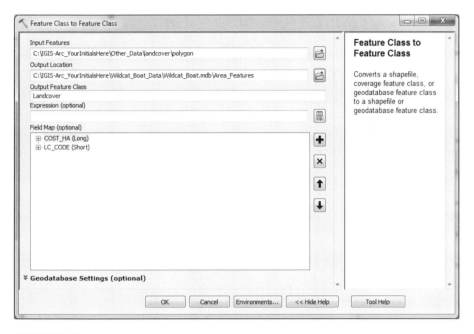

FIGURE 1-37

Looking at the Landcover Personal Geodatabase Feature Class

____ **20.** Navigate to the Area_Features personal geodatabase feature data set. Click it. Within it you should see Landcover feature class that you created from the Landcover coverage. Select it. Click the Preview tab. You see a map of the landcover you worked with in Assignment 1-1. In the drop-down menu in the Preview text box, select Table. How many land cover polygons are there? _____ Check this against the number of polygons in the Landcover coverage that you wrote down previously.

____ **21.** Examine the Landcover feature class table. Notice the fields Shape_Length and Shape_Area. These attributes are the perimeters and areas of the polygons. Sort the Shape_Area column into descending order. The largest polygon is a portion of the lake, with a land cover code of 500. What is the perimeter of the polygon with the largest *land* area (to the nearest meter)? _____ meters. What is the area of the smallest polygon (record all digits)? _____ square meters. Contrast this "double precision" number with the one you previously recorded for the same polygon in the Landcover coverage.

___ **22.** Find the Landcover feature class attributes and their properties with a right-click on the Landcover icon. Then, in the Properties window select Fields. Fill out the following table, except for Width:

Field Name	Data Type	Width
_____	_____	_____
_____	_____	_____
_____	_____	_____
_____	_____	_____
_____	_____	_____
_____	_____	_____

___ **23.** Cancel the personal geodatabase Feature Class Properties window. Find the names of the attributes by using the metadata, check them against what you wrote above, and add the Width. (Hint: It's in ESRI Fields and Subtypes. To repeat: when you click text with an arrow pointing down beside it, you see more text and the arrow points to the right. Clicking again or clicking the Hide button with an upward-pointing arrow contracts the text.)

Further Examining the Wildcat Boat Facility Area Data Sets

Do the following steps using the datasets in

        ___IGIS-Arc_YourInitials

___ **24.** In the Catalog Tree, click Soils feature class, within the Area_Features feature data set. Click the Preview tab.

This feature class uses polylines to enclose areas (polygons). Each of the polygons is considered to be homogeneous in a particular type of soil.

___ **25.** Using the Identify tool from the Geography toolbar, click a polygon near the center of the map. The polygon becomes momentarily highlighted, and the Identify Results window appears. You get the plane area of the feature in square meters in the field Shape_Area. You also get, in the field Shape_Length, the sum of lengths of the polylines (in meters) that enclose the polygon. In the attribute field named SOIL_CODE, you are shown a code that identifies the soil type, such as Tn4. There is also an attribute (SUIT) that indicates whether the soil is suitable for the construction of a building: 1 for unsuitable, 2 for moderately suitable, or 3 for suitable. Click some other polygons. Click the largest polygon in the northeast (That's water—note the SUITability and the SOIL_CODE: _____, ____). Dismiss the Identify Results table.

___ **26.** Look at the attribute table for the Soils personal geodatabase feature class. Obtain statistics for the Shape_Area column of the table. What is the smallest area (to one decimal place)? _____ The largest? _____Close the Statistics window and sort the table from smallest to largest and verify the minimum and maximum numbers. What is the area of the largest polygon that is not water? _____.

___ **27.** In the Catalog Tree click on the Soils icon. Right click the icon. Then select Properties > Fields. Fill out the following table, except for Width.

Field Name	Data Type	Width
_____	_____	_____
_____	_____	_____
_____	_____	_____
_____	_____	_____
_____	_____	_____
_____	_____	_____

___ **28.** Cancel the personal geodatabase feature class Feature Class Properties window. Find the names of the attributes by using the metadata, check them against what you wrote above, and add the Width.

___ **29.** Click the Contents tab. In the Catalog tree select the Line_Features personal geodatabase feature dataset. Check out the presentation using the four display options on the standard toolbar: Large Icons, List, Details (scroll to see them), and Thumbnails.

___ **30.** In the personal geodatabase feature dataset Line_Features, select Roads. Make a thumbnail of the northwest portion of it. (You may need to check your Fast Facts File to recall how to make a thumbnail.) Click the Contents tab and note the thumbnail and some general information about the feature class. What is its projection?[38] _____.

___ **31.** Click the Description tab and quickly glance through the overall Description, the ArcGIS Metadata > Resource Identification as well as FGDC Metadata > Identification. As with much metadata you will find, there is a lot of missing information. Under FGDC Metadata > Metadata Reference, note the metadata date the contents were last updated: _____. (Cryptically the date is given YYYY-MM-DD and times are often given HH-MM-SS-OO.)

___ **32.** Look at the FGDC Metadata > Spatial Reference of the data. What is the geographic coordinate system name? _____ What is the projected coordinate system name? _____ To the nearest hundredth of a degree, what are the geographic coordinates of the northwest corner? _____, _____. The southeast corner? _____, _____.

---

[38]If the projection is not revealed, it may be that ArcCatalog reverted to its default options. Reset the options. (How to do this should definitely be in your Fast Facts File.)

____ **33.** In the projected coordinates, to the nearest meter, what are the coordinates of the northwest corner _____, _____. The southeast corner? _____, _____. What are the **units** of Roads? _____ (Hints: Check under ArcGIS Metadata > ESRI Spatial Information. Ctrl+F allows you to search metadata text that is expanded.)

____ **34.** Look again at the Preview of the Roads geography. Zoom to full extent. Then look at the table.

Here again, each row in the table corresponds to a geographic feature. In this case, each feature is a polyline, which is a series of straight-line segments that each run between two points. Among the fields of data you see a Shape_Length, which is the sum of the segments that make up each polyline. This is also information that you "get for free" when you use a GIS. There is also an attribute named RD_CODE, which indicates something about the road, like paved or not, width, who it is maintained by, and so on. (You will see a much more comprehensive discussion of polylines and segments in the Overview of Chapter 4.)

____ **35.** Again look at the geography of Roads. Zoom up on the northernmost road. Notice that it is not a smooth curve but a series of straight lines. These lines begin and end at Cartesian points, which have x-y coordinates. Use the Identify tool to see the particular attribute values for the polyline that represent this road. Be sure that, as you click the polyline, the feature is briefly highlighted. What is the length of the road to the nearest meter? _____. Each polyline is numbered with an OBJECTID. What is the OBJECTID of this road? _____. The road code for this polyline is _____. Dismiss the Identify Results window. Zoom to the full extent of the Roads personal geodatabase feature class.

____ **36.** Find the Roads attributes and their properties. Fill out the following table, except for Width:

Field Name	Data Type	Width
_____	_____	_____
_____	_____	_____
_____	_____	_____
_____	_____	_____

____ **37.** Dismiss the personal geodatabase Feature Class Properties window. Find the names of the attributes by using the metadata and check them against what you wrote above, and add the Width.

____ **38.** Look at the Sewers personal geodatabase feature class. How many polylines are there? _____ What are the diameters of the various sewer pipes (inches)? _____, _____ Make a thumbnail of sewers and examine it with the Contents tab.

____ **39.** Find the Sewers attributes and their properties. Fill out the table below except for Width.

Field Name	Data Type	Width
_____	_____	_____
_____	_____	_____
_____	_____	_____
_____	_____	_____
_____	_____	_____

____ **40.** Cancel the personal geodatabase Feature Class Properties window. Find the names of the attributes by using the metadata, check them against what you wrote above, and add the Width.

____ **41.** Look at the Streams personal geodatabase feature class. Make a thumbnail. Use Statistics to determine the total length in *kilometers* of all the Streams in the study area. _____ What is the longest stream polyline in *meters*? _____

____ **42.** Fill out the following table, except for Width:

Field Name	*Data Type*	*Width*
_____	_____	_____
_____	_____	_____
_____	_____	_____
_____	_____	_____

____ **43.** Cancel the personal geodatabase Feature Class Properties window. Find the names of the attributes by using the metadata, check them against what you wrote above, and add the Width. Dismiss ArcCatalog.

Exercise 1-7 (Project)

Looking at Wildcat Boat Data with ArcMap

We will be using the Wildcat Boat Data a lot in the ensuing exercises in the book. You may have elected to skip the steps that converted the Landcover coverage to a geodatabase. Or that process may not have worked correctly. In any event, just to be sure you are working with the proper data, I'm going to ask you to delete the Wildcat_Boat_Data folder you have been working with and replace it with one that is on the DVD that comes with the book.

____ **1.** Using the operating system (Windows or whatever) navigate to the folder

[____] IGIS-Arc_AUX\Wildcat_Boat_Data

to be sure you have it.[39] Within that you will find Wildcat_Boat.mdb.

____ **2.** Assuming that you do have

[____] IGIS-Arc_AUX,

the next step will be to delete the old Wildcat_Boat_Data folder that you have been working with. Navigate to the folder

____ IGIS-Arc_*YourInitials*\Wildcat_Boat_Data

_____

[39]The Folder IGIS-Arc_AUX is on the DVD that came with the book. If it is not on the computer's hard drive in the [____] path, you should copy it there with the operating system.

and select it. Press the Delete key. Confirm that you want to delete it after making sure that it is only the Wildcat_Boat_Data folder in the

> ___IGIS-Arc_*YourInitials*

folder that you are deleting.

___ **3.** Navigate to the data set

> [___]IGIS-Arc_AUX\Wildcat_Boat_Data

and select it. Press Ctrl-C to copy the folder onto the clipboard

Navigate to the folder

> ___IGIS-Arc_*YourInitials*

and select it. Press Ctrl-V. Verify that your personal folder now contains a Wildcat_Boat_Data folder. Start ArcCatalog. Make a Folder Connection to the Wildcat_Boat_Data folder. Open the folder. Expand it completely. Verify that Wildcat_Boat.mdb exists, and that it contains Area_Features, which contains both Landcover and Soils.

___ **4.** Launch ArcMap. Under New Maps click My Templates. Click on Blank Map. Click OK. Add the Catalog Tree, using an icon on the Standard toolbar. From your personal

> ___IGIS-Arc_*YourInitials*

folder drag the Sewers personal geodatabase feature class from Catalog Tree into the ArcMap map window.[40] Now drag the Streams personal geodatabase feature class in. Put in the Landcover personal geodatabase feature class. Make ArcMap occupy the full monitor screen. Because we started with the Sewers feature class and it has less extent than Streams or Landcover, we only see a part of these other two feature classes. The solution? Click the Full Extent icon on the Standard Toolbar.

Here you see one of the major bonuses of a GIS: the ability to see data of different types and from different sources easily represented on the same map. An additional advantage is that you can immediately choose what is shown and what is not. Notice that the Landcover data set is displayed "under" the Sewers and Streams—both in the Layers pane and, in a different sense, in the map window.

___ **5.** Experiment with displaying and un-displaying the different personal geodatabase feature classes by clicking the check boxes next to the data set names.

___ **6.** Remove both Landcover and Streams by right-clicking the data set name and choosing Remove.[41]

---

[40]If you run into a "schema lock" problem, dismiss ArcCatalog and load the data set into ArcMap directly with Add Data.

[41]You can, by holding down the Ctrl key and clicking, highlight both data sets at the same time.

___ 7. Sewers should now be displayed by itself. Open the attribute table for Sewers by right-clicking Sewers and, from the menu that appears, selecting Open Attribute Table. (Is the procedure to open an attribute table in your Fast Facts File? If you didn't remember how to open an attribute table, it should be.) Reduce the dimensions of the attribute table and position it so that you can see it and the graphic image of the Sewers as well. On the Tools toolbar (which resembles the Tools toolbar in ArcCatalog—zoom, pan, etc.), find the Full Extent Icon and click it to be sure you are seeing the entire Sewers map. Then, press the Select Features icon (not the Select Elements icon, which is also on the Tools toolbar). Now, click the northernmost sewer line. It should turn cyan *and the relevant row of the attribute table should be highlighted*. Pick another line and observe. Sort the records, in ascending order, according to Shape_Length. Click the shortest arc you can find on the map and verify its length in the table.

___ 8. In the table, click in the small box to the far left of the last record. The selected record becomes highlighted *and so will the equivalent line*.

I went through the last few steps to reinforce the point (that maybe you now understand so well you don't want to hear it anymore): A GIS is a special case of an information system whose database is the marriage of a (geo)graphic database and an attribute database.

Next, you look very briefly at the capacity of ArcGIS to let you add a completely separate database table to the Table of Contents, view such a table, and then join it together with a feature class attribute table so that the information in the separate table becomes part of the attribute table.

___ 9. Click the New Map File icon on the Standard toolbar. You want to open a blank map. Don't save changes. Add the Landcover feature class from

    ___`IGIS-Arc_YourInitials\Wildcat_Boat_Data`

    `\Wildcat_Boat.mdb\Area_Features.`

___ 10. Open the Landcover attribute table. Notice that the table contains the column LC_CODE but no indication of what the codes mean. Undock the table if it is docked, and then close the table.

___ 11. Click the List By Source button at the top of the Layers window. Add, using the ArcCatalog tree (in the side pane), the database table

    ___`IGIS-Arc_YourInitials\Other_Data\LC_Code&Type.dbf.`

Right-click LC_Code&Type and Open the table. What are the columns of the LC_Code&Type Table?

    _____, _____, _____.

Dismiss the table window.

___ 12. Right-click Landcover in the Table Of Contents. Choose Joins and Relates > Join to bring up a Join Data window. You want to join attributes from a table. In box number one—the field in the layer Landcover that the join will be based on—select LC_Code from the dropdown menu. Such a field is called a *key field*. In box number two, the table you want to join to the layer is, of course, LC_Code&Type. In box number three, which may have been automatically filled in for you, the field in the LC_Code&Type table is also named LC_Code.[42] Click OK. (You will be asked if you want to create an index for the join field. When you are processing thousands of records, this can ultimately save time. For our minor demonstration, you can ignore the offer. Select No.)

Seeing the Results of the Join

_____ **13.** Dismiss the LC_Code&Type table. Using the Identify tool cursor, click a polygon in the map display. Notice that the amount of information you get is substantially greater than before. Specifically, you see not only the land code (LC_CODE) but also the type (LC_TYPE) that is associated with the code. The important thing is that you have added the contents of one table to another. Dismiss the Identify window.

_____ **14.** Open the attribute table of the Landcover layer. Notice that all the information has been put together. Sort the LC-type. How many polygons consist of "Barren" land? _____

A note of warning: The product of the preceding process—the table with information gleaned from two tables—will persist only as long as ArcMap is running and Landcover remains as a layer. If you wanted to save the new table together with the geography of Landcover, you would have to take additional steps. One simple way to not lose your work is to save the map.

_____ **15.** Dismiss ArcMap. When asked if you want to save the map indicate Yes, giving it the name "Map with Joined Landcover Table" and placing it in ___IGIS-Arc_*YourInitials*.

Exercise 1-8 (Project)

Understanding the ArcGIS Help System

The Help system in ArcGIS has many facets. There are several ways to access different kinds of Help. Knowing how to access the various pieces of documentation will make your use of ArcGIS more efficient and less frustrating. You will start with a nifty mechanism for getting help on tools and buttons.

A Button for Instant Help: "What's This?" (for ArcGIS Desktop version 10.0 only)

_____ **1.** Start ArcCatalog. Click the "What's This" tool on the Standard toolbar, then click the Connect To Folder icon on the Standard toolbar and read all about it. Since you know most of this information, I am mainly trying to show you what you can expect from the Tool. Click any blank area (or press Esc on the keyboard) to dismiss the information box.

Press Shift-F1. What happens? _____. Click the Search window icon on the Standard toolbar. Dismiss the text explanation.

---

[42]The key fields in the two tables do not need to have the same name. In this case they happen to.

Click again the "What's This" tool. Now click it again and read about the tool itself.[43] Try using Shift-F1 on a content menu item by first moving the cursor down to View > Refresh, and then pressing Shift-F1. Read the box.

Getting Instant Help for a Tool or Command (for ArcGIS Desktop version 10.1)

As you have seen previously, when you "mouse over" a tool, icon, or command you get a description from a ToolTip. In version 10.1 this capability substitutes for the What's This tool, which is no longer available.

The Help System and Documentation

Not only is it unlikely that you will ever learn all of ArcGIS, you may not even learn all there is to know about the Help system. You should make a major effort, however, because whatever question you may have about the system probably has an answer somewhere in Help. There is an immense amount of information available. However, finding it can seem somewhat like a scavenger hunt.

___ **2.** Launch ArcCatalog. On the ArcCatalog Main menu click Help. Pick About ArcCatalog. From the first line you can tell what version of ArcCatalog you are running: _____. From the second line, if it is not blank you can determine what, if any, service packs are installed. From the third line, you can find out the License Type: _____. The Esri Web site is: _____ _____. Click OK.

___ **3.** For ArcGIS Desktop 10.0: Place the cursor in the right pane and click. Press F1. A window opens named ArcGIS 10 Help. Click GIS glossary and you will be served up with definitions of GIS terms that reside locally with your software. It is extensive. To access its search capability, press Ctrl-F to bring up a Find window. Find "geodataset". What is the glossary text entry for this term?_____

_____

_____

_____

Dismiss the GIS glossary.

For ArcGIS Desktop 10.1: The glossary is no longer supported. Instead you can go on the Internet and point your browser at

http://www.esri.com/what-is-gis/overview#glossaries_panel

From there you can select the Esri Online GIS Dictionary. Search for "geodataset". What is the text entry for this term?_____

_____

_____

---

[43]The reference to the Table Of Contents has to do with ArcMap, which you explore in detail in the next chapter.

Return to ArcCatalog

___ 4. The remainder of the help system under Contents is a hierarchical nested set of "books." When you get down to the bottom of a given hierarchy, you will find a help document. For example, in version 10.0, use the "+" icon in front of Professional Library, then Geoprocessing > Introduction and finally the text What is Geoprocessing. For version 10.1 go straight to the Geoprocessing book, then Introduction and finally the text What is Geoprocessing. Read the first page or so of the document and then scroll to the bottom of the page where you will find Related Topics. To get back to previous screens, click the Back arrow at the top of the window.

___ 5. Check out What's new and, within that, A quick tour of what's new. If you continue to use GIS after you finish this textbook, you will undoubtedly be faced with newer versions the software.

___ 6. Right-Click any entry in the left pane. Click Open All. Scroll down through the pane. It may be daunting to see how much information there is, because of its implication for the size of the software package. That's why they will pay you the big bucks when you know how to use it. Right-click again and select Close All.

___ 7. Click the topmost entry: Welcome to . . . Read the section.

___ 8. Look at the buttons across the top of the help window. List them:

_____, _____, _____, _____,

_____, _____.

___ 9. To do this step, you may have to download the Windows 7 help file viewer, which, unfortunately, doesn't come with ArcGIS. You might require assistance from your system administrator or instructor.

Press the Search tab. Type

"catalog tree" thumbnail.

Click Ask. About how many topics are found? _____. Add the phrase

"large icons"

and press Ask again. Open A quick tour of ArcCatalog

and scroll through until you see the Standard Toolbar section. Review the material to refresh your knowledge of that. Close the ArcGIS Desktop Help window.

ArcGIS Help across the Internet

___ 10. Click Help on the ArcCatalog Main menu. Click the ArcGIS Resource Center. If your computer is connected to the Internet, you will get to resources.esri.com, which provides interesting information. Spend some time looking at the Web page. Click on the tabs at the top of the page, and look at the drop-down menus. With a browser look at support.esri.com.

___ **11.** Look for the Knowledge Base heading. This entire site changes frequently, so don't be surprised if this is a difficult step to follow. If you want to get a very extensive GIS dictionary, you will see GIS Dictionary as an entry within the Knowledge Base tab. Click it.

Look up "geodatabase." Look up "file geodatabase." Look up "datum." Look up "relational database." Look up "projection." Look up some other term that you might have been wondering about.

What's the difference between the GIS Dictionary and the Glossary? Just that they were put together by different people at different times. Understand that Esri evolved from a single consultant working out of his home into a huge organization. Esri was not planned from the top down; it grew from the bottom up. So, you can expect lots of duplication.

In this text, I will not define many terms. Nor will I include a glossary. With all the capabilities you have to get definitions on demand, I decided it would be better to save a few trees. If you want a printed dictionary you can obtain *A to Z GIS: An Illustrated Geographic Information Systems*. You may also search online. I also highly recommend the *Glossary of Mapping Sciences* from a joint committee of American Society of Photogrammetry and Remote Sensing (ASPRS), American Congress on Surveying and Mapping (ACSM), and American Society of Civil Engineers (ASCE) which can be found at the following Web site: http://www.geoworkforce.olemiss.edu/tools/glossary_asprs/index.php

There are other parts of the help documentation that you will encounter later, when you use ArcToolbox. Esri has an extensive training Web site: training.esri.com which offers both free and priced seminars and tutorials.

Exercise 1-9 (Dull Stuff)

Using ArcCatalog for Mundane Operations

ArcCatalog serves as an operating system (like UNIX or Windows) for geographic data sets. In this brief exercise, you will copy, paste, rename, and delete a feature class. In copying the feature class you will also automatically copy the feature data set and part of the geodatabase that houses it.

___ **1.** With ArcCatalog running, highlight

      ___IGIS-Arc_*YourInitials*

in the Catalog Tree. Select File > New > Folder and name the folder Housekeeping_Stuff. (Never, ever accept the proffered name "New Folder" because it has a blank in it. ArcGIS may complain—and fail—later if a space is used in a path name. Never use a file or folder name with a blank character in it!) There are other file naming "gotchas," such as dashes, so it is good practice to be simple yet descriptive with uppercase and lowercase characters, numbers, and the occasional underscore.

___ **2.** Navigate to

      ___IGIS-Arc_*YourInitials*\Village_Data\Water_Resources.gdb

and highlight it. Select: Edit > Copy to place the Water_Resources.gdb onto the ArcCatalog clipboard. (In place of Edit > Copy, you could press Ctrl-C, or you could press the Copy button on the Standard toolbar.)

___ **3.** Highlight the word Folder Connections in the Catalog Tree and select: View > Refresh (or just press F5). This ensures that the Catalog Tree properly reflects all data sets and puts them in order in the Tree. It is not always necessary to do this, but it doesn't cost much (only a bit of time and computer processing), and it will save you on occasion from wondering where something went that was supposed to be there.

___ **4.** Navigate back to Housekeeping_Stuff, and highlight it. Select Edit > Paste (or press Ctrl-V, or press the Paste button on the Standard toolbar). Refresh the Catalog Tree again, expand Housekeeping_Stuff, and display the Geography of this copy of the Fire_Hydrants feature class.

___ **5.** With Fire_Hydrants highlighted, right-click on the name. Choose Rename. (Alternative ways to start the renaming process: press F2, or left-click on the highlighted name.) Type "NEW_Fire_ Hydrants" to change the name of the feature class. Click somewhere away from the name to end the renaming process. Preview NEW_Fire_Hydrants with the Preview tab.

___ **6.** Delete the feature data set Hydrants by highlighting it and selecting File > Delete, or pressing the Delete button on the Standard toolbar, or by selecting Delete on the drop-down menu that you receive by right-clicking the highlighted name. Also delete the folder Housekeeping_Stuff. (You may have to delete its contents, before you will be allowed to delete the folder.) Close ArcCatalog.

ArcCatalog operates pretty much like Windows Explorer or the regular operating system file browser when it comes to copying, moving, renaming, deleting, and so on. *But always use ArcCatalog when dealing with geographic data sets, whether they be geodatabases, coverages, or shapefiles.* You may, however, use Windows to move or rename *entire folders* that contain geographic data sets.

Exercise 1-10 (Review)

Checking, Updating, and Organizing Your Fast Facts File

The Fast Facts File that you are developing should contain references to items in the following checklist. The checklist represents the abilities you should have upon completing Chapter 1.

Name: _____

My User or Logon Identifier:_____

My password is written down in this secure location: _____

The hard drive location of this FastFactsFile is _____

The FastFactsFile is backed up on _____

The date of the last update of the FastFactsFile is _____

The date of the last reorganization of the FastFactsFile is _____

The path to IGIS-Arc, associated with [____], is _____

The path to IGIS-Arc_*YourInitials*, associated with _____, is _____

Operations using ArcGIS Desktop software:

- ❏ To initiate ArcCatalog
- ❏ To determine the level of ArcGIS (ArcView, ArcEditor, or ArcInfo)
- ❏ To see properties of an entry in the Catalog Tree
- ❏ To set options for ArcCatalog
- ❏ To check on and install Patches and Service Packs
- ❏ To expand or contract entries in the Catalog Tree
- ❏ To copy, paste, delete, rename, make a new something (folder, shapefile, geodatabase feature class, and so on), search, or reveal the properties of the entry in the Catalog Tree
- ❏ To see the entire path of an entry in the Catalog Tree
- ❏ To connect to a folder so that it appears as a single entry in the Catalog Tree
- ❏ To disconnect from a folder
- ❏ To turn toolbars off and on
- ❏ To move toolbars around
- ❏ To turn on ToolTips
- ❏ To get specific help on a tool or command
- ❏ To launch ArcMap or ArcToolbox from ArcCatalog
- ❏ To examine metadata
- ❏ To see the attribute table of a geographic data set
- ❏ Ways of selecting a row in a relational database table as the current record
- ❏ To change the appearance of a table cosmetically
- ❏ To get statistics on the numeric values in a table column
- ❏ To sort the values in a column
- ❏ To arrange the records order using both primary and secondary sort
- ❏ To move a table column
- ❏ To search for values in a table
- ❏ To make a column always visible
- ❏ To look at the graphics of a geographic data set

❏ To look at the items in a table that relate to a graphically represented feature

❏ To see the coordinates that the cursor is pointed at

❏ To see the characteristics and parameters of the attributes of items in a table

❏ To magnify the graphics of the area being examined

❏ To move around on a map that is being examined

❏ To make thumbnails of geographic data sets

❏ To explore metadata: description, ArcGIS, and FGDC (if present)

❏ To select among various metadata stylesheets

❏ Three ways to get data sets into ArcMap are

❏ Three ways of initiating the data set copying process in ArcCatalog are

❏ Three ways of initiating the data set renaming process in ArcCatalog are

❏ Three ways of initiating the data set deleting process in ArcCatalog are

❏ Ways of searching for geographic data sets with ArcCatalog are

❏ To find the properties of a geographic data set

❏ In ArcMap, to highlight both a selected feature and the corresponding row in the table

What's Next?

In this chapter, you used ArcCatalog to find and explore geographic data sets and to install those data sets in ArcMap. In the next chapter, you begin working with ArcMap extensively—looking at examples of the wide variety of GIS data types available.

CHAPTER **2**

Characteristics and Examples of Spatial Data

OVERVIEW

IN WHICH you look at a variety of geographic datasets involving vector, raster, triangulated irregular networks, and terrains. Datasets of shapefiles and geodatabases are considered, and you are introduced to Esri's ArcMap.

The Original Form of Spatial Data: Maps

Thirty years ago spatial data meant maps. A single map is a spatial database and, for many purposes, a very good one. For hundreds, perhaps thousands, of years, almost all of the information used to support land-related planning and management, and myriad other activities such as navigation, has come from maps. Mapmaking became a well-developed activity. The piece of paper on which the map is drawn is a continuum that can represent the quasi-two-dimensional surface of the Earth in an obvious way. ("A picture is worth a thousand words.") Many times when people are asked "How do I get to . . . ?" they respond, "Let me draw you a map."

Why, then, should we spend millions of dollars on databases composed of discrete symbols when maps are available? Among the answers are that maps alone are extremely hard to use for many of the analyses that human activity requires. There are precious few ways to combine graphic information with other graphic information. Decisions involving the space we live in are becoming more difficult all the time because of the larger number of factors that must be considered. Physical techniques have been evolved for combining maps, such as overlaying one transparent map with another and looking through the composite, but these methods are tremendously time-consuming and have substantial limitations in terms of useful output.

The reason spatial databases composed of discrete symbols (numbers, letters, and special characters) are overtaking map use is that, in the last 30 years, we have learned a great deal about handling discrete symbols, and we have developed both techniques and equipment (primarily digital computers and software) that can manipulate symbols efficiently and quickly.

Thus, an approach that seems basically less appropriate to the task does in fact serve us well—especially since high-quality two-dimensional maps, that is, analogs of the landscape with their innate advantages in conveying information—can now be produced by computers, at desired scales.

Moving Spatial Data from Maps to Computers: Forces for Change

Force for Change #1: There Are Difficulties and Limitations with Using Maps for Decision Making

Maps can depict things beautifully and usefully, so for many applications, a paper map is exactly what is needed. But for many purposes maps are hard to use, for these reasons:

❏ A map is a compromise between a storage function and a display function. As more and more information is stored on a map, it becomes more cluttered. At some point, it becomes unreadable. A map that stored every theme of interest to everyone would be black. Aeronautical charts are a good example of this problem. The aeronautical charts of the 1950s were pretty simple affairs, showing terrain, prominent features, and some airport information. As new regulations came into effect, and new communication facilities were established (whose radio frequencies were placed on the map), as new types of official airspaces were defined, as new military training grounds were introduced, the map had to depict more and more. As a consequence, without careful study (not an activity that can easily take place in the cockpit of an airplane, where the primary activity should be piloting), it is easy to misread such a cluttered map. In a GIS, the storage function and the display function are separated. When display is required, a map can be constructed of only those themes wanted by the user.

❏ It is difficult to analyze a map. Consider a map that shows highways. Suppose you are interested in knowing the distance from city A to city B. What is meant by "distance"? How about straight-line distance from city center to city center? Obtaining an approximation of a straight-line distance isn't too hard. The map has a scale indicating that a certain linear distance on the map corresponds to a certain number of miles or kilometers along an idealized "Earth" with no bumps. You only need a way to measure a distance on the map and compare the measurement to the map's scale. Perform a little arithmetic and you're done. But since the map is a projection of the spherical Earth, the distance won't be exact. The scale varies over the map's surface, so the scale applies exactly only in very few places on the map. Another reason the straight-line distance is not exact, even if you could follow the straight-line distance over the surface of the Earth, you would probably encounter hills, which add to the distance.

❏ Let's make the problem harder: You want distance from "A" to "B" along a highway route. Depending on the type of map, you may get some help in this analysis. If you are looking at an oil company map or an automobile association map, it might have one of those triangular matrices that indicates distances between selected cities. Here, part of the analysis has been done for you—provided your origin and destination are represented on the chart. Another approach could be used if the map has numbers printed beside segments of the road that you can add up to get the total distance. Again, some of the analysis has been done for you, but you still have a bit of work to do. If the map doesn't have these features, then you could use the scale of the map to approximate distances along the route. Based on what the graphics of the map show, you could determine the sum of all those curvy road segments. The more the road curves, the more arduous the task. Also, even when you can sum the segments up, you have to remember that the road curves in three dimensions. Going up and down hills adds miles to the distance a car travels, over and above the distance that would be traveled by simply following the two-dimensional line. Once data are in a GIS, such distance calculations are trivial—though they are still an approximation. So the process of finding the distance—one of the simplest answers you might want from the map—is not simple, and is guaranteed to be imprecise.

❏ If you are unimpressed with the difficulty of analyzing distances on a map, let's move to a harder problem. You have a map of a county that has several parks. Suppose your map shows parkland as green areas. You would like to know the area—in acres, square miles, or square kilometers—that the parks occupy. How do you determine that? You might use the linear scale of the map to make a two-dimensional grid of squares on some transparent material, lay this over the parkland polygons, count squares, and do some arithmetic to estimate the area. Or you could divide each polygon up into triangles and calculate the area of each triangle.[1] Or you might obtain a remarkable device called a planimeter—a gadget that is made to measure the area of a graphically represented planar region—and run its stylus around the boundary to get an approximate value. (You'll find the planimeter in a museum, next to the slide rule, which is next to the abacus.) Or you could paste the map down on a thin sheet of aluminum, use tin snips to cut out the green areas, and compare the weight of all the cutouts to that of a known area of the aluminum sheet. Again, none of these processes is easy, or particularly accurate. The point is this: Maps are hard to analyze. In a GIS, once the data are in the computer, an excellent approximation of the area is a quantity that you get for free.

❏ It is difficult to compare maps. If I didn't convince you how difficult it is to analyze a single map, look at the issue of comparing maps, or analyzing multiple maps to get information from the combination of them. Suppose a municipality wants to build an airport to serve its region and you are to advise the government on how to find adequate locations. What factors must be considered regarding the several hundred square kilometers around the municipality? Here is a partial list:

❏ Topography (of the potential site and of the surrounding area)

❏ Geology

❏ Environmentally sensitive areas

❏ Soil characteristics

❏ Land cost and land availability

❏ Access to ground transportation facilities

❏ Weather patterns (e.g., tendency for fog to occur)

❏ Obstructions in the airspace (e.g., towers, wires)

❏ Existing land use

❏ Surrounding structures and their heights

❏ Habitat of endangered species

❏ Proximity to populated areas

and several more.

---

[1]Knowing the three sides of a triangle, you can calculate the area as the square root of $s(s-a)(s-b)(s-c)$, where a, b, and c are the lengths of the sides and s is the semi-perimeter: $(a+b+c)/2$. I mention this because it seems now to be lost knowledge as far as high-school geometry teaching is concerned.

You are not likely to find all these features on a single map. If you have multiple maps, you certainly have two major problems, and you probably have three.

❑ First, just inspecting several maps for the right combination of factors is quite difficult. How would you determine where to put the airport? You could pin the maps side by side on a wall and look at them. "Here's a location with nice large, flat area on Map A." "Whoops, no. Map B shows the soil wouldn't support a runway." Trying to look at several maps in a serial fashion is not easy.

❑ The maps you will be able to get will be different shapes, at different scales, using different projections and different units, and will cover different areas.

❑ Maps get out-of-date. A U.S. Geological Survey topographic map covering part of Lexington, Kentucky, was produced in 1965. It was updated in 1993. During that time interval, the city grew by 100,000 people. It takes a lot of time to produce a good map. Even as soon as a map is published, it is out-of-date. Of course, part of this problem occurs because it is impractical to survey and record all significant changes just as they occur. But another major problem is that once a map is printed, the publisher cannot change it. If the accuracy of a map is critical, as with aeronautical maps, an updated version can be reissued frequently. With GIS, changes can be made to maps as soon as information about changes becomes available.

On the other side of this coin is the considerable value that maps, as they have been made for the last several decades, have for many sorts of activities:

❑ Maps provide a visually intuitive reference to the features and activities of an area of interest. They connect us with our environment with a level of directness and lack of ambiguity that one may not get with the "black box" and small screen of the computer.

❑ Maps are easily portable. They don't weigh you down. They display information you can read out of doors—something your laptop computer or tablet may not do. They don't need batteries or charging; thus they don't expire in the field at inopportune times.

❑ Maps usually give you large display areas. Yes, you can zoom to great levels of detail and pan on a computer screen or a tablet screen. But there are times when you need both a reasonable level of detail and a large view. It's hard to beat six square feet of map in those instances.

❑ Many maps are basically honest. Those produced by the U.S. National Mapping Program adhere to rigid standards. (Look at http://nationalmap.gov/standards.) Frankly, the ability to provide users with measures of accuracy and quality control is something that all GIS programs fail at because vendors haven't truly gotten to it yet. The move to associate metadata with GIS datasets goes in the right direction, but when you combine GIS datasets, there is no hint as to how good the resulting datasets are.

Force for Change #2: The Need for Better Resource Allocation and Environmental Protection Became Evident

Before the 1960s, it was the view that the source of things humans wanted, such as land, resources, energy, air, and water, was independent of the sink where we put things we no longer wanted, such as garbage, heat, sewage, and combustion products. (See Figure 2-1)

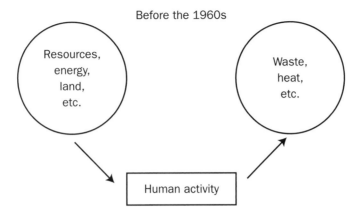

FIGURE 2-1

In the decade of 1960s, perhaps beginning with the publication of *Silent Spring* by Rachel Carson, we began to understand the implications of our freewheeling use of resources and our disposal habits. At the end of the decade, the National Environmental Policy Act (NEPA) was passed and the Environmental Protection Agency (EPA) was established. Some, at least, began to understand our situation as depicted in Figure 2-2: The source and sink are connected, and in a way that has serious implications for our future. Stuff moves back from the sink to the source. If, following the path of the dashed arrow, that movement occurs naturally, we call it pollution. If the transfer, according to the solid arrows is back through human activity, we call it recycling. In any event, we have to deal with the fact that our source and our sink are connected. GIS can help.

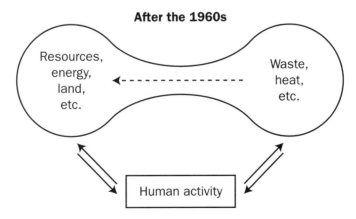

FIGURE 2-2

Force for Change #3: The Evolution of the Garden-Variety Computer System Has Been Amazing

To compare computers, we might use this formula as a gross measure of computing power (P):

P ~ (S*M)/$

to be read as P is proportional to "computing speed" times "primary memory size" divided by "machine cost." Let's look at the change of computing power in the past 50-odd years (since the second-generation machines came out, using transistors instead of electron tubes).

❏ A popular machine in 1960 (the IBM 1620) had a memory of 20,000 bytes. A popular machine today could have 4 billion bytes (4 gigabytes). So today's machine has on the order of 200,000 times as much memory.

❏ The IBM 1620 had a clock speed of 50 kilohertz. Today's machine might run at 2.5 gigahertz. That's another factor of 50,000.

❏ The cost of the IBM 1620 in 1960 was about $100,000 or, considering inflation $500,000 in today's dollars. Today a popular machine might cost $2,000. This gives us a factor of 250.

So one could say that, considering cost, today's machine was 200,000 × 50,000 × 250 times as powerful as one in 1960. That's a factor of 2.5 trillion. To put the idea of a "factor" in perspective, consider a factor of "2." Let's say your income was doubled (or halved). That's a factor of 2. Quite an effect, no? Or consider a factor of "10"—say, the speed of a car compared to a human running speed, or the speed of an airliner over a car. A factor of 10 changes the nature of whatever is being considered. What does a factor of 2.5 trillion do? Boggles the mind, that's what. Of course, much of that increase is used up with graphics, poor programming, and nonuse (computers which are only typed on are loafing along at about 1 percent of their capacity, no matter how fast you type). Still, when the chips are down—so to speak—today's machines are quite amazing.

In addition to the sheer increase in power of today's computers, the hardware has become much more reliable, so that the failures are quite rare, even over years. However, this is much more than offset by the fact that software has become less reliable. Software vendors tend to want to get their products on the market quickly, whether they are buggy or not.[2]

Spatial Data

When location or position is used as a primary referencing basis for data, the data involved are known as *spatial data*. For example, the elevations, in feet, of the landmarks Clingman's Dome and Newfound Gap in the Great Smoky Mountains National Park are data. If the primary referencing basis for these data is the "Great Smoky National Park," or *x* miles south of Gatlinburg, Tennessee, on U.S. Highway 441, or *p* degrees latitude and *q* degrees longitude, then the elevations could be referred to as spatial data.

---

[2]It is my opinion that 30 percent of those who fund and supervise computer programming should have to be examined, remedially trained where necessary, and then certified. The other 70 percent should be taken out and shot. (I'm joking, of course—but there are major problems with the software produced in the last decade or so.)

Spatial data, then, are discrete symbols (numbers, letters, or special characters) used to describe some entity; these data are organized according to the location of that entity in the three-dimensional world. They are data that pertain to the space occupied by objects. It includes cities, rivers, roads, states, crop coverages, mountain ranges, and so on.

Normally, when it is desirable to describe things in the real world by spatial data, the objects are abstracted into some geometrical or mathematical form, as discussed in Chapter 1. For example, a fire tower might be represented by a point, a stream by a set of connected straight lines, and a lake by a polygon boundary.

Limiting the Scope

Spatial data (again, facts about the real world organized by locational coordinates) can be used to describe molecular structures, a human central nervous system, positions of books in a library, or stars in the universe. Since this book relates to geography, we now exclude several categories of spatial data. Usually not considered are data that relate to:

❏ Conditions that change quickly in time—in matter of hours, days, or even weeks. Current pollution levels, weather, and ties will not be included, although average pollution at a point, climate, and ranges of times could be included. Exceptions to this rule include using sensors to collect immediate data about conditions and put those data out to Internet sites for display on the Web.

❏ Objects that move about in space—such as automobiles, animals, or people. However, data about flows of these objects past a certain point at a certain time might well be included. Exceptions to this rule include using GPS to keep up with trucks and cars, or to track animals in the wild.

❏ Circumstances in which the locational identifiers must be more precise than 1 decimeter (a tenth of a meter) to ensure that the related data are useful or valid. The smallest distance separating two adjacent entities that can be distinguished from one another is called the *resolution distance*, or simply, resolution. If the separation in distance is less than the resolution, the data cannot be used to resolve any differences in the condition or situation. Again, there are exceptions: Surveyors are making increasing use of GIS, and they make measurement to within a centimeter.

Spatial Data for Decision Making

We have said spatial data relate to conditions, facts, and objects in three-dimensional space. Most spatial data that now exist use a two-dimensional referencing scheme such as latitude/longitude (or projections thereof) or street addresses. Unless the dataset is specifically one that considers the matter of altitude, this third coordinate is either included as part of the attribute data (rather than part of the locational identifier) or is implied by the nature of the data. For example, if the data describe soil characteristics, one understands that the top few feet of the Earth's crust, regardless of altitude, are being described.

Data types that might be part of spatial datasets are exemplified in Table 2-1.

Sets of Spatial Data: The Database

We have discussed databases in general (a medium containing numbers, symbols, or graphics organized according to some scheme). And we have commented on the idea of spatial data (data describing entities in the three-dimensional world where the location of the thing being described is an integral part of the description.) A spatial database, then, is a collection of spatial data, organized in such a way that the data can be retrieved according to their locational identifiers and in other ways as well.

TABLE 2-1   Example Data Types Included in Spatial Data

Soils	Types, physical and chemical properties
Vegetation	Species composition, age
Wildlife habitat	Types, carrying capacity
Hydrology	Ground and surface water, volume, flows
Geology	Rock types, minerals and ores, physical and chemical properties, oil and gas deposits
Physiography	Elevation, slope, aspect
Land use	Activity types, structure types, zoning
Land cover	Types, facts about what covers the surface
Transportation facilities	Types, capacity, schedules, condition, age
Utility distribution systems	Service areas, capacity, historical features and landmarks—importance, condition, ownership, use
Census districts	Population, housing, other demographic information
Fire districts	Equipment rating, insurance rating
Zip code zones	Delineation of areas
Centers of employment	Types, work hours, number of employees, industrial classification
Locations of police stations	Area of jurisdiction, facilities
Pollution sources	Types, duration, occurrences
Land parcel information (Cadastral)	Owner name, address, value of land, value of structures, tax information

This presentation of a list of data types or variables whose data might be stored in a spatial database does not imply that satisfactory storage schemes are easy to determine, nor that each variable will be stored in the same manner. Whether the geometric abstraction for a data type is best selected as a point, line, area, or volume can be an important consideration in a particular storage scheme.

The development of spatial databases to be used for analysis and decision making is both an art and a science. Thus, when any data-handling program is being developed, a key point to remember is that there is no single best way.

Spatial Databases: Inherent Difficulties

In addition to all the problems one has in building, maintaining, and operating any large database, spatial databases have their own peculiarities and challenges. Here are some that need to be considered.

Size

An airplane pilot once said (in pre-radar days) that the thing that kept ATC (Air Traffic Control) from folding up completely in its attempts to keep airplanes from colliding was that "God packs a lot of

FIGURE 2-3   An orthophotoquad image of a portion of northwestern Michigan

airspace in three dimensions." Anyone who has worked planning or management related to the land knows that God also puts a lot of surface area in two dimensions. Thus, any spatial database used for land and resource considerations will either (a) not cover much area, (b) not include much detail, or (c) be very big. Very big databases, regardless of their simplicity, are expensive to build and maintain.

Spatial data in general use up a lot of computer memory and disk space. For example, Figure 2-3 is an orthophoto image of a part of northern Michigan (around Frankfort and Pilgrim) and Lake Michigan.[3] It represents an area of about 30 square miles.

The image consists of square (picture elements, or pixels) that are 1 meter on a side and can be displayed white, black, and 254 shades of gray. The file underlying the image, represented in the most basic form, binary, looks like this:

```
11111110101010101011111111101000000101010000010100000000000001010101111010101011001110001
01010100010011101010001000001000010000111101000100100101001000110010001001110010111
11110101010101011111111101000000101010000010100000000000001010101111010101011001110001010
101000100111010100010000010000410000111101000100100101001000110010001001110010111111
11010101010101111111110100000010101000001010000000000000101010111101010110011100010101 0
10001001110101000100000100001000011110100010010010100100011001000100111001011111110
10000001010100000010100000000000001010101111010101100111000101010001001110101000100 00
```

---

[3]All Figures in this textbook are on the DVD that accompanies the book. They are in the folder Color_Figures. The size, and in some cases, color of these figures may provide you with additional insight.

```
0100001000011110100010010010100100011001000100111001011111110101010101111111010000
0010101000001010000000000001010101111010101100111000101010100010011101010001000000 10
0001000011110100010010101111111010101000000000001010101010101011101001001001011001 00
1011100111111101010101011111111010000001010100000101000000000001010101111010101100
1110001010101000100111010100010000010000100001111010001001001010010001100100010 0111
0010111111101010101011111111010000001010100000101000000000001010101111010101100 111
00010101010001001110101000100000100001000011110100010010010100010011001000100111 001
0111111101010101011111111010000001010100000101000000000001010101111010101100111000
1010101000100111010100010000010000100001111010001001001010010001100100010011100101 1
1111101000000101010000010100000000000101010111101010110011100010101000100111010100
0100000100001000011110100010010010100100011001000100111001011111110101010101111111 11
0100000101010000010100000000000101010111101010110011100010101010001001110101000 10
0000100001000011110100010010111111101010100000000000101010101010101111010010010010 1
1001001011100111111101010101011111111010000001010100000101000000000001010101111010
1011001110001010101000100111010100010000010000100001111010001001001010010001100100 0
1001110010111111101010101011111111010000001010100000101000000000001010101111010101
1001110001010101000100111010100010000010000100001111010001001001010010001100100010 0
1110010111111101010101011111111010000001010100000101000000000001010101111010101100
1110001010101000100111010100010000010000100001111010001001001010010001100100010 0111
0010111111101000000101010000010100000000000101010111101010110011100010101000100111
0101000100000100001111010001001001010010001100100010011100101111110101010101 1
1111101000000101010000010100000000000101010111101010110011100010101010001001110 10
1000100000100001000011110100010010111111101010100000000000101010101010101110100100
10010110010010111001111111010101010111111110100000010101000001010000000000001010101
1110101011001110001010101000100111010100010000010000100001111010001001001010010001 1
0010001001110010111111101010101011111111010000001010100000101000000000001010101111
0101011001110001010101000100111010100010000010000100001111010001001001010010001 1001
0001001110010111111101010101011111111010000001010100000101000000000001010101111010
1011001110001010101000100111010100010000010000100001111010001001001010010001100 1000
1001110010111111101000000101010000010100000000000101010111101010110011100010101000
1001110101000100000100001000011110100010010010100010011001000100111001011111110 1010
1010111111110100000010101000001010000000000010101011110101011001110001010101000 100
1110101000100000100001000011110100010010111111101010100000000000101010101010101110
10010010010110010010111000101010101010101010101011010010101000100100110010001000 1010
1010101010101111111101000000101010000010100000000000101010111101010110011100010101
0100010011101010001000001000010000111101000100100101001000110010001001110010111111 1
0101010101111111101000000101010000010100000000000101010111101010110011100010101 01
0001001110101000100000100001000011110100010010010100010011001000100111001011111110
10101010111111110100000010101000001010000000000010101011110101011001110001010101 010
0010011101010001000001000010000111101000100100101001000110010001001110010111111110 1
0000010101000001010000000000010101011110101011001110001010100010011101010001000000
1000010000111101000100100101001000110010001001110010111111101010101011111111010000 0
0101010000010100000000000101010111101010110011100010101000100111010100010000001 00
0010000111101000100101111111101010100000000000101010101010101110100100100101100100 1
0111001111101010100000000000101010101000101010000010100000000000101010111101010 11
0011100010101000100111010100010000010000100001111010001001001010010001100100010 0111
0010111111101010101011111111010000001010100000101000000000001010101111010101100 111
0001010101000100111010100010000010000100001111010001001011111110101010000000000 010
1010101010101110100100100101100100101110011111010101000000000001010101010101011 1010
0100100101100
```

This goes on for about another 630 pages, which the publisher has, understandably, declined to include. Our point is, it takes a lot of bits to represent even a small photo.

Continuous Nature of the Referencing Basis

In most databases, a particular, unique key points to a unique thing. For some examples, a given auto license number identifies a particular car; a name or Social Security number tags an individual person; a house number and street constitute a pointer to a residence. Spatial phenomena do not enjoy any such autonomy, however—they are a mixture of discrete and continuous. That is, there is no natural and completely satisfying one-for-one correspondence between spatial locators and the related data. A virtually infinite amount of data is potentially available about even the smallest area of the real world; we can store only a small part. Thus, by choosing a particular technique for organizing the continuous into the discrete, we are screening out or "throwing away" an infinite amount of potential information. Clearly, it takes some sophistication and forethought to select a technique to represent the coordinates that apply to the continuous real world and have a database that will be useful in solving problems.

Continuous Nature of Data

In addition to the continuum of two- and three-dimensional space just mentioned (i.e., the fact that our basic referencing scheme potentially has infinitely many points in it), there are also problems with the continuous nature of the data themselves. Soil type is probably a good example. Just as no two snowflakes are alike, no two soils are exactly alike. Soils must be categorized into groups and a judgment made about which group a particular soil belongs to. In naturally continuous variables, such as elevation, the parallel issue of precision comes in: Do we measure (vertically) to the nearest meter? To the nearest millimeter?

Abstraction of Entities

The simplest reference that can be made in a spatial database is to a point, but no material entity is ever just a point. Many of the things we deal with are either linear features, areas, or volumes, so the referencing scheme becomes more complicated. Where is a house? Well, it's a lot of places when you get right down to it. Do you define it by its corners in plan view? Do you select a single point, a "centroid," and define the house to exist at that point? Do you simply say it exists in town "X," with many other houses? There are many fundamental variations in the way the "real world" is and can be referenced. These varying methods can be incompatible, precluding any easy transfer of data or techniques for manipulating data.

Multitude of Existing Spatial Coordinate Systems

Many spatial coordinate systems exist. Most of those used for planning and resource management rely on the use of flat projections of curved surfaces. Many of the datasets that will be used to build a multivariable spatial database will come from data recorded with distorted and dissimilar methods of representation. Matters of units, datum, spheroid, and projection must be addressed. A single state may use many coordinate systems in its various agencies. Examples are latitude and longitude (both NAD27 datum and NAD83 datum), UTM (both NAD27 datum and NAD83 datum), a state plane coordinate system (one or more zones), road miles, river miles, a special coordinate system for particular features (e.g., oil and gas wells), and so on.

Existing but Inappropriate Data

While it is true that considerable data of the types important to this discussion have been collected, many of them are not directly usable in a spatial database. This occurs principally because these datasets, collected by groups or agencies with specific missions to serve, have been assembled in nonuniform categories or have been interpreted in a specific manner for a particular purpose. For example, early soils data categories may not contain the necessary information that will enable measurement of some environmental effects of land use activities.

Effort Required for Development

The data in our base won't develop as a natural consequence of some already ongoing process. Other database developers are more fortunate. As a clerk processes applications for auto license tags, he or she may type the pertinent information about the car, owner, and tag directly into a database. Thus, the database develops as a result of the tag-selling process that must occur anyway. Spatial databases about the environment have not evolved as consequences of other processes; the work starts almost from scratch in most cases.

The Changing Environment

One cannot get an entire spatial database of any size developed before part of it is incorrect because some of the values in the real world will have changed over time. Land use is an example of a variable whose data values are changing in many places on a daily basis. Houses are built. Roads are paved. Even such stable phenomena as topography change drastically over time. For example, the Mississippi River was about 1300 miles long when LaSalle floated down in his canoe. When Mark Twain wrote about it 200 years later, it was less than 1000 miles in length. Not only that, very little of what was wet in LaSalle's day was still river in Twain's. And to further illustrate the futility of any attempt at a "permanent" spatial database, Ole Man River has moved at least two towns from one state to another by its meanderings. The moral is that some data values of all variables in a spatial database are going to change over time. Some procedure for updating the base must be developed or the value of the base will be degraded by time. Further, different variables are of different value to the analysis and decision-making process and, of course, change at different rates. In some cases, the efficient thing to do is to note changes as they occur; in other cases, replacement of all data related to a particular variable is in order. Either way, there are difficulties and costs.

Multiple Paradigms for Storing Geographic Data

Ingenious ways of taking the continuous, virtually infinite environment and storing its important features in a discrete computer have been developed. Spatial data stored in one scheme are not easily converted to another, and one almost always loses information in such a transfer.

For example, take the matter of representing the elevations above sea level of a geographic area. To begin with, one is dealing, in theory, with an infinite number of values. If you establish the elevation of a given point, then, depending on the precision of your measurements, a nearby point, say 1 meter to the north, will have a different elevation. Elevations of an area may be thought of as a continuous surface, potentially different at every latitude and longitude position. Unless we can model this surface with a mathematical equation (and, with the average mountain or cow pasture, this is usually out of the question), we are stuck with having to select a set of specific points, determine their elevation, and make assumptions about the elevations between those points.

Several ways of representing elevation in GIS have been used. Three are as follows:

❏ Contours

❏ Digital elevation models

❏ Triangulated irregular networks

These will get more detailed treatment later, but let's look at their essential characteristics.

Contours are familiar to you from you experience with topographic maps. Each contour line represents a given elevation. That is, if you walked along the path depicted by the contour line, your elevation would not change. Recalling that a GIS is a marriage of geographic database and an attribute database, you see that the geographic points along the path form the geographic part of this partnership, while the elevation of the line is the attribute datum. To obtain an estimate of the elevation at a point between two contour lines interpolation might be used.

Digital elevation models (DEMs) rely on the idea of a raster. A *raster* is a set of equal-sized squares, arranged in rows and columns, which cover (tessellate, if you want a highbrow word) the plane. Think of a chessboard—or square tiles on a kitchen floor. The geographic position of each square (e.g., its center) can be calculated by its row and column number. Each square has an attribute that might be its average elevation (or, if the DEM were being constructed for aircraft pilots, it would be better if the attribute were the maximum elevation!) The reported elevation is, therefore, constant within a given square and usually changes at each edge of each square, making for a rather lumpy representation. Obviously, DEMs with more, and smaller, squares potentially represent the surface better. Of course, the value obtained from a DEM for a given position is almost guaranteed to be somewhat incorrect.

Triangular irregular networks (TINs) represent the surface of a geographic area by a set of triangles whose vertices are points of known elevation. Since three points determine a plane, the computer can come up with an estimated value of elevation for any requested point among those points. Of course, the Earth's surface is not made up of triangles, and poor selection of the points of known elevation (e.g., midway up a hill rather than at its top) could dramatically, negatively affect the accuracy of the TIN. Again, the more known points, the more triangles, and the (potentially) better representation of the true environment. The use of LIDAR from aircraft sensors can result in elevations of points that are perhaps only two meters apart horizontally. So high accuracy is possible.

Information Systems

This issue of "processing" data in a map form (analog) versus symbol form (digital) brings us to the matter of processing or handling data in general. The conceptual model we will use is that data are processed to produce information. Actually, the terms are not absolute, because what is information to a person filling one role may be data to someone else filling another. However, the idea of a before–after concept, distinguishing the two states as data and information, turns out to be useful, so we employ it.[4]

---

[4]The idea that data precedes information might occur in the following context sequence: Existence, Awareness, Observation, Measurement, *Data*, *Information*, Knowledge Understanding, Wisdom.

An *information system*, in the context of this material, is a set of steps, or set of processes that is executed by a "device" to produce information. We choose to call the symbols that are input to the process by two names: data and parameters. "Data" we have discussed along with its formation into bases. "Parameters" we consider to be information, which the user of an information system supplies at the time of use of the system. Such parameters might specify which data in the base are to be used, how they are to be combined, what the format of the resulting information—output—is to be, and other specifications and/or constraints.

For an example, a professor may assign a student the task of compiling a bibliography of the works of Shakespeare. The database might be a library catalog; the parameters are such descriptive terms as "bibliography." "Shakespeare," "word-processed"; the device is the student, who, with his eyes, pencil, word processor, and so on produces the information. Figure 2-4 is a diagram of a generic information system.

A Garden-Variety Information System

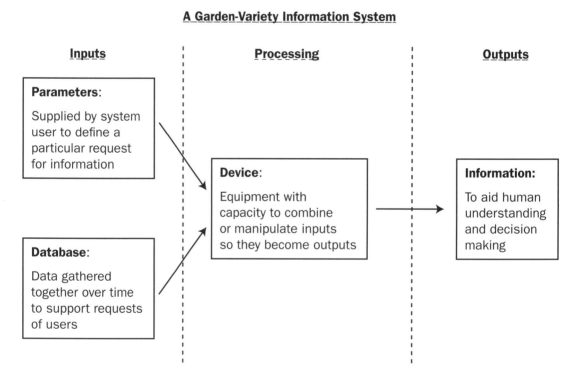

FIGURE 2-4   Schematic of a general information system

In Summary

Based on the preceding discussion of spatial data, databases, and information systems, we can offer yet another definition of a GIS: A geographic information system is an information system that has its primary source of input a base composed of data referenced by spatial (or land, or geographic) coordinates. The system accepts parameters, examines its database, and provides information for decision making and resource management. In an automated spatial information system, a major part of the device that does the processing is an electronic digital computer. Much of the database it uses is stored on computer hard drives in servers or PCs.

Uses for a Geographic Information System

The following list of examples of the use of a GIS is hardly comprehensive, but as you read through it you may happen on ideas that will apply to your areas of interest or study.

Land and Its Use

Discounting the possibility of sudden catastrophe, the strongest factor in how things will be tomorrow is how they are today. A planner or manager who fails to provide himself or herself with information about the current state and characteristics of the environment will probably misplan and mismanage.

Perhaps the most important variables in a geographic database are as follows:

1. What now exists on the land (land cover and resources)

2. How the land is employed (land use and human-oriented activities)

3. What is legally permitted to happen to the land (zoning and legal control)

Once the present state of the environment, or the portions of it with which we are concerned, is recorded in a form amenable to processing, we can begin to make decisions about its conversions to some other use. A geographic information system can be useful in dealing with at least three general categories of issues:

❏ Determining the effect a particular activity or land use will have at a particular location (sometimes called environmental impact analysis).

❏ Given a particular activity, with its characteristics known, determining a set of locations where it might be placed (sometimes called locational analysis).

❏ Given a particular location or site, determining a set of land use activities that might well be placed there (sometimes called site analysis).

Let us now briefly look at a variety of specific areas in which geographic information systems could have an impact. The thrust of presentation is mostly by example and is far from comprehensive. The format we will use, for the most part, is as follows:

1. Examples of types of spatial (and other) data that might be stored

2. Advantages that might accrue by careful use of those data

An important factor to notice in the following is the degree of overlap among variables of different areas of concern.

The Natural Environment

Knowledge of the natural environmental state of the land is central to a determination of what should be preserved, what should be enhanced, what activities could be supported, what impacts are likely to occur from given uses, and a host of other questions.

This change—the realization of "spaceship Earth"—has been occurring since the 1960s and has had many profound and far-reaching effects. As a partial result, many of the first attempts to use spatial information systems to support decision making have had the storage of natural science information as their basis.

103

Storage of geographic (and other) data about the following:

Climate

Bedrock

Surficial geology

Physiography

Hydrology

Soils

Vegetative cover

Wildlife habitats

helps us

Identify, delineate, and manage areas of environmental concern

Analyze land-carrying capacity

Write environmental impact statements

Energy

Energy potentials begin their service for humankind in many forms—oil, gas, coal, hydraulic head, wind, tide, sun, fission, fusion—and always wind up in the same way: *heat.*

The problems associated with the efficient and useful transfer of energy potential to heat are myriad; much of the time we throw away large amounts of energy because it does not serve a particular process—for instance, nuclear power plants discharge vast quantities of heat into rivers to the detriment of the fish and the loss to human beings paying for gas to heat their home. A geographic information system is not the answer to the sensible use of energy, but it is a tool that can help reduce energy waste. Spatially distributed data on energy sources, energy movements, and energy use of all kinds could lead to a greater understanding of our wastefulness and how to prevent it. These datasets could lead to discovery of new energy sources and how to tap them And could alert us to dangers to water supplies from methods of gas and oil extraction, such as hydraulic fracturing of sub-surface rock, (fracking).

Storage of geographic (and other) data about the following:

Potential energy sources

Location

Size

Cost of extraction or tapping

Surrounding environment

Access

Process capability

Energy distribution systems

   Location

   Paths

   Capacities

   Intermediate storage facilities

   Types of energy conveyed

   Degree of hazard in extraction

Energy use patterns

   Industrial

   Residential

   Peak usage

   Distribution by user characteristics

may lead to information allowing analysis of

   Costs of moving energy

   Remaining available energy reserves

   Efficiency of different allocation schemes

   Waste

   Heat pollution

may lead to information for delineating

   Areas of danger to humans

   Environmental impact

may lead to information for developing

   New distribution lines

   Resource allocation schemes

Human Resources

It is for people that we operate our governments. It is primarily people who use the land, the energy, and the resources, and, in part, it is people who feel the effects of its ill use.

The vast amount of data about people is not stored in spatial format for at least two reasons:

❏ They move around—day to day and year to year.

❏ We protect their privacy to a considerable extent.

However, the storage of information about human resources and conditions in a spatial context offers two major advantages:

❏ It allows us to deal in a very direct manner with our primary concern: humankind.

❏ Many sets of data have been developed, largely by the Bureau of the Census, in such a way as to permit relatively easy loading into a spatial database, even though—for reasons of privacy and reasons related to the mission of the census—the "grain" of such two-dimensional information storage is very coarse for most applications.

Storage of geographic (and other) data about human beings:

Where they live

How much they consume

How much they earn

How old they are

What they discard

Where they play

What crimes they suffer

What mishaps befall them

What facilities are available for their employment, shopping, and learning

may lead to information:

To plan for

Mass transit

Recreation areas

Police unit allocation

Pupil assignment

To analyze

Migration patterns

Population growth

Crime patterns

Welfare needs

To manage

Public and government services

Sustainable lifestyles

Areas of Critical Environmental Concern

Areas of critical environmental concern are those geographic areas that are important to the needs of humans. Not only do they perform functions related to health, safety, and welfare of the general public, but they may also serve economic and educational needs as well. Areas become of critical environmental concern when natural resources become scarce or are threatened through the actions of human beings, or when the areas themselves present a threat to the human population.

Storage of geographic (and other) data about the following:

> Agricultural lands
>
> Natural and scenic resources
>
> Soil
>
> Aquifers
>
> Geology and geologic hazards
>
> Wildlife habitats
>
> Vegetation
>
> Floodplains
>
> Wetlands
>
> Scientific areas
>
> Wild and scenic rivers
>
> Cultural activities
>
> Transportation networks

may lead to information to facilitate

> Identification of unique resource
>
> Management of designated areas
>
> Determining relative importance of kinds of resources

Water

Water is the most important resource to the functions of natural environmental processes and human activities. It is a dynamic resource—its movement, both as surface water and groundwater, and in the atmosphere, creates a very broad management problem. Through information in the spatial context we can better analyze and manage our water resources.

Storage of geographic (and other) data about the following:

> Natural bodies of water
>
> Supplies
>
> Flooding

Use patterns

Recreation needs

Climate

Watersheds

Elevations

Industrial locations

Settlement locations

may lead to information about

Floodplains

Availability of clean water

Irrigation

Pollution (potential and existing)

Natural Resources

Natural resources are both finite and necessary for the survival of humans and the maintenance of quality of life. Some natural resources are renewable with proper management, while others will simply run out. A continuing supply of information is necessary for a proper evaluation of how we should use our resources wisely.

Storage of geographic (and other) data about the following:

Forests

Mineral sources

Energy sources

Rivers, streams, lakes

Wildlife and fish

Agriculture

Harbors

Geology

may lead to information to facilitate

Timber management

Preservation of agricultural land

Conservation of energy resources

Wildlife management

Market analysis

Resource allocation

Resource extraction

Resource policy

Recycling

Resource utilization

Agriculture

The production of food has received increasing attention with a growing world population and shrinking agriculturally productive lands. Demands for grain and other crops has increased tremendously as North America has drawn closer to the world marketplace. The need for good agricultural management becomes more obvious as we try to meet the needs of others. Data demands will also increase as we seek solutions to this growing problem.

Storage of geographic (and other) data about the following:

Land conversion

Soils

Geology

Crop productivity

Climate

Hydrology, water supply

Irrigation

Erosion

Crop disease, blight

Insect control

Pesticides

Fertilizers

may lead to information to facilitate

Crop management

Protection of agricultural lands

Conservation practices

Prime agricultural land policy and management

Sustainable agriculture

Crime Prevention; Law Enforcement; Criminal Justice

The Criminal Justice System has many potential applications for geographic data: to assist the system in predicting likely points of criminal activity and to enable efficient allocation of resources through systematic identification of locations warranting increased manpower, analysis, or resource allocation. Although spatial data are routinely collected by law enforcement agencies, frequently it is done in a non-uniform manner that lacks sufficient precision to be tactically useful or to enable ready comparison of location information from occurrence to occurrence.

Storage of geographic (and other) data about crime:

> Where (specifically) crime occurs
>
> Where stolen property is recovered
>
> Where arrests are made
>
> Where high risk businesses are located
>
> Where arrestees live and were schooled

may lead to information to plan for

> Selection of sites or premises for target-hardening attention
>
> Procedures for establishing risk ratings for particular locations
>
> Tactical and strategic patrol allocation
>
> Selection of particular locations for detailed crime prevention analysis
>
> Crime pattern recognition
>
> Selection of areas or schools for delinquency prevention attention

Homeland Security and Civil Defense

Homeland Security and Civil Defense agencies have been established to respond to natural or human-caused accidental disasters or those caused by war or terrorism. As such, we need plans for both short-term and long-term aid to communities across the nation.

Storage of geographic (and other) data about the following:

> Population distribution
>
> Sources of food
>
> Geologic activity (earthquakes)
>
> Transportation
>
> Military installations
>
> Public facilities

Medical facilities

Rescue equipment

Terrorist cells

may lead to information to facilitate

Alternative disaster relief plans

Need for stockpiling of foods and medical supplies

Evacuation plans

Proper designation of disaster relief areas

Communications

Communications represent the way in which humans stay in touch with occurrences around them and through which they transmit information. The physical requirements of communication systems have considerable impact on the natural environment. In order to maintain harmony between the two, information on geographic data must be kept before the decision makers.

Storage of geographic (and other) data related to the following:

Communication stations and antennas

Population

Terrain

Power sources

Current events

Technical information

may lead to information to facilitate

Siting of transmission lines

Location of cellular equipment

Education

Transportation

The movement of people and materials for economic, social, and recreational reasons requires consideration of spatial data. Analysis is required on levels ranging from small-scale local transportation to the national and global scales.
Storage of geographic (and other) data about the following:
Highways, roads, interchanges, and so on

Rapid transit

Airports

Seaports

Railroads

Origins and destinations of travelers

Population shifts

Centers of employment

Commercial traffic

may lead to information to facilitate

Alternative transportation plans

Locational analysis

Mass transit

Energy conservation

So, given the applications and benefits alluded to above, the comparison of GIS with the elephant in Chapter 1 is not that farfetched. It is to be hoped that decision makers are not as blind as the six who were investigating the elephant. And that the readers of this textbook will attempt to broaden the use of GIS for the betterment of society.

Characteristics and Examples of Spatial Data

Kentucky River Authority

1994
Ky. River Clean Sweep

Appreciating Geographic Space and Spatial Data

For this exercise, you will need a notebook and may need a calculator and a long tape measure. This exercise is couched in English units (feet). Your instructor may want you to use metric units (meters), which make a lot more sense but are not as common in the United States.

_____   **1.** Carefully measure the length of your stride. (A procedure that might help with this is to take several steps across a floor with fixed-width tiles or under a ceiling with uniform length panels.) To the **nearest tenth of a foot**, your stride is _____ feet.[1] How many of your paces would constitute 100 feet? _____.

_____   **2.** Find an area of landscape that is a square, roughly 210 feet on each side, and walk its perimeter, examining its interior as you go. Aside from giving you an idea of what an acre[2] is, this activity will probably let you view the complexity that can be contained in a small bit of ground—perhaps the land use, soil, rocks, pavement, vegetation, crops, buildings, fire hydrants, parking meters, street lights, pipes, wires, and other features.

_____   **3.** How many acres constitute a square mile? (Use the exact definition of an acre in the earlier footnote; the area in square feet of a square mile is 5280 times 5280.) _____. The land area of the United States is about 3,500,000 square miles. How many acres would that be? _____. An average-sized state would be about one-fiftieth of that. You can see that recording the data for a state for just one simple theme could use up a lot of computer storage.

_____   Open your Fast Facts File.

_____   Open the Color Figures file for this chapter, so you can see the illustrations in more detail.

---

[1]To convert inches to feet (with tenths), divide by 12; for example, 30 inches is 2.5 feet.
[2]Defined approximately as the area enclosed by a square that measures 208.7 feet on a side, or defined exactly by a rectangle 1 foot wide by 43,560 feet (about 8 miles) long. In other words, an area of 43,560 square feet.

Exercise 2-2 (Setup)

ArcMap Toolbar Examination and Review

1. Start ArcMap. Examine the Getting Started dialog depicted in Figure 2-5. You have a choice of Existing Maps and New Maps, among others. Under New Maps, click on My Templates. This should bring up a Blank Map icon. Click that, then OK. (Or just click OK. Or just click Cancel.)

You will see an Untitled blank map that looks pretty much like Figure 2-6. Below the menus and toolbars, the map window is divided into two panes, as you know from Chapter 1. On the left will be the Table of Contents (T/C from now on). (If the T/C does not appear, go to Windows on the Main menu and turn it on.) On the right will be the area in which the map you construct will appear.

Enlarge the ArcMap window so it occupies the full monitor screen. Note the choices on the main menu bar: File, Edit, View, Bookmarks, Insert, Selection, Geoprocessing, Customize, Windows, and Help. Using Customize > Toolbars, turn on the following:

3-D Analyst

Draw

Editor

Layout

Spatial Analyst

Standard

Tools

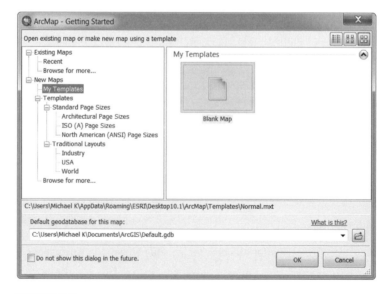

FIGURE 2-5

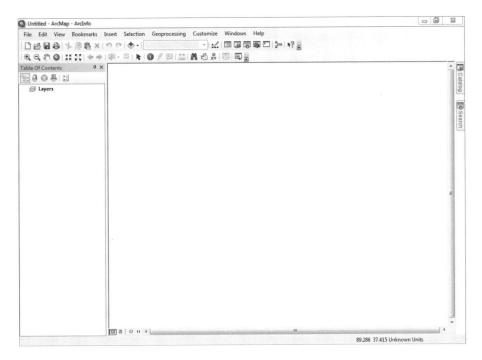

FIGURE 2-6

—— **2.** These toolbars will appear—either docked or free floating. For each docked toolbar, hover the mouse cursor over the toolbar handle (four tiny dots—arranged vertically—at the left end of each toolbar, as shown in Figure 2-7) until your cursor turns into a four-sided arrow. Drag the elements around until they resemble Figure 2-8.

Double-click in the title of the T/C which makes the window leave its docked position and become freestanding. Maneuver it, again looking at Figure 2-8.

When you use ArcMap to display data sets, the T/C will contain the names of those data sets and information about symbology used to display them.

As you saw when you selected the toolbars to turn on, ArcMap has a lot of toolbars, which suggests a lot of capability (and complexity). The preceding toolbars are those you will probably use most often. I will describe each tool later as you come to it, but for now simply record the names of the icons on the Standard toolbar and the Tools toolbar in your Fast Facts File. To determine the name, position the cursor over each and look at the ToolTips. (In ArcMap version 10.0, you can look at the Status Bar for further description; in version 10.1 the description will appear in the ToolTip.) As you write the toolbar name and look at the description, think about what operation would occur when the icon is clicked.

Write the names of the icons of the Standard toolbar in your Fast Facts File.

      Done? ( ) Yes. ( ) No

Write the names of the icons of the Tools toolbar in your Fast Facts File.

      Done? ( ) Yes. ( ) No.

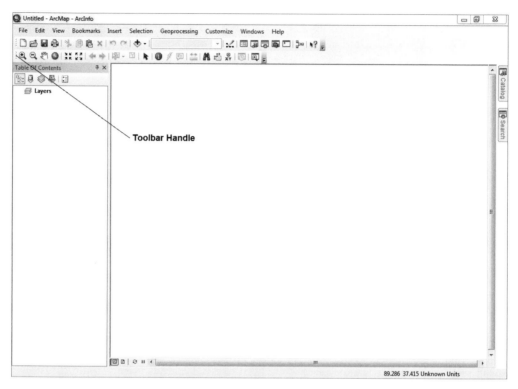

FIGURE 2-7

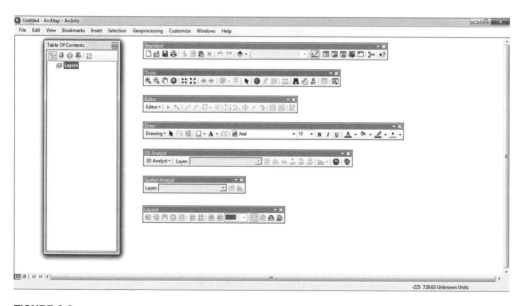

FIGURE 2-8

—— **3.** Dismiss the 3D Analyst, Spatial Analyst, Editor, Draw, and Layout toolbars. Double-click the title bars of the remaining toolbars (Standard and Tools) to restore them to their original positions. If the result doesn't look like Figure 2-6, drag the toolbars around by their handles until it does. Click on the T/C title bar and start to drag the window back to its original location. Notice that multiple blue arrows appear, which provide options for dock locations. Experiment with docking the T/C in various locations (move the cursor over a blue arrow) When through demonstrating this docking procedure to yourself, pick the arrow that is the furthest to the left to return the T/C to its original location. Dismiss ArcMap.

Exercise 2-3 (Major Project)

Exploring Different Types of Geographic Data

The Basic Difference between ArcCatalog and ArcMap

The most general statement that can be made about ArcCatalog and ArcMap is this: ArcCatalog deals with the exploration, examination, and finding of geographic data sets; ArcMap uses those data sets to form layers that display maps and allows analysis of the underlying spatial data.

Exploring Data from the NAVSTAR Global Positioning System (GPS)

A GPS receiver, using the U.S. Department of Defense NAVSTAR system of about 30 satellites, collects positional information—in the form of latitude, longitude, altitude, and time fixes—and stores these coordinates in its memory. Computer files of these points can then be made into Esri data sets.[3]

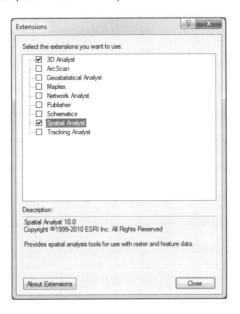

FIGURE 2-9

---

[3]See my textbook: The Global Positioning System and ArcGIS, 3rd Edition, Taylor & Francis – CRC Press, 2009.

In November of 1994, students and faculty of the Department of Geography at the University of Kentucky participated in a cleanup of the Kentucky River. They took a GPS receiver on their trek; the antenna was mounted on the roof of a garbage scow (originally built as a houseboat).

One file the students collected, along the river from a marina to an "island" in the river, was C111315A. SSF (designating the 11th month, 13th day, 15th hour). Using post-processed differential correction, the files were adjusted to yield greater accuracy. The file was then converted to an Esri shapefile and renamed Boat_SP83.shp. The file is called Boat_SP83, since the dataset was projected to **State Plane** (Kentucky North Zone [1601] coordinates, using the North American Datum of 1983 (NAD**83** datum).

Preliminaries

—— **1.** Start ArcCatalog. Use Connect To Folder to place the

[___]IGIS-Arc\River

folder in the Catalog Tree. Launch ArcMap from ArcCatalog. You want a Blank Map.

"Extensions" are additional software packages that extend the capabilities of the main software. If you have the authorization to use an extension, then it was probably loaded on the hard disk with the rest of the ArcGIS Desktop software.

—— **2.** Make the ArcMap window occupy the full extent of the screen.[4] Enable the 3D Analyst and Spatial Analyst extension with Customize > Extensions.[5] See Figure 2-9. Close the Extensions window.

[___]IGIS-Arc\River

Make sure that the List By Drawing Order button (rather than List By Sources, List By Visibility, or List By Selection buttons) is active at the top of the T/C pane.

Seeing the GPS File in ArcMap

A GPS file is basically a file of individual three-dimensional spatial points, or fixes. The fundamental form of each point is a latitude value, a longitude value, and an altitude value. When this file was converted to a point shapefile, the latitudes and longitudes were converted to Cartesian northings and eastings. Each altitude became represented as an attribute value associated with the appropriate point.

—— **3.** ***Add the layer Boat_SP83.shp:***[6] To add the file, select File > Add Data > Add Data. In the Add Data window, find the Contents View Type drop-down button. Select Details. Repeatedly press the Up One Level button until it is no longer active. Activate the drop-down menu for the Look in field. Highlight Folder Connections and click. Navigate to the River folder you connected to earlier and pounce.[7] Find and highlight Boat_SP83.shp. The results will look something like Figure 2-10. Press the Add button. (Ignore any warnings.) In the left-hand pane (the T/C, as

---

[4]Sometimes the mouse cursor comes up as the pan tool when you didn't order that. If that happens simply change it back to the Select Elements tool on the Tools toolbar.

[5]To conserve computing resources (time, memory space), extensions are not loaded into the fast memory of the computer unless the user takes specific action.

[6]When the first sentence of a step is in bold italics, like this one, it means that you should read over all of the text of the step before you attempt to follow the directions in the step.

[7]A "pounce" is either a double-click or a single-click (to select) followed by pressing Enter on the keyboard.

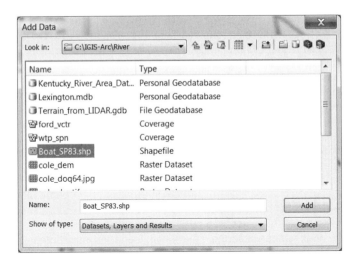

FIGURE 2-10

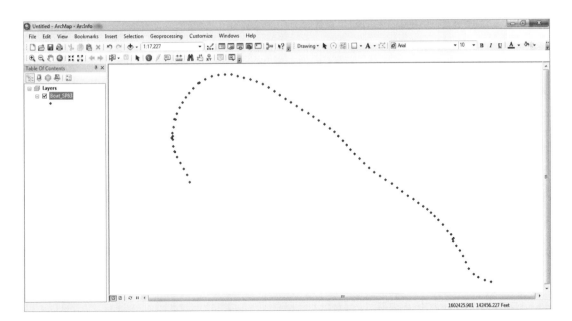

FIGURE 2-11

you saw previously), you will see the Boat_SP83 entry. In the right-hand pane, which is formally called a *data frame*, you should see a GPS track composed of a plotting symbol that looks like a diamond or a dot. The track resembles a fishhook. Compare with Figure 2-11.

___ **4.** *Make the color of the plotting symbol bright red:* The plotting symbol color has been chosen randomly by the software, but you can change it. To do so, right-click the *symbol* itself in the T/C to bring up a window-ette containing a hundred or so colors. If you pause the cursor over a color, its name is revealed as a ToolTip. You might use Mars Red here—click it. The process for

119

changing the color of a plotting symbol is something you might want to record as a Fast Fact. Find a tiny text box next to the plotting symbol and click it (or click just to the right of it). Type "GPS point" (just to identify it for yourself in the future) and press Enter.

5. **Set the map units and display units:** If you move the cursor around the map, you will see numbers appear in the Status bar at the bottom of the screen. These are display units and are presently shown as Feet. Right-click the data frame to bring up a menu. Click Data Frame Properties to see the Data Frame Properties window. Click the General tab. In the Units area, you will see Feet in the Map text box. The box is grayed out, so you can't change it. The reason is that the data set has its projection defined already (Kentucky North Zone [1601]), which has survey feet as its unit of measurement.

6. Notice that you can change the display units in Data Frame Properties. They are set as feet, by default—the same as the map units. Place the cursor in the Display text box, click, and press Shift-F1. If things are set up properly, the software cooperates, and you are lucky, context-sensitive help will appear; read the information, then click elsewhere in the window to dismiss it. From the Display drop-down menu select Miles for the Display text box. See Figure 2-12. Click Apply. Click OK. You may note, as you slide the cursor around the map, that now the status bar shows the display units to be Miles. These are the number of miles, in the X and Y directions, respectively, from the origin of the coordinate system.

7. **Understanding the difference between map units and display units:** It used to be the case that you could get a GIS glossary by clicking the data frame, pressing F1 > Contents tab > GIS Glossary. Unfortunately, this seems to have disappeared from later versions of the software. You might try the procedure to see if the glossary has been restored. You will get the ARC GIS

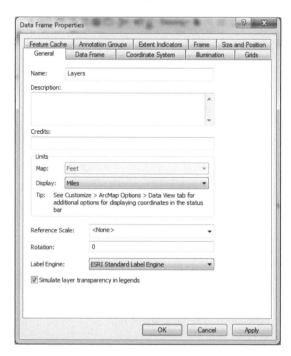

FIGURE 2-12

Help Library, which is sometimes helpful and sometime not. A better bet in this case is to use a search engine with the keywords "esri" and "dictionary". Look up display units there. You might want to highlight the definition (drag the cursor across it), copy it (Ctrl-C) to the clipboard, and paste it (Ctrl-V) into your Fast Facts File. Now look up Map Units. You need to know the difference between map units and display units. Dismiss whatever windows you had open in the search for the definitions.

____ **8.** Change the Display Units to UTM. As you move the cursor around the map you are told that the GPS track is almost 4.2 million meters from the equator. (The other number – the easting coordinate – requires more explanation, which will come later.) Being able to identify a point in Universal Transverse Mercator coordinates can be very handy at times. Change the Display units back to Feet. As you now move the cursor around the data frame, you can see the coordinates of the tip of the cursor. Notice that the "easting" (leftmost number, the x-coordinate) increases as you move the cursor directly to the right. The "northing" number (y-coordinate) increases as you move the cursor directly up. What are the approximate coordinates (just to the nearest foot) of the eastmost GPS fix? Easting (shown first) _____ feet. Northing _____ feet.

____ **9.** *Bring up the attribute table of Boat_SP83.shp*: Right-click the text Boat_SP83 and pick Open Attribute Table from the drop-down menu. Note that there are 83 records in the table, numbered with feature identifiers (FIDs) of 0 through 82, as you can tell by scrolling from top to bottom. See Figure 2-13. You should see columns containing the local time each fix was recorded and its height above mean sea level in feet. By scrolling (vertically) through the records in the table, you can note the beginning and ending time for the data collection run. How many minutes elapsed between the start and end time of data collection? _____ minutes.

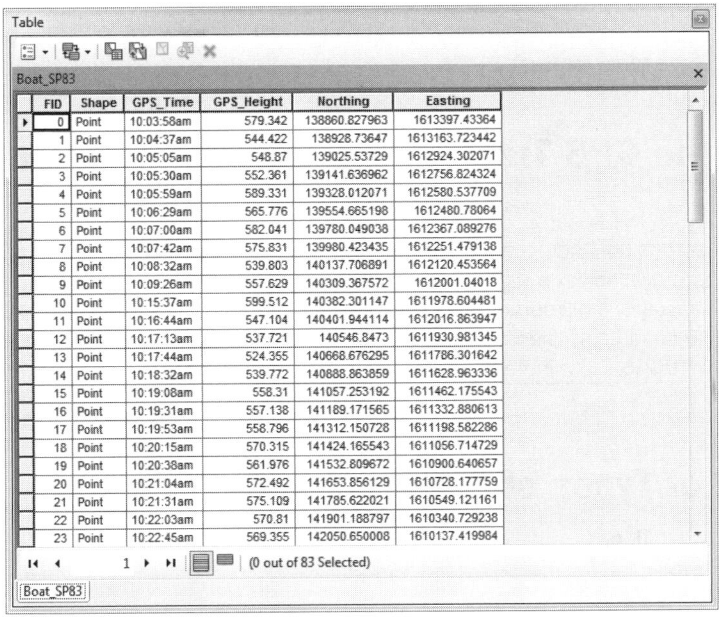

FID	Shape	GPS_Time	GPS_Height	Northing	Easting
0	Point	10:03:58am	579.342	138860.827963	1613397.43364
1	Point	10:04:37am	544.422	138928.73647	1613163.723442
2	Point	10:05:05am	548.87	139025.53729	1612924.302071
3	Point	10:05:30am	552.361	139141.636962	1612756.824324
4	Point	10:05:59am	589.331	139328.012071	1612580.537709
5	Point	10:06:29am	565.776	139554.665198	1612480.78064
6	Point	10:07:00am	582.041	139780.049038	1612367.089276
7	Point	10:07:42am	575.831	139980.423435	1612251.479138
8	Point	10:08:32am	539.803	140137.706891	1612120.453564
9	Point	10:09:26am	557.629	140309.367572	1612001.04018
10	Point	10:15:37am	599.512	140382.301147	1611978.604481
11	Point	10:16:44am	547.104	140401.944114	1612016.863947
12	Point	10:17:13am	537.721	140546.8473	1611930.981345
13	Point	10:17:44am	524.355	140668.676295	1611786.301642
14	Point	10:18:32am	539.772	140888.863859	1611628.963336
15	Point	10:19:08am	558.31	141057.253192	1611462.175543
16	Point	10:19:31am	557.138	141189.171565	1611332.880613
17	Point	10:19:53am	558.796	141312.150728	1611198.582286
18	Point	10:20:15am	570.315	141424.165543	1611056.714729
19	Point	10:20:38am	561.976	141532.809672	1610900.640657
20	Point	10:21:04am	572.492	141653.856129	1610728.177759
21	Point	10:21:31am	575.109	141785.622021	1610549.121161
22	Point	10:22:03am	570.81	141901.188797	1610340.729238
23	Point	10:22:45am	569.355	142050.650008	1610137.419984

1 ▶ ▶I (0 out of 83 Selected)

Boat_SP83

FIGURE 2-13

Right-click the gray box at the left end of Record 0. Click Identify on the drop-down menu. The resulting Identify window is a quick way to see all the values of a record.

What is the GPS height shown in this record? _____.

In the same way, identify the last record in the table.

What is the GPS Height shown in this record? _____.

Dismiss the Identify Results window.

Note the wide variation of altitudes; since the river is usually very placid and almost level, this gives you an idea of what you can expect in the way of vertical accuracy for individual GPS fixes—even those that have been adjusted through a process called differential correction. On the other hand, this data set was created a long time ago—things have improved considerably, with the addition of more and better satellites, and more advanced receivers, but GPS vertical measurements will always be worse than horizontal ones.

____ **10.** Note the icons across the top of the table. Referring to ToolTips, type the names of the these icons in your Fast Facts File.

Done? ( ) Yes. ( ) No.

____ **11.** If any records have been selected, click the Clear Selected Features button on the Tools toolbar. (Alternatively, you can select the Table Options drop-down button and pick Clear Selection.) Engage the capabilities that let you obtain statistics and see a graph (the same capabilities that you used in ArcCatalog) to determine the average elevation (to the nearest foot). _____ feet. Look at the graph to see where most elevation values are clustered. Dismiss the Statistics of Boat_SP83 window. What is the difference between the highest elevation recorded and the second highest? _____ feet (Hint: use sort). Dismiss the attribute table for Boat_SP83.

Looking at the GPS Track in the Context of a Variety of GIS Data

A GPS receiver gives position, not location, information. To recognize a location, you need contextual information. You have available, on the DVD, part of which you copied to your personal folder, several digital maps and images of a portion of the Kentucky River plus a couple of vector datasets of a few arcs from two USGS quadrangle maps. The county to the north of the river is Fayette; that to the south is Madison. Of the two USGS 7.5-minute quadrangles that cover the area, the westernmost is COLETOWN; the other is FORD.

A Potpourri of Types of Geographic Data

The data exist in two folders:

[___]IGIS-Arc\River

[___]IGIS-Arc\Kentucky_wide_data

You will explore a dozen or so data sets.[8] Look over the following. The terminology won't mean much to you at this point, but as you add these data sets to the map you can refer back to this list for more detailed information.

- A personal geodatabase (PGDB) named Kentucky_River_Area_Data containing two PGDB feature datasets: Quadrangle_Data and Country_Streams, described here:

    - Quadrangle_Data contains three vector-based PGDB feature classes:

        - The soil types in the part of Fayette County that is covered by the Coletown quadrangle: cole_soil_polygon.

        - Geologic (surface rock) data in the Coletown triangle: cole_rock_polygon.

        - A vector (line) PGDB feature class of the elevation contour lines for the COLE quadrangle: cole_contours_line.

    - County_Streams contains two vector (line) PGDB feature classes that have been derived from TIGER/Line files:

        - The streams of Fayette County, Kentucky: Fay_Tiger.

        - The streams of Madison County, Kentucky: Mad_Tiger.

- A line *shapefile* containing a few features digitized from the Coletown, Kentucky Triangle: cole_vctr.shp.

- A line *coverage* containing a few features digitized from the Ford, Kentucky quadrangle: FORD_VCTR.

- A *coverage* with both arc attribute table (AAT) and a polygon attribute table (PAT), showing the Kentucky counties and county boundaries—that is, both the areas of the counties (polygons) and the lines that separate them (arcs) are depicted: CNTY_BND_SPN.

- A digital raster graphics file scanned from the USGS Coletown topographic quadrangle: COLE_DRG.TIF.[9]

- A small orthophoto image: COLE_DOQ.TIF.

- A personal geodatabase named Lexington that contains a single, freestanding PGDB feature class:

- A set of line with labels showing Lexington-area vehicle transportation system, derived from the 2002 TIGER/Line files for Fayette County: Roads.

---

[8]None of the data names in Esri software is case-sensitive. That is FORD_VCTR is considered the same as ford_vctr. However, blanks are not permitted in dataset names, nor in folders or in the paths to data set names. Dashes may also cause problems in filenames but not folder names. Not knowing all this causes trouble at times.

[9]A regular TIFF file that is used to portray geographic areas requires a separate world file that provides the geographic coordinates of the TIFF. A GeoTIFFs, on the other hand, contains the relevant world file—it is embedded. ArcGIS may use either setup, but not all software does. You may convert a GeoTIFF to a TIFF and a world file with and the ArcToolbox conversion tool Raster To Other Format (multiple) (or the older ArcInfo command CONVERTIMAGE, if you have access to that software).

- ❏ A digital elevation model (DEM) in the form of an ArcInfo raster consisting of square pillars or posts of elevation that are approximately 30 meters on a side: COLE_DEM.

- ❏ A triangulated irregular network (TIN) showing elevations derived from the DEM, in the form of an ArcInfo TIN: COLE_TIN.

- ❏ A raster showing Fayette County land landcover: fyttelc06alii.

- ❏ A raster showing Kentucky land landcover: MSU_SPNorth.

- ❏ A File Geodatabase: Terrain_from_LIDAR.gdb.

    - ❏ \KY_SP_North_NAD83\River_Bend_and_Water_Plant_Terrain.

In the steps that follow, you will add these feature-based, grid-based, image-based, and TIN-based layers to the map.[10] The photograph in Figure 2-14 is a newspaper photo of a part of the area that you will be looking at.

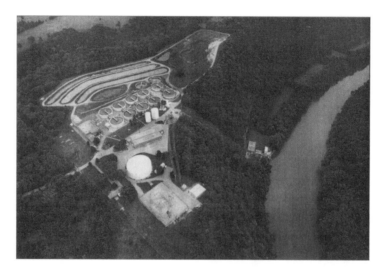

FIGURE 2-14   Courtesy of the Lexington Herald-Leader, Lexington, Kentucky

___ **12.** Enlarge the map window a bit so that it occupies almost the entire ArcGIS window, which should itself be set to occupy the full monitor screen. If you are really short on screen space, making the T/C auto-hide will increase the real estate of the map window. Under Customize > Extensions, make sure 3-D Analyst and Spatial Analyst have checks beside them.

---

[10]Raster and image files may be contained as freestanding entries in file and personal geodatabases. TIN files stand on their own directly in folders.

Displaying Layers from Vector-Based Datasets

The **coverage** FORD_VCTR and the **shapefile** cole_vctr.shp[11] contain a few arcs digitized from the USGS 7.5-minute topographic quadrangles Ford and Coletown. In particular, the arcs trace the banks of the Kentucky River and some highways and Interstates from the quad sheets. The datasets have been converted to Kentucky State Plane coordinates (survey feet) in the NAD 1983 datum.

Since we will be adding several layers to the data frame, you should be aware of the different ways to accomplish this. As a reminder, first, you know of the File > Add Data sequence. Also, there is an Add Data button on the standard toolbar. Also, if you right-click on the word Layers, you can select Add Data. In each of the preceding cases, an Add Data window appears. The most efficient way to add data is by dragging the feature class from the partial ArcCatalog window that shows up (without its Preview section taking up screen space) when you click the Catalog icon on the ArcMap Standard toolbar. Note these four ways in your Fast Facts File.

> Done? (   ) Yes. (   ) No.

_____ **13.** Use Connect to Folder to place the

> [____]IGIS-Arc\River

> folder in the Catalog Tree. Add the cole_vctr.shp shapefile from

[____]IGIS-Arc\River. Lines appear in the data frame, some of which outline the river and are parallel to the set of points of Boat_SP83 (which will also be referred to as the GPS track). Also, a small island is depicted.

To the left of the window you may see either the T/C or the Catalog—one obscuring the other. Click the tabs at the bottom of the pane to select the one you want to view at the moment. If you want to see them both at the same time you can separate them by dragging the tab of either one of them away from the other. To restore the view where one overlays the other, double-click the title bar of the one you moved, or slide its title bar back over the other.

_____ **14.** **Add the arc component of the vector coverage FORD_VCTR from [____]IGIS-Arc\River:** To add the arc component of the coverage, bring up the Add Data window, find the coverage name (FORD_VCTR), **double-click the name**, click "arc", then click Add.

You should see the GPS track, a few arcs from the Coletown topographic quadrangle (topo sheet) depicting the river's banks, and a few arcs from the Ford topo sheet.[12] Both (the coverage and the shapefile) are vector datasets, but they have different structures in the memory of the computer and on the disk drives, as you will see in Chapter 4. The starting point of the trip is in the southeast; at the other end of the track, you can see the polygon outlining a tiny island.

---

[11]Why are FORD_VCTR and Cole_Vctr.shp in different formats? Simply to acquaint you with two of the different data-representation structures (coverage and shapefile) that ArcGIS is capable of.

[12]Yes, this short trip crossed the boundary between two quad sheets. Not only are you learning about GIS, but you are also confirming the First Law of Geography: Any area of interest, of almost any size, will require multiple map sheets to represent. GIS, as its displays evolve toward seamlessness, is set on making that law unimportant.

___ **15.** ***Make the FORD_VCTR coverage line symbol a green line of width 2 points:***[13] Earlier you changed just the color of the symbol, using a right-click on the symbol itself. Since you now want to change the color **and** size, you have to open the symbol selector window, which gives you a great deal of control over how features are displayed. Click (left-click, that is) on the ***line symbol*** that is under the text name "ford_vctr arc" to bring up the window. All you need to do here is to enter a 2 in the Width box (by clicking or typing) and select a color (say, Medium Apple) by clicking the little Color patch drop-down menu to bring up a pallet of colors. While you are looking at the Symbol Selector window, check out the possibilities for line symbols: everything from highways to abandoned railways to aqueducts. Not only is there the set of symbols you see in the window, but if you can't find exactly what you are looking for you can search for more or load more Style References (from the "Style References…" button). In addition, if you want, you can edit symbols with the Edit Symbols button. The possibilities are limitless. Now back to the main issue at hand: Click OK. Observe.

___ **16.** Now change the symbology of the cole_vctr shapefile. Make the line symbol some bright red color, width 2.

___ **17.** Layers may be displayed or not (turned on or off) by clicking the box next to the layer name. Experiment by turning the three layers off and on; leave them all on when you are done.

___ **18.** ***Make a group layer:*** Be sure the List By Drawing Order[14] button at the top of the T/C is pressed. Click any of the three feature class names in the T/C. Then, with the Ctrl key held down, click each of the other two layer names to highlight them. Then, right-click any one of them and click Group from the resulting menu. All three are now subsumed under the name New Group Layer. In a group layer, the entire set of layers may be turned off all at once. If the group is turned on, whether a particular layer is displayed depends on whether its box is checked or not. In other words, you can turn these layers on and off together, or you can turn individual layers on and off. Again, experiment, leaving everything on at the end.

___ **19.** ***Change the name of the group layer:*** Right-click the text New Group Layer. Select Properties > General. In the Layer Name text box, replace, by typing, New Group Layer with GPS_and_ Vectors. Click Apply, and note that the name has changed in the T/C. Click OK. Write in your Fast Facts File how to change the name of a group layer. This name-changing technique also works for a plain layer, though it's usually a good idea to keep the names of layers the same, or close to the same, as the names of the data files that created them.

___ **20.** Experiment with collapsing and expanding the entries in the T/C by clicking all the boxes containing minus (−) and plus (+). You can reduce the T/C here to a single entry (Layers). Alternately, you can see each constituent layer, and each of those with or without its legend. At the end, expand everything.

___ **21.** ***Zoom to the full extent of all data:*** Find the Full Extent button on the Tools toolbar. Examine the map. Those lines that don't represent riverbanks represent highways in the general vicinity.

___ **22.** ***Zoom back to the GPS layer:*** Right-click Boat_SP83, then click Zoom To Layer.

---

[13]There are 72 points in an inch.

[14]Esri software seems to insist on capitalizing prepositions sometimes. If you are like me, after a while it won't bug you so much.

Housekeeping: Saving and Restoring a Map

As you work it is a good idea to save your work. While you could easily reconstruct what you have done so far, you have some time invested that you would rather not lose.

___ 23. On the File menu, find and click Save. A Save As window appears. Navigate to the ___IGIS-Arc_*YourInitials* folder so that it appears in the Save In box. For a filename, which currently reads Untitled.mxd, use River_Map. The Save As Type field should read ArcMap Document. The file extension MXD designates the file as an ArcGIS Map document. See Figure 2-15. Click Save. Note that the title bar of the ArcMap window has changed to River_Map and that the ArcMap button on the Windows taskbar also reflects that change when you mouse over it.

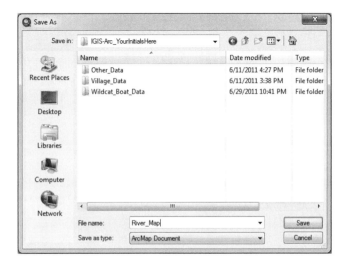

FIGURE 2-15

___ 24. Dismiss ArcMap. Using the Windows operating system, look at the contents of the folder ___IGIS-Arc_*YourInitials*. Notice it contains the file River_Map.mxd.[15] How large, in kilobytes, is the file? _____ KB. Pounce on River_Map.mxd. ArcMap will restart and load the map you just saved.

It is vital that you understand what an MXD file is and what it is not. While it contains a great deal of information about **_how_** the data it refers to is represented (symbols, colors, line widths, and so on), **_the MXD file does not itself contain data_**. The MXD file contains **pointers** to the data—that is, the map file tells ArcMap the names of the folders where the datasets are located. You could not, for example, send only an MXD file to a colleague and expect her or him to be able to see the map with only ArcGIS desktop. You would also have to send any relevant geodatabases, shapefiles, and coverages. (You could, however, send a Layer Package—to be discussed later.)

---

[15]If the extension name "mxd" doesn't appear after the name River_Map, find out how, with Windows, to turn off the checkbox in front of "Hide extensions for known file types."

Selecting: Both Map Data and Attribute Data

In a geographic dataset file (cole_vctr.shp, for example), each feature in the map corresponds to a row in the attribute database. Here the features (lines) have lengths, in the unit of measurements of the shapefile (feet). Also, in this case, a description has been added to each line.

___ **25.** Look at Figure 2-16 as you work this step. Make sure you are zoomed to the extent of the GPS layer (Boat_SP83). Open the attribute table of cole_vctr. Find the WHAT column. Find the attribute value Island. At the left end of that record you will see a gray box; click inside the box. The ***record*** becomes selected—shown by the fact that it is highlighted. At the same time, the ***line*** that delineates the small island at the end of the GPS track will also become selected— again shown by the fact that it is highlighted. Again, see Figure 2-16.

Whenever a record in a layer's attribute table is selected, the corresponding feature in the data frame will be selected. If the layer is being displayed, the feature will be highlighted.

___ **26.** Click the box of the record that has the WHAT attribute value of Right Bank. Notice that a line bounding the river is highlighted. Hold down the Ctrl key and click the box related to Left Bank. Now both records are highlighted, as are both bounding lines. At the bottom of the table,

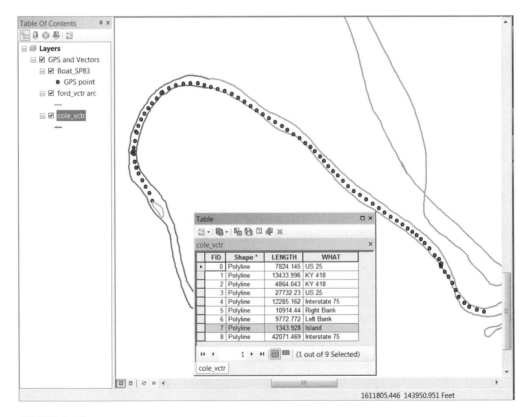

FIGURE 2-16

press the Show Selected Records (a graphic with only blue lines in it) button. The two records highlighted in cyan will appear at the top of the table by themselves. A new group of buttons appears at the top of the table, but they're grayed out. Now click the gray box to the left of the Right Bank record. The record turns yellow (indicating a selection within a selection), as does the associated line. The new buttons are now available, so check out their ToolTips. The lesson: If only selected records are shown, you can highlight records and features **within** selected records and features, using a different color. Show all records. Find the Table Options button at the top of the table. With the Table Options button, press Clear Selection.

____ **27.** Zoom to the extent of the cole_vctr layer. Again select the records representing the banks of the river. Click the Table Options button. Choose Switch Selection. Now the records and lines (of cole_vctr) that were selected are not, and those records and lines that were not selected now are. Now press Table Options, pick Select All, and observe the results. Find and click the Clear Selection button at the top of the table. Zoom back to the previous level of magnification with the Go Back to Previous Extent button on the Tools toolbar. The GPS track should again dominate the window. Keep the Attributes of cole_vctr table open.

The ArcMap Tools toolbar contains some additional buttons not found on the ArcCatalog Tools toolbar. One of these is a Select Features button that lets you graphically select features or areas on a map.

____ **28.** On the Tools toolbar, find and press the Select Features icon. The ToolTip will explain what it does and give some hints for use. Next to the Select Features icon is a drop-down menu arrow. Click it and note the possible ways you can select features. Write the names of the icons of the Standard toolbar in your Fast Facts File.[16]

Done? (   ) Yes. (   ) No.

Experiment with the various Select Features options to understand how they work on the GPS and Vectors layer. Note that each option has a different cursor.

____ **29.** Make sure Select Feature by Rectangle is active. If any selections remain from your experimentation has selected anything, use the Clear Selected Features button on the Tools toolbar. Use the cursor to point at the line that delineates the island and click. The line again becomes highlighted as does its associated record.

So here is the main message: You can select either records or features, and the selection is carried through to the associated features or records.

____ **30.** Hold down the Shift key and click again on the line of the island. Note that it becomes unselected. Select it again. Now hold down the Shift key and click one of the other red arcs; it is now selected as well.

Select a third red line. (The strange fact that you hold down Ctrl to select multiple records in a table but use the Shift key to graphically select multiple features on a map is something that you might want to write down in your Fast Facts File.)[17] To clear all selections: click the

---

[16]Version 10.0 operates somewhat differently but the capabilities are the same.
[17]Okay, for the most part, I'll quit reminding you about the Fast Facts File now. Either you are in the habit by this time or you aren't.

Clear Selection button at the top of the attribute table window. What are other ways to clear selections? _____.

___ **31.** Drag a small box across the river just north of the island. Notice that all points and lines that have any part within the box are selected. This is one case of Select by Location; there are many more. Clear all selections—this time by clicking the Clear Selected Features on the Tools toolbar, not the table.

The Interactive Selection Method in the Selection menu allows you to set up the operation of the Select Features tool.

Zoom to the extent of cole_vctr by using the Go To Next Extent button in the Tools toolbar (which points to the right and is next to the leftwards pointing Go To Previous Extent button). In the attribute table, select the Island, Left Bank, and Right Bank records. Now click Selection (on the main menu) > Interactive Selection Method. What are the interactive selection methods? See Figure 2-17. Write them in your Fast Facts File.

Done? ( ) Yes. ( ) No.

Experiment with adding and removing graphic features with the Interactive Selection Method. Finish by choosing Create New Selection. Clear all selections. Close the attributes table of cole_vctr. Zoom to the GPS track.

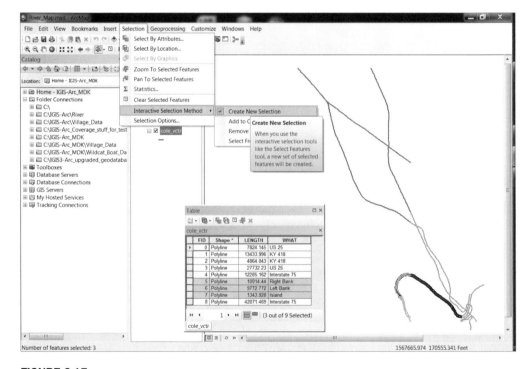

FIGURE 2-17

Using the Measure Tool and the Identify Tool

Also available from the ArcMap toolbar is the capability for measuring distances (using an icon that looks like a ruler), according to how you have set some display parameters.

____ **32.** *Measure the length (the distance along the GPS track) of the trip using Measure from the Tools toolbar:* Activate the measure tool by pressing the ruler icon; a Measure window appears. Read the contents. Pause the cursor over each of the buttons in the window and write the ToolTip in your Fast Facts File.

Done? (   ) Yes. (   ) No.

____ **33.** Use the button that indicates Choose Units and set the Distance units to Feet. Move the Measure window out of the way of the GPS track if necessary. Using the plus (+) sign as the reference point of the cursor, click on the easternmost point of the GPS track, move the cursor, and observe the Measure window contents. Then, trace the GPS track, clicking at points along the route of the trip, using points closer together when the route bends so that you can more nearly estimate the true length of the trip. At the last point double-click (or just keep the cursor on the point and press the Esc key) to get the trip Length total, which, to the nearest tenth of a foot, is _____ feet. Close the Measure window.

____ **34.** Click the Identify icon on the Tools toolbar. Press F1 on the keyboard to look for the help file for the Identify tool. (You may find it and you may not, depending on how your ArcGIS has been set up. In any event you know how to get help through other means.) Observe it, then dismiss it—noting that F1 is sometimes a good way to get help for a tool. Click somewhere on the map where no features exist. Set the layer in the dropdown menu "Identify from" as Boat_SP83.

____ **35.** Using the Identify tool, look at the beginning and ending times of the trip and determine the average speed of the boat in feet per minute. _____.

What is the average rate in miles[18] per hour? _____ Dismiss the Identify Results window.

____ **36.** *Change the viewing area and measure some other distances:* Zoom in on the lines around the island. What is the length of the island (_____ feet), and the greatest width of the river, including the island, near the island? (_____ feet). Dismiss the measurement window.

____ **37.** Zoom back to the previous extent by pressing the left-arrow icon on the Tools toolbar. (If things don't look right, zoom to the extent of the GPS track.)

County Boundaries and Polygons

____ **38.** Click the Add Data button on the Standard toolbar. In the Add Data window, find (yet another instance of) the Connect To Folder icon and press it. Use the Connect To Folder window to select the

[___]IGIS-Arc\Kentucky_wide_data folder.

See Figure 2-18. Click OK. This action will add the folder to the Catalog Tree.

---

[18]There are 5,280 feet in a mile.

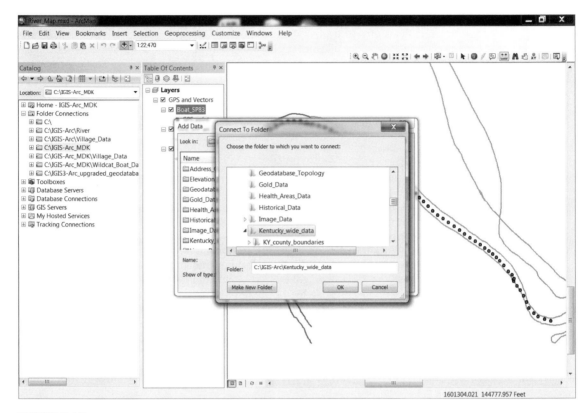

FIGURE 2-18

_____ **39.** Load as a layer the *arc* component of the CNTY_BND_SPN coverage from

[____]IGIS-Arc\Kentucky_wide_data\KY_county_boundaries.

Make its symbol a black line of width 2. How would you describe the boundary between Fayette and Madison counties with the course of the Kentucky River. _____.

_____ **40.** Load as a layer the *polygon* component of the CNTY_BND_SPN coverage from

[____]IGIS-Arc\Kentucky_wide_data\KY_county_boundaries.

By right-clicking the color patch of the layer, make its color Yucca Yellow. The yellow areas of the screen now consist of polygons that represent counties. Use the Identify tool to determine the name of the Area Development District (ADDNAME) that Fayette County is in. (If this doesn't work at first, you probably didn't change the layer that the Identify tool looks at to cnty_bnd_spn polygon. Try again.) _____ What State Plane (SP_ZONE) zone is it in? _____ What is the name of the county that you see to the east of Fayette? _____. Dismiss the Identify Results window.

_____ **41.** Zoom to the full extent of the CNTY_BND_SPN polygon coverage to see all the Kentucky counties. Open the attribute table, sort by NAME, find Fayette County, and select the record. By how much did Fayette County's population grow from 1970 to 1990? _____. Note that a highlight outlines the county on the map. Use the Identify tool to point at the highlighted county. As you noted before, there is exact correspondence between the items in the Identify Results window and the attribute table column heading names. Dismiss the Identify Results window and the table.

____ **42.** Figure out how to remove the polygon layer from the T/C and do so. Then put it back. ArcGIS may have picked a different color when the layer is returned. Click the color patch and pick Beige. Zoom back to the GPS track. Right-click cnty_bnd_spn polygon in the T/C and pick Label Features from the drop-down menu. Labels from the NAME field of the attribute database should appear: FAYETTE, MADISON, and CLARK.

TIGER/Line Files

TIGER/Line files are a product of the U.S. Bureau of the Census. TIGER stands for Topologically Integrated Geographic Encoding and Referencing. (You really should write this identifier in your Fast Facts File, unless your memory is better than mine. I can barely parse it, much less remember it. You have to wonder which came first, the name or the acronym.[19]) TIGER/Line files primarily contain data related to streets—name, street numbers, census tracks, zip codes, county and state codes, and the geographic coordinates of these features. They were designed as the framework on which to hang the census data, and for the use of census takers. A bit later in this chapter, you will look briefly at TIGER/Line files that represent streets, roads, and highways.

But TIGER/Line files also include other types of data that can be represented in linear form, such as political boundaries, railroads, and streams. Fay_Tiger and Mad_Tiger are feature classes in the feature dataset County_streams, in the personal geodatabase Kentucky_River_Area_Data, in the folder [___]IGIS-Arc\River.

The Fay_Tiger and Mad_Tiger datasets are not TIGER/Line files, per se, but were derived from them.

____ **43.** In the T/C, click Layers to select it. Be sure that you are zoomed to the GPS layer and that List By Drawing Order is active at the top of the T/C. Click Add Data and navigate to [___]IGIS-Arc\River. Pounce on it. Pounce on Kentucky_River_Area_Data.mdb. Pounce on County_Streams. Select Fay_Tiger and click Add. Display the line as a bright blue color and a size of 2. Pan (using the "hand" icon on the Tools toolbar) so that you can clearly see the two streams that come into the Kentucky River along the GPS track. What are their names? _____, _____. Notice also, using the Identify tool and the feature's line color, that the Fayette county side of the river is considered a Fayette county stream. Zoom back to the GPS track.

____ **44.** A portion of the dividing line between Clark county and Fayette county (move the map around to see it) is visible in the northeast of the map. Zoom in on it. The line of demarcation between the two counties is given by two different datasets: the county boundaries (cnty_bnd_spn) and the streams (Fay_Tiger), which you will see as separate lines that cross and recross several times. Zoom in again, looking at the place where the two lines seem most divergent. Use the Identify tool to verify

---

[19]The forerunner of TIGER files were DIME files. DIME stood for Dual Independent Map Encoding – reflecting the idea that two very different types of data were included: latitude/longitude on the one hand and postal addresses on the other hand. That acronym made a lot more sense.

the name of the stream: Boone (as in Daniel) Creek. The legal description of the line that separates the two counties is the centerline of that creek. But as you can see, the geographic description of the dividing line is depicted in two different ways that do not exactly agree. Measure the greatest distance between the two lines. _____ feet. Dismiss the Identify and Measure windows.

The lack of complete agreement between two geographic datasets can pose annoying problems for those trying to do spatial analysis. There exist several ways of coping with this problem—either by eliminating the slivers that are formed between the lines or by using the topological features that come with geodatabases. For now just be aware that data from different sources, which each attempt to represent the "real world," will frequently disagree at some level. This sort of data error something to remember when you begin working independently with your own data.

You can see why the quality of GIS datasets must be matched to the needs of the user. These county boundaries might be fine for those delineating watersheds. But surveyors would roll their eyes at discrepancies of 50 feet. They work with accuracy of inches, and very few of those. Land ownership must be precisely defined. Local wars have been declared over distances of less than a foot.

_____ **45.** Change the cursor to Select Elements. Zoom back to the Boat_SP83 layer. Add Mad_Tiger to the map, in the same fashion as Fay_Tiger, but use a pale blue color of width 3 points. Make a group layer of the two TIGER files. (How? Check your Fast Facts File.) Click in a blank area of the T/C, and this time change the New Group Layer name by clicking it and waiting a second or two, and clicking again. You can then edit the field. Type the new name (call it TIGER_Streams) and press Enter. To get an idea of the extent of the streams, zoom to the extent of the group layer. Then return to the GPS track.

The Table of Contents: Display vs. Source vs. Selection

So far we have looked at shapefiles, coverages, and vector-based personal and file geodatabases. You will learn the distinctions between them in detail in Chapter 4. The geodatabase will be the data model of choice as this century progresses, but all three are important, mostly because, as you use ArcGIS in your work or other classes, you may find (and be asked to use) extensive datasets in all three forms. The coverage has been around for decades, the shapefiles for years. The geodatabase, particularly the file geodatabase, is the database of choice for now and the future. One of the major differences between the types is the way in which they are stored on the hard disks of computers. If you look at the T/C of ArcMap, with the List By Drawing Order button active, the different types are pretty much represented in the same way. If you look at the T/C with the List By Source button active, there are some obvious differences.

_____ **46.** Toggle back and forth between Source and Drawing Order, using the buttons at the top of the T/C. Turn off Auto-hiding if it is on and widen the T/C pane horizontally by dragging its right edge so you can see all the text. See Figure 2-19. When List By Source is pressed, the most obvious difference is that you not only see each dataset that contains each layer but also the path all the way to the hard disk. Layers are grouped by their type (shapefile, coverage, or geodatabase) and by their hard drive location. Note, for instance, that while FORD_VCTR arc and cole_vctr are both located in the same place on the [___] path and were made into a group layer, they show up in different places on the List By Source T/C, because one is a shapefile and the other is a coverage. Neither of the group layer names (TIGER_Streams or GPS_and_Vectors) is anywhere to be found, since you are now looking at underlying datasets.

The List By Selection button of the T/C lets you determine which layers are "selectable" and which cannot have features selected from them by clicking on them. The T/C also reports on which layers actually have selections made from them.

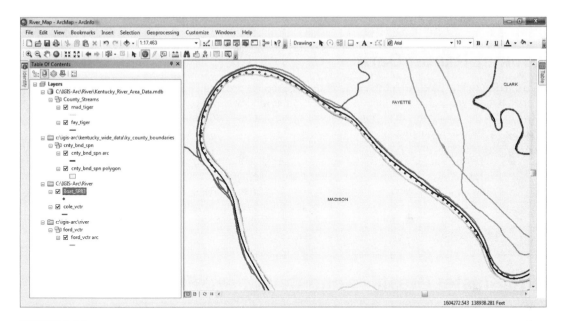

FIGURE 2-19

___ 47. Press the List By Selection button at the top of the T/C window. Use the Select Features tool and, while holding down the Shift key, select in the map window each of the three counties visible in the CNTY_BND_SPN polygon and note the results. The window divides into two sections: selected layers and selectable layers. Experiment by selecting and deselecting other features, such as several of the points along the GPS track, and a stream or two, noting the change in the T/C. You can make layers un-selectable (that is, features cannot be selected) by clicking an icon to the right of the layer name. Using ToolTips determine the name of that icon. _____. Make CNTY_BND_SPN polygon un-selectable, observe the change in the T/C, and then attempt to select a county polygon with the Select Features tool. As you would expect, nothing happens. Click the toggle icon (for CNTY_BND_SPN) again to move the layer back to the Selectable area. In summary, the T/C can show you layers (feature classes) that are:

❏ Selected (containing selected feature elements)

❏ Selectable (but not containing feature elements)

❏ Not Selectable[20]

Press List By Drawing Order icon at the top of the T/C.

___ 48. Using Save As on the File menu, save the map as River_Map_2.mxd in your *Yourinitials* folder, where you originally stored River_Map.mxd. Close ArcMap.

_____

[20]Actually a layer can be made "not selectable" and yet be under the Selected heading. This happens if you toggle the selectability off after some features have already been selected. So no more selections are permitted but the already selected features remain selected.

A Look at Raster Data

Here's a dichotomy: Be aware of the difference between (1) opening a map and (2) adding data to a map. To open a map, you use File > Open to get to an Open window, which then allows you to browse for an MXD file. On the other hand, if a map is already open, you use Add Data (which you can get to a variety of ways) to put an additional dataset on the map.

___ **1.** Start ArcMap. If you have not done anything else in ArcMap since completing Exercise 2-3, River_Map_2.mxd should conveniently available for selection in the Existing Map section of the Getting Started window. Highlight it and click Open. (If River_Map_2.mxd is not available, be sure Existing Maps > Recent is selected on the left hand side of the Getting started window. If River_Map_2.mxd still isn't available, select Existing Maps > Browse for more and navigate to the location of River_Map_2.mxd. Click Open.)[21]

Digital Raster Graphics and Cell-Based Files

Shapefiles, coverages, and geodatasets are all capable of storing points, lines, and polygons. These three are illustrative of the data model we have called "vector." Now we turn to a completely different method of representing geographic data: "raster."

Digital raster graphics (DRG) files are images (photographs, pictures) of the 7.5-minute topographic quadrangles (topo maps) produced by the United States Geological Survey (USGS). They have been scanned into a graphic image format (TIFF—Tagged Image File Format) and provided with information that fixes them in geographical space, making them GeoTIFFs.

___ **2.** *Display a layer based on the digital raster graphic (DRG) image of the Coletown quadrangle:* Set the T/C to List By Drawing Order. Remove the layer CNTY_BND_SPN *polygon*. Zoom to the GPS track. In the T/C, right-click Layers. Add COLE_DRG.TIF from

[___]IGIS-Arc\River.

(If asked, do not build pyramids but read the explanation.) Zoom the view to the full extent of the DRG to see what a USGS DRG image looks like. A part of the city of Lexington is in the northwest corner; the river datasets of interest are in the southeast. Zoom back to the area of the GPS track and notice that only part of the GPS track is in the area covered by the Coletown quadrangle. (See Figure 2-20.)

When (and only when) List By Drawing Order is selected, you may rearrange the order in which entries appear in the T/C. This is important because the T/C order determines the order in which the layers are drawn on the computer monitor. To change the order, you simply drag a layer's name to where you want it to be in the T/C. In general, you should put point-based layers

[21]When ArcMap is already running you can open a map by (a) pressing Ctrl-O, (b) Choosing File > Open, or (c) clicking the Open An Existing Map (see Status Bar or ToolTip) icon on the Standard toolbar. Opening a map requires that you do something (either save or discard) with the map that is presently displayed.

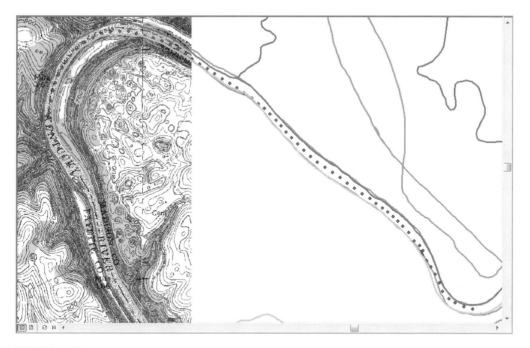

FIGURE 2-20

at the top of the T/C; line-based layers below them; and polygon, image-based, or grid-based layers at the bottom. ArcMap generally attends to this on its own, but sometimes you have to help out.

___ 3. *Experiment with the drawing order:* The title COLE_DRG.TIF probably appeared at the bottom of the T/C. Drag it to the top. Notice that it blocks out the display of all the other layers in its vicinity. Drag COLE_DRG.TIF back to the bottom of the T/C and zoom back to the GPS track, if you strayed from it.

___ 4. By panning and zooming, look at the part of the DRG between the island and the southern border. Observe the contour lines. Obviously the elevation changes sharply, since the contour lines are close together. Read the map to verify the river name and the names of the counties on each side. Zoom back.

Zoom in on the bend in the river—the most northwest quarter circle of it. What is the NORMAL POOL ELEVATION "printed" on the COLE_DRG.TIF? _____. If you don't see it on your map, rearrange things so that the map looks like Figure 2-21.

___ 5. *Since you might want to return to this level of zoom, set a spatial bookmark:* From the main menu select Bookmarks > Create Bookmark[22] and type River_Bend in the Bookmark Name text box. See Figure 2-21. Click OK. Check that it works by zooming to the entire DRG, then select Bookmarks > River_Bend.

---

[22]It's just "Create" in Version 10.0.

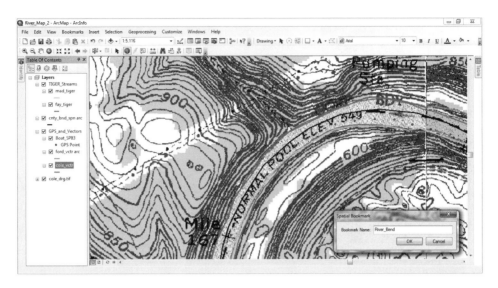

FIGURE 2-21

   6. Pan so that you can see the human-built artifacts (shown in purple) and text just north of the bend of the river. What would you say was the function of this facility? _____. Zoom back to the Boat_SP83. Make a bookmark of the GPS track view, so you can get to it more quickly. Call the bookmark whatever you wish. Name it here: _____.

   7. Pick the colorful area near the island and zoom way, way in on the DRG, so that you can see individual square pixels. In the T/C, click the plus (+) sign in front of the COLE_DRG.TIF. All 256 symbols[23] that could make up the image are now revealed. Actually, only the first dozen or so are used.

   8. Open the attribute table for the DRG. What you see here is the number of pixels (picture elements) of each value (color) on the map rather than information about the geography being represented. See Figure 2-22.

   9. Click the Identify tool icon. Click on the map window to bring up the Identify Results window. Look at the possibilities in the drop-down menu labeled "Identify from". See Figure 2-23. You can ask for results to be shown in the

     Topmost layer

     Visible layers

     Selectable layers

---

[23]Why 256? Because the index number of the colors is stored in a single byte, which as you know is 8 bits, which allows two to the eighth power combinations, which is 2*2*2*2*2*2*2*2, which is 256.

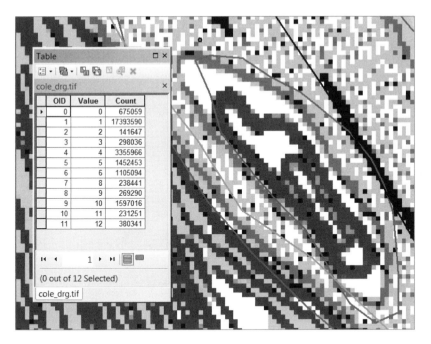

FIGURE 2-22

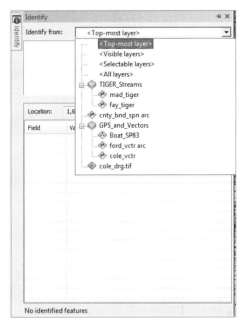

FIGURE 2-23

All layers

{Named layer}

{Additional named layer}

{Group name}

{Grouped named layer}

{Additional grouped named layer}

{Additional named layer}

{And yet another named layer}

{And so on}

___ **10.** Select COLE_DRG.TIF in the "Identify from" text box. Click a pixel on the map and note its Color Index. Match this number with the value in the T/C and note the adjacent color symbol. Select a couple more pixels. Altogether, this should give you some idea of how a DRG TIFF image is put together. For a more detailed description do Steps 11 through 13 below. If you aren't interested in this topic, close the Identify window and skip to step 14.

A Look (Optional) at How DRG Color Values Are Put Together

(Skip to Step 14 if you don't want to do this section.) A DRG is simply a color picture made up of pixels (pixel elements). It may be interesting to you to see how this works. Each color used on the picture is made up of specific quantities of three primary colors: red, green, and blue.[24] Imagine that you have three flashlights, one of each color, focused on a single spot (pixel) and that the amount of light coming from each flashlight can be controlled, with 0 indicating none and 255 indicating brightest. Of course, when all three lights are off, you get black. When all three lights are on at their brightest, white light is produced. These two conditions are represented in the table as Value 0 (black) and Value 255 (white).

___ **11.** Click a brown pixel on the map. Its color index is 4. The Identify window Color field will tell you that the maximum value for any color value is 255, and that this color is made up of intensities 131 (red), 66 (green), and 37 (blue). (Incidentally, recall that you previously viewed the attribute table for the DRG. This Identify window represents one of the few cases where the values shown in the Identify window are not simply the values in the attribute table.)

---

[24]You may have learned that the primary colors were red, yellow, and blue and that you could use these to make other colors. You perhaps verified this by mixing paints that went on paper. Actually, these paints absorbed wavelengths of light for certain color ranges. Paints give off no light on their own. A paint that absorbs red and blue wavelengths from the white light that falls on it and reflects the rest appears as yellow. "Red, yellow, blue" is a subtractive or reflective color model. Light is only reflected, not generated. However, when you look at a computer monitor, you are seeing colors from an emissive or additive color model. Light is being generated by the interaction of the particles coming from the electron gun and the special phosphors on the inside of the monitor screen (or by some more modern whiz-bang technology that generates colored light). Here, the best combination of pure colors to generate the visible spectrum was found to be red, green, and blue. These aren't the only color models. For example, most color printers use magenta, yellow, and cyan (plus black separately) as their primary colors.

___ 12. Right-click the yellow rectangular symbol (7) in the T/C, and click More Colors to open the Color Selector window. Notice that the Red, Green, and Blue (RGB) values are approximately 254, 234, and 0. Move the slider on the Green bar (G) back to 0. Note the resulting color in the lower-left corner of the Color Selector window: pure Red. Click OK to change the yellow color symbol to that color. This change is also reflected in the number 4 symbol in the T/C as well as on the map. Experiment by setting several symbols to different colors. To see the pure primary color Green, set the Green color bar to 255 and the other two to 0. Do the same for the other primary colors. Notice also that the intensity of the resulting color (shown at the lower left of the Color Selector window) is controlled by the sliders. Click OK and note the change in the DRG. Change the white pixels to black and the black pixels to white.

___ 13. When you have finished experimenting, remove COLE_DRG.TIF from the map. This nullifies any changes you have made to the color symbols. Add COLE_DRG.TIF back to the map; expand its legend to verify that you haven't done any permanent damage to the default color scheme.

___ 14. You don't really want every color symbol cluttering up your T/C, so collapse the dataset legend if it is showing. Initiate a zoom out with the Fixed Zoom Out button on the Tools toolbar, and click it repeatedly. Observe how visually meaningless pixels combine to form an image.

Experimenting with Different Ways of Seeing Data

___ 15. Zoom to the GPS track using the bookmark you made previously. Click the Windows menu and click Magnifier. When the Magnifier window appears, drag it, by its title bar, so that the cross hairs are centered over the Normal Pool Elevation number and verify its value in the magnified view. _____. Then, drag the window over the island and compare the COLE_VCTR lines with the elevation lines depicted by the topographic map. Notice that you can resize the Magnifier window by dragging its sides and corners; also you can change the level of magnification with a drop-down menu. Right-click the title bar of the Magnifier window and click Viewer to lock the view, so you can move the Magnifier window—now Viewer window—and not change its contents. Open another Magnifier window and check out its menu options by clicking the little right-pointing triangle and selecting properties from the resulting drop-down menu. Note the current properties and then experiment with them (you may have to move the properties windows if it has appeared directly on top of the magnifier window) before returning them to the original settings and dismissing the properties window. Notice that you can "see through" the Viewer window with the Magnifier window. See Figure 2-24. Dismiss the Viewer window and the Magnifier window.

___ 16. Zoom to Boat_SP83. Select Windows > Overview. In the Layers Overview window, right-click the title bar and click Properties. In the Overview Properties window, find the Reference Layer text box and, if COLE_DRG.TIF is not selected, pull down the menu and click COLE_DRG.TIF. Now click Apply, then OK. Resize and drag the Layers Overview window so that it fits comfortably in the lower left of your screen. See Figure 2-25. The Data Frame window shows the GPS track, part of the DRG, and other data. The Layers Overview window shows the entire DRG. Within that window is a box (gray with a red outline) that indicates whatever is displayed in the data frame.

___ 17. Dragging the outlined box in the Layers Overview allows you to pan around the COLE_DRG.TIF layer. Move it over to the city of Lexington. Then move it back to the river bend. Also, if you use the Zoom controls or the Pan control on the map in the data frame, the new view is reflected by changes in the size and location of the outline box in the Layers Overview window. Use Zoom

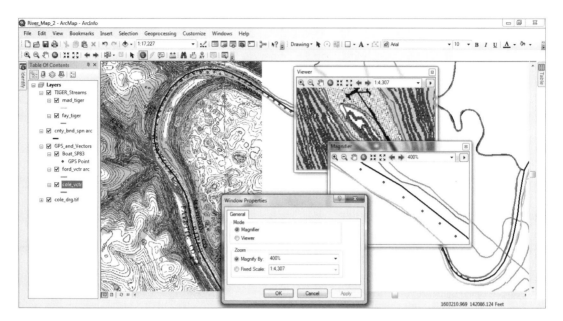

FIGURE 2-24

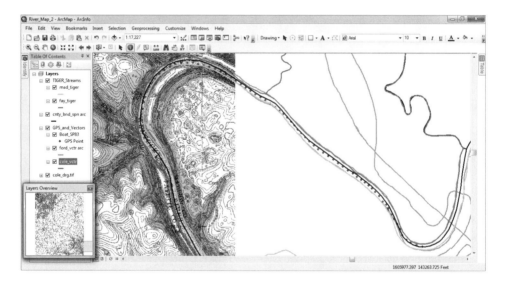

FIGURE 2-25

controls on the data frame so that the scale box in the Standard toolbar indicates a scale between 1:13,000 and 1:15,000.[25] Drag the outline box to the larger blue patch at the top of the Layers Overview window, about in the center. What is the number of the Lexington Reservoir that you see? _____.

---

[25]You can also type a value in the scale box, say 14000.

___ **18.** You may have multiple Magnifier and Overview windows open at a particular time. Before you experiment with these features together, save your map (like, River_Map_3), so you can get it back if something goes wrong and ArcMap closes. What you are about to do is a sort of high-wire act for the software. And frequent saves are a good idea anyway.

Open another Layers Overview window, resize it, and place it above the other one at the left of the screen—obscuring the T/C. Make its Reference Layer GPS_and_Vectors. Open a Magnifier window and plunk it into the data frame. See Figure 2-26. Spend some time looking at the capabilities here. Zoom and pan the data frame with the usual tools. Then pan by using the outline areas in the Layers Overview windows. Move the Magnifier window as well.

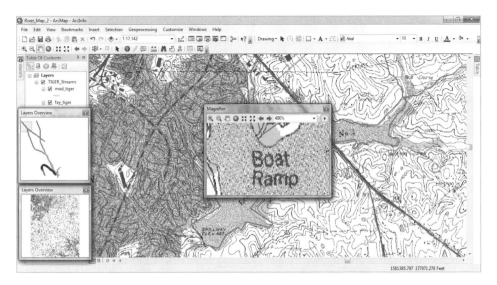

FIGURE 2-26

___ **19.** When you are through playing (and being impressed), close all magnifier and overview windows.

___ **20.** Be sure that the scroll bars are turned on. Select View > Scrollbars if they are not. You can zoom to the full extent of all the datasets in the T/C by pressing the Full Extent icon on the Tools toolbar. Do so now. Notice that the slider bars at the right and bottom of the data frame are fully extended. Zoom to the extent of the DRG. Now you can shift your view of the data frame (a different way of panning) by dragging the scroll bars. Experiment. Close all extra windows that you have opened.

Digital Orthophotos

A digital orthophoto is an aerial photograph that has been rectified (adjusted) so that it may be used as a map with (an almost) constant scale throughout. (Fixing aerial photos so that they become orthophotos is a complicated process, beyond the scope of this book. Consult remote sensing texts or the Internet if you are interested.)

___ **21.** *Add a layer from an orthophoto "TIF" file:* Zoom to the GPS track. Navigate to [___]IGIS-Arc\ River and add a layer to the map based on COLE_DOQ.TIF. (Ignore any warnings about projecting layers. If asked, do not build pyramids.) If necessary, drag the COLE_DOQ title next to the bottom of the T/C (with only the DRG layer below it). Zoom to the extent of the orthophoto image layer. See Figure 2-27. You can get an interesting perspective on the images by flipping COLE_DOQ.TIF off and on. Notice the water filtration plant and other artifacts on both the DRG and the DOQ. Zoom up on the area of the island. Observe that you see the island on the DRG (and see it outlined by vectors) but on the DOQ the island has disappeared. It may well have been under water at the time this photo was taken. A 20-foot rise in water level on this river during floods is not uncommon.

FIGURE 2-27

___ **22.** Experiment with different levels of magnification on the DOQ. Notice that when you are zoomed in very tightly, the "meaning" contained in the image "disappears" in that you see only squares (pixels) of various colors. Zoom so that there are about 10 pixels across the screen. How many feet across is each square? _____. What length is this in meters?[26] _____. This is the resolution of the dataset. Crudely, the resolution is a measure of the limit of how much detail you can wring out of the data. Zoom back to the extent of the DOQ.

---

[26]There are 39.37 inches in a meter.

___ 23. If it isn't expanded already, expand the COLE_DOQ.TIF entry in the T/C so that you can see the three bands that make it up: Red, Green, and Blue. Right-click on the layer name. Notice that Open Attribute Table is grayed out, indicating that the TIF has no attribute table. However the Identify tool can provide some information about each "color". Make the Identify window show the COLE_DOQ layer, and then click on various places on the photo. Query values of several pixels. You will notice that the colors are made up of intensity values. Zoom to the water filtration plant in the northeastern corner of the DOQ. With the Identify cursor, pick a light square on the map and note the values. Do the same for a dark square. Note that the lightest color has the most of red, green, and blue. In the T/C click on each legend color box and uncheck Visible. The image disappears. Now turn each on separately and note the image. Look at them in pairs to get an idea of how equal amounts of different colors produce different results. Turn all bands back on to restore the map image. Close the Identify window.

___ 24. Zoom to the full extent of the DOQ. Find the northern boundary of the DOQ, where it meets the DRG in a forested area. Zoom way up on this area so that you can see the pixels of each layer. What distance on the ground does the DRG pixel width cover? _____ feet. Zoom to the extent of the COLE_DRG. Dismiss the measure window.

___ 25. By right-clicking the data frame and choosing the Data Frame Properties > General, change the display units to Miles. Use the Measure tool, which you should also set to miles, to obtain the height of the DRG in miles (to the nearest one-hundredth). _____, _____. Measure the DRG across the top. _____. Measure it across the bottom. _____. How many square miles is this? _____.

A topo map, as represented by this DRG, covers an area bounded by 7.5 minute boundaries—both north-south and east-west. The fact that it is taller than it is wide in miles demonstrates that a minute of longitude is less than a minute of latitude in this area, as well as almost everywhere else. As one looks at higher latitudes—further north—the seven and a half minute quadrangles get skinnier and skinnier.

___ 26. Change the display units back to Feet. (Under Data Frame Properties again, let me remind you that you cannot change the map units. They are locked in by virtue of the fact that you originally set up the map with a coordinate system that used survey feet as the basic map unit.) Close the Measure window.

More TIGER/Line Files

___ 27. Zoom in on the northeast corner of the DOQ, so that the water filtration plant occupies most of the window. You will notice a road that starts in the northwest of the image and traverses the area towards the east-southeast.

By using the Roads personal geodatabase feature class, contained in the Personal Geodatabase Lexington, you will be able to determine the name of that road. This feature class is somewhat different from the feature classes you saw previously, in that it does not exist within a PGDB feature dataset. Rather, it is found directly within the geodatabase. A feature class that exists directly within a database, without an intermediate feature dataset, is called "freestanding."

___ 28. *Add a layer based on the freestanding feature class Roads:* Navigate to

[___]IGIS-Arc\River\Lexington.mdb\Roads

FIGURE 2-28

and add it to the data frame. Make it bright green, width 2. You should see a line crossing the DOQ. You will notice that it doesn't fit the road in the image very well—which is sometimes characteristic of TIGER/Line data. That census application is less concerned with geographic accuracy than with the topology—that is, the main emphasis is on what connects to what. Try to ignore the fact that the TIGER-depicted road takes a little detour through the plant machinery. See Figure 2-28.

_____ **29.** Start the Identify tool. Click somewhere in the data frame to bring up the Identify window. Select ROADS from the "Identify from:" drop-down menu box. Click on the green line to determine the name of the road. _____. What would you say the zip code for the plant was? _____. The TIGER/Line files contain several attribute fields, which we will explore later. Dismiss the Identify window.

_____ **30.** Just to get an idea of the volume of data in Roads, zoom to that layer. Use the Zoom In tool to look at the level of detail. Lexington has a (mostly) limited-access "ring road" that circles the central part of the city; its "diameter" is about 5 to 7 miles, depending on where you measure. Can you determine its name? _____. Zoom in or out as necessary.

_____ **31.** Zoom to the DRG, then to a small square in its northwest corner that includes the southernmost interchange of New circle Road. Look at the fit, or lack of fit, of the Roads feature class with those shown on the DRG. See Figure 2-29.

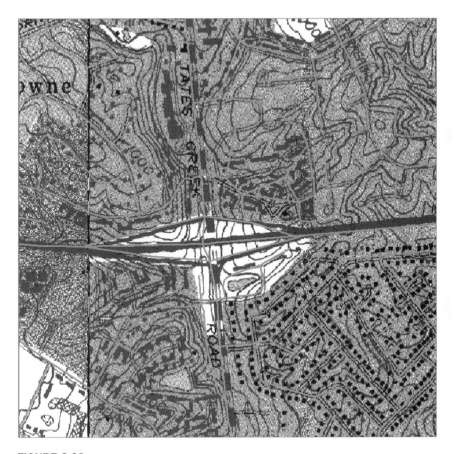

FIGURE 2-29

____ **32.** Open the attribute table of Roads and expand it horizontally. Notice that, at the bottom of the window, where the number of records is shown, the 2000 has an asterisk (*) in front of it. This means that there are more than 2000 records and ArcMap doesn't want to load all of them unless asked. To force ArcMap to display all of the records, look the bottom of the table window and find the buttons for moving around among the records of the table. Check out the ToolTips. You will see a text box in the middle of the buttons. Click the rightmost button (ToolTip: "Move to end of table.") It will take you to the last record of the file. How many records are there in the feature class? _____.

____ **33.** Sort the table into ascending order by feature name (FENAME). Scroll to the top of the table, then down a bit so as to get past all the road segments that have a blank FENAME. You will notice that many road names are repeated. That's because each road segment, between intersections, has its own record. Since TIGER/Line files exist for every bit of real estate in the United States, you can see that the whole thing is a monster-sized database. It has been available in shapefile format, in fairly reasonable file sizes relating to specific geographic areas, using File Transfer Protocol (FTP):

ftp://ftp2.census.gov/geo/tiger/TIGER2010/

or through the web page:

http://www2.census.gov/

I am not suggesting that it is particularly easy for the uninitiated to find what they want through these web sites, but if demographics is of interest to you, the census site is a great source of data.

Use Windows Explorer to determine the size of the geodatabase (called Lexington.mdb in the River folder), which, recall, contains only Roads. _____ megabytes. Dismiss the attribute table, then remove the Roads layer from the map, using the T/C.

Another Tie between Attributes and Geographics

_____ **34.** Zoom to the GPS track. Click the Find button, whose icon looks like binoculars, on the Standard toolbar. Make sure the Features tab is active. Move the Find window to the right, past the DRG. Type Island in the Find text box. The text box below that should say <Visible layers>. Press Find. At the bottom of the window, you should see that a reference to Island was found in the cole_vctr shapefile table, in the field WHAT. See Figure 2-30.

After you have "found" a feature with Find, you have a remarkable number of options for examining that feature. Here are some:

FIGURE 2-30

_____ **35.** While looking at the island on the map,

❏   Click the text Island in the Find window to flash the feature.

❏   Right-click the text Island and click Select on the menu that drops down.

❏ Right-click the text Island and click Identify. Look at the Identify window, and then dismiss it.

❏ Right-click the text Island and click Create Bookmark. Select Bookmarks > Island. Observe the result, temporarily moving the Find window as necessary. Then select Go Back To Previous Extent.

❏ Right-click the text Island and click Zoom To. Zoom back.

❏ Right-click the text Island and click Unselect.

❏ Dismiss the Find window.

Have you added to your Fast Facts File lately?

More Housekeeping: Shutting Down and Restarting ArcMap

____ **36.** ***Shut down ArcMap:*** Use File > Save (or its equivalent: Ctrl-S) to save your map file. Now click the "X" in the upper-right corner of the ArcMap window (or press Alt-F4) to close the ArcMap Program.

____ **37.** Restart ArcMap. Unless someone has changed some settings, you will be given a choice of recent maps and near the top should be River_Map_3.mxd. Select it. You should see what you had before you terminated ArcMap.

____ **38.** ***Stop and restart ArcMap using River_Map_3.mxd to invoke the program:*** Again click the "X" in the upper-right corner of the ArcMap window. Now using Windows Explorer, navigate to

___IGIS-Arc_*YourInitials*\River_Map_3.mxd

Pounce on it and ArcMap should open with the map you saved.

So you see that you can easily save your work, and easily restart ArcMap so that it will pick up where it left off.

Digital Elevation Model Files

A digital elevation model (DEM) data file is a digital representation of elevations, using a raster format. A DEM consists of an array of elevations for many ground positions at regularly spaced intervals. In this dataset they appear as square cells, or, visualizing them three dimensionally, "square posts" whose flat tops represent the height of the land or water at the particular geographic positions.

____ **39.** ***Display a layer from the Digital Elevation Model (DEM) that is in the form of an ArcInfo Raster:***[27] From [___]IGIS-Arc\River add a layer based on the Coletown Quadrangle digital elevation model named COLE_DEM. What is the range of elevations shown in the

---

[27]In order to see this raster, you will need the Spatial Analyst extension or 3D Analyst extension. As indicated earlier, you may need to tell ArcGIS to install the extension(s) in the current session. To do this, go to the Customize menu and click Extensions. Put check marks by Spatial Analyst and 3D Analyst.

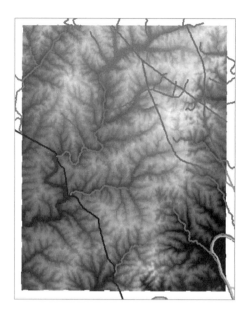

FIGURE 2-31

T/C? _____ feet to _____ feet. Drag its title to the bottom of the T/C. Turn off the two image layers above it. Zoom to the extent of the DEM layer. You can get an idea of the topography by looking at the DEM. White areas are the highest elevations; dark areas the lowest. See Figure 2-31. Select the Identify tool on the Tools toolbar. Click a light area that doesn't have other features (such as vectors) nearby. The Identify window will show the elevation in two places: under cole_dem and as the Pixel value. (You will also see a stretched value number, which should be close to 255. This indicates the intensity of the "color" at that location. Pure white would be 255 which corresponds to a byte with all bits set to 1s, that is, 2 to the 8th power.) Set the "Identify from" text box to COLE_DEM (scrolling the drop-down menu as necessary) so that you no longer have to avoid other features when clicking on the DEM. Dock the Identify window to the left of the screen, next to the T/C. Shrink its horizontal size; again zoom to the extent of the DEM. Now use the Identify tool on a dark area. Here the stretched value should be near zero. Click on the DEM at the river. Record the Pixel value, representing the elevation, here: _____.

____ **40.** Zoom in on an area where there is a large variance in elevation. Zoom in some more. At some point, you will see the individual pixels that make up the DEM. Now, with the display units set to Feet, measure the pixel size to the nearest foot, by measuring across 10 pixels. _____ feet. This number, rounded to the nearest foot, is one of the standard cell sizes for DEMs.

____ **41.** Zoom to the GPS track. Yes, the appearance of the DEM is unimpressive—mostly black, due to the way the heights of the grid posts are classified. To fix this, right-click the DEM layer name in the T/C and click Properties. Under the Symbology tab in the Layer Properties window, select Classified in the Show area. Select a color ramp that goes from Yellow to Dark Red. You can do this either by (1) looking at the actual color choices in the drop-down menu or (2) right-clicking the Color Ramp bar and turning off the Graphic View, so that the drop-down menu shows you text descriptions. When all this is set up, click Classify. Look at the Classification Statistics.

How many DEM posts are there? (see Count) _____.

What is the height of the shortest one? _____.

What is the height of the tallest one? _____.

What is the height of the average post to the nearest tenth of a foot? _____.

___ **42.** For Method specify Natural Breaks (Jenks); also specify 30 classes. Examine the histogram. Specify 40 columns for it. Your Classification window should look like Figure 2-32. Note that the great majority of the elevations lie around the 1000-foot level. However, there are some at around the 550-foot level, which you may recall from the DRG, is near the normal pool elevation of the river. Click OK. Click Apply. Dismiss the Layer Properties window by clicking OK. Note the range of elevation values, and the associated color symbols, by widening and scrolling the T/C. Then zoom to the GPS track again.

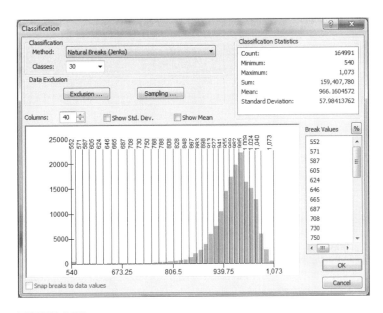

FIGURE 2-32

___ **43.** Zoom in on the part of the GPS track that is within the Coletown DEM, including the island. Click again on the Identify tool and unhide the Identify window if necessary. Notice how quickly the elevation changes in the vicinity of the river. Getting the results from the COLE_DEM, what is the elevation of the surface of the river? _____. Record an elevation taken on the island. _____. Does the DEM river elevation agree with what you wrote before for the normal pool elevation? How much different? _____. Let the Identify window auto hide again.

___ **44.** A faster way to see the values of the DEM posts is to right-click the DEM layer name, then click Properties and switch to the Display tab. Check the Show MapTips check box and then click Apply and OK. Now when you move the mouse cursor around the map, even if the Identify

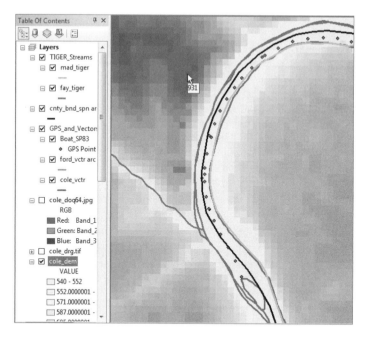

FIGURE 2-33

tool is not being used, you will see the elevation of the DEM post when you pause over a given geographic location. See Figure 2-33. Using Measure and MapTips, find the steepest average slope you can—where the land drops _____ feet in elevation over a distance of _____ feet.

___ **45.** Examine the COLE_DEM entry in the T/C, after widening the pane enough to see all the text. Collapse the COLE_DEM entry, to hide all the values and their symbols. Open the attribute table for COLE_DEM. It is called a value attribute table; it has an extension of "vat". The number of different heights that are represented is 529. The Value field contains the height of the grid "post" in feet. The Count field indicates how many posts there are of that height. How many posts are there of height 1000? _____.

___ **46.** Since this table is more vertical than horizontal, let's dock it on the right-hand side of the screen and resize it so that we can just see all three columns. Select the records whose elevation lies between 901 and 1000, by clicking the box to the left of the record with elevation 901, scrolling with the scroll bar to 1000 (being careful not to click on any other boxes), and then holding the shift key down and clicking the record that shows elevation 1000. (Alternatively, you can click to select 901 and drag your mouse to 1000. You may find this to be somewhat difficult because the table can scroll very quickly. So, if this doesn't work very well because the table scrolls by too quickly, do it in stages, using the vertical scroll bar to scroll the list; hold down the Ctrl key while clicking to select each successive set of records.) Once those 100 records are selected, run statistics on the Count attribute. Recall that you will get results based on the selected records only. Use the Selection Statistics of cole_dem.vat window to determine how many posts (the Sum) there are with elevations between 901 and 1000. _____. How much surface area in acres would that be? _____ acres.

How many square miles?[28] _____ . You may find the numbers that are presented a little confusing. In the attribute table, count refers to **the number of posts** of a given height. In the selection statistics window, count refers to **the number of records** being considered, while sum refers to the total number of posts represented by those records.

___ **47.** Dismiss the Statistics window and the DEM attribute table. Zoom to the extent of the DEM. The selected areas (bright blue) show those elevations between 901 and 1000 feet, which, as you will note, is almost all the quadrangle. Clear selected features with the appropriate button on the Tools toolbar.

___ **48.** Again experiment with various levels of magnification, including full extent of the DEM. Notice that in most places the TIGER streams seem to flow down the valleys shown by the DEM, but sometimes they don't. Again, GIS data is like other data: wide variations in quality—perhaps even within a given dataset. Data sets from different sources are not always consistent. Pick a stream and attempt to determine its direction of flow by looking at elevations. Some stream designations in the southwest area of the map stop abruptly at the black line. Can you recall why? _____ (Hint: Turn off cnty_bnd_spn arc.)

Comparing the DEM and the DRG

___ **49.** *Compare the elevations "printed" on the DRG with those values given by the DEM:* Zoom to the bookmark River_Bend. Turn off the DOQ layer if it is on. Arrange the DRG and DEM layers so that the DEM is at 'the bottom and the DRG is just above it. Turn both layers on. The DRG image will obscure the DEM. Unhide the Identify window again and activate the tool if it is not already. Set up "Identify from layer" as COLE_DEM. Use the Identify tool and MapTips on the topographic quadrangle, pointing at contour lines where the text on the map shows elevations (e.g., 600, 700, 850, and 900 feet, zooming out as necessary) near the bend in the river. You visually see the elevations on the DRG, but the Identify results and MapTips come from the DEM. Note the agreement (or, sometimes, lack thereof) between what is printed on the topo map and the values of the underlying grid. Zoom out somewhat and try this exercise again. You may notice better agreement between the two datasets as you get further away from the steep slopes around the river. Auto-Hide the Identify Results Window.

___ **50.** Zoom to the extent of the DRG. Locate Lexington Reservoir No. 4 in the north central part of the quad. Zoom in on it. By turning the DRG on and off, flip between viewing the DEM and the DRG. Determine, from the DEM, the elevation of the surface of the water in the reservoir. _____ feet. According to the topo map, what is the elevation of the spillway? _____ feet.

Contour Line Files

Another way of representing elevation on a map, familiar to you from your examination of the topographic map, is through "isolines" or "contour lines." For any given such line, every position on that line is supposed to represent the same elevation. The contour lines you have seen so far have been "printed" on the DRG, but a GIS can represent these lines as data as well.

---

[28]There are 43,560 square feet in an acre. There are 640 acres in a square mile.

____ **51.** While still zoomed up on the reservoir, turn off all layers except COLE_DEM. Add the feature class cole_contours_lines from

[___]IGIS-Arc\River\Kentucky_River_Area_Data.mdb\Quadrangle_data

using a Fire Red line of width 1. Start Identify tool and set its window menu so it shows cole_contours_line. Adjust the Identify Results window so that you can see all the attributes. (MapTips will still show the DEM values.) Click the map. Investigate the contours around Reservoir 4 by clicking individual contour lines. See Figure 2-34. What is the name of the field in the attribute table that specifies the elevations of the contours? _____. Notice how it is helpful that the feature you click is highlighted briefly. What is the contour interval (the number of feet measured vertically between adjacent contour lines)? _____ feet. Turn off the DEM.

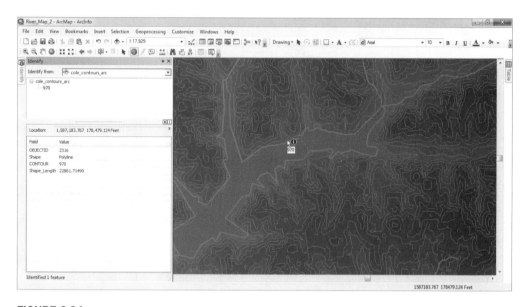

FIGURE 2-34

____ **52.** Turn on the DRG layer. Use the Identify tool to compare the elevation values of the red contour lines with those shown on the DRG. Notice that the contours from cole_contours_line and the contours from the DRG match pretty closely in this area. Zoom in some more on the areas where you see some discrepancies. Using the Measure tool, find a place where the difference between two corresponding contour lines seems excessive. How far apart are they horizontally? _____ feet.

____ **53.** Autohide any tables that might be open to maximize the size of the image. Zoom to the River_Bend. Look again at the differences between the two corresponding contour lines data representations. The process of comparing contour lines may work better if you pan to a location away from the river. Near the river the contour lines are so dense that it is hard to see what is going on. Pan the map so that you can see more of its western part. You will notice that the contour lines from the two different datasets are close, but certainly not congruent. Knowing the elevation at every point in an area is tricky (actually impossible) task. These two datasets were derived in very different ways. They actually show a fair degree of agreement.

____ **54.** Turn off the DRG and turn on the DEM. Again compare the DEM elevations with the contour lines. Use MapTips to see the DEM elevations while using the Identify tool to check the contour lines. Auto-hide the Identify Results window.

____ **55.** Save the map as River_Map_4.mxd in your *Yourinitials* folder, where you originally stored River_Map.mxd. Close ArcMap.

Exercise 2-5 (Project)

Triangulated Irregular Networks

____ **1.** Start ArcMap with River_Map_4 (in your ___IGIS-Arc_*YourInitials* folder). Under Customize > Extensions, make sure 3D Analyst and Spatial Analyst are active.

TINs (triangulated irregular networks), in addition to contour lines and digital elevation models, constitute another way of representing surfaces (particularly topography) in a GIS. A TIN represents elevation by tessellating[29] a surface with triangles. Each triangle is positioned above[30] a horizontal plane such that each of its three vertices is at a known elevation. The result, then, is a surface made up of triangular plates, connected at their edges and positioned at different angles. Each such plate, therefore, has a lot of information associated with it: elevation of its centroid and vertices, elevation of any point on the plate, slope of the plate, and the compass direction of a line perpendicular to the plate (called aspect).

____ **2.** *Add a layer based on a Triangulated Irregular Network (an ArcInfo TIN):*[31] Navigate to [___]IGIS-Arc\River and add a layer based on the COLE_TIN dataset. Drag its title to make it the second from the bottom entry in the T/C, with the DEM below it, turned on, and with its MapTips still active. Turn off any layers above it, including cole_contours_line, so you can see the TIN without the interference of the contour lines. Zoom to the TIN layer. Make sure the cole_tin is expanded in the T/C so that you can see the colors associated with the elevation ranges. Look at the TIN, comparing the colors on the map with the elevations shown in the T/C. Turn all entries in GPS_and_Vectors on. Now zoom in, using the River_Bend bookmark. Zoom in more—on the southeast quarter—to see the construction of the TIN—it is composed of triangles. (Under some conditions of elevation slope, and aspect, you will not see the line separating them, so some of the figures will not appear as triangles.)

---

[29]Tessellated is a 50-cent word that means "completely covering an area with geometric figures without gaps or overlaps." Actually, it is the projection of the triangles on the plane that tessellate it, since the sum of the areas of the actual triangles is more than the area of the plane—as the actual triangles that make up the surface are not parallel to the plane.

[30]Or below, or in the vicinity of the horizontal plane.

[31]In order to see this TIN, you will need the 3D Analyst extension.

_____ **3.** Zoom back to River Bend. To get a better idea of the changes in elevation, pick a monochromatic color ramp—say, Yellow to Green to Dark Blue—by selecting COLE_TIN's Properties, the Symbology Tab, and making a checkmark next to Elevation in the Show area. Right-click Color Ramp, unselect Graphic View, scroll to the desired ramp, and click it. Then, using Classify, pick Defined Interval as a method and 25 as the interval size. Press OK. Make sure the box next to "Show hillshade illumination effect in 2D display" is checked. (This produces a neat presentation that lets you see the individual triangles.) Click Apply, then OK. See Figure 2-35.

FIGURE 2-35

_____ **4.** *Use the Identify tool to notice that, with a TIN, the elevations are calculated rather than stored in an attribute table:* Zoom in so that you can view a dozen triangles or so in the southeast corner. In the Identify window, set "Identify from" to COLE_TIN. Pick a triangle and get the elevations close to its vertices. Since the triangle is tipped, you will get different values for Elevation when you use Identify. MapTips will show you the value of the elevation of the underlying DEM. Now pick a vertex where three or more triangles come together, and get the elevations of the adjacent corners of those neighboring triangles. Those values should be about the same, since the vertex has a single elevation.

_____ **5.** The Identify tool also presents the slope of the triangle (in degrees) and the aspect (the direction in compass degrees that the face of the triangle is inclined towards). In Figure 2-36 the particular triangle identified has a slope of about 17 degrees and an aspect of approximately southeast. As illustrated shortly, you may also view the TIN at oblique angles, rather than from simply directly above. You could spend a full day exploring the different ways to display TINs. Come back to this if you are interested. Get guidance from the ArcGIS Help files.

FIGURE 2-36

If you try to open an attributes table of a TIN dataset, you will discover that it has none. In most geographic datasets the values produced by the Identify tool come from an attribute table. In the TIN, however, the values placed in the Identify Results window are calculated on the fly immediately—based on where you have placed the cursor and clicked.

___ **6.** Collapse the legend for the TIN in the T/C. Zoom to the GPS track, and pan to fill the window with TIN. Again, with the MapTips of the DEM active, compare the DEM elevations with some TIN elevations as revealed by the Identify tool. Under the Display tab in the Properties window for the TIN check Show MapTips, which will take the MapTips away from the DEM and assign them to the TIN. Slide the cursor around the display looking at elevations, created by the TIN MapTips. Click on a few points to verify that the MapTips agree with the elevation shown by Identify.

___ **7.** Change the Primary Display Field of the TIN (in Layer Properties, under the Fields tab) so that MapTips shows Slope. Click Apply, then OK. Experiment by sliding the Select Elements cursor around on the map. See Figure 2-37. Near the river you can find very steep slopes, since the river basically runs through a canyon it has cut into the rock. What is the greatest slope you can find near the river? _____ degrees. The slope you see is expressed in degrees measured from the horizontal plane.[32]

---

[32]You have to be careful when dealing with the numbers that specify slope. Sometimes slope is expressed as a percentage: the rise in elevation divided by the run along the surface of the Earth, times 100 to make it a percentage. If you want the angle of a given percent slope, divide the percent slope by 100, and find the angle whose tangent is that value. Such an angle is called an arctangent, defined as "the angle whose tangent is x."

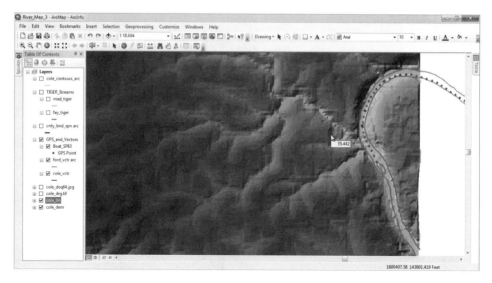

FIGURE 2-37

---

    8. Now make TIN MapTips display Aspect. The aspect is the compass direction (zero degrees to 360 degrees) of a vector[33] that is perpendicular to the surface of the triangle. So an east-facing slope would have an aspect of 90 degrees. Experiment again, looking near the river where you know the directions of the slopes. If you look approximately at the middle of the River_Bend you will find the triangles on the northwestern side facing southeast, and vise-versa.

---

    9. Hide the Identify Results window and close any open, docked tables. Turn off all layers except the GPS and Vectors group layer. Save the map as River_Map_5.MXD in ___IGIS-Arc_YourInitials. Dismiss the Identify window. Dismiss ArcMap.

TINs Are Three-Dimensional Datasets

---

    10. To look at a TIN in another way, and to get a glimpse of the 3D viewing capability of GIS, we go back to ArcCatalog. We need the full ArcCatalog capabilities to do this, so launch ArcCatalog from the All Programs menu or an icon on the desktop. Make ArcCatalog occupy the full monitor screen. Under the Customize menu in ArcCatalog, activate the 3D Analyst extension. Find COLE_TIN in [___]IGIS-Arc\River and click it. Select the Preview tab and observe. In the Preview text box, select 3D View. A different-looking cursor appears as well as a map being viewed from an angle. The position of the cursor is initially unimportant, but if you move it around, slowly, holding down the left mouse button, you can dramatically affect the view of the TIN. Play with it. Then, (gently) do the same with the right mouse button. What is the effect? _____ You can achieve the same effect by rotating the wheel on your mouse. To move the map on the screen press both mouse buttons (or press the center wheel on the mouse) and move the cursor.

---

[33]To be precise, it is the direction of the horizontal component of that vector.

___ **11.** Now make sure Customize > Toolbars > 3D View Tools is turned on (checked). Put the toolbar somewhere out of the way. Use it to zoom in on the area of the river, using the magnifying glass icon. Now use the Navigate cursor (left end of the toolbar) to look at the gorge of the river, from the eastern edge of the TIN and somewhat above. You can take your own little helicopter ride around the area. See Figure 2-38. Play.

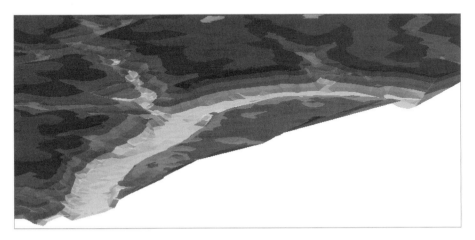

FIGURE 2-38

___ **12.** Most of the 3D View Tools icons are familiar to you. Experiment with those that aren't. At the right end of the toolbar is an icon to launch ArcScene.

ArcScene is like ArcMap, but for mapping in three dimensions. It is a 3D viewing application with many capabilities, including letting you drape raster and vector data over surfaces. I discuss it in Chapter 9 and show you here only enough to whet your appetite.

___ **13.** Launch ArcScene, with a blank scene, from the 3D View Tools of ArcCatalog. Add Data: Boat_ SP83 (use a red dot) and COLE_DRG.TIF. Make sure the Tools toolbar is active and note the tools it contains. Experiment. Zoom in on the area of the river. (Try out the zoom control that you activate with the right mouse button.) Now add the TIN layer.

With the addition of the TIN layer, things become a bit tricky. The GPS path and the DRG appear to be under the TIN—as, upon reflection, you would expect. The GPS track has an assumed elevation of zero—recall that its elevation data was simply recorded as an attribute. Likewise, the DRG is simply two-dimensional. But the TIN is truly 3D!

___ **14.** Use the Navigate cursor to view the image in a horizontal state from the eastern edge, so that you see the TIN layer floating above the other two. Then look around the area. Find a view that you like and save the scene as River_Map_6.SXD (note the ArcScene file extension SDD) in ___IGIS-Arc_*YourInitials*. Dismiss ArcScene. Dismiss ArcCatalog.

Elevation Based on Massive Sets of Data: The Esri Terrain

Most computer-based models of elevation are based on relatively few measured points, usually with considerable distance between them. With the development of airborne LiDAR, knowledge of surface elevation has increased dramatically: Elevation points on the surface (x, y, and z) can be spaced horizontally approximately two or three meters apart! GIS LiDAR (Light Detection and Ranging)[34] datasets are collected by aircraft using light from lasers that is bounced off the earth's surface and returns to the aircraft. The time it takes for this round trip is precisely recorded and used to determine the distance from the aircraft to the ground. Combined with other information, the elevation of the point on the ground can be calculated.

In the following exercise, you will add a LiDAR dataset to your map and look at a few of its properties. LiDAR is not an image or a photograph but resembles both because, in addition to obtaining elevation data, different surfaces return different amounts of the laser beam, so that objects or conditions can be discerned from a LiDAR dataset.

___ **15.** Start a new, blank map in ArcMap. From the [___]IGIS-Arc\River folder add as data the GPS track. Click Add Data again and, in the same folder, you will find the File Geodatabase Terrain_ from_LIDAR.gdb. Double-click it. Add

KY_SP_North_NAD83\River_Bend_and_Water_Plant_Terrain

to the map. Zoom to the extent of this File Geodatabase terrain.

What you are seeing is a great number of triangles that indicate elevation, slope and aspect. The colors of the triangles indicate elevation. However, because of the density of the data, you are not seeing all the triangles that could be generated by the data. The creators of the LiDAR terrain representation rightly assumed that, given the relatively small scale, you don't need, can't see, and can't use all the information that a complete TIN of the area could display, so there is no reason to try to display a high level of detail. This terrain is based on more than 440,000 points!

___ **16.** Use the Identify tool to sample a few locations within the terrain. Write down a few elevations, slopes and aspects:

Elevation	Slope	Aspect

___ **17.** What does the scale text box in the Standard toolbar indicate? 1:_____.

Immediately under the name of the terrain layer in the T/C there is a designation preceded by an asterisk (*).

---

[34]Contrast LiDAR with the more familiar concept of RADAR (Radio Detection And Ranging), developed around the time of World War II, used to display aircraft positions and weather. RADAR uses radio waves in much the same way as LiDAR uses light (laser) waves.

What is that designation? _____.

The way ArcGIS copes with the immense size of LiDAR datasets is by showing only a reasonable amount of information, based on the scale of the display. In what follows, you will experiment with viewing the terrain layer with various levels of scale.

____ **18.** Highlight the contents of the scale text box in the Standard toolbar, type 5000, and press Enter, which changes the scale to 1:5000—or rather will change the scale to 1:5000 after a bit of time, depending on the speed of your computer; a lot of computation is taking place. If you now look at about the middle of the terrain layer, you will see the water treatment plant, which you did not see before because of the smaller scale.

____ **19.** Look at the T/C. Notice that, where before it said *Overview Terrain, it now gives a Z Tolerance of 4. This value basically indicates that only those elevations that are different by 4 feet are represented in the image.

So, this is the mechanism that ArcGIS uses to cope with the massive amount of data: When a relatively large area is shown the level of detail is reduced. As the scale gets larger, more detail is shown, but the amount of computation needed does not greatly increase because the area shown is smaller.

____ **20.** Change the scale to 1:3500. Pan the image so that the water plant view is in about the center. What is the Z Tolerance now? _____.

____ **21.** Record the Z Tolerance for the scale 1:2500 and examine the map. _____Then 1:1500, still keeping the water plant near the center by panning. _____. Finally use 1:500. The Z Tolerance has become zero, indicating that every data point is being used to generate the TIN, regardless of difference in elevation. Dismiss ArcMap without saving changes.

Exercise 2-6 (Project)

Geodatasets of Soils, Rocks, and Land Cover

The type of soil that exists at a location is important for agriculture, for assessing the effects of erosion, and for other reasons. The rock below the soil is of great interest to those who investigate the resources of the planet. In the database Kentucky_River_Area_Data.mdb, we have a Personal Geodatabase Feature Dataset called Quadrangle_Data, which contains feature classes cole_soil_polygon and cole_rock_polygon.

In the soils data, the different classifications appear in a field named MINOR1 of its relational database table. The different classifications of surficial rock types are identified in the NAME field of its table. The origin of these databases is not given, nor the meaning of the classifications. While these are real datasets, they are not valid for display or analysis, other than in a tutorial sense.

____ **1.** ***Display a soils personal geodatabase feature class***: Work with River_Map_5 in ArcMap. In [___] IGIS-Arc\River, find the MDB (Microsoft Access Database) file named Kentucky_River_Area_ Data. Pounce. Find the Feature Dataset Quadrangle_Data. Double-click. Add cole_soil_polygon to the map. Drag its title to the bottom of the T/C. Make sure all image layers, the DEM, and the TIN above it are off. Turn everything else off as well except the GPS and Vectors group. Zoom to the extent of cole_soil_polygon. You will note two blank areas—one to the southwest and a little one in the crook of the river. These occur because this type of soils feature class is not compiled primarily by quadrangle, but by quadrangle *within* county—in this case the part of Fayette County that is in the Coletown quadrangle.

____ **2.** Open the attribute table for cole_soil_polygon. Undock it by dragging the Table title bar into the center of the map, without dropping it onto one of the many blue arrows. You want the table to be freely floating within the map. How many records do you find? _____. This type of data set is sometimes called "detailed soils"; you can probably see why. Run Statistics on the field MINOR1, which designates the soil type. How many polygons are there? _____.

The Summarizing Procedure

You can't get much out of the other statistics, because the numbers in MINOR1 are "nominal," meaning that the numbers are used just as names. Under these circumstances, you can't do any mathematical operations other than to compare for "equal" or "not equal." Even the histogram (Frequency Distribution) doesn't give any definitive information because pairs of adjacent numbers (e.g., 34 and 35) are clumped together in the graph, so this only tells you the total number, not how many polygons are of type 34 and how many of type 35. You can use Summarize to tell you how many polygons of each MINOR1 type there are.

____ **3.** Dismiss the Statistics of cole_soil_polygon window. Click the field MINOR1 to highlight its column. Right-click the field name. Click Summarize. The Summarize window has three input boxes. The first, at the top, should indicate that MINOR1 is the field to summarize. Ignore the second one (in the middle). The third and last should indicate that a table named Sum_Output will be placed in the default geodatabase. To change both the name and location click the Browse Folder button and locate.

____IGIS-Arc_*YourInitials*.

Pounce. Change the name to Summary_MINOR1 in the Name box. From the drop-down box Save as Type, change it from File and Personal Geodatabase tables to dBase Table. Click Save. Click OK. When the Summarize Completed window appears, agree to add the table to the map. In the T/C, note the appearance of Summary_MINOR1 and the fact that the List By Source button is now active. Press the List By Display Order button and notice that the reference to the table goes away. Click the List By Source tab again. The lesson here: You can only see the reference to ancillary tables such as this List By Source is active. Right-click the Summary_ MINOR1 entry and select Open. The Attributes Of Summary_MINOR1 appears, showing the number of records (and hence polygons) with each different value of MINOR1. Notice that there are tabs at the bottom of the Table window. This enables you to switch between cole_soil_ polygon and Summary_MINOR1 easily. Grab the Summary_MINOR1 tab and drag it into the tab window, docking it on the right side so that you can see the information in cole_soil_polygon and Summary_MINOR1 at the same time. Open the Table Options drop-down menu from the top of the window and select Arrange Tables, and within that Move to Next Tab Group, which should return the table to the tabbed view. What is the count for the MINOR1 value of 34? _____. How about 35? _____. Close the tables. Press the Display tab.

____ **4.** Make sure that List by Drawing Order is selected in the T/C. Zoom in on the part of the GPS track that lies within the soils layer. You see lines outlining the polygons, but the polygons are all the same color. You can use the Identify tool (be sure to set the proper layer) to see the attributes, including the value of MINOR1. Note that a polygon "flashes" when you click it. From your knowledge of what is where, determine the "soil type code" for water? _____. What is the "soil type" for an area with no data? _____. Hide the Identify Results window.

5. ***Label each polygon with the number that indicates soil type:*** In general, you can automatically place a label in each polygon. The label, which comes from the attribute table, is a value from the record associated with the feature.[35] Bring up the Layer Properties of cole_soil_polygon. (Recall: To get to Layer Properties, you may either double-click the entry name or right-click it and choose Properties.) Click the Labels tab. Turn on Label Features in this layer by clicking the check box next to it. Click Apply. A number appears in each polygon, but for each it is 999. That's because these are the values found in the MAJOR1 field of the attribute table. By default, the software picks the leftmost "real" field (or a field titled "NAME") in the attribute table to use in labeling. Not what we want. In the Text String Label Field dropdown menu change the text MAJOR1 to MINOR1 and press Apply again. Now you should see the two-digit value for the soil type in each polygon. In the Layer Properties window, press OK. See Figure 2-39.

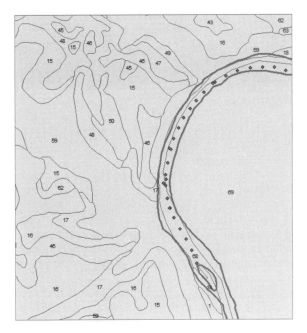

FIGURE 2-39

6. ***Display the feature class with different colors based on the field Minor1:*** In order to more easily distinguish between polygons, you can change the symbology so that the polygons are displayed in different colors. Bring up the Layers Properties of cole_soil_polygon. Click the Symbology tab. Notice that you can Show:

❏   Features

❏   Categories

---

[35]You saw labels briefly earlier when we labeled counties.

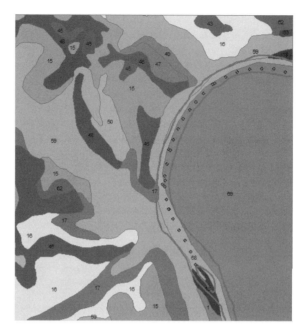

FIGURE 2-40

❏ Quantities

❏ Charts

❏ Multiple attributes

Pick Categories, and within that, Unique Values. This setting will ensure that when you leave this window the software will assign a color to each polygon depending on the value of a field in the attribute table. For the Value Field select MINOR1. For the Color Ramp choose one that has strongly different colors, say Basic Random. Click the Add All Values button. Click Apply, then OK. See Figure 2-40.

____ **7.** In this step, turn on other layers as needed. Determine the MINOR1 value of the soil type classification (1) of the island, _____, (2) near the stream that flows southeast into the river, _____, (3) that is used to indicate water, _____, and (4) for areas not in Fayette County _____. Use the Identify tool to verify your answers. Also check that the color used for the island polygon corresponds to the color next to its number in the Layer pane (in the T/C)— you will have to scroll down to see the values and associated colors. Collapse the list of colors by clicking the minus sign next to cole_soil_polygon. Turn off cole_soil_polygon.

Some Geological Data

____ **8.** Also in the Quadrangle_Data feature dataset you will find cole_rock_polygon, which shows the classification of the surface geology. Add it as a polygon layer, just as you previously added the soils feature class.

9. Run through the same exercises with cole_rock_polygon as you did with cole_soil_polygon, looking at different levels of zoom:

❏ Summarize (on the NAME field) calling the resulting table Summary_NAME

❏ Identify some polygons

❏ Label polygons (use the NAME field)

❏ Display the polygons with different colors

Using the Identify tool, indicate what sort of rock the river bed is made of _____.

10. In the Layer Properties window for cole_rock_polygon, click the Labels tab and change the size of the labels to 10 points. Experiment with different scales by typing into the text box on the Standard toolbar. (You only need to type the denominator of the scale fraction—the software will understand the scale you mean.) Notice that with scales smaller than about 1:30,000 (like 1:50,000), the labels are pretty much useless. And annoying! Select the Labels tab from Layer Properties, and then click the Scale Range button. Activate the radio button for "Don't Show Labels When Zoomed. In the Out Beyond text box, type 30000. Click OK, Apply, and OK. Now notice that a zoom level of 1:30000, you see labels. At 1:30001, you don't.

The concept of showing details of layers only between specified scale levels applies to feature as well as labels. Suppose that you didn't want to see the streams when the data frame was zoomed out beyond 1:80,000.

11. Turn on all TIGER streams. Right-click TIGER_streams and click Properties. Under the General tab, in the Don't Show Layer When Zoomed Out Beyond field, type 80000. Click Apply, then OK. Using the Magnifying glass icons and/or the Fixed Zoom In and Out tool on the Tools toolbar, zoom in and out on the data frame. Observe the scale text box for the scale and observe the map for the presence (or absence) of streams. Also notice that when the streams aren't being displayed, the on-off check box next to TIGER_Streams changes to checked but grayed out.

Rasters of Land Cover Data

Of considerable interest to urban and regional planners is what is called "land cover" (or the allied but somewhat different concept of "land use"): the vegetation, water, natural surface, human activity on the land surface, and so on. In the following, you will load a data set that describes land cover in Fayette County Kentucky at a high-level of resolution.

12. Load as a layer fyttelc06alii, which is a raster, you can find in [___]IGIS-Arc\Kentucky_wide_ data. If asked, do not build pyramids. Turn all other image and polygon layers off. Zoom to the full extent of the data set to get an idea of the amount of data. Zoom to the GPS track. See Figure 2-41. Based on the color of the cells and looking at the legend, what would you say the code for open water is? _____. What is the code cultivated crops? _____. Verify each of these using the Identify tool. Zoom further in so that you can measure the width of a cell. What are the dimensions of the cells in this data set? _____ feet. Which is _____ meters. How much area in each? _____ square feet. An acre would contain how many cells? _____. As I said, this is a high-resolution data set.

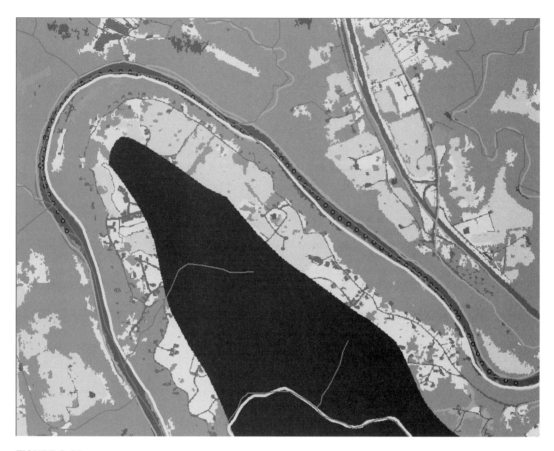

FIGURE 2-41

_____ **13.** Open the attribute table of fyttelc06alii. Based on the value of Count, how many cells represent open water? _____. Therefore, how many square miles of water would you calculate cover land in Fayette County? _____. How about cultivated crops? _____.

For another look at land cover, you will load a data set that describes land cover in the state of Kentucky at a generalized level. You are being shown these two different data sets to illustrate that different organizations select widely varying categories into which to place the activities and conditions on Earth's surface.

_____ **14.** Load as a layer MSU_SPNORTH, which is a raster found in [___]IGIS-Arc\Kentucky_wide_Data\ KY_Landuse_MSU\MSUdata. If asked, do not build pyramids. Turn all other image and polygon layers off. Zoom to the full extent of the data set to get an idea of the amount of data. Zoom into the GPS track. See Figure 2-42 (bearing in mind that your colors will be different, since the software assigns colors at random). Based on the color of the cells and looking at the legend, what would you say the code for water is? _____. What is the code for roads? _____. Verify each of these using the Identify tool. Zoom further in so that you can measure the width of a cell. What are the dimensions of the cells in this data set? _____ feet. How much area in each? _____ square feet.

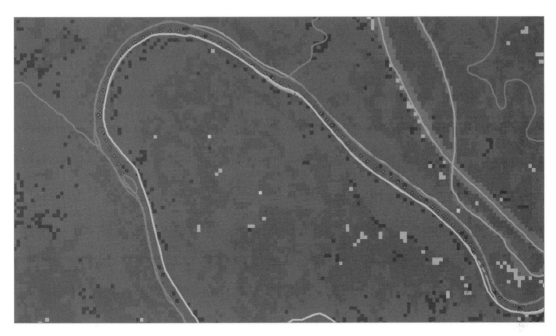

FIGURE 2-42

____ **15.** In the Kentucky_wide_data folder you will find the text file KY_Landuse_MSU\MSUdata\msu_lut.txt. Open this file(lut stands for land use table) with a text editor or word processor. Verify that your answers about water and roads (transportation) above were correct.

____ **16.** Open the attribute table of MSU_SPNORTH. Based on the value of Count, how many cells represent water? _____. Therefore, how many square miles of water would you calculate cover land in Kentucky? _____. How about transportation? _____. Compare. _____.

You Are Not Alone (Assuming you have an Internet connection)

____ **17.** Make a new blank map. Add as data Boat_SP83.shp. There is a dropdown menu besides the Add Data button, which you probably haven't noticed. Click it. Click Add Basemap. Assuming you have an Internet connection, you will see a dozen or so thumbnails of basemaps that are available over the web, courtesy of Esri. Select National Geographic. Click Add. Wait. (If nothing shows up after a few seconds, click the pan tool and move the map very slightly, to encourage the software to create the display.) You will see topography (by means of shading) and some roads. Note Interstate 75 for future reference. Type 50,000 in the scale box and then Pan, following I-75 north, to Lexington. Find Lexington Reservoir No 4 to the west of I-75, as you begin to get into the more developed area of Lexington. You could pan all the way to Toronto if you wanted to. This is a world-wide basemap. Type 4,000,000 into the scale box. Find Lake Ontario and Toronto, Canada. At a scale of 1:500,000 you can see that Toronto has an airport south of the city on an island in Lake Ontario. At 1:25,000 you can read the name of the Toronto Airport. _____.

___ **18.** In the T/C click on the word Basemap, wait a second, and click again. You can now change its name to Basemap_NatGeo. Click the minus sign in front of the name to collapse the entry.

___ **19.** Zoom to the GPS layer again. Again add a basemap—this time picking Imagery. In the T/C change its name from Basemap to Basemap_Imagery and collapse the entry. In the T/C, drag the Basemap_Imagery layer above the Basemap_NatGeo layer. The imagery layer will block the NatGeo layer. Pan the image down slightly so that you can see the water treatment plant.

___ **20.** Suppose you would like to see both layers at the same time. Right-click the Basemap_Imagery layer and pick Properties > Display. Set the Transparency to 70%. Now you can see names of the roads that are present on the NatGeo layer as you look through the imagery layer. What is the name of the road to the south of the river that appears to parallel the river? _____.

___ **21.** Start a new map. Add the GPS track. Using the dropdown menu next to the Add Data button click on Add Data From ArcGIS Online. Experiment with some of the basemaps, starting with USA Topo Maps. That's right, all of the more than 55,000 7.5 minute topographic quadrangles of the United States are available to you! You could use combinations of these basemaps to find an area of interest and see considerable detail of any place in the U.S. For example, you could zoom way out on a DeLorme map (add from ArcGIS Online), locate an area of interest (your home town?), and zoom in on it—getting a good topo image of the area at a scale of 1:4000 and a scale of 1:2000 on an imagery basemap. The same approach could be used to look worldwide, depending on the data available.

In Chapter 3 you will learn how to add your own data to basemaps like these and publish the results so that others can call them up.

Next Steps on Your Own

Of course, there is a long story behind each of the types of data that we have looked at briefly here. If you are interested in a particular type or source of data, I suggest that you look first at the Web for sources and then study Web pages or texts for more detail.

Exercise 2-7 (Review)

Checking, Updating, and Organizing Your Fast Facts File

The Fast Facts File that you are developing should contain references to items in the following checklist. The checklist represents the abilities to use the software you should have upon completing Chapter 2.

❏ To initiate ArcMap:

❏ Square feet in an acre; acres in a square mile

❏ To move toolbars by double-clicking

❏ To turn toolbars on or off

❏ To include extensions in the software

- ❏ The T/C contains _____ when List By Drawing Order is selected.

- ❏ The T/C contains _____ when List By Source is selected.

- ❏ The T/C contains _____ when List By Visibility is selected.

- ❏ The T/C contains _____ when List By Selection is selected.

- ❏ To change just the color of a plotting symbol

- ❏ To change the color and size of a plotting symbol

- ❏ To set the map and display units

- ❏ Map units are

- ❏ Display units are

- ❏ To open an attribute table

- ❏ To dock and automatically hide attribute table(s)

- ❏ A personal geodatabase (PGDB) is

- ❏ A PGDB feature dataset is

- ❏ A PGDB feature class is

- ❏ A PGDB freestanding feature class is

- ❏ A coverage is

- ❏ A shapefile is

- ❏ A DRG is

- ❏ A DOQ is

- ❏ A DEM is

- ❏ A TIN is

- ❏ Vector means

- ❏ Raster means

- ❏ TIGER refers to

- ❏ To group layers together

- ❏ To rename a layer

- ❏ To zoom to the extent of a layer

- ❏ To zoom to the extent of all layers

- ❏ The extension of a map document is

- ❏ A map document contains

- ❏ To save a map document

❏ The difference between Add Data and Open is

❏ Ways of selecting records are

❏ Ways of selecting features are

❏ To find a text string in a database

❏ To set a bookmark

❏ To return to a previous zoom level

❏ To measure distance on a map

❏ To see the values of all attributes of a given feature

❏ To connect to a folder in ArcMap

❏ TIFF refers to

❏ To change the drawing order in the Table Of Contents

❏ The additive colors are

❏ To set layers for the Identify tool

❏ To dock and automatically hide the Identify tool results window

❏ To collapse and expand entries and legends in the Table Of Contents

❏ The Magnifier window

❏ The Overview window

❏ Ways of panning the image are

❏ To set MapTips for a layer to be displayed

❏ A TIN does not have an associated attribute table because

❏ To see a 3D data set with ArcCatalog

❏ ArcScene is

❏ To run statistics on an attribute

❏ To summarize an attribute

❏ To compare tables

❏ To label features

❏ To symbolize features with different colors

❏ To bring up a layer's Properties window

❏ To change whether or not labels are shown, depending on scale

❏ To change whether or not layers are shown, depending on scale

❏   To classify features

❏   To get basemaps from the Internet

❏   To look through one layer at another you can

The Next Chapter

In Chapter 3 you will learn how to make real maps, virtual maps, and publish maps on the Internet.

Products of a GIS: Maps and Other Information

OVERVIEW

IN WHICH you investigate the relationship between GIS and map-making, look at a variety of types of output from a GIS, and learn the rudiments of mapmaking with ArcGIS.

GIS and traditional cartography share a fundamental idea: Depict a part of the real world so that human beings can understand it and use the knowledge for navigation, planning, resource management, or other forms of decision making. The traditional cartographic product, the map, is a single entity that is usually the result of a major project. For example, a USGS topographic map may take several years to produce. Decisions about the font and placement of textual information, colors used, and so on are carefully considered. It has fixed size and scale. To many, making maps is a work of art.

GIS is the worst thing that ever happened to cartography.[1]

Unattributable
Circa 1990

GIS and Cartography— Compatibility?

Earlier, in Chapter 2, we discussed the differences between maps and GIS. Here I want to illustrate a dramatic, seldom recognized difference between published maps and GIS-developed maps. It turns out that, no matter how good the tools in GIS become in facilitating the production of paper maps, there is a fundamental incompatibility between GIS data and even a fairly large-scale paper map. In addition to the previously discussed advantages of GIS over maps is the fact that cartographers sometimes have to distort the positions of features in order to let the map convey important information. For example, suppose that in a mountainous region there is a populated valley. Running through this valley may be several linear features: a stream, a road, a power line, a gas main, and a railroad track. These elements may have to exist within a corridor only a couple of hundred feet wide. Further, the road may cross the stream and the railroad. A distance of 250 feet

---

[1]Probably said because the tools weren't there to make high-quality maps and because it lets amateurs like the author, with all the artistic ability of a can of spray paint, make maps.

translates to only an eighth of an inch on one "2000-scale map," where 1 inch corresponds to 2000 feet. (The scale is 1:24000 – a standard scale for US topo maps.)

An eighth of an inch certainly does not supply enough space to show these features and the relationships between them. If the symbols for these features are shown in their correct spatial locations, they will appear on top of each other. What is the cartographer to do? Fudge! Widen the valley and exaggerate the distances between the features, which means putting some of them in places where they, in fact, do not exist. "Generalize" is a term that is used for this. The cartographer rightly deems it more important to show the correct relationships among the features than to be precisely spatially accurate.

One can correctly argue that such adjustments should not be a part of a GIS database. The different uses, at different scales, of GIS data suggest that everything placed in the database should be recorded as accurately and precisely as reasonably possible. This means that if a map is made or data displayed at the 1:24000 scale from such a database, you would not be able to distinguish among the features. To see the features in their proper relationships, you would need to zoom to a larger scale.

This incompatibility between the paper map and GIS will be exacerbated as more and more detail is introduced into GIS databases. The incompatibility, along with the lack of GIS tools for good mapmaking (a problem that is actually fast disappearing), is why many cartographers convert GIS maps into drawing programs like Freehand and Illustrator. In any event, the virtual map or the Web-based map, will not push the paper map into antiquity, but the trend will be in that direction. After all, what is usually desired is to have spatial data and analysis tools that will create products for decision makers.

Products of a Geographic Information System

Let's shift focus from learning GIS to thinking about the desired end results. By this I mean the decisions taken by human beings to change, navigate, or protect the environment.

C. P. Snow wrote, "To be any good, in his youth, at least, a scientist has to think of one thing deeply and obsessively for a long time. An administrator has to think of a great many things, widely, in their interconnections, for a short time."[2] It is to the attention of these administrators that products of a GIS are mostly directed. You, as a producer of information, need to be mindful of limits on their time.

Overall Requirements for Utility

To be useful in decision making, products of a GIS must meet several criteria:

1. The decision makers must know it is available.

2. They must be able to understand it.

3. They must have some reason to believe that it is worth their time to determine how to use it.

4. Assistance to aid the decision makers' understanding of the product must be available.

5. The product must be available at the time it is needed.

6. It must be relevant to the area of concern.

---

[2]Snow, C.P., 1961. *Science and Government*. London: Oxford University Press.

7. It must have considerable accuracy and integrity; if the product lets decision makers down, a long time will elapse before they depend on such information again.

Classification of GIS Products

Products from a GIS might be classified in several ways. I will use the terms media, format, purpose, and audience.

Media: I use the term "media" to denote the physical carriers of the information presented to the decision maker. Common media are paper, photographic materials (opaque ones like photographs and translucent ones like slides and films), and electronic visual devices like computer monitors. Three-dimensional electric displays activated by laser beams—called holograms—may be available in the future, but more conventional products are now available that can meet more important, if less exotic, criteria. Almost all products of existing GIS are designed to respond to the sense of vision in some manner.

Format: While the number of visual media that carry information is limited, the number of forms or formats that information can assume is without limit. An (almost) infinite variety can be obtained with characters—the 26 letters of the alphabet, Arabic number symbols, and some special symbols. This type of information is called character-based.

Character-based information can appear in the form of text, tables, lists, formulae, and so on. The way in which information is organized has a tremendous impact on whether or not it will be useful. Character-based information can be processed by an individual in serial fashion (like a reader "processing" this line of text) or in search mode—a procedure in which a person examines unconnected groups of characters in order to find desired information. Looking up a number in a telephone directory exemplifies use of the search mode—a procedure in which a person examines unconnected groups of characters in order to find desired information—followed, of course, by serial mode.

For purposes of mental model building, the best products allow a user of character-based information to quickly grasp two things: the overall scheme of organization of the information (revealed by tables of contents, materials on "how to use this information," executive summaries, etc.) and the subject of the information itself (illustrated by introductions, table titles, lists of parameters relating to the information, etc.). Development of products that can meet these criteria is an art and a science.

Graphic information—pictures, photographs, drawings, maps, displays, graphs, diagrams, and so on—is also as versatile as character-based information.

Simplistically, character-based information is read, while graphic information is viewed. Both can help a decision maker form a more complete mental model of an issue, but each provides information in different ways.

It is rare that any information is either totally character-based or graphic. Combinations of the two are the most effective (graphs have descriptive headings and designations; reports have diagrams and illustrations), though the process of "marrying" the two is not always straightforward, particularly when computers are used.

Purpose: Another classification that might be considered during design of a product from a GIS are the purposes of that product or information. Some of the possible purposes include the following:

- ❏ Inventorying

- ❏ Analyzing

- ❏ Explaining

- ❏ Documenting

- ❏ Designating

- ❏ Defending

- ❏ Managing

- ❏ Forecasting

- ❏ Monitoring

The design of the product is frequently more effective if the purpose is kept well in mind. For example, if the major purpose is monitoring, the most appropriate product might be one that reflects a change over time rather than the production of two documents, each of which shows the situation at a given point in time. This sounds elementary, but the amount of effort that has been spent in trying to compare two similar documents, side by side, to ascertain the differences between them is staggering. A GIS has the capability to generate the "difference" and that capability should be used.

Other purposes will be most appropriately met by differing formats. The Important point to consider, for each product, is *how that product will be used.*

Audience: A good GIS will be capable of producing many sorts of information products at varying levels of detail and sophistication. An additional classification for these products might be the audience for whom the product is intended.

Attempts to develop a "superproduct" should be avoided. As various products evolve, this becomes a strong temptation. Those responsible for system and product design and evolution keep adding more bells and whistles, which the designers, of course, understand completely.[3] But a person charged with making a decision, who is looking at the product for the first time, may find that an elaborate demonstration interferes with his or her understanding of—and use of—the information.

One approach designed to avoid this problem is the development of an information product series—several forms of information of similar origin, media, format and purpose. This type of series might show different levels of detail that can be used by anyone exploring all perspectives of an issue. As a product series is used, the first priority should be development of a product appropriate for the needs of the decision maker. That product should be supplemented by others of the same general form that provide more detail or addition information. For example, if information on the limits of the 50-year floodplain for a reach of river is needed, the information product should not show the 20-year plain, the 100-year-plain, the normal yearly

---

[3]This phenomenon is well known to those who use successive versions of software packages and see them become increasingly complex and difficult to learn and use. Nice, compact, simple programs become nightmares of complexity as bells and whistles are added. True improvements become obviated by changes in the graphical user interface. The problem is related to the need for corporations to bring out new versions that will make older versions obsolete—and hence create income for the businesses.

range of bodies of water, or the expected annual rainfall, and so forth. Instead, it should clearly show the 50-year floodplain, with a notation that more detailed or sophisticated information is available for other floodplain limits in roughly the same format. If the system can produce a complicated map, it should also have the ability of producing less complicated ones.

Documenting Products

Developers of GIS products sometimes become so caught up in the task of providing primary products to decision makers that they neglect a second but vital component of the operation: providing explanatory, documenting, and context-setting information. When the information relates to data sets, we call it, as you know, metadata. Products need equivalent attention. Some of the information about products can be obtained from the metadata; some cannot. A partial list of defining information for a report, document, or map might include the following:

- ❏ A title

- ❏ A descriptive paragraph on the content of the document and how it is to be used

- ❏ The geographic area covered

- ❏ The date the information was produced

- ❏ Identifiers that allow the user to determine the defining information about the data that support the information in the report—references to the metadata, perhaps

- ❏ Statements about the precision of the information in the report

- ❏ Estimates of the accuracy of the information in the report

- ❏ The variables that went into the production of the report

- ❏ A name, phone number, e-mail and postal address of a person (or agency) to contact for information about the report

- ❏ All of the parameters that were supplied by the user in the production of the report

- ❏ Any warnings to users

- ❏ Identification of agencies and individuals responsible for the report

Not all of this information needs to appear in the same format in the same place. Some of it may be generated by the computer and appear on the printout: date, data identifiers, parameters, precision and accuracy information, and so forth. Various descriptive information might better be printed separately and attached to the computer-produced report. Regardless of the methods of production or dissemination, a product from a GIS should be a complete package. And, of course, pointers to the metadata of the constituent data sources must be provided.

It may be determined that each user of an institution's GIS should be given a user's manual of the system that mainly describes any customization of an off-the-shelf GIS such as ArcGIS. Such a manual could set the context of the entire system and then describe each product series. Such a manual should be loose-leaf and modular; it should also be available online. An updating scheme should be thought out carefully.

When defining information about a report is contained in the user's manual rather than in the product itself, the product must contain references to the user's manual; the loss of the link between the report and its defining information can prevent acceptance and use of the system.

Thoughts on Different Types of Products

A major reason to have a GIS is to either:

❏ provide new information or

❏ provide information in new forms.

The obvious sort of information that comes from a GIS is in graphic form: the map. However, there are many other ways to convey information about the environment—some more attuned to the style of the decision maker than the casual user.

Don't Ignore Character-Based Information

Character-based information is the sort that many decision makers are most used to using but graphic information is the form that most planning professionals who advise decision makers deal with in formulating their recommendations.

Some humans, perhaps innately, are better at dealing with graphic information and some with character-based information. There is also physiological evidence to suggest that the two different types are processed by different hemispheres of the brain. One might draw a parallel with left-handedness and right-handedness.

It is not my purpose to suggest (or deny) that each person has an inherent dominance of ability to process character-based or graphic information but to point out the danger that an individual may well naturally opt for information in a particular form—just as her or she might naturally use a screwdriver with his or her left hand—when another form might be more appropriate in helping the person gain the necessary insights needed.

GISs can provide both character-based and graphic information, and any GIS that is used to provide only one type may be missing a good bet.

Don't Hesitate to Sort Information

For example, suppose some 348 "areas of critical environmental concern" have been nominated and an identification code attached to each; your GIS has the information to calculate a factor between 1 and 100 that suggests the degree of danger to which each is subject. The output you envision is a list, in order of identification code, with the "danger factor" printed out in an adjacent column. As you begin to use the output, you discover yourself looking through the list to find the area with the highest factor, then the next highest, and so on. At this point, while all that you want is there, it clearly could be in a better form: in order of the danger factor rather than (or in addition to) the identification code order.

Consider Hard Copy

Consider the use of a GIS in some applications to have it print out lists, catalogs, or tables of numbers—of which 99 percent are never viewed. This type of output can be replaced on a periodic or as-needed basis. There is also the option of putting such information on the Internet, but that may make it less accessible in some instances.

While printing a lot of paper sounds wasteful, it may be cost-effective. Consider the example of a telephone book. Despite duplication, paper, and distribution costs—and the fact than an information service is provided over the phone—it is less costly to organize and provide mostly unwanted information to each customer in a region rather than respond dynamically to the customers' need for information at a particular point in time. On the other hand, the phone company does not provide a list of subscribers for the entire nation. The key point is that the issue of product utility and cost must be looked at in a comprehensive way—not just in terms of time, materials, or human effort alone.[4]

Consider Balance in Product Content

In the design of a product, there should be a balance between simplicity and generality. In one sense the best product is the one that speaks directly to the decision maker, respecting her or his particular abilities, relating to the issue he or she is dealing with. On the other hand, it is nice to have a product of such good design that it can serve the decision maker, his or her advisor, those in other areas, perhaps the public and the courts. Mapmakers, of course, are well aware of this design problem. A map's usefulness is increased by adding a new type of information and yet decreased at the same time because the map becomes more cluttered. A map that "shows" everything shows nothing, because it is black.

Elements of Product Design

Product design is an art that marries what is possible with what is needed. Many factors go into successful design:

❑ *Examination*—Of products from other systems

❑ *Innovation*—The ability to conceive of more meaningful ways to display information within the constraints of the devices which put the information together

❑ *Refinement*—Of products that the GIS is already producing by obtaining, and heeding feedback from users of the product

❑ *Knowledge*—Of what data are required (and the other characteristics of those data) to produce the necessary information

❑ *Lack of bias*—Toward either character-based or graphic information and an ability to provide information in the best format for the given customer

---

[4]The availability of the Internet has changed this somewhat, although looking up a number in a paper-based directory is sometimes preferable to dodging advertisements and dealing with sites that want to sell you the information. Many people view the Internet as a way to make money and they don't mind inconveniencing users if they think they can achieve that aim.

Units, Projection, and Scale

In Kentucky, eight different coordinate systems for spatial information are in use by state agencies alone. No doubt, other states share this problem. How should distances on a map be presented—metric or otherwise? What shows up on the product should take into account the nature of the audience. If you can, avoid cluttering products with multiple coordinate systems. If the user doesn't mind the maps looking weird, you might make the map using latitude and longitude coordinates. Usually though, users like maps whose scales are consistent in all directions.

Thoughts on Resolution and Scale

Resolution is, basically, the smallest length in ground units at which the identity or characteristics of something can be resolved by looking at the product. For example, what is the diameter of the smallest object that can be seen in an aerial photograph? Clearly, the answer depends on several factors besides the diameter of the object. It involves the reflectance of the object compared with the surrounding ground, the quality of the vision of the person looking at the photo, and the scale of the photo. An 8 × 10 photo of Arizona might allow only an object the size of Phoenix to be visible. Through enlargement of the same photo, however, it might be possible to increase the resolution so that an object 100 feet across could be identified. If further enlargement—no matter how extensive—does not allow identification of objects with diameters smaller than 100 feet, then the resolution of the photo is said to be 100 feet. Thus, it is obvious that resolution and scale are closely linked—the larger the scale, the greater the resolution—up to a certain limit. Beyond that limit, no further information can be obtained from the product simply by enlarging the scale. In previous exercises you have had the experience, looking at raster images, of watching the information disappear as you zoomed in.

If a specific ground area is to be covered, designers must then weigh the relative factors of size, scale, and resolution to determine the most appropriate method of producing an information product. Too often, these determinations are made without sufficient thought, or are influenced by habit and conference room table size. The issue of appropriate scale must be considered early in the process because some GIS products utilize overlays and base maps of other materials to make them meaningful. Obviously, a nearly exact physical match must occur. Further, the size, scale, and resolution of both base maps and overlaid products must be appropriate. It is worth mentioning here that reproductions of paper map output, from any source, are not always the same size as the original—and hence do not have the same scale as the original. Unfortunately, statements regarding scale printed on the map, such as 1 inch = 1000 feet, are reproduced along with everything else, creating a built-in lie.

Making Sure There Is a Base Map

Your GIS will include, probably, a multitude of layers. The existence of a base map—a cartographically or photographically produced product of, usually, great accuracy, high precision, and great detail to which all other products of the GIS can be referenced—is essential. To attempt to build GIS products without first developing (or otherwise obtaining) a geographic base map is folly. But be aware: To use primarily GPS to create GIS data sets intended to serve the base map function is a monstrous undertaking.

Measure of Quality Assurance

One of the values of a GIS is its capability to combine different types of information that have the same geographic base. For example a user might need a map displaying a single variable derived from a combination of soil type, bedrock type, and depth to bedrock. An important aspect of the defining information of a report combining these three variables is the confidence one can place in the accuracy of the resultant information. Each of the constituent individual variables is stored in the database and each has its own characteristics of precision and accuracy—measured in terms of geographic coordinates and the values of the variable. Therefore, just as the values of the three individual variables are combined to produce a single variable, the precision and accuracy attributes of these variables should be combined—according to appropriate numerical techniques—determine the accuracy of the final result. Statements of accuracy should be included as part of the resultant report.

The process of assessing accuracy is not always easy, nor are the results always encouraging. For example, in a study done by the Australian Commonwealth Scientific Industrial Research Organization a number of years ago, involving three variables, an analysis was made of the output. When particular points on the Earth's surface were picked out and "ground truthed," it was found that at least one of the constituent variables was incorrect more than 60 percent of the time. The implications for the accuracy of the combined report are obvious.

Frankly, the lack of sufficient quality reporting of derived products is a major shortcoming of all commercial GIS software.

The Decision Maker–Product Interface

The act of a decision maker sitting down at a computer terminal is one that cuts out virtually all the "people buffers" between the decision maker and the computer. Many decision makers will not want to spend the time necessary to either learn or operate the software. There is also a certain element of justified fear involved for even the ablest individual in doing something new with others looking on.

A person familiar with the product (including the assumptions underlying it and other products that might be useful to the decision makers) should be present when the product is used. People charged with making decisions have a way of asking questions no one thought they would ask. Anyone who presumes to provide them with new information in new forms had better be ready.

The person charged with the responsibility of understanding and explaining a document is also in an ideal position to recommend changes in the document's structure or information content based on conversation with the users of the product. The dissemination of a product containing information is very much a two-way street and relies on user feedback for its successful continuation.

Often, user needs are not correctly perceived by product designers. Further, user needs change. These and many other factors suggest that a continuing dialogue between the providers and users of information products must exist.

It may be that, instead of having a person assigned to a particular set of products as the interface between the decision maker and the product, personnel will be assigned as liaisons to various departments using the products. Whichever scheme is chosen, personnel charged with the function of providing an interface between products and decision makers should meet among themselves regularly to aid in improving the effectiveness of the GIS operation.

In Summary

You can see that the number and diversity of potential information products from a GIS are almost unlimited. These products, however, are useless if they don't fill a need or if they are inappropriate for their intended audience. The song, "Alice's Restaurant Massacree" by Arlo Guthrie contains the line: ". . . and the judge [with the seeing-eye dog] wasn't going to look at the twenty-seven 8 by 10 color glossy pictures with the circles and arrows and a paragraph on the back of each one. . ."

Design of an information product must, of course, proceed from an enlightened view of the data used to support it. Equally important, however, is a clear understanding of the needs of decision makers or administrators who will be using it.

Products of a GIS: Maps and Other Information

—— Open your Fast Facts text or document file.

—— Open the Color Figures file, so you can see the illustrations in more detail.

Up until now the maps you have worked with have fundamentally been portrayals of geographic data. However, if you look at any map that is produced commercially or by government sources, you will notice that the spatial data reside in a context of other text and graphics. For example, you will probably find a title of the map, an arrow indicating the north direction, a legend showing the scale, and so on. Also, a single map sheet might consist of several maps at different scales or maps showing different data from the same geographic area. While you can print a map directly from the sort of ArcGIS view you have been working with, ArcMap has capabilities that let you produce a map with the additional elements that form sophisticated cartographic products.

While the thrust of this book is to lay groundwork so that you can use GIS to do geographic data analysis, synthesis, and modeling, I would be less than candid if I didn't let you know that the most popular use of GIS currently is to *display*, in map format, geographic information. This chapter gives you the beginnings of how that is done. Of course, when you do use GIS to do analysis, you will need to display the results, so what follows is essential. Before we launch into how this works you need to know some terminology:

❑ *Feature*—A representation of a real-world thing, like a house, a city, or a pipe

❑ *Object*—A point, line, network, or area that represents a feature

❑ *Layer*—A set of objects that represent a number of features

❑ *Data frame*—One or more layers, each displayed in a particular way (scale, style, etc.)

❑ *Layout*—One or more data frames, with optional tables, graphics, and so on—the finished graphic and text product that will become a map sheet.

The Data View and the Layout View

To use ArcMap to make maps, you need to be aware of two distinctly different ways of displaying geographic data: the Data View and the Layout View. The Data View is the view that you are familiar with. The main ArcMap window shows the datasets you have selected; they fill the window pane.

Chapter 3

In the Layout View you see an image of a map sheet meant to be displayed or printed. In this view you "lay out" the map elements, including the images from datasets that make the map a true cartographic product. A lot is involved in transforming geographic data into a map worth the name. You discover how to begin the process in the exercises that follow.

Exercise 3-1 (Warm-up)

Templates

A template in ArcMap is somewhat like a preprinted sheet of drawing paper. It can be almost blank or can contain graphics and/or data. We start by looking at some examples:

—— **1.** Start ArcCatalog. Copy the folder Trivial_GIS_Datasets from [___]IGIS-Arc to ___IGIS-Arc_ *YourInitials*. Make a connection to this new folder. Under ___IGIS-Arc_*YourInitials* create a new folder named Map_Making. Make a connection to this folder also. Launch ArcMap from ArcCatalog. (By the way, folder connections are easy to make, and even easier to get rid of. So, disconnect folders when it's likely you won't need them again soon. This will keep things more organized.)

—— **2.** When the initial ArcMap dialog box—Getting Started—comes up (assuming it does[1]) select My Templates from under New Maps within the left pane. One template (probably the only one) will be called Blank Map. It is based on Normal.mxt—the one you get when you start fresh with ArcMap. It is a file whose location on disk is given below the selection areas of the Getting Started window. The file extension of all templates is .mxt; look for it. Write down the location of (the path to) the file Normal.mxt.[2] Do **not** click OK.

—— **3.** Find the Traditional Layouts under Templates (which is under New Maps). Click. In the upper-right corner of the window you will find that you can choose to see the Traditional Layouts in a List, as Large icons, or Thumbnails. Choose Thumbnails. Among your choices of templates here are LandscapeModern. Single-click the thumbnail. See Figure 3-1. (Stretch the window if necessary to see three columns of Layouts.) Write down the location of the file LandscapeModern.mxd. Notice the file extension is MXD and not MXT. (This file is sort of a cross between a template and a map, because it contains information that make it easier to start a published map, once you put your data in it.) Do **not** click OK.

---

[1]The dialog box is optional and may have been turned off. Or ArcMap may be already running. If so, choose File > New to get the New Document window. It may appear to be slightly different from the Getting Started window depicted in Figure 3-1 and referenced in Step 3. (To turn on the startup dialog for the future, you may choose Customize > ArcMap Options, then on the General tab, make sure Show Getting Started dialog has a check mark in its check box.
[2]The path name may be too long for the text box. If so, place the cursor in the text box, click once, and press arrow keys to see the rest of it.

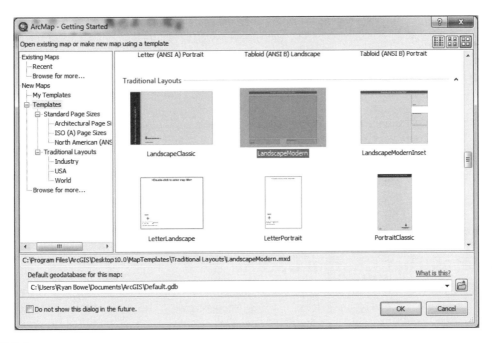

FIGURE 3-1

4. In general, all ESRI-supplied templates, except Normal.mxt, are stored together in a MapTemplates folder. By using the up and down arrow keys, you can see miniatures of several templates. Look at some of the other templates. Do **not** click OK.

5. Make sure that LandscapeModern.mxt is highlighted. Now you may press OK. This makes LandscapeModern the template for the current ArcMap document, as you can see from the map display area window.

6. In Customize > Toolbars make sure there is a check mark beside Layout. Locate the Layout toolbar. It will look something like this:

Some of the icons will seem similar to those in the Tools toolbar, but note that most include the image of a page as well. These tools are for working with map layouts and they operate somewhat differently, as you will see shortly.

7. ***Examine and use the map template:*** There is some tiny text in the lower-right box of the template. Zoom in on it (using the + magnifier on the **Layout** toolbar) by dragging a box around the text. What does it say? _____ Click the Select Elements tool on the **Tools** toolbar and double-click the text you zoomed in on. A Properties window pops up with the Text tab active. In the Text box, type Just Playing (replacing the text that is already there) and press Apply. (Move windows around as necessary to see relevant parts of the screen.) In the Properties window press Change Symbol, which brings up a Symbol Selector window; make the text red, bold, Arial, 36 point. Click OK, then Apply, then OK.

____ **8.** Use the Zoom Whole Page button[3] in the Layout toolbar to see the entire layout again. If necessary, click again on the Select Elements tool on the Tools toolbar. Using what you learned above, change the text you put in previously to More Play and click Apply, then OK. When the cursor is inside the selection box, you should see a cursor that has four arrows. Using this cursor, you can now drag the text around. Drag the text to the center of its box.

____ **9.** *Now (on your own):* In black, underlined, bold, italic, 40-point Arial, enter a title at the top of the layout that reads I'm Done Here. Center the title.

Exercise 3-2 (Project)

Templates That Contain Data

____ **1.** Using File > New,[4] in ArcMap, pick and click either the USA template category or the World category, depending on where you live. By flipping between the three buttons at the top of the New Document window, under the red dismissal X, you can look at the possibilities in different ways. Pick a template that contains your region. Click OK. If asked, don't save your previous layout. What template did you pick? _____.

Notice that the titles of the Layers in the Table Of Contents (T/C) reflect your choice and that a number of data sets—Capital Cities, Rivers, etc.—are represented. Notice also that the map window contains a title, a scale bar, and a legend. The *thin* black line represents the boundaries of the page that may be printed. You should also see rulers along the top and left side.[5]

____ **2.** *Switch between the Layout View and the Data View:* From the View menu, pick Data View. You will see the sort of map display you are used to seeing, without titles, scales, and so on. Now pick Layout View from that same menu. For a shortcut way of switching between the two views, locate the four icons to the left of the horizontal scroll bar at the bottom of the window.[6] (See Figure 3-2.) The leftmost one, an Earth symbol, is the Data View. The next, a page, gets you the Layout View. The third is a refresh button, should the display not look quite right and you want it "repainted." The fourth pauses the drawing. Flip back and forth between the two sorts of views.

FIGURE 3-2

---

[3]Find it with ToolTips. If ToolTips is not on, select Customize > Customize Mode and on the Options tab check Show ToolTips on toolbars.
[4]File > New brings up the New Document window, which is different from starting ArcMap fresh, where you receive the Getting Started window and have the option to open recent maps as well as templates. The New Document window can be accessed in additional ways: the New Map File button and Ctrl + N.
[5]If the rulers aren't showing, right-click the layout page outside the thin black line, then select Rulers > Rulers.
[6]If the scroll bar isn't showing, make a check by the option View > Scroll Bars.

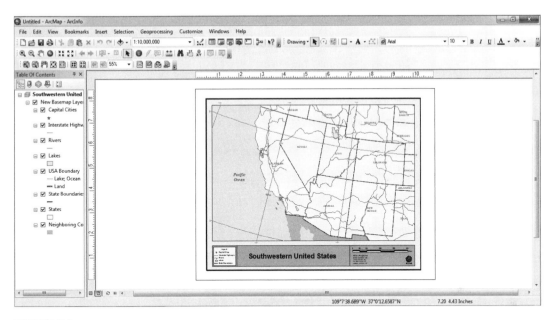

FIGURE 3-3

——— **3.** Select File > New again. In the New Document window, locate the text " What is this?" relating to the Default geodatabase for this map. Click it. Read the first two paragraphs. If later you are confused about the "Home" folder and default locations you can come back and read more. Dismiss the Help window.

——— **4.** Click the title bar of the New Document window, and then pick the USA template category (New Maps > Templates > Traditional Layouts > USA). Click SouthwesternUSA, and then click OK (not saving changes from the previous map). The map pane comes up in Layout View. (See Figure 3-3.) Observe it, then switch to Data View.

——— **5.** Since geographic data sets are being shown, there is likely to be some underlying coordinate system. Determine what it is: Right-click the map, then in the menu that presents itself, click Data Frame Properties to bring up the Data Frame Properties window. Select the Coordinate System tab. What is the coordinate system? _____What is its Central Meridian? _____ degrees. What is the Latitude of Origin? _____ degrees. Look under the General tab; what are the Map Units? _____ Check that the Display Units are Miles. Click Apply if necessary, then OK.

——— **6.** Using the buttons at the lower-left, switch back to the Layout View. When the cursor is on the map proper (the area where geographic features are being shown), two types of coordinates are displayed: geographic (map) coordinates and page coordinates. Place the cursor at the lower-left corner of the map proper. What are the approximate coordinate values?

Map: Easting _____ Northing _____ in _____ (units)

Page: X-coordinate _____ Y-coordinate _____ in _____ (units)

Do the same for the upper-right corner of the map proper:

Map: Easting _____ Northing _____ in _____ (units)

Page: X-coordinate _____ Y-coordinate _____ in _____ (units)

By the way, it is easy to confuse latitude and longitude. It is also easy to confuse easting and northing. Usually latitude is given first when coordinates are in degrees, but usually the x-coordinate (easting) is given first when you are operating in a Cartesian coordinate system. Note that these orders are the reverse of each other.

Now check out the lower-left and upper-right coordinates of the entire rectangular map page, as delimited by the thin black lines.

Lower-left page: X-coordinate _____ Y-coordinate _____ in _____ (units)

Upper-right page: X-coordinate _____ Y-coordinate _____ in _____ (units)

Controlling Your View of the Map: Zooming

The two sets of zooming and panning controls take some getting used to. Let's look at zooming first. If you are in the Data View, the Layout toolbar is inactive. However, in the Layout View, both toolbars may be employed; their tools have different effects.

_____ **7.** Go into Data View. Using the familiar Zoom In tool, drag a box around Colorado, leaving significant margins around the state boundary. You will be able to make out the word Colorado and may be able to see the green star symbol representing the capital city, Denver. Zoom up on that area by dragging about a 1-square-inch box around it. Notice that the place names and the symbol get no larger, but you see more detail of the highways. Notice also that the state name Colorado is visible, having moved from where it was before into your view. Zoom in again tightly on Denver and its symbol. Again you see "Colorado." And again, the symbols stay the same size.

_____ **8.** Click the Go Back to Previous Extent button until the entire state of Colorado is shown again.[7] Switch to Layout View. Now using the Zoom In tool on the _**Layout**_ toolbar, zoom in on Denver. Notice now you don't see "Colorado," that Denver's star is larger, and that the roads are shown with wider symbols. The lesson: When you use the magnifier on the Layout toolbar, the Zoom In feature really does act like a magnifying glass; that is, in addition to seeing a smaller area in more detail, the graphics and symbols get bigger. Further, the text symbols do not relocate as they do in the Data view. In the Layout toolbar, click the Go back to extent button so that the entire state is visible.

In the Layout View, the zoom buttons simply let you view the map—geographic elements, title block, text, and so on—with different levels of magnification. Zooming and panning do not move the labels and symbols of the map.

_____

[7]If at any time you run into difficulty and feel that you would like to start over, do. Just choose File > New, pick USA template category, click SouthwesternUSA, and click OK. If asked, don't save your previous layout.

____ **9.** ***Experiment with the buttons on the Layout toolbar:*** Zoom using the Zoom In tool on the Layout toolbar, so you can see the Denver area (green star) and the highways surrounding it. Notice again how place names, highway detail, features, and boundaries become larger as you zoom in. Again recall that this is different from your previous experiences with the Data View zoom control. Now use the Identify tool to click on the highway that goes out of Denver toward the west. What is the route number of that Interstate? _____. How long is it? _____. Click a tab at the bottom of the window so you can see the T/C again.

____ **10.** ***Determine the use of the Zoom to 100% button (labeled and depicted 1:1) on the Layout toolbar:*** Use the What's This (?) button (in version 10.0) or mouse over the 1:1 button in version 10.1. Try the button. Notice that the text box on the Layout toolbar reads 100%. Then press the Go Back to Extent button. What does the text box read now? _____.

____ **11.** Click the Zoom Whole Page button. The entire layout returns, title block and all. (This button will be your good friend in a number of circumstances—use it often.) Try the other Zoom buttons, the Go Back to Extent and Go Forward to Extent buttons, and the Zoom Control text box. Experiment with panning the map sheet at different levels of zoom. Notice that the information provided by the rulers changes as you zoom and pan.

____ **12.** Observe the scale text box on the Standard toolbar. Press the Zoom Whole Page button again. Notice that the scale on the Standard toolbar does not change when you change the zoom level with the Layout toolbar, because you are merely looking around a "paper" map, studying different parts of it with a magnifying glass.

____ **13.** Experiment by flipping back and forth between the two view types at different levels of magnification. Remember, you can alternate views easily by using the two mini-buttons at the left of the horizontal scroll bar. Go into Data View. The Data View of this template includes a lot of territory, as you can see by panning. On the Tools toolbar, press the Go Back to Previous Extent button (repeatedly if necessary) to get the overall map of Colorado again. Return to Layout View, and use Zoom Whole Page. If the map does not show exactly what you want, alternate between Data View and Layout View, panning and zooming, until you get the desired result. (For each of the following operations, read the Status bar (version 10.0) or the ToolTip (version 10.1) to see what the button is intended to do.)

Understanding the Panning and Other Controls

____ **14.** To eliminate any confusion after all that map manipulation, let's start over: Click File > New, pick the USA template category, click SouthwesternUSA, and click OK. Again, if asked, don't save your previous layout. In Data View, zoom in on Colorado, as before. Change to Layout View. (For each of the following operations, read the ToolTip (and maybe the status bar, depending on the ArcGIS version you are using) to see what the button is intended to do. If you want more information remember that the What's This? tool could be useful.) Using the "hand" icon on the Tools toolbar, pan over to Utah. Get back with the Go Back to Previous Extent button. Now pan the map by dragging it using Pan on the Layout toolbar. Notice this simply moves the entire layout (and adjusts the rulers so that the same ruler numbers intersect at the same geographic features). Return to the original image by using the Go Back to Extent button on the Layout toolbar. Pan again. Go back this time with Zoom Whole Page.

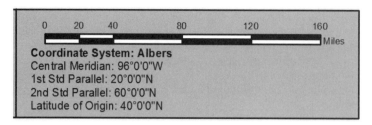

FIGURE 3-4

___ **15.** The image of Colorado looks crooked on the page. The Albers projection does this for areas not near its central meridian—in this case, 96 degrees west. On about what meridian does Denver lie? _____ degrees. (Hint: check the Status bar). The boundaries of the state are in fact parallels and meridians of the latitude and longitude coordinate system, so it would be nice if they ran east-west and north-south. Fix the crooked image by clicking Customize > Toolbars and make sure that Data Frame Tools is checked. Then, type the desired number of degrees of *counterclockwise* rotation into the text box in the Data Frame Tools window that appears. Try 3 degrees and press Enter. Not enough? Clear Rotation and try 6 degrees. (If you wanted a clockwise rotation, you would use a negative number.) Still not satisfied? Try the Rotate Data Frame interactively tool. Finish up by using a rotation of 6 degrees. Dismiss the Data Frame Tools tool bar window.

___ **16.** *Since you have changed the scope of the data, the title on the map is now incorrect, so change it:* In Layout View, press the Select Elements button on the Tools toolbar. Click various places on the map proper and on the other parts of the page. Notice the blue rectangles and "handles" that appear. Note that multiple clicks in the same location sometimes result in different selections. You could resize these elements and move them around—but don't. Zoom in on the area of the title of the map, using the Layout toolbar zoom. With Select Elements active, click the title of the map. A faint blue line should appear around the text of the title; no handles should be seen. (If this doesn't happen the first time you try, keep trying.) Right-click and select Properties. Under the Text tab, type Colorado. Press Apply, then OK.

___ **17.** Pan and zoom (using the Layout toolbar) to the left end of the title box. Check out the Legend.

___ **18.** Pan to the right end of the title box. Check out the ESRI logo (in version 10.0), scale bar, and the Coordinate System. See Figure 3-4. Verify the Central Meridian and the Latitude of Origin. _____.

Adding Other Map Elements

The layout has no north arrow. Let's put one in. We'll erase the Coordinate System information and put the north arrow there.

___ **19.** With the Select Elements tool, click the Coordinate System text so that it gets a selected border. Tap Delete on the keyboard.

___ **20.** Hmmm . . . on second thought maybe erasing that information isn't such a good idea. An "undo" command is in the Edit menu, but let's try the universal undo: Ctrl-Z. Bingo! (You *will* put that in your Fast Fact File, won't you?)

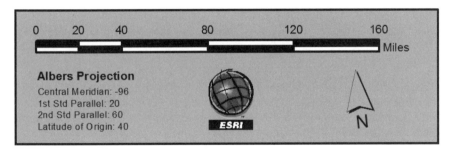

FIGURE 3-5

_____ **21.** If you are running version 10.0 you will see an ESRI logo. Delete that to make room for a north arrow.

_____ **22.** Click North Arrow on the Insert menu. Pick ESRI North 6 and click OK. It appears, selected, with handles, on the Layout. If you place the cursor over one of the handles, the cursor changes to a double-headed arrow. If you drag the handle with this cursor, you can resize the selected element. Resize the arrow that appears and drag it into the legend/scale area. Press the Esc key, (or click somewhere away from the arrow on a blank area) to unselect the element and, thus, turn off the handles. Then, move the elements around to make things look neat. Examine the result. See Figure 3-5.

Well, that's annoying. Upon inspection it appears that the North Arrow doesn't seem to be pointing quite north. We went to some trouble to line Colorado up so its meridians were as north-south as possible,[8] and now apparently our actions to rotate the map 6 degrees are reflected in the direction the North Arrow is pointing. In general, that is probably a useful feature (rotate an image and the North Arrow reflects the change), but here it is causing us a problem.

_____ **23.** Select the arrow and right-click it. Select the Rotate or Flip button from the menu that appears. The options to Rotate Right or Rotate Left appear, but they are grayed out. Use Esc to get rid of the menus.

_____ **24.** Delete the North Arrow. Let's try Insert > North Arrow ESRI North 6 again to see if there are any options to change the angle of the arrow. Maybe look at Properties. Calibration Angle looks promising. Since we rotated the image 6 degrees, perhaps we should also rotate the North Arrow by 6 degrees. Try that.

_____ **25.** That's better. Resize and position the arrow.

The preceding digression, in which we explored putting in a North Arrow that looked right, is illustrative of how you sometimes must use ArcMap. Very few people know all the software well enough to apply it in all situations. Sometimes you just have to tinker with things. Generally there is a way to do what you want to do—it just may be hidden in cascading menus.

---

[8]Of course, no north arrow is exactly right for the whole map. On the eastern edge of the state, the boundary points north, as does the western boundary. As you know those two lines are not parallel—they converge at the North Pole.

___ **26.** Zoom in on the scale bar. Select it. Fatten it slightly (vertically). Shorten it. Notice that as it gets shorter the numbers change. These numbers reflect what the actual scale of the map would be if it were printed out on paper. Play with the scale bar until it shows 100 miles. Center it. Press Zoom Whole Page.

___ **27.** Begin, but don't complete, the process of saving by using File > Save As. Notice that ArcMap wants to save this as an MXD (map) document rather than a template. (You could save it as a template, but you would want to be sure to give it a new name—so as not to overwrite the default template.)

One point of this exercise is to show you that you can use the template capability of ArcMap to access, and save for yourself, already developed data sources.

___ **28.** Save the map as Colorado.mxd in

___IGIS-Arc_*YourInitials*\Map_Making.

Exercise 3-3 (Major Project)

Data Frames

A *data frame* is a "visual container" for layers in ArcMap. Up to now you have been working with data applied to a single data frame—named "Layers" by default. But Arc lets you work with multiple data frames. Among other advantages, this lets you make a map that shows different datasets and also data at different scales on the same map document. In the Layout View, multiple data frames may be seen. In the Data View, only the datasets in one data frame (that is, the active data frame) are displayed.

___ **1.** Use Ctrl+N to bring up the New Document window In ArcMap. Select New Maps > Traditional Layouts > LetterLandscape. Open it by clicking OK.

___ **2.** Now make certain that you are looking at a Data View, not a Layout View (View > Data View).

___ **3.** *Make two new data frames:* Select Insert > Data Frame, and then Insert > Data Frame again. Note that, in the T/C, the name of the last data frame you create appears in bold type. This is the active data frame. There can be only one active data frame at a given time.

___ **4.** Change the name of the active data frame from New Data Frame 2 to LINES by right-clicking the name, selecting Properties, pressing the General tab, and typing in the new name in the Name text box of the Data Frame Properties window. Press Apply, then OK.

___ **5.** Make New Data Frame active by right-clicking the name and clicking Activate. Now click the layer names LINES to highlight it. Notice that New Data Frame is still in bold, while LINES is not. *New Data Frame is the active data frame; LINES is the selected data frame*.

___ **6.** Change the name of New Data Frame, using a different way than you did previously: click on the name, wait a full second, and click its name again. You will be able to type the new name—call it ALL. Press Enter.

___ **7.** Highlight the data frame that is labeled Layers and change its name to POINTS.

Adding Data to Data Frames

You have to be careful here. If you use the Add Data option from the File menu, or use the Add Data button on the Standard toolbar, a data set is added to *the active data frame*. However, if you right-click a data frame name and get the menu that says Add Data (showing the Add Data icon), the data will be added to the *selected* data frame, regardless of whether it is active or not. In other words, Add Data operates differently depending on the source of the command.

___ **8.** Make POINTS the active data frame. Then add the *shapefile*

___IGIS-Arc_*YourInitials*\Trivial_GIS_Datasets\some_points.shp

to the POINTS data frame. Now right-click the name some_points in the T/C. Examine the Properties of the shapefile named some_points. What is the Projected Coordinate System (under the Source tab)? _____.
Using the Measure tool, determine the approximate distance covered by the data set in an east-west direction to the nearest tenth of a meter. _____ meters.

___ **9.** Activate the LINES data frame. Add the geodatabase feature class some_lines_arc, found in

___IGIS-Arc_*YourInitials*\Trivial_GIS_Datasets\Carto.mdb

to the LINES data frame. Change the name to just "some_lines".

___ **10.** From ___IGIS-Arc_*YourInitials*\Trivial_GIS_Datasets, add the *polygon component* of the *coverage* some_polygons to the ALL data frame. Change the name to just "some_polygons".

___ **11.** Switch to the Layout View. Something of a mess appears in the map pane. All three of the data sets, albeit at different sizes, are put in the same place. Make the Select Elements pointer on the Tools toolbar active. Move the pointer to the approximate middle of the pane and click. A box with eight blue handles appears. Click again, and again. What is happening here is that you are clicking inside all the data frames at the same time, so different data frames are being made active. Keep clicking (slowly—don't double-click), noticing now the data frame names in the T/C. With each click a different data frame becomes active, as you can tell from the bold font.

The data frame boxes will not all be the same size, nor will the images inside be the same scale. We will fix the scale problem now and worry about the aesthetics of making the boxes the same size later.

___ **12.** Make ALL the active data frame by clicking the map area until the name ALL in the T/C becomes bold. Within the Data Frame you should see a cursor that has four arrows as well as the pointer. Using this cursor, you can now drag the data frame around. Drag the ALL data frame to the lower-right corner of the layout. Make the LINES Data Frame active and drag it to the upper-right corner of the layout.

If you place the cursor over one of the handles, the cursor changes to a double-headed arrow. If you drag the handle with this cursor, you can resize the data frame and the image inside it. The image's dimensions remain proportional—which is good, since it contains geographical features and you wouldn't want to change their proportions.

_____ **13.** Make the POINTS data frame active. Drag the handle in its lower-right corner towards the upper-left corner, reducing the size of the data frame. Make its location the upper left portion of the layout, below the title.

_____ **14.** Drag a box around the legend, north arrow, scale bar and text, selecting all of those elements together. Drag the whole works to the lower-right bottom corner of the map sheet, pretty much on top of the ALL data frame. Select the ALL data frame and drag it to the lower-left corner of the layout.

_____ **15.** Using the tools you just learned about, move and resize the three maps so that the boxes around the data look approximately like Figure 3-6. Don't worry if the geographic elements don't seem to be the right size. (If a data frame doesn't appear to contain all the data, right-click on its dataset and zoom to layer.)

Recap: When you see several data frames in the Layout View, you may click in any one of them to make it active. Your choice will be reflected by the bold font in the T/C and the look of the data frame (with blue handles or dashed lines) within the layout space.

_____ **16.** Click the POINTS data frame, in the map pane, to make it active. Check the T/C for the bold font. Notice also that the Map Scale text box on the Standard toolbar reflects the scale of the map in the data frame.

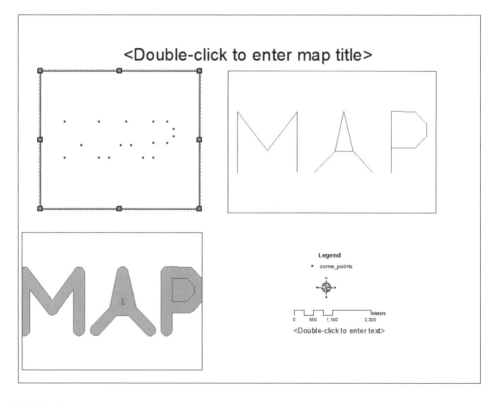

FIGURE 3-6

A Summary of the Graphic Indicators

❏ A blue dashed-line rectangle, with or without handles, indicates a selected element or data frame. Multiple elements or data frames can be selected by clicking while holding down the Ctrl key, or by dragging a box.

❏ A line of black dashes around a data frame indicates that it is the active data frame.

❏ Just for completeness: A hash-mark border around the active data frame indicates that the frame is "in focus," which means you can edit what is in the data frame in the Layout View. (Normally you can only edit within a data frame when you are in Data View. This is a subject for later.) If you double-click a data frame, you put it in focus.

Tinkering with the Map—Scale Bars

_____ **17.** Make sure POINTS is the active data frame. Using the Layout toolbar, zoom in on the scale bar. What number appears at its left end? _____. Right end? _____. Flip to the Data View.

_____ **18.** Now select the other Zoom In tool—the one on the Tools toolbar. Drag a box around the points that make the "P." Go back into Layout View and look at the scale bar. Now what are the numbers at the ends of the scale bar? _____ _____. Now select Zoom Whole Page. Notice that you have made the "P" in the POINTS data frame much larger, and that the change was reflected in the legend of the scale bar.

_____ **19.** Flip back to the Data View and zoom to the some_points layer. As you return to the Layout View you can see that the scale bar has changed back to about the original number.[9]

_____ **20.** Click on Zoom Whole Page. In the Layout View, select the POINTS data frame. Drag the box larger. The box becomes larger but the contents stay the same size. Move the borders of the box back. Now use the Zoom In tool on the Tools toolbar—making the "A" larger. Note that the scale on the Standard toolbar changes. Also the scale on the map legend changes.

There are three ideas to be understood here:

❏ The scale bar dynamically keeps up with the true map scale. It may not appear so on the screen, but it will be pretty close to right when the map is printed out or when you select the Zoom To 100% (1:1) button.

❏ The _scale_ of the geographical elements shown on the map is controlled by the Tools toolbar buttons and by the size of the data frame on the layout, while the _portion_ of the map shown on the screen is controlled by the Layout toolbar buttons.

---

[9]If at any time the geographic elements disappear from a data frame, click on the Refresh View button (next to the Layout View button). If that doesn't restore the image, drag the dividing line between the T/C and the map viewing area well to the right over all the data frames, let it go, and then drag it back to its original position. Another trick to get around this problem is to fetch the properties of the data frame and click on a tab or two. This problem appears to have been fixed in version 10.1

❏ The scale bar is keyed to the POINTS data frame—because POINTS was the original, default data frame (originally named Layers, which you changed to POINTS) that came up with the template. If you wish, you can go through steps 17 to 20 above with one of the other data frames and notice that the scale bar does not change.

___ **21.** Make sure you are in Layout View, whole page. Adjust the three data frames so they are about the same size. Since the scale bar applies only to the POINTS data frame, click the scale bar and drag it inside that data frame.

___ **22.** Activate the LINES data frame. On the Insert menu, pick Scale Bar. Choose Stepped Scale Line since it most resembles the scale bar on the POINTS data frame. Under Properties make the Division Units Meters. OK. Position it in the LINES data frame, shortening it as necessary. Zoom in on the Layout so that you can see both the scale bars better.

Well, heck. Unfortunately the two scale bars don't agree. That is because the data sets, although they cover about the same territory, are shown at different scales. Let's use a common scale (e.g., 1:50,000) that will let the geographies fit comfortably in each data frame.

___ **23.** Zoom to whole page, make the Select Elements active, click the POINTS data frame, and type 50000 into the Map Scale text box on the Standard toolbar. (ArcMap knows you mean "1:50000" so that is what appears when you press Enter.) Do the same with the LINES data frame.

___ **24.** The scales of the two data frames are now the same, but the scale bars differ in appearance and length. And since the data really beg to be displayed at the same scale, it seems unnecessary to have multiple scale bars. Select the LINES scale bar, right-click, and from the context menu, select Delete. Drag the POINTS scale bar back where it came from.

___ **25.** Fix the scale of the ALL data frame to match the other two.

___ **26.** Add some_points and some_lines (changing the name as before) to the ALL data frame. Save the map as PntLnPlygn_1.mxd in ___IGIS-Arc_*YourInitials*\Map_Making.

Legends

___ **27.** Zoom in on the Legend and notice that it applies only to the POINTS Data Frame. Zoom to the whole page and slide the Legend into POINTS Data Frame's lower-right corner.

___ **28.** Make the LINES data frame active. Select Insert > Legend. A Legend Wizard window appears. Arc will let you pretty much click through (Next > Next > Next) and come up with a legend that is right for the data frame. But, as you can see as you do this, the user has lots of control over the appearance of the legend. Slide the legend into the LINES data frame.

___ **29.** Make the ALL data frame active. Insert > Legend. Because ALL contains three different layers, you have some options about what the legend will contain. The arrow keys let you move layer names from one pane to another. Arrange it so that the Legend Items pane has only

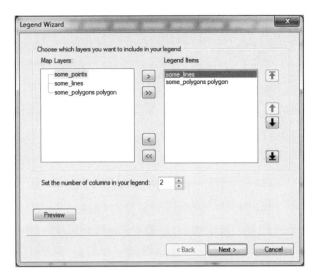

FIGURE 3-7

some_lines and some_polygons in it, as in Figure 3-7. Set the number of columns to 2. Finish the wizard. Drag the new legend to the ALL data frame.

___ 30. In the Layout View, zoom so you can see the legend of the ALL data frame. In the Data View, change some symbology (color, width) of polygon and line features in the ALL data frame. Check the Layout View. Notice that the legend has automatically changed. If you want a more drastic demonstration of the automatic change, change symbology within the T/C and watch the Layout View update. Zoom Whole Page.

___ 31. In Layout View Double-click the text at the top of the map to enter a title. Call it Exercise 3-3: Data Frames. Notice that you could change a lot of the characteristics of the text string used to title the map. Click OK.

___ 32. Beneath the scale bar, where it says Text, change it to display your name. Also put your organization's name below yours by going Insert > Text, which will bring up a tiny text box somewhere on the map. Click somewhere away from the word "Text" and then drag the word to a spot underneath your name. If you then double-click the word a Properties window will appear that will let you enter the name of your school or organization. Also, with Change Symbol you can modify the size of the text; use 12 point type. OK, Apply, OK.

___ 33. Using File > Save As, save the map as PntLnPlygn_2.mxd in

___IGIS-Arc_*YourInitials*\Map_Making.

Use File > Map Document Properties to bring up a window that describes the map. There you could add information about the map, as well as, get information regarding it. See Figure 3-8.

FIGURE 3-8

Looking at the Plethora of Mapmaking Tools and Options

The ability of GIS systems to make maps has evolved over a period of some four decades. It has become quite sophisticated. With that sophistication has come complexity. I wish I could tell you that there is some underlying grand theory of automated mapmaking, but there isn't. The tools you are about to look at are a monument to ad-hockery.

In this exercise, you get a look at only the links to the vast array of options that the serious GIS mapmaker has in ArcMap. What follows will seem something like busy work, but it will serve as a reference for you later, should you become more interested in making maps.

—— **1.** With PntLnPlygn_2.mxd in ArcMap Layout View, click the ALL data frame to select it and make it active. Now right-click the data frame. A menu of many possibilities appears, all of which apply to this particular data frame. List them in your Fast Facts File, starting with Add Data, thinking briefly about each and what it might do.

Done? Yes _____ No _____.

Run the cursor over each possibility that has a triangle pointer mark to note the options within them. We will use a couple of them later.

____ **2.** Click Properties to bring up the Data Frame Properties window, which you have seen before. The window has several tabs. List them in your Fast Facts File, starting with General, clicking on each one to see what comes up.

Done? Yes _____ No _____.

____ **3.** Close the data Frame Properties window. Click somewhere away from any data frame. Now right-click in that same place. A menu with many possibilities appears. These choices relate to the map sheet as a whole. List the choices in your Fast Facts File. (Zoom Whole Page, etc.)

Done? Yes _____ No _____.

____ **4.** Click ArcMap Options to bring up an ArcMap Options window. You should see several tabs. Click on each one while listing them in your Fast Facts File.

Done? Yes _____ No _____. Close the ArcMap Options window.

You get the idea: Myriad possibilities for actions on your part. None of them particularly complicated, as it turns out, but mind-boggling when taken all together. What we have just done is to let you take a first cut at seeing just the names that suggest the possibilities that lie underneath.

In what follows, you will experiment with some of the data frame properties and layout options.

____ **5.** Right-click the layout page, somewhere outside all data frames, then select the Layout View tab of the ArcMap Options window. You don't want the contents stretched when the window is resized (for geographic data this is an unwelcome distortion). Show the horizontal and vertical guides. Show a dashed line around the active data frame. Show the rulers, but not the grid. On the rulers make the smallest division one-tenth of an inch. Snap elements to the guides, but nothing else. Click Apply (if you changed anything), then click OK.

____ **6.** Narrow the T/C and close or hide any other windows, so you have a good view of the layout. Zoom Whole page. Right-click the horizontal rule at the 1-inch mark. Click Set Guide. Notice the little arrow that appears and the light blue vertical guide line. With the right-click menu, clear that guide. Set it again. If the guide isn't exactly on the 1-inch mark, use the double-headed cursor and drag it over. (A "RulerTip" will tell you the current position.)

____ **7.** Actually, if you just click a point on the ruler, a guide will be set. Set additional guides on the horizontal ruler at 5 inches, 6 inches, and 10 inches. Set guides on the vertical ruler at 1 inch, 3.5 inches, 4.5 inches, and 7 inches. You will make the data frames fit into the rectangles created by these guides.

You may have noticed that every time you change the size of a data frame, the scale changes. If you want to keep, say, a constant 1:40,000, you have to keep typing it in the scale text box. There is a way around this.

____ **8.** Pick a data frame and open the Properties window. Click the Data Frame tab. Under Extent in the Data Frame Properties window select Fixed Scale from the drop-down menu and type 40000 in the text box. Click Apply, then OK. Notice that the Standard toolbar now shows 1:40,000, but the number has a gray background—fixed at that value by the Properties window. Prove this by greatly enlarging the data frame. Its content will not change size. Return the data frame to a reasonable size. Set the scale on the other two data frames to 1:40,000 as well.

9. Grab a corner of a data frame and drag it close to an intersection of guidelines, and release the mouse button. Notice that the boundaries of the frame snap to the guidelines. Fix up all three data frames so that they fit the guides.

10. Use Pan on the Tools toolbar to slide the geographics up toward the top of each data frame—center it as well as you can. (Notice that while Pan is still an active tool, the zoom controls are disabled—because you fixed the scale.)

11. Use Select Elements to adjust the positions of the legends of each data frame. If you are a perfectionist, you can fix up the sizes of the legends texts so that they are all the same. To do this, select a legend, choose Properties to bring up a Legend Properties window, and change virtually anything you want about the appearance of the legend. Cancel the Legend Properties window if you brought it up.

In the lower-right corner of the layout, you have a north arrow, scale bar, and a place for your name and organization. In the next two steps, adjust the position of each of those, so the layout looks like Figure 3-9.

12. Zoom up on the lower-right quadrant of the map. Click the Select Elements tool. Click the north arrow to select it. Right-click there, then click Nudge > Nudge Left. The graphic moves a few pixels to the left. Actually, if you simply tap the arrow keys on the keyboard, you can move the graphic more quickly. Place the north arrow where you want it.

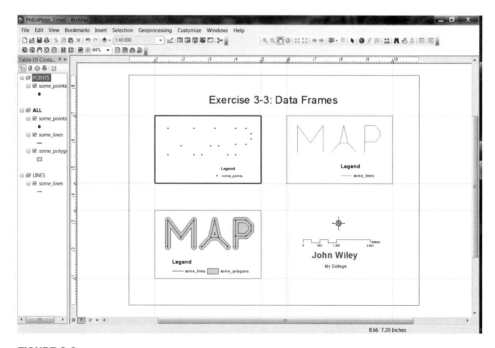

FIGURE 3.9

____ **13.** *Modify the name field:* When you select text, you do not get the sizing handles. Text sizing is done by a properties window, which you call up by double-clicking the text (or by right-clicking and selecting Properties). Do that, and then click Change Symbol. In the Symbol Selector window, make your name bold in 24-point type, with the color red. OK everything and observe the results. Move your name around so that it appears properly placed.

____ **14.** Zoom to the whole page and observe the finished product. Save it as PntLnPlygn_3. Print the map (Sometimes there are problems with printing in ArcGIS—and even problems with viewing pages before printing. If you have difficulty go to www.esri.com/support to search for a solution.) Close ArcMap.

As you can see, many options for mapmaking exist within ArcMap. We have but scratched the surface.

Exercise 3-5 (Major Project)

Making a Map of the Wildcat Boat Datasets

____ **1.** Using the techniques and principles that you learned previously in Exercises 3-3 and 3-4, make a comprehensive layout, showing the data for the Wildcat Boat project. Use the geodatabase in

___IGIS-Arc_*YourInitials*\Wildcat_Boat_Data.

The layout you create, which should look very much like the one in the previous exercise with the "MAP" data frames, should contain *three separate data frames*. Call those data frames:

❑ SOILS

❑ LANDCOVER and ROADS

❑ SEWERS and STREAMS

Snap the three data frame edges to the guidelines as you did in the previous exercise. Use a consistent scale throughout. Use distinctive symbols to indicate features. Use labeling as appropriate.

____ **2.** Save the map as Wildcat Boat Overview in

___IGIS-Arc_*YourInitials*\MapMaking.

Print the map.

Exercise 3-6 (Major Project)

Publishing Maps on the Internet

Publishing maps on the Web can provide a rich opportunity for getting your information out to those that you want to see it. This can be very involved, as when your school, company, or agency maintains a library of dynamic maps on a server. Or it can be quite simple, through use of Esri's cloud.

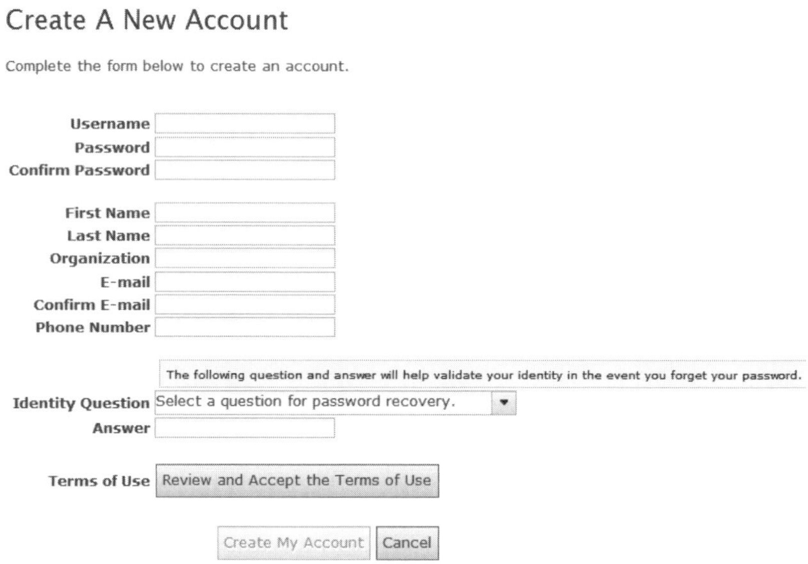

Create A New Account

Complete the form below to create an account.

Username [ ]
Password [ ]
Confirm Password [ ]

First Name [ ]
Last Name [ ]
Organization [ ]
E-mail [ ]
Confirm E-mail [ ]
Phone Number [ ]

The following question and answer will help validate your identity in the event you forget your password.

Identity Question [Select a question for password recovery. ▼]
Answer [ ]

Terms of Use [Review and Accept the Terms of Use]

[Create My Account] [Cancel]

FIGURE 3-10

In this exercise, you will use a service provided by Esri to make and store a map that can be accessed over the Internet by anyone. This Exercise requires an Internet connection and that you copy [___]IGIS-Arc\ River to your personal folder

___IGIS-Arc_*YourInitials* if you have not already done so.

I will make this exercise as simple as possible—asking only that you put the well-known shapefile Boat_SP83 on some imagery that Esri provides—and then publish the resulting map. Be aware, however, that the Esri cloud will allow you to upload all sorts of data and other elements, including layer packages (covered in the next exercise).

___ **1.** *Create an Esri Global Account:* Open an Internet browser (Internet Explorer, Firefox, or other— no guarantees that any given browser will work, given the variety of them and the fact that they change all the time). Point the browser at

www.arcgis.com/home/signin.html

You should know that things on the Web change frequently. What worked at the time this book was written may not work for you now. So you may have to improvise—both in getting access to an Esri Global Account and in putting your data in the cloud. However, unless something has drastically changed, you will be able of put data on a base map that Esri supplies.

___ **2.** Find the Create a Personal Account button and click it. Fill out the form (See Figure 3-10.) Use the User Name[10]

---

[10]If you already have an Esri Global Account create another one with this name. If you want you may delete it after completing this Exercise.

___IGIS-Web_*YourInitials*.

Decide on a password. Note well from the Esri web site: User names are **6** to **24** characters in length. Passwords are **4** to **14** characters in length. Use letters and numbers only for both fields. Both User Names and Passwords are case-sensitive, so be careful. It is a good idea to use a password different from any others you are using—if someone learns your global account password it's probably no big deal, unless it happens to also be the password you use at your bank! It's also a good idea to note the user name (somewhere) and password (somewhere else) so you will have them the next time you access your Esri Global Account. Provide the other information asked for in the form. Review and accept the Terms of Use, and click on Create My Account. (You will have up to 2 gigabytes of storage available to you.)

_____ **3.** On the screen that appears click Edit my profile. If you want to, you may add information to your account. Be aware that this is public information, so be cautious about what you write. (If you go into the edit screen, you have the opportunity to make the information Private—but, as you probably know, nothing that you put into electronic form is guaranteed to be private.) Click Save.

_____ **4.** Click ArcGIS at the left side of the ribbon at the top. Click Learn More and browse the material. Dismiss that tab.

_____ **5.** Find the Sign Out button and do so. Close the Browser.

_____ **6.** Go to www.arcgis.com in a browser, and click on Sign in, in the top-right corner. Click on Register your Esri Global Account. Provide the User Name and Password that you used in creating your Esri Global Account. Click Register. Accept the Terms of Use.

_____ **7.** Click on Gallery to view some interesting maps. When done, click the Back button on your browser to return to the Gallery.

_____ **8.** Click Map. Click Basemap. For a base map click Imagery. Examine the image. Type the street address of a place that you know well. Observe. Click the back button on the browser. Minimize the browser.

_____ **9.** Using Windows Explorer, navigate to

___IGIS-Arc_*YourInitials*\River

Locate the files that begin with Boat_SP83, and highlight them all together. (There will be a few of them; if there is one that ends with "zip" delete it.) Right-click on the block of highlighted file names and choose Send to > Compressed (zipped) folder. A file named Boat_SP83.zip should appear in the River folder. Press enter to accept the filename. This is the file you will send to ArcGIS.com.

_____ **10.** Restore the browser with the arcgis.com window. Click Map. Click the arrow next to Add to get a dropdown menu and select Add Layer from File. In the window that appears click Browse. Navigate to the zip file that you created from the Boat_SP83 shapefile and select it. Your window should look something like Figure 3-11. Click Open.

_____ **11.** The Add Layer from File window should reappear. Click on the radio button for Keep original features and press Import Layer. Wait while the file is imported.

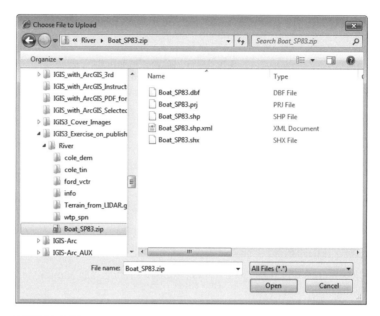

FIGURE 3-11

____ **12.** The GPS track should appear, superimposed on the color orthophoto basemap. Click with the mouse pointer on the last fix (the one near the island) of the GPS track. What is the FID number of this fix? _____. What is the northing (feet)? _____ What is the easting? _____ You can slide the map around by dragging.

____ **13.** Determine how to measure the length of the GPS track in Miles. (While measuring you might have to be patient between clicks along the GPS track, since this process is happening interactively, over the Internet.) _____ miles. Also under the Measure button you will find Location. Click on it and choose Degrees. Note that, as you slide the pointer cursor around the map, you get a display of the location of its tip. Click on the center of the last GPS fix. Latitude _____. Longitude _____.

____ **14.** Drag the map so that the last fix is approximately in the center of the map. Use the vertical bar at the upper left of the map (or the mouse scroll wheel) to zoom in as far as you can, while still being able to see the imagery. Recheck the latitude and longitude (click Location again). You will see some differences in the less significant digits, because zooming in gave you greater precision.

____ **15.** Click Save. Click Save from drop-down menu to place the map in your ArcGIS.com account. In the Save Map window use the title

____IGIS-Web_*YourInitials*_First_Map.

For tags put IGIS, River, and the identifier of the course you are taking (all comma separated, no blank spaces). For Summary, put in any text that seems appropriate. Click Save Map.

____ **16.** Click Share. In the share window, put a check by Everyone. Ignore the other options on the window, but write down the Link to this map. _____ Click Close.

___ **17.** Find the sign out button. Sign out. Close the browser.

___ **18.** Open a browser and go to

`explorer.arcgis.com`

Click on Featured, and look at the various types of maps that are available.[11]

___ **19.** Click on My Content. Be prepared to sign in with your Global Account Name and password. A thumbnail of your map should appear. Click it. Experiment with the various controls in ArcGIS Explorer Online by using the little circle at the lower left of the window and the controls at the upper left of the map: zoom out and in, zoom to full extent, zoom to layer, pan, select, display details, show coordinates, identify, measure, and so on. Fast pan is dragging with the left mouse button; fast zoom is the wheel on your mouse. In general the controls are different from ArcGIS desktop, but you will be able to sort them our since you know what sorts of capabilities to look for. Log Off or Sign Out (if you find a way to), then close the browser.

___ **20.** Ask another student to log on to her or his www.arcgis.com account and search All Content (under Show) for[12]

___IGIS-Web_*YourInitials*_First_Map.

A thumbnail of your map should appear on this other person's display. When the thumbnail is clicked, the map should open. Click on Open this map in ArcGIS Explorer Online. Explore. Close the browser.

___ **21.** Reciprocate by opening the other person's map by searching for

___IGIS-Web_*TheirInitals*_First_Map.

___ **22.** Sign out of both accounts and close all browsers.

In this Exercise, what you have shown is that you can

(a) make use of a basemap (we chose Imagery but a number of others were available),

(b) add your own shapefile to it,

(c) save the resulting map in "the cloud" on ArcGIS.com,

(d) publish the map so that it is available to others with ArcGIS Explorer Online and at www.arcgis.com.

You have done about the simplest Web publishing job imaginable, but at least you now have the concept of putting your maps on the Internet. The programs that make this ability available change rapidly, as do the techniques. You can put maps on mobile devices. You can make the maps dynamic. And on. And on.

---

[11]You might have to install something called Microsoft Silverlight to make use of Explorer.ArcGIS.com. It's pretty straightforward for those with administrative rights to the computer. Others will have to contact their system administrator.
[12]If you can't find someone else to help you with this create another arcgis account. Plan on deleting it when you finish this exercise.

This is a rapidly growing area of endeavor. However, the software to do this will evolve rapidly so if you are going to be occupied with Internet map publishing in a couple of years from now you might want to wait to learn the details.

Enhancing Communication: Styles, Layer Files, Layer Packages, Reports, Charts, and Graphics

Somewhere in the conceptual space between raw data and finished maps lie the ideas of styles, layer files, and map templates. We've already looked at templates, which may or may not have data associated with them.

Layer Files

Layer files are based on raw data files. Basically, a layer file tells ArcMap how to draw a data file—what symbols and colors to use. As you know, if you add a raw data file in ArcMap, the software makes random choices as to how feature are drawn. Sometimes this is satisfactory; more often it is not, if you have serious intentions of examining and analyzing the data. Let's look at an example, in which you can see an obvious advantage to choosing how features are symbolized.

1. Use ArcCatalog to copy the shapefile KY_Streams_spf from

[___]IGIS-Arc\Kentucky_wide_data

to

___IGIS-Arc_*YourInitials*\Map_Making.

2. Start ArcMap with a Blank Map. Add

___IGIS-Arc_*YourInitials*\Map_Making\KY_Streams_spf.shp

to the map, using Data View

This is a fairly large dataset (about 60 megabytes) that contains information about the streams of Kentucky, from the largest (Order 8) to the smallest (Order 1). When two streams of the same order (e.g., Order 1) come together, they make a stream of the next highest order (i.e., Order 2). However, if two streams of different order come together (e.g., Order 6 and Order 5), the output is just a stream of the higher order (i.e., Order 6).[13] Therefore, "stream order" cannot be considered true ordinal data, in terms of stream size, volume, rate of flow, and so on, even relative to those streams that flow into it. That is, the Order 5 stream mentioned previously might have a greater flow volume than the Order 6 stream it converges with.

---

[13]This is according to the Strahler method of stream order analysis. In another method, Shreve, headwater streams are also assigned an order of 1. But when two or more streams converge, then the stream downstream of the confluence is assigned an order equal to the sum of the orders of the upstream streams. Stream analysis is discussed in Chapter 8.

___ **3.** Open the KY_Streams_spf attribute table. Dock it at the bottom if it is still at the side to make it easier to read while docked. Shorten it vertically and zoom the image to Full Extent. How many stream segments are there? _____.

___ **4.** All the KY_Streams_spf are shown with a single color.[14] Suppose that you want to see the smaller streams in a lighter blue and the larger ones in a darker color. The attribute table has a column labeled ORDER_, which has values from 1 to 8. So, let's change the way the streams are drawn: Right-click the shapefile name in the T/C, then click Properties > Symbology > Categories > Unique Values. In the Value field, select ORDER_, then click Add All Values.

___ **5.** Double-click the symbol for Order 1 streams. Pick the color Sodalite Blue with a width of 1 to symbolize this stream. For Orders 2 through 8, use width values of 1.33, 1.67, 2, 2.33, 2.67, 3, and 3.33, respectively. Pick colors of blue that are darker for higher orders, ending with Dark Navy for order 8. Click Apply. Click OK. The result will look something like Figure 3-12.

You probably noticed that it took some time and concentration to symbolize the map in this way. Should you remove KY_Streams_spf from the T/C, all that work would be lost. A layer file (extension LYR) will preserve the symbology.

___ **6.** Right-click the name KY_Streams_spf and select Save As Layer File from the menu. Place KY_Streams_spf.lyr in

___IGIS-Arc_*YourInitials*\Map_Making.

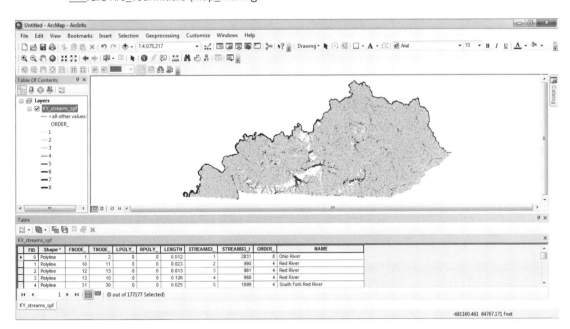

FIGURE 3-12

---

[14]You might think, looking at Kentucky streams, that there is a "U" shaped swath of missing data. Not so. There just aren't any streams there, because of the karst geological formations. Rainwater water soaks in rather than running along the surface. There are some Kentucky roads on which you can drive many miles without crossing a bridge.

_____ **7.** While we are at it, let's also make a simple layout of the data and save it as a map named KY_Streams.mxd in the same folder: Map_Making. Go to Layout View, then click File > Save. Type in a filename, and click Save again.

_____ **8.** Click on the New Map File icon, bringing up the New Document window. You should not be asked to save anything. Go to Data View. Add KY_Streams_spf.*shp* from

    ___IGIS-Arc_*YourInitials*\Map_Making.

Notice that the KY_Streams data set is drawn all in a single color, with no size differentiation. Remove the shapefile.

_____ **9.** Click the New Map File icon, specifying a blank map, and saying no when asked about saving. You should be in Data View, but go there if you aren't there already. Add KY_Streams_spf.lyr from

    ___IGIS-Arc_*YourInitials*\Map_Making.

Notice that the KY_Streams data set is drawn as you symbolized it.

There is an important caveat to be mentioned here. The layer file does not contain data; it only contains the instructions as to how the data set is to be drawn.

_____ **10.** Use Windows Explorer to navigate to the folder

    ___IGIS-Arc_*YourInitials*\Map_Making.

Ask for details of the files listed there. What is the size of:

KY_Streams_spf.shp _____

KY_Streams_spf.dbf _____

KY_Streams_spf.lyr _____

KY_Streams.mxd _____

The SHP file is the geographic feature data. The DBF holds the attribute data. Together they constitute more than 60 million bytes. The LYR file, in contrast, occupies a mere 9,000 bytes. Obviously, the layer file does not contain the KY_Streams data. Also, the map file (mxd) is way too small to hold the actual data.

_____ **11.** For a more dramatic illustration that layer files do not contain the data, do the following. In ArcMap, start a new blank map file without saving the changes to the current file. In ArcCatalog, click the words Folder Connections and press F5 to refresh the catalog tree. In ArcCatalog, delete KY_Streams_spf.shp from ___IGIS-Arc_*YourInitials*\Map_Making. (You can do this either with the Delete key or by right-clicking the selection and clicking Delete.)

Now try to preview the Geography of KY_Streams_spf.lyr by clicking its name. What is the message you get? _____
_____
_____

___ **12.** Try to add data: KY_Streams_spf.lyr in ArcMap. Interestingly, you see the T/C with the symbolization you created. However, the map drawing area—both data and layout—are blank because the underlying dataset is missing. Next, try to open the file KY_streams.mxd. Again, nothing to see.

Note that both the layer file and the map file have been ruined because you made the data on which they depend inaccessible. (If you recopy KY_Streams_spf.shp back into Map_Making, all is forgiven and both the layer file and the map will work again.)

In summary, layer files can be useful in several ways. You can make the data available to others— through a network or e-mail—and be sure that the data will be portrayed as you have prescribed. I will warn you, however, that this can be tricky. Obviously, you have to send the data along with the layer file. Just as important, however, the layer file must be able to know precisely where the underlying data set is. Suppose that both the data seta and the layer file resided in C:\Some_Folder and you sent them to someone who loaded both in D:\Some_Folder; the layer file might not be able to access the data set. There are things you can do (fairly easily—check the help files for a discussion of relative paths) to solve this problem. However, you have to be careful to preserve or remake the linkages between data sets and layer files. A better solution comes in the form of a Layer Package, discussed next.

Layer Packages

We can go a step further in making a complete unit out of a set of layers.

Another way to get an entire dataset, with its symbology, geographic data, and attribute data put into transferrable form, is the Layer Package. In ArcMap, in addition to Save as Layer File you will also find an option to Create Layer Package. A Layer Package is a powerful way to share a dataset, symbolized the way you want it. It wraps everything up together into a single file that you can move to another computer or to www.arcgis.com. Making a layer package (file extension LPK) is a little involved but well worth learning about if you want to send someone a complete, symbolized dataset or want to post such a dataset on the Internet. I discussed earlier how to put a shapefile on www.arcgis.com so it can be viewed with ArcGIS Explorer Online (at www.arcgis.com/explorer) . You can also put a layer package there. The layer package can be made in ArcMap, ArcScene, or ArcGlobe. (Note: The following instructions are for ArcGIS version 10.1. Version 10.0 operates a little differently, with less stringent requirements, but you should be able to see through the differences easily.)

___ **13.** Start ArcMap with a new, blank map. From ___IGIS-Arc_*YourInitials*\River add as data Boat_ SP83.shp, cole_doq64.jpg, cole_dem, Roads (from Lexington.mdb), and ford_vctr\arc. So you have here a map made up of a shapefile, a couple of different rasters, a geodatabase feature class, and a piece of a coverage. This is sort of hodgepodge but the idea is to demonstrate that you can put any sort of layer into a layer package.

___ **14.** A little work is required. Since a layer package might go onto the Internet the software insists on some descriptive material, so the map may be retrieved later by someone to whom you have given access information. In particular, for starters, each layer must have some sort of description. Bring up Layer Properties for Boat_SP83. In the Description field (under the General tab) type River GPS Track. Click OK. In the same way, provide each of the other four layers with at least a cursory description. Possibilities for descriptions are Ford Quad Vectors, Lexington Roads, Cole Tiny Grayscale DOQ, and Cole Quad Digital Elevation Model.

____ **15.** Hold down the Ctrl key and click on each of the five layers, so they are all selected. Right-click on one layer and select Create Layer Package. In the window that emerges verify that your five layers are Included. Click Save package to file (rather than uploading it to ArcGIS). Browse to ___IGIS-Arc_*YourInitials*\River and provide the file name First_Layer_Package_*YourInitials*. lpk. Click Save. In the Layer Package window, click Item Description. For the Summary type Hodgepodge_*YourInitials*. For tags type GPS, Ford, Roads, DOQ, DEM. For Description type IGIS Exercise 3-7. Click Analyze and hope for no errors. If there are any, you can click the Error symbol and take the appropriate remedial action. Dismiss the Prepare window. Click Share. Wait until the progress window says Succeeded. Click OK.

____ **16.** Start ArcMap with a new, blank map, without saving. In ArcCatalog (either the sidebar or the program) look for First_Layer_Package_ *YourInitials*. (You may have to refresh the Catalog Tree: Highlight ___IGIS-Arc_*YourInitials*, right-click on the folder, and click Refresh.)

____ **17.** Getting a layer package into ArcMap is, candidly, a little weird. You can't Open it. You can't add it as data. But you can drag it from ArcCatalog into ArcMap—either the T/C area or the map area. Once you do you will see a map of the five layers, at its full extent. The attribute data is there as well. Open the Attribute table of Boat_SP83 to check it. Close the table. Dismiss ArcMap, without saving. Obviously there is no need to save anything, since you have it all wrapped up in a layer package.

The layer package you have just made and tested is simply a file with an LPK extension. You can email it, put it in your favorite cloud, transfer it to a flash/thumb drive—in other words treat it as just a file that you can transfer to anyone who can run ArcGIS Desktop. That person can access all the information, using the symbology that you set up. Basically the problem of easily transferring ArcGIS data is solved!

Styles

Styles basically let you draw maps using colors, symbols, and patterns developed by other people and organizations. When you have drawn maps before, you have been using a style developed by ESRI. In fact, it is difficult to separate the software, which basically lets you draw points, lines, and polygons— admittedly in myriad colors—from the ESRI predeveloped symbols. Let's start your understanding of styles by eliminating all of them.

____ **18.** In ArcMap, enter Data View and click the New Map File icon on the Standard toolbar to start with a Blank Map. Add the **_polygon component_** of the coverage SOME_POLYGONS, which you will find in the

    ___IGIS-Arc_*YourInitials*\Trivial_GIS_Datasets.

    folder. Set the T/C tab to List By Drawing Order.

____ **19.** Set the software so that it uses no styles at all, by clicking Customize > Style Manager to bring up the Style Manager window. Click the Styles button to bring up the Styles References window. Clear all the boxes you can. Which one can't you clear? _____. Click OK. Click Close.

____ **20.** In the T/C, right-click the polygon symbol. You may recall that that usually that brings up an array of distinct colors (e.g., Medium Apple, Sodalite Blue, Mars Red, and so on). However, those are part of the ESRI style. At the bottom of the box, click More Colors. Now what you have is the Color Selector window, which you met in Chapter 2. It lets you select any color the computer is capable of producing—millions of them[15]—but without benefit of being able to name the color or easily select it again. The Color Selector window lets you define a color in the most basic way. You can move the R, G, and B sliders to determine the amount of red, green, and blue, each on a scale of zero to 255, that go into making up the color that will wind up on the polygons. As you move the sliders, the lower-left rectangle in the window shows the new color. Adjacent to it is the current color that is to be changed. In the area of the window just up from the bottom is a box showing, as a continuum, all the colors. Clicking or dragging in this box is also a way to select a color. Try this out, watching the text boxes and the slider bars. Move the cursor both horizontally (to change colors) and vertically (to change brightness). Finally, pick a garish yellow by typing in the boxes—say, R255, G255, B99. Click OK.

___ **21.** Click the polygon symbol in the T/C. This brings up a Symbol Selector window. Here, you can modify the outline width and the outline color. Make the color a light green with an outline color red. You could also open the Symbol Properties window by clicking the Edit Symbol... button, and there you will have access to another bunch of options. In fact, you could doubtless spend half a workweek exploring the possibilities that ArcMap gives you in the color arena. For now, just cancel the Symbol Property Editor, and press the Style References button.

This menu shows you the different symbol sets that come with ArcMap. It is simply another way to get at the list of Styles available to you.

Adding and Using a Style

___ **22.** Clear off the menus and windows.

Add the shapefile some_points from the

___IGIS-Arc_*YourInitials*\Trivial_GIS_Datasets.

folder. You get the generic dot. Look briefly at the attribute table for some_points.shp. Note that TYPE_ is keyed to the polygon the point is in: M, A, or P. Hide the table. Bring up the related Style References window again. Let's go for something really ridiculous: click 3D Trees and OK. This now gives you the capacity to replace the generic dot with elements from a style sheet called 3D Trees. Click Close on the Style Manager window. In the T/C, click the some_points symbol. A Symbol Selector window appears with a plethora of tree images. Slide down through the list, taking a botany lesson as you go. How many trees are there?[16]

___ **23.** Pick a Rocky Mountain Maple, either by scrolling down in the alphabetized list or by typing it in the text box which says "Type here to search."[17] Click OK. It's pretty hard to see that you have made any difference in looking at the Data View. Return to the Symbol Selector window. Use the maple again but change the size to 50 points. The symbol has taken on some form. To see it in all its pixilated glory, change to the Layout View and zoom in on a point, using the zoom on the Layout toolbar.

___ **24.** Zoom to the whole page and go back to the Data View.

---

[15]256 * 256 * 256 – you do the math.

[16]Just kidding.

[17]If this is the first time you have used the search, it may take a while to initialize and you will see a message that says "Updating index database..."

_____ **25.** All the capabilities you had with the software before you have now. Only the symbols you may use have been curtailed. Let's use different symbols for the points in the M, and the A, and the P.

_____ **26.** Bring up the Layer Properties window for some_points. Click Symbology > Categories > Unique Values. Make the Value Field TYPE_, and Add All Values. Double-click the symbol next to the M to bring up the Symbol Selector window. At the top of that window, you see two radio button options: All Styles and Referenced Styles. (This is a different kind of category than on the Layer Properties window.) Pick All Styles and search for Plant. You get several possibilities, including nuclear plants. We want a botanical type of symbol, so pick the option, of a Century Plant and again make the size 50. Click OK, and OK again. Now put the Jade Plant in the A, and then search for SUCCULENT, and pick Cereus for the P. Observe the results.

The object of these preceding steps is to show you the large number of already developed styles that are available to you. Something to think about before you spend time developing your own symbols.

_____ **27.** Click Customize > Style Manager to open the Style Manager window, and then click the Styles button to bring up the Styles References. Turn off 3D trees and turn on ESRI. Close the Style Manager window. Dismiss ArcMap, saving changes if you want to in Map_Making.

Reports

As useful as an attribute table is, its format leaves a lot to be desired. It is seldom reasonable to print out a large table in regular form. What is very useful at times is a summary of the information in the table. ArcMap gives you the ability to generate textual reports from an attribute table. You will see this ability demonstrated in a three-step process. First, you will create a second table by using the Summarize feature, available by right-clicking a column in a table. Secondly, you will make a report from the second table. Finally, you will put that report on a map.

Report making requires design and usually a pair of tasks: manipulating the information on a mockup and then looking at the results of the manipulation, which take the form of the actual report. You will be alternating between an editing window—in which you will use a Report Designer—and "running" the report to examine the results of your design.

You may recall that the Wildcat Boat data contained a personal geodatabase feature dataset named Sewers. It consisted of a few linear segments representing lengths of sanitary sewer pipe of two different diameters: 60 inches and 45 inches.

Assume that you need a map of the sewers and want to place a report on that map showing the total lengths of each diameter of pipe. You might proceed as follows.

_____ **28.** In ArcMap, start a new map, using the Letter (ANSI A) Portrait. This will automatically put ArcMap in Layout View. Add as data the Sewers feature dataset from

___IGIS-Arc_*YourInitials*\Wildcat_Boat_Data \Wildcat_Boat.mdb\Line_Features

_____ **29.** Open the attribute table of Sewers and cut the table down to reasonable size. Notice that there are four lengths of pipe 60 inches in diameter and two lengths 45 inches in diameter. Just as in Data View, you can use the Select Features tool (in the Tools toolbar) to graphically select features in the Layout View and see the selections reflected in the table. Try it. Also you can select records in the table and see the results highlighted in the map. Using Ctrl-click on the table, highlight the four pipes of diameter 60 inches. Right-click the Shape_Length column heading and pick Statistics. From the Selection Statistics of Sewers window, determine the total length, to one decimal place, of 60-inch pipe. _____ meters. Dismiss the window. On the table window using ToolTips, find the Switch Selection icon. Press it. What is the total length of 45-inch pipe? _____.

Let's examine how you can get this information into a report and then onto the map.

___ 30. Clear all selections using the Clear Selection button on the table menu. Right-click the DIAMETER heading and pick Summarize. In Box 1 the field to summarize should read DIAMETER. Skip box 2. Accept the default output table name, Sum_Output.dbf, but double-check that the output location path is your Map_Making folder.[18] Click OK. When asked if you want to add the results table in the map, choose Yes. Open the table.

___ 31. The new table, Attributes of Sum_Output (the Sum refers to summary, not to sum, as in total), tells you the numbers of segments but little else. Let's try again. Close the table and remove it from the T/C.

___ 32. Make sure the Sewers attribute table is open. Again, right-click over the DIAMETER heading and pick Summarize. This time, in box 2, expand Shape_Length and check Sum. Continue as before, add the table to the map, and open it. This time you see that you get an additional field: Sum_Shape_Length. Check to see that the numbers you wrote above are the same as those in the table. Write the name of the table here: _____.

___ 33. Notice that the sum (total) of the Shape_Length column numbers gives the value of the lengths to thousandths of a meter. While quite precise, this seems unlikely to be accurate and is certainly useless. We can fix that by chopping off the decimal part with something called the Field calculator. Right-click the Sum_Shape_Length column and pick Field Calculator. Ignore the warning (since you can always reproduce the table if you make a mistake). In the Field Calculator window click INT(). The expression INT() should appear in the codeblock box at the bottom of the dialog. Your cursor should be inside the brackets of the INT expression. (The INT function produces integers by removing the decimal fractions of the argument.) Double-click Sum_Shape_ in the Fields box. When you click OK, the numbers in that column will become integers.

Now that you have a table that contains the needed information in a reasonable form, you may make a report. ArcMap has considerable report making capability. We will create only the most elementary example. When you finish this section, you will at least know that report generation capabilities exist.

___ 34. Click View > Reports > Create Report to bring up a Report Wizard. As the wizard progresses you will be presented with a series of six questions:

❏ Which fields do you want on your report?

❏ Do you want to add any grouping levels?

❏ Which fields do you want sorted on your report?

❏ How would you like to layout your report?

❏ What style would you like?

❏ What title do you want for your report?

---

[18]Note: if you have to change the path, be sure to Save as type "dBase Table." The default is a "File and Personal Geodatabase table" and since you're outputting to a folder, a File and Personal GDB table won't work.

_____ **35.** Starting with the first question: In the Layer/Table drop-down menu, pick the table name that you wrote in the blank in the step before the last two steps. Move all the available fields to the Report Fields area by clicking the right-pointing double arrow. Now move the OID field back by highlighting it and using the left-pointing single arrow, since you don't want it in the table. Click Next.

_____ **36.** We don't want to add any grouping layers (used for more complex data), so we will skip the second question by clicking Next. The third question is on sorting. Select DIAMETER from the drop-down menu and make sure the Sort column reads Ascending (which is the default). Click Next.

_____ **37.** The defaults in the Layout query are acceptable. Click Next. Select a simple style, such as the aptly named Simple. Click Next. The last question asks us to title our report. Call it Sewer Pipe Lengths. Select Preview the report and click Finish.

_____ **38.** The results are underwhelming. The title is a little large, and the field names are cut off although there is still plenty of room on the page. You will save it and work on the saved file. Find the "Save report output to file" icon and press it. Save the report as PipeLengthReport.RDF in your

   _____IGIS-Arc_*YourInitials*\Map_Making.

   folder, if you are using version 10.0. With version 10.1 the extension will be RLF.

_____ **39.** We clearly have some editing to do—so click the Edit button to enter into a Report Designer interface. The number of options here is extensive. An entire course could be built around the Report Designer. For now, I only show you how to get the most rudimentary report on the map.

In the Report Designer you have the items you can add to the report: "Design Elements" on the left. On the right, there are the Properties of the document or of the elements selected. Then, in the center of the Report Designer, you have a mockup of the report, which is divided into sections. At the top, you will find the Report Header—basically, the title of the report. Next down are the titles of the columns of the report. Below those are the references to the data in the table. Suppose that we want to generate the most elementary report to display the length of the two diameters of pipe. See Figure 3-13.

_____ **40.** Click on the title Sewer Pipe Lengths. Note that it appears in bold on the right side under Data > Text. Double-click the name. When it becomes highlighted, you can type in a new report title: Pipe Lengths. Once you press Enter, the title changes in the report area.

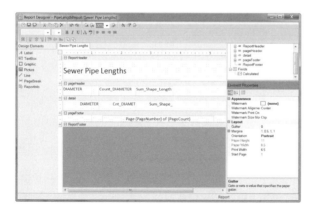

FIGURE 3-13

Let's change the headings of the columns DIAMETER, Count_DIAMETER, and Sum_Shape_Length with the names Diameter, # of Pipes, and Length. Click on DIAMETER under page header. Find where it shows up under Data >Text, double-click, and press Enter. Change its value to "Diameter".

____ **41.** Now change heading Count_Diameter to # of Pipes. Change Sum_Shape_Length to Length. Now look at the results of these changes. Click on the Run Report icon (or press F5).

____ **42.** The report looks somewhat better. We have improved column names but the numbers don't line up under the headings. So, click the Edit button to bring back the Report Designer. Under the various headings we can move the locations of the title, the column headings, and the columns. Shift these around, alternating between Edit and Run Report until you have a report that seems reasonable to you. Select Save on the Report Viewer page (which creates a file with an "rdf" extension) and use the name PipeLengthReport.rlf. To show the capability of producing PDF files, switch to the Report Viewer export it to a PDF naming it PipeLengthReport.pdf and save it into the Map_Making folder in your

____IGIS-Arc_*YourInitials* folder.

____ **43.** In the Report Viewer, print the final report. Find and click the "Add report to ArcMap Layout" button, bringing up the Add to Map dialog box. Uncheck "Add Page Borders" because we will only be using part of the report page in the layout. Click OK. Close the report editor and examine the results. Note that since you saved the RLF file, you can load or run the report and make modifications to the report.

____ **44.** The report is full page, so it may be as big as the Layout. You should use only a part of it. Using Select Elements, select the report (it will have four handles, not six) and slide it down so the text fits above the bottom margins. Size it so that it fits between the vertical map outline limits. Use the Pan control on the Tools toolbar to slide the geographics up toward the top of the page. Label each segment with the diameter of the pipe with 24-point bold type.[19] The results should look something like Figure 3-14.

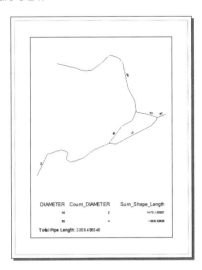

FIGURE 3-14

[19]If this doesn't label each of the six features, go to the Labels tab in the Layer Properties window. Under Placement Properties, choose Place One Label Per Feature.

_____ **45.** Display the map at 100 percent of the size it would appear on a page. Use the scroll bars, look around the map to see that things are about the right size. Title the map with your name. Using File > Print, print the map. Save it in the Map_Making folder as Sewers_Specs.mxd.

Charts and Graphs

Another form of communication—neither text nor map—can also be created by ArcMap: the chart or graph. The software is capable of producing graphs and charts of both two- and three-dimensional appearance.

_____ **46.** In ArcMap Click File > New. Select Letter (ANSI A) Portrait. Switch to Data View. Add the personal geodatabase feature class named Soils from

_____IGIS-Arc_*YourInitials*\Wildcat_Boat_Data\ ... you know the drill.

Once the map has appeared in the window, symbolize the SUIT categories using Unique values.

_____ **47.** Open the Soils attribute table and summarize the suitability (SUIT) column, including the sum from Shape_Area. Accept the default name and save it in Wildcat_Boat.mdb. Agree to add the table to the T/C of the map. Dismiss the Soils attribute table. Open the Sum_Output_x table. It should look about like Figure 3-15.

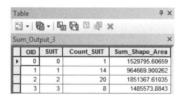

FIGURE 3-15

_____ **48.** Select View > Graphs > Create Graph to bring up the Create Graph Wizard. Click the drop-down menu of Graph type, and look at the many types of graphs you can create. Pick Vertical Bar. In the Layer/Table field choose Sum_Output_x (x is some digit, depending on how many tables you have created in this session). For the Value field, you will want Sum_Shape_Area. Make the "X field" SUIT. Uncheck Add to legend. For the Color use Pallet (Excel). Leave the rest of the options at their default values. Click Next. Change the title to "Square Meters of Soil Suitabilities". In Axis properties make sure the Visible boxes of the Left and Bottom axes are checked. Give the Left axis the title "Totals of Areas"; use the title "Suitabilities" on the Bottom axis. Both Right and Top should be blank. The axis titles will appear on the graph. Click Finish.

_____ **49.** A Graph should appear. Make it taller and narrower. Slide it over on top of the Sum_Output_x table so you could get to either table by clicking on its header.

_____ **50.** Assuming that the graph looks as it should (see Figure 3-16) slide your cursor to it.

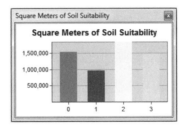

FIGURE 3-16

____ **51.** Right-click on the title bar of the graph, and click Add To Layout. Dismiss the graph window as well as the Sum_Output_x table. In the Layout View, drag the graph to the bottom of the page. Using the handles on soils_polygon, resize and move it so that it doesn't conflict with the graph. Click Zoom whole page. Adjust each map element so that a reasonable amount of layout space is devoted to the data and the graph. It should look somewhat like Figure 3-17. Title the map Soil Suitabilities.

(Optional) It would be helpful if the soils map showed the polygons with the same colors as the graph. Unfortunately, the best way to do this is by changing the individual colors on the feature class. If you want to take the time, try changing the random colors assigned in the T/C so that they more or less match those of the graph. Start by clicking the color patch next to the zero value and making it the same color as the zero column on the graph (using the fill color in the color selector window. Or, to be meticulous, click More Colors...). Do the same for values one, two, and three. Click Apply, then OK.

Save the map in

___IGIS-Arc_*YourInitials*\Map_Making.

with the name Soils_with_Graph_1.

Suppose that you want to see the relative amount of each suitability. A pie chart will serve this function.

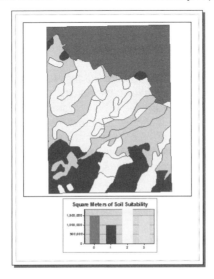

FIGURE 3-17

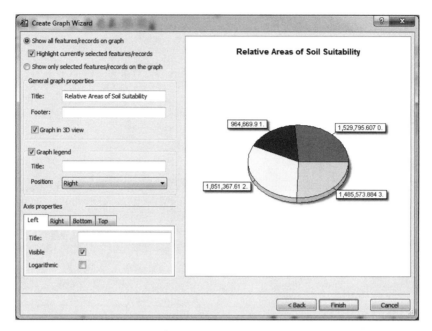

FIGURE 3-18

____ **52.** Right-click on the graph part of the layout and select Properties to bring up the Create Graph Wizard again. Change the Graph type to Pie. For the Value field select SUIT. For Color pick Palette (Excel). Turn on Show labels (marks). Click the Appearance tab. Make the title Relative Areas of Soil Suitabilities. Click Graph in 3D view. See Figure 3-18. Click Apply, then OK. Zoom in (Layout Toolbar) on the chart to get a better look.

____ **53.** Zoom Whole Page. Save the map with the name Soils_with_Graph_2. Print the Layout if you want.

____ **54.** Notice that water is shown as a portion of the pie chart. That didn't bother us so much when it was on the bar graph, but here it skews the results. On the Sum_Output_x table select records with SUIT values 1, 2, and 3. Start the Graph Wizard and proceed as before, this time check the Use Selected Records box. Now the graph shows that only three soil suitabilities of actual land area are represented. If you want the color on the graph to correspond to the colors on the map you have to correct them on the map. The colors of the graph cannot be easily changed. Place the graph on the Layout and save as Soils_with_Graph_3.

Graphics

As a last topic: you can put graphics or text information into a data frame directly from a variety of sources. This subject really gets us away from our intended goal—preparing you to do analyses with GIS—but it is a major feature of the software that you should know about, so we will look at it briefly.

Placing ancillary information on a data frame is done primarily with two sets of controls. The first is Insert on the Main menu. The second is the Drawing toolbar.[20]

_____ **55.** In ArcMap start a new map with Letter (ANSI A) Portrait. In Data View, add the shapefile

[____]IGIS-Arc\River\Boat_SP83.shp

to a new map.

_____ **56.** Also add the following data sets, in this order:

[____]IGIS-Arc\River\wtp_spn (the point component, water treatment plants)

[____]IGIS-Arc\River\cole_drg.tif

[____]IGIS-Arc\River\cole_doq64.jpg

[____]IGIS-Arc\Kentucky_wide_data\KY_Streams_spf.shp

Make the point symbol for wtp_spn a bright red square of size 10. Make the Boat_SP83 symbols bright green circles of size 8. Pan and zoom the image until all of the DOQ and the GPS track are in. Your data frame should look something like Figure 3-19. Save this as

____IGIS-Arc_*YourInitials*\Map_Making\LWP1.mxd.

_____ **57.** Turn your attention to the Drawing toolbar. Run your cursor from left to right over each of the buttons while reading the ToolTips and/or the Status bar. On those buttons with drop-down menus (little triangle symbol just to the right of the button), look at the options.

Suppose that you want to put some identifying text on the data frame, pointing to features.

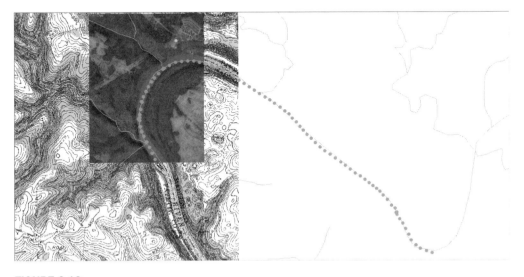

FIGURE 3-19

---

[20]If the Drawing toolbar is not on the ArcMap window, choose Customize > Toolbars and make sure there is a check mark next to Draw.

____ **58.** On the Drawing toolbar, find an "A" (for new text—add text to the map by typing it in) and access the menu next to it. Find the callout box, and click. With the Callout cursor, click the easternmost point of the GPS track. In the text box that appears type

Beginning of GPS Track

and press Enter. Drag the box below the point so that all the text is within the data frame. Click away from the box to deselect it.

____ **59.** Repeat the procedure of adding a callout box—this time referencing the last point of the GPS track. However, instead of typing in the text box, click somewhere off the box, then double-click Text to bring up a Properties window. In the Text area, type:

End of GPS Track

then click Apply and OK. Drag the callout box to the right, off Cole_DRG.

Deselect it.

____ **60.** Make a callout box that says "Filtration Plant" pointed at the facility in the northeast corner of the Cole_DOQ.

____ **61.** Use File > Save As to save the data frame as a map named LWP2.mxd.

We have available an oblique aerial photo of the water plant. No geographic coordinates come with it. It is just a picture. However, we can add it to the data frame.

____ **62.** Choose Insert > Picture and navigate to

[____]IGIS-Arc\River\Lexington_Water_Plant.JPG

Open. Drag the photo to the upper right corner of the data frame.

____ **63.** Bring up the menu next to the callout icon (it changed from the "A" earlier, while you were looking at something else). Find and click the icon for inserting Text. Click the data frame in the lower right quadrant. Type Lexington Water Plant and press Enter. Start changing the Properties of the text by double-clicking it. Press the Text tab and then select Change Symbol. Make the text red, 20 points, Arial, bold, and underlined. Click OK, and OK again.

____ **64.** Change to Layout View. You probably would like to rearrange some of the elements, given the change in format, but you notice that you cannot select any element except the entire data frame. ArcMap doesn't let you edit a data frame in a Layout unless you put that data frame "in focus." Do that by double-clicking the data frame. Note the hashed border around it. (You can also toggle focus on and off by right-clicking the Data Frame and selecting Focus Data Frame or by clicking on the Focus Data Frame button on the Layout toolbar).

____ **65.** Select the picture and move it up in the layout. Enlarge it by dragging its corners. Select the text title and move it down, centered, but still inside the Data Frame. (If you go outside the Data Frame the title disappears; you can get it back with Edit > Undo, or with Ctrl-Z.) Click somewhere away from the title to unselect it.

____ **66.** Since the title crosses some features, perhaps you want to give it a background. Click the Rectangle icon button on the Drawing toolbar and draw a rectangle over the title, covering it up. Click the word Drawing on the Drawing toolbar, pick Order, and send the selected rectangle to the back. Pull up the menu for fill color and make the color of the rectangle Lapis Lazuli.[21] Change the color of the text in the box to White. If you have any cartographic design experience, shake your head over what a wretched mapmaker the author is, and save the map as LWP3. mxd, after fixing it up to suit yourself.

In what follows you will experiment with some of the drawing tools on a blank data frame. Feel free to vary the process and to experiment.

____ **67.** Click the New Map File icon, choosing Letter (ANSI A) Landscape, and go into Data View.

____ **68.** On the Drawing toolbar the icon that starts out as a rectangle becomes whatever object you choose from the dropdown menu. We could call it the new shape icon: Rectangle, Polygon, etc. Try out the different possibilities. With Rectangle, Circle, and Ellipse, just click and drag. With Polygon, and Line, just click to make successive vertices; double-click to end the graphic. Curve is particularly fun. Again, just click to make successive vertices; double-click to end the graphic. With Freehand simply drag the cursor around.

With any of these graphic elements, you can, after selecting them on the Drawing toolbar with Select Elements, right-click and change their positions by rotating 90 degrees or flipping around an axis. You can also bring up a window to get information about and/or change their properties.

____ **69.** Pick a graphic that you have made and select it (with either the Select Elements pointer on the Tools toolbar or the Select Elements pointer on the Drawing toolbar. Click Zoom To Selected Elements on the Drawing toolbar. Grab one of the cyan handles to get a two-headed arrow, then drag the handle to shrink the graphic to about half its original size. Zoom back to full extent. Select another element and experiment with the Rotate tool (on the Drawing toolbar). First click the Rotate tool button, then click the feature. Press the "A" key to type in a number of degrees of counter-clockwise rotation.

____ **70.** Select another shape. Toward the right end of the Drawing toolbar, change the Fill Color and/or Line Color to whatever you want.

____ **71.** Create some text. Change its font to Courier New, 16 point. Experiment with the Text options. Circle Text, for example, will, let you create a circle and then place lines of text within it. Once the text is no longer selected, the circle's bounding square disappears.

____ **72.** Click New Map File, continuing to use Letter (ANSI A) Landscape. Find the button for New Splined Text and press it. Make a spline (like a snake) starting in the northwest corner of the data frame: Click, move an inch or so to the right, and click again; continue until you have made a spline that looks somewhat like Figure 3-20. Double-click to end the spline. Type the following into the text box, without using the Enter key except at the very end:

A quick move by the enemy may jeopardize six fine gunboats. Now is the time for all good men and true to come to the aid of their party. The quick brown fox jumped over the lazy dog.

---

[21]The color of an exotic gem. Also University of Kentucky Wildcat Blue.

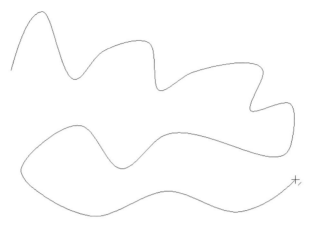

FIGURE 3-20

____ **73.** When you press Enter you should see the text following along the first part of the spline you made. The text should remain selected. Change the font color. Change the font size to 16. Click away from the text to clear the selection. Observe. See Figure 3-21. Note that where the curve has extreme bends the text suffers.

____ **74.** Click New Map File keeping the Letter Landscape template. Go into Data view. Make a polygon of several vertices, using the Polygon option. Click Edit Vertices in the Drawing toolbar. Drag the little cyan squares around with the four arrow-head only cursor to reshape the polygon. In version 10.1 right click on a vertex and delete it. Right click on a line of the polygon and add a vertex. Experiment with doing both. Notice the difference between reshaping and resizing, which we discussed earlier.

____ **75.** Create a new curve. Edit Vertices again. By dragging the cyan boxes around, you can change the locations of vertices. By dragging the purple boxes, you can change the shape of the part of the curve that goes into a vertex. By dragging a point on the curve that is not a vertex, you can move the entire curve.

FIGURE 3-21

_____ **76.** Click New Map File. Add the Layer Sewers from Line_Features of

_____IGIS-Arc_*YourInitials*\Wildcat_Boat_Data\Wildcat_Boat.mdb.

Make sure you are in the Data View. Draw a Rectangle so that it covers some of the northern pipe. So you can see through it, double-click the rectangle to bring up a Properties window, click the Symbol tab, and make its fill color No Color. Click Apply, then OK. Be sure the rectangle is selected—you will see the eight blue handles. Now from the Selection menu pick Select By Graphics. The northern pipe should become highlighted. If you open the sewers attribute table, you will see its record highlighted as well. In the Selection menu, click Selection Options. Note the various ways you can define the selection process. In the Selection menu, select Clear Selected Features. Dismiss the attribute table.

Making Graphics out of Geographic Features

_____ **77.** Right-click Sewers in the T/C. Click Convert Features to Graphics. In the window, specify Convert All Features of Layer Sewers. Only draw the converted features. Click OK. With Select Elements on the Draw toolbar click a line of sewer pipe to select it. You will find that you can now move on or change it around like any other graphic. Dismantle the sewer system, and pile the pieces up in the southeast corner of the data frame.

_____ **78.** Use Ctrl-O to open an existing mapfile. Open the map file

_____IGIS-Arc_*YourInitials*\Map_Making\LWP3.mxd.

In Layout View, put the data frame in focus. With the Layout toolbar, zoom up on the DOQ. Pick Splined Text from the Drawing toolbar. Make a spline paralleling the curve of the river, about an inch to the east of the GPS track, starting with the fix at the end of the track and continuing up to the edge of the DOQ. In the text box, type "Kentucky River" and press Enter. Change the text color to White. Change the text size to 16. Slide and rotate the text until it fits nicely in the bend of the river. If you don't like the result, delete it and try again. Zoom to the whole page, and save the map as LWP4.

As you can see, ArcMap has a remarkable number of tools that aid you in making maps. Admittedly these tools are not as extensive as those in various drawing programs, but don't forget: you retain the advantage of having a dynamic map with all the "intelligence" that GIS gives.

_____ **79.** Just to demonstrate that a GIS map is really different, use the Identify tool to click the red square that represents the Lexington Water Plant. As a result, you will see an extensive amount of information on this particular plant. Now right-click wtp_spn point in the T/C and select Zoom To Layer.

These are the water treatment plants in the state. Open the wtp_spn point attribute table and observe the amount of information available. Try that with your drawing program! Close ArcMap.

Exercise 3-8

Checking, Updating, and Organizing Your Fast Facts File

The Fast Facts File that you are developing should contain references to items in the following checklist. The checklist represents the abilities to use the software you should have upon completing Chapter 3. ***Important Note:*** This checklist is on the DVD that accompanies the book. It is available in Microsoft Word format. Rather than typing or writing by hand the text that follows, you can copy and paste it into your Fast Facts File from the DVD file.

❏ The Layout View (contrasted with the Data View) is

❏ A map template is

❏ The file extension of a map template is

❏ The name of the map template that is the basis for a blank document is

❏ To get a variety of map templates

❏ A major toolbar used to produce a map is

❏ The map templates that contain data are located

❏ Ways of changing from Data View to Layout View are

❏ To determine the coordinate system of data in the data frame

❏ The types of coordinates available in the Data View are

❏ The types of coordinates available in the Layout View are

❏ Zoom controls in the Data View and the Layout View

❏ To rotate the map display

❏ Elements of the map that may be added are

❏ To rotate the north arrow

❏ Care has to be taken saving a map created from a template with data because

❏ A data frame is

❏ To make a new data frame

❏ To make a data frame active

❏ Two ways to change a data frame name are

❏ The difference between the active data frame and the selected data frame is

❏ Care must be taken when adding data to a data frame because

❏ The projected coordinate system could be found under Properties through the T/C under this table:

❏   If data frames overlap selecting a particular one of them may be done by

❏   To move a data frame within the Layout

❏   To select elements on a layout

❏   The active data frame appears in the T/C with

❏   A blue dashed line indicates

❏   A black dashed line indicates

❏   A hash-mark around the active data frame indicates

❏   The scale bar will reflect the true map scale when

❏   A scale bar is keyed to only one data frame. It is the

❏   A map scale can be set by typing in the

❏   The Map Document Properties enables you to manipulate

❏   The Legend is tied to the T/C

❏   To make different data frames line up one can use

❏   A layer file is related to a data file

❏   The extension of a layer file is

❏   A layer package is

❏   To get a layer package into ArcMap

❏   If one erased a data file and then tried to draw the associated layer file

❏   You set the software to use particular styles by

❏   The Color Selector window

❏   To add and use a style

❏   The purpose of Summarize is

❏   To use the report writing capability of ArcMap, first create

❏   The sections of the Report Designer are

❏   A report can be added to a layout by

❏   To get to the Graph Wizard

❏   Ancillary information can be placed on a data frame in two ways:

❏   To add a call-out box

❏   To edit data in a data frame while in layout view one must

- ❏ To create text along a spline

- ❏ To make graphics out of geographic features

- ❏ Two ways of putting a layout "in focus" are

- ❏ The website ArcGIS.com allows you to publish

Structures for Storing Geographic Data

OVERVIEW

IN WHICH you explore the ways geographic datasets are stored in the memory and on the disk drives of a computer. You also learn the rudiments of using ArcToolbox.

Why Is Spatial Data Analysis So Hard?

Spatial (that is, geographic) datasets are notoriously difficult to analyze. In other fields of human endeavor, most of the datasets one wants to analyze are naturally made up of numbers. What is the history of the stock market's up and downs? Numbers. What are the statistics relating to the grades of students in the sophomore class? Numbers. How many parts-per-million carbon monoxide molecules may be safely tolerated by different types of air-breathing animals? Numbers. But the chief way of *storing spatial data* for most of human history has been the map—whether paper, Mylar, or computer image.

Numbers and text are composed of nicely behaved discrete symbols. Each symbol may be represented by a bit of ink or by a few pixels on a computer screen that fit neatly into a square roughly an eighth of an inch on a side. And, in English, there aren't very many different symbols: 10 digits, 26 letters uppercase, another 26 lowercase, and a bunch of special symbols—in total a maximum of 256. Maps use symbols also, but they are not nearly so well behaved. For example, symbolizing a road may result in a wavy line 2 feet long.

As discussed in Chapter 2, maps are difficult to analyze, and it is hard to compare maps. Also, the map has been the primary way of *both storing and displaying* spatial data—an idea we discussed earlier. One of the major advantages of a computer-based GIS is that we separate the storage function from the display function.

A physical method of comparing maps involves a set of, initially, clear plastic sheets, one for each theme in the study area. Each map is darkened in certain areas to indicate the lack of suitability of that theme (for some activity or structure) in the location. A completely clear area of the map might mean a completely suitable area on the ground. A totally black area would indicate a total lack of suitability. Other levels of suitability could be indicated by lighter or darker (grayscale) areas. For example, suppose that you were searching for a site for an airport. On one sheet, expensive land would be created as darker, less expensive as lighter. On another sheet, areas where structures would have to be demolished might be made black. A third sheet would show a very flat area as clear. Assuming that all these maps were made the same size, shape, scale, projection, and so on (quite a chore in itself), you could then line them up and place them on a light table, making sure that equivalent geographic areas properly registered (lined up) with one another, and look through them to perceive the resulting image. Using this "map overlay" technique,[1] the lighter a resulting area, the more suitable that area would be. You can probably think of several reasons why this method is pretty inexact (relative importance of different factors, for one—are weather patterns as important as topography?), but the overlay method was one way used to analyze spatial data sets that come from several map sources.

How the Computer Aids Analyzing Spatial Data

Computers can aid in spatial data analysis and synthesis in a variety of ways. First off is speed. It helps that computers can add and compare numbers billions of times faster than you can. (Computers, while stupid, are fast and accurate. Humans are smart, but slow and sloppy.) Further, a computer is capable of doing repetitive tasks (read: boring) for hours or years on end. You probably would not want to know a person with this capability. A third virtue of computers in GIS is the ability to store very large datasets.

A vital factor in using a computer to analyze spatial data is the paradigm or schema (data model, data structure) that is used to store the data in the memory of the machine. While the issues about the format in which to store data are not unique to GIS, lots of other fields have much less of a problem. Usually when one stores data in a computer, the questions that arise are ones like the following:

❏   Should I use integers or numbers with decimal points?

❏   Is the number likely to be very big or very small?

❏   Would it cause problems if I used a text string to store a numeric value?

Such sets of numbers usually exist in simple lists, databases, or perhaps in matrices.

Complexity of Spatial Data

With spatial data the problem is much more complex than with numbers or text. The natural and human-made environment we want to work with

❏   Is virtually infinite in detail

❏   Is a mixture of continuous and discrete phenomena

❏   Needs to be considered at different levels of detail

---

[1]A method given prominence by Ian McHarg in his 1969 book *Designs with Nature*.

A computer, on the other hand, is finite (small, really) and discrete to a fault (made up, at its most fundamental level, of things, i.e., bits, that either are or aren't, i.e., 1s or 0s—there is no middle ground).

So the question is this: How can we extract significance from the complex, virtually infinite, multidimensional natural and human-made environment and, using only numbers, letters, and patterns of bits, make the computer form a "map" that can be easily analyzed and compared with features that make up the environment we are interested in. Put another way, we need to find a way of structuring the geographic data in the computer's memory so that we can derive answers to queries we might make.

Structures for Spatial Data

What are the principles, fields, ideas, tools, and techniques that are in play in the development of spatial data structure? There are several:

Geometry. A branch of mathematics that deals with the measurement, properties, and relationships of points, lines, angles, surfaces, and solids. With plane geometry we can define a set of polygonal areas with line segments. We can overlay one polygonal set with another, using geometry to calculate where line segments intersect and make new polygons.

Topology. Loosely, a branch of mathematics concerned with the properties of geometric configurations that are *unaltered* when positions of points, lines, and surfaces are altered. (Classic joke: A topologist is a mathematician who can't tell the difference between a coffee mug and a doughnut [since each is a solid objects with a single hole].)

Look at the three plane figures composed of lines connected to nodes (see Figure 4-1). Nodes are shown by heavy dots. While configuration "A" and configuration "B" appear to have a lot in common cosmetically, configuration "A" and "C" are topologically identical and Configuration "B" is different from both. "A" and "C" have the same number of lines and nodes as each other, and you can find equivalences in the connections of the nodes in those two configurations. However, you cannot "map" "B" on to either "A" or "C". If you don't see this, assign letters to the nodes and numbers to the lines in all three. Make a table for "A," "B," and "C," showing what node is connected to what node with what line.

Idealization. Easily manipulated symbols are substituted for actual, three-dimensional real-world objects. All physical objects exist (over time) in three-dimensional space. If the object's measure in

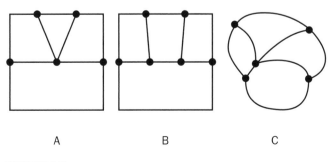

A   B   C

FIGURE 4-1

one or two dimensions is quite small compared with other dimension(s), we may be able to safely ignore a dimension or two. For example, we tend to think of a single sheet of paper as a two-dimensional object, but of course it has thickness as well. We might think of a fire hydrant (depicted on a map as a dot—just a geometric point) as a zero-dimensional entity, but it is, of course, a three-dimensional artifact. (Ask the engineer who designed it, the workpeople who installed it, the firefighters who use it, or yourself, should you try to lift it.) Just as we idealize objects depicted on maps, we do so in a GIS. We say that the fire hydrant exists at a location specified by a single latitude and longitude pair, when in fact parts of it exist at an infinite number of latitude-longitude pairs—all, admittedly, close together but different nonetheless.

Aggregation. Entities having similar characteristics are put together. For example, saying that an area has x acres where corn is grown and y acres where soybeans are grown is a statement of aggregation. Information about where respective acreages of crops are located may or may not be detailed.

Interpolation and extrapolation. We probable-ize. We assume. Data points with a believed high degree of accuracy are interpolated or extrapolated to obtain new information. If we know that the altitude of a certain point on the Earth's surface is 900 feet and that the altitude of another point very close by is at 910 feet, we might interpolate between the two to say that the altitude of a point half-way between them is 905 feet. To get a better estimate, we might also consider the 890-foot contour and the 920-foot contour. In any event, the elevation of such an unknown point is probably known to be not less than 900 feet nor more than 910 feet. Thus, in some cases, there are bounds on the error introduced by the process of probablization.

Categorization. We categorize when we break up a continuous set into a number of discrete sets. For example, we might subsume slopes of $0°$ to $1°$ in category A, slopes of greater than $1°$ up to $3°$ in category B, and so on.

Storage Paradigms for Areal Data

Now we turn to looking at the specifics of the different data structures used by ArcGIS. Representing "almost zero-dimensional objects" (e.g., parking meters) and "essentially one-dimensional objects" (e.g., narrow streams) is relatively simple. If an object is, for our practical purposes, just a point then a simple, single coordinate pair will suffice. If a feature can be represented by a sequence of line segments, then just a sequence of coordinate pairs does the job. Representing areas, however, is a much less straightforward problem.

Fundamental Bases of Geographic Data Mode

Figure 4-2 is a orthophotoquad showing a picture of a piece of Earth's surface. It shows houses, green space, warehouses, roads, trees, railroad, parking lots, a horse race track, and so on. Suppose that you have been given the task of determining the area occupied by each of the feature types: x square feet of housing, y square feet of highway, and so on. Information about where these various land uses exist is also desired. Suppose further that the year is 1960 and you have a computer available to use for the project. If you use the computer, your employer insists that you store the information so that whatever you do can be verified by someone else.

What approach would you take? Basically, to use the computer, you would have to transform the "picture" into numbers and symbols (which the computer would transform into bits). For a given theme (such as land cover) these numbers and symbols must answer two questions at the same time:

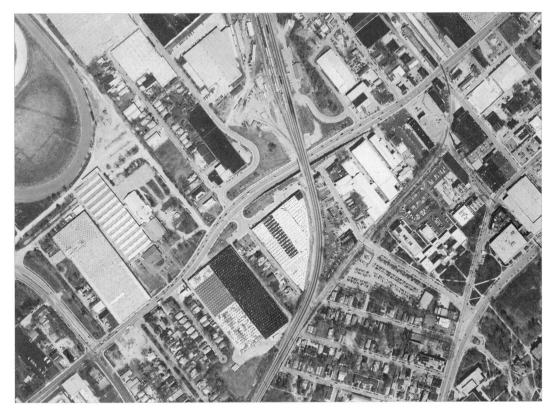

FIGURE 4-2   An orthophotoquadrangle of part of Lexington, Kentucky

❏   WHAT? (entity or quantity)

❏   WHERE?

I don't know how *you* would do this. If you think about it, you may come up with a viable, effective, and efficient scheme that no one else has thought of. If so, head for the patent office.

Here are approaches that others have come up with:

❏   Systematically divide the overall area up in a regular way into a large number of equally sized subareas (e.g., small squares). Record what is in each subarea. Have a reference scheme so that you know where each subarea is. This technique falls under a broad category called *raster* (or *grid* or *cell*). Almost always, a raster may be viewed as a rectangular space composed of rows and columns. A given cell is at the intersection of a given row and a given column.

❏   Completely delineate each of the features—"delineate," in this case, is a real, physical delineation. It means: in the two-dimensional plane, draw a series of straight-line segments around each area. Develop a method for determining where the lines are and for giving each segment a direction. This is often referred to as a *vector*[2] approach, since a directed straight-line segment is a vector.

❏ Just to exhaust the fundamental types of GIS storage methods, although it doesn't help solve this particular problem: Partition a surface that is above (or below, or both) the area of interest into irregular triangles. Except for the periphery, each triangle shares sides and vertices with an adjacent triangle. The triangles approximate the height of the surface (e.g., elevation), the slope, and the direction (e.g., aspect). This sort of dataset is known as a triangulated irregular network, or TIN, which you met briefly in Chapter 2.

The Raster Data Model

One way of systematically dividing up an area of interest is shown in Figure 4-3. Here, regularly spaced horizontal and vertical lines, like those that generate the squares of a chess board, make a *grid* that creates relatively small areas called *cells*. In the past, and sometimes currently, the practice was to index each cell by a row number and a column number. Generally, the top (north-most) row was numbered one (1) and the left (west-most) column was numbered one (1). More recently the indexing has shifted to strictly geographic coordinates. In this case, the coordinates of the center of the southwest-most (lower-left) cell are specified by the easting and northing of (usually) the center of that cell. The horizontal and vertical lines are parallel to the x- and y-axes of the coordinate system. Since the cell size is known, the coordinates of the center of any cell may be easily calculated.

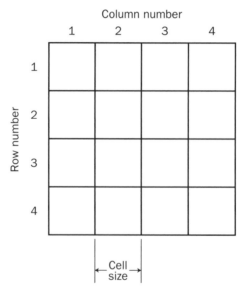

FIGURE 4-3   A basic raster that allows storage of categorical data

---

[2]A vector is a mathematical or physical entity that has magnitude (in this case length) and direction (it has a starting point and an ending point, and is, therefore, an arrow pointing in a geographic direction).

If the person deciding on the spacing between the grid lines has done a good job, and the overall area being depicted is cooperative, the user will frequently be able to know, for each cell, what feature or condition most occupies the cell, thus answering the "what" question. Actually, several issues, to be addressed later, come into play in determining the "what," when, as will frequently be the case, more than one feature, or condition of the particular theme, appears in the area covered by a cell.

The determination of "where" in the raster case is, on the surface, quite simple. As I indicated before, if (1) the horizontal grid lines run east-west, (2) the location of a specific cell is known (e.g., the upper left (northwest-most) cell is known or the lower left (southwest-most)), and (3) the cell size is known, then the geographic location of any given cell is a simple calculation based on the row and column number of the cell.

Although a raster of squares (or "almost squares," if the dimensions of a cell are couched in latitude-longitude terms) is a set of discrete areas, the fact that they are regular in nature, and that each one has the same configuration of four nearest neighbors and four next-nearest neighbors, makes it a fairly good model for representing continuous surfaces, where each cell probably contains a different value indicating, perhaps, elevation. (see Figure 4-4).[3]

A sequence of rasters is also an excellent way to represent, analyze, and predict phenomena that change quickly over time, such as the spread of an oil spill or a forest fire.

The raster approach can also represent discrete areas, albeit "lumpily" with straight vertical and horizontal lines separating nonhomogeneous areas (see Figure 4-5). Here three different areas, designated A, B, and C, are represented, indicating, say, three different soil types.

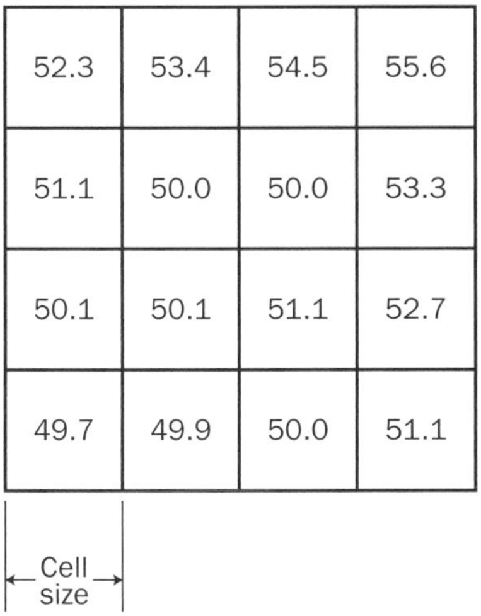

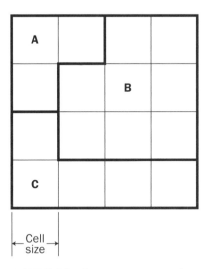

52.3	53.4	54.5	55.6
51.1	50.0	50.0	53.3
50.1	50.1	51.1	52.7
49.7	49.9	50.0	51.1

← Cell size →

FIGURE 4-4   A basic raster that allows storage of continuous data

← Cell size →

FIGURE 4-5   Raster representation of areas

---

[3]A raster does not represent a continuous surface as well as a TIN. A surface represented by a raster (such as a DEM) has discontinuous breaks; in a TIN, the surface representation is continuous but not differentiable in places. If you don't know or care what differentiable means, ignore it.

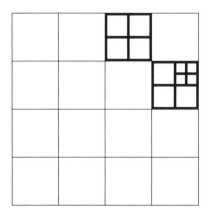

FIGURE 4-6   The quadtree technique for storage of raster

The computation of "where" becomes more complicated if the matrix of cells is not oriented along Cartesian grid lines or the graticule formed by meridians and parallels, but, after all, computation is something that computers are good at. A more subtle complication occurs if the area covered by the raster is large in a north-south direction because of the issues related to projecting the curved earth onto a flat plane. However, for the most part, the matter of location using a raster approach is easily handled.

If a raster cell contains more than one type of area, as many on boundaries between areas do, there is an approach called "quad tree" in which the raster cell is divided into four subcells, as in Figure 4-6. If a subcell is homogeneous in the feature value, then it is left alone. If not, it is redivided into four more sub-subcells, and the process is repeated. This redivision continues until all the subcells are homogeneous, or they become too small to make further subdivisions reasonable.

Vector Data Model

Using the vector approach, the level of difficulty of determining "what" and "where" is somewhat reversed. The "what" is relatively simple. Each unique area is enclosed inside a polygon, so the content of the polygon is homogeneous—containing a unique value (or a unique set of values) related to the theme. Contrast this with the raster approach, where several features or conditions may occur in one cell and (usually) one feature or condition is picked for recording in the database.

The "where" with the vector approach is a bit more problematical but usually allows greater precision. Whereas with a raster approach, the location of each area (cell) is but a simple calculation, with the vector approach, each vertex of each polygon has an explicit geographic location. Therefore, the "where" question may have thousands to millions of answers, in terms of coordinate pairs for a large area or one with many polygons that have complex boundaries. Of course, the issue with respect to projection, datum, and so on are present—for all the coordinate pairs.

A Multiplicity of "Storadigms"

ArcGIS supports, or at least recognizes, several different spatial storage paradigms:

1. File and Personal Geodatabases[4]

2. Shapefiles

3. Coverages

4. Computer-aided design (CAD) files

5. Vector Product Format (VPF) datasets

6. Raster (GRID) datasets

7. Triangular irregular network (TIN) datasets

8. Terrain datasets

Numbers 2, 3, 4, and 5 of these are based on the concept of the vector. We will examine and work with the first two extensively. Number 6 is the raster format already briefly discussed and covered in detail in Chapter 8. Number 7, as briefly discussed earlier, is a technique for storing data where there is an independent, continuous variable (e.g., elevation) whose values are based on the two dependent variables x and y, frequently longitude and latitude. That is, a TIN is used to represent a surface. Terrains are based on TINs covering areas with closely spaced data points.

Ideally, there would be only a single storage paradigm. We would store all spatial data in this way, and when we made a query of the database, or asked for a map of a given area and scale, it would be provided. One problem with this approach is that different sorts of data—representing different aspects of the environment—have distinctly different characteristics. Pick a point on Earth's surface. It has elevation. If there is soil there, it has physical characteristics. Someone or some entity probably owns it. At a certain moment it has a certain temperature, and over a year period it has an average temperature. It may be suitable for growing some kinds of crops, but not others. It has a particular slope and aspect. The vertical distance down to bedrock has a certain value, and that bedrock is of a certain type. There may be a volume of coal or oil under the surface. The atmosphere above it contains some pollutants. We want to represent these facts with datasets.

Some of this information is more easily and effectively stored in one paradigm or format, other information in a different format. Further, a lot of inventing and creating has gone into the problem of representing the infinite, continuous environment as numbers, letters, and symbols. So we wind up with a lot of different paradigms for storing spatial data, or, to coin a term, "storadigms." One advantage of using Esri software is that you may easily convert from one method of storing data to another.

Vector-Based Geographic[5] Datasets—Logical Construction

Geodatabases and coverages are the two most sophisticated data structures used by Esri software. They are sophisticated in different ways, as you will see later. The coverage concept dates back many years and was the foundation of ArcInfo–lines that represented linear features and that separated polygons

---

[4]Actually, geodatabases can include, or will in the future, other paradigms in this list.
[5]Or, should you hail from the United Kingdom: Geographical instead of Geographic.

were call "arcs." The geodatabase is a more recent development. You will spend a fair amount of time and effort understanding and working with geodatabases. The coverage data structure is, frankly, rapidly becoming obsolete. My recommendation is that GIS professionals of today convert their coverages (and, admittedly, there are a lot of them around) to the File Geodatabase structure. Another data structure, the middle-aged shapefile, is relatively simple, but because multitudinous data sets exist in this form, nationally and internationally, it is also important. Understanding geographic data structure is vital to being able to do some forms of analysis with GIS. Please note that you can convert any of the three of these dataset forms into any of the others.

The primary elements of vector-based datasets are points (zero-dimensional entities), lines (one-dimensional entities), and polygons (two-dimensional entities). Terms used with all Esri data models are described in the sections that follow, and then when we look at particular data models. Figures that graphically show the entities follow in the detailed discussions of geodatabases and shapefiles.

Zero-Dimensional Entities in a Two-Dimensional Field: Points

While vital to the functioning of a GIS, a zero-dimensional thing (generically a point) is pretty dull from a geometric view. It is basically a pair of numbers (x- and y-coordinates, or perhaps, latitude and longitude coordinates) stored as single- or double-precision numbers.[6]

The concept of a point is used in a variety of ways in GIS to represent features, as end points of lines, as vertices in sequences of line segments, vertices of triangles, as reference points tying to the feature set to the real world, as locations to hang labels on, as centroids of areas, as junctions and nodes in geometric networks, as centers or corners of raster cells, and others.

We work mainly with points in a two-dimensional arena, but of course they exist truly in three-dimensional space. ArcGIS will let us add information about this third dimension, sometimes as what amounts to an attribute (a "z" value) and sometimes (for example, in a TIN) as a measure in the true third dimension. Even when a true 3-D point is represented, the units of measurement of the vertical may not be the same as those used in the horizontal plane. Some of the uses of points are described below.

Points representing features—In these cases, a point has associated with it a row in a relational database table that identifies the point and allows the user to add other (attribute) information about the feature the point represents. You became acquainted with points representing features in the fire hydrant example of Chapter 1. A point may be used to represent a feature that is too small to have a meaningful area. How does the concept of a point fit into a vector system? A point is simply a vector with zero magnitude and an unimportant direction.

Multipoints—A multipoint is a collection of points that share the same attribute values (e.g., several gas wells which have the same characteristics and the same owner). The collection of points is represented by a single row in a table.

Vertices—Sets of coordinates where two line segments are joined or where a line segment ends. Also considered vertices are the corners of a triangle in a TIN. Usually, no database table row is associated with a vertex.

---

[6]To really locate a point in real (3-D) space, one would need a trio of numbers—but vector GIS either assumes that the third dimension is the elevation of the surface of the Earth, or is not relevant, or is stored as attribute data. An exception to this is a TIN, which stores coordinate triples.

Labels—A point carrying textual information about what the point represents or the polygon that the point resides in.

Junctions—Points in geometric networks where ends of lines (edges) are joined. Discussed when we cover geometric networks in Chapter 9.

Lattice points—Set of points in a raster, usually defined to be at the centers of cells.

One-Dimensional Entities in a Two-Dimensional Field: Lines

You saw examples of lines representing streams and sewers in Chapter 1. Lines were also used there to delineate boundaries of polygons. A line may be used to represent a linear feature that is too narrow to have a meaningful area. As with zero-dimensional entities, vertices on a line may have a "z" (e.g., altitude) value.

In GIS Terms, a line is a simple geometric entity that consists of a sequence (that is, an ordered set) of vertices, which are simply coordinate pairs or triples. Between each adjacent pair of vertices there is a segment. A segment is frequently simply a straight line, but in vector-based geodatabase feature classes, it can also be a part of a circle or ellipse, or it may be a spline (called a Bézier[7] curve). A line that consists of multiple straight-line segments connected at vertices can approximate a curve. Lines can therefore be used to represent curvilinear features, such as roads and streams. The segments of a line are not allowed to intersect each other.

Paths—A path is a line as described previously, composed of a sequence of connected segments (or a single segment). The term "path" is used in vector-based geodatabase feature classes.

Polyline—A polyline is made up of one or more paths. If there are multiple paths, the paths may be connected or disjoint. Even if a polyline representing a feature consists of multiple paths, it has only one row in the attribute table. If polylines are used in shapefiles, the segments of the path must be straight lines.

Rings—In a geodatabase vector feature class, when a path encloses an area (polygon), the path is called a ring. A ring starts and ends at the same place. A ring is a sequence of nonintersecting segments that form a closed loop. Its primary purpose is to enclose areas. If the segments are directed straight lines (vectors), then the area enclosed is a polygon, in both the mathematical and GIS sense. If the segments are curvilinear elements (arcs of circles or ellipses, or splines), then the enclosed area is a GIS polygon, but not a geometric one. A ring has an unambiguous inside and outside. The length of a ring is automatically stored in the associated attribute table.

Arcs—Used in coverages which are not discussed here. (The arc concept is the basis for the designation "Arc" of ArcInfo. An Esri arc is basically a vector. "Info" was the name of the original brand of database that the system used.)

Routes—Routes are subsets or supersets of polylines. They allow the user to define collections of linear features (e.g., the parts of a road system that constitute bus route #99), or measured distances along a linear feature (where a stream changes from clear to turbid). Routes make use of an *m* (measure) number that specifies the distance along a feature to a location at which something changes.

---

[7]Developed by Pierre Bézier in the late 1960s for computer-aided design (CAD) and computer-aided manufacturing (CAM) operations for the Renault automobile company. It involves an anchor point at each end of a segment.

Two-Dimensional Entities in a Two-Dimensional Field: Polygons

Let's look quasi-philosophically at definitions of plane areas and the lines that define them. Lines define polygons, but sometimes it's not really clear what is meant by some basic terms.

A common definition: A polygon is a closed plane figure bounded by three or more line segments. This implies that the area inside the segments is included and could be calculated. For example, a standard definition of a triangle is that it is a three-sided polygon. But what is a triangle? It is the metal frame of a truss or the sail of a boat? Which of the two figures below would you consider a triangle?

 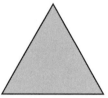

Probably you would say both, although one is three line segments while the other is an area. If the one on the right, which is an area, is called a triangle, then what is the figure on the left to be called?

Also one could ask: what is a circle? Is it the curvilinear line? Or is it the plane figure like a coin? That is, is it the locus of points at a distance "d" from a single point "c" (the center), or is it the locus of points at a distance "d" or less than "d" from the single point "c." "Circle" is used in English both ways.

Look at these figures—which do you consider the circle?

You still might say both, but probably the one on the left fits the usual definition. The one on the right, with the area included, is more properly called a disk (or disc). On the other hand, we speak of the "area of a circle" and have a formula for it. But based on the definition of a circle shouldn't we refer to "the area within a circle"?

With ArcGIS we can make a definitive statement. A figure is a polygon if it has an associated area. A sequence of line segments, although it may be closed, is not considered a polygon. That being said, be aware that the ArcGIS geodatabase definition of a polygon completely butchers the mathematical definition of a polygon. The following are considered polygons:

 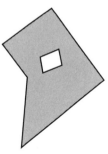 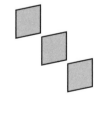

A polygon in ArcGIS, however, is almost always considered a plane figure rather than the line elements that bound it.

Polygons—In geometry a polygon is a plane figure with three or more straight-line sides. The sides may not cross. Perhaps you think of a polygon as a square or hexagon—and you are correct. However, there is a finite but no small limit to the number of sides a polygon may have, nor do the sides have to be of equal length, as they are in a *regular polygon* where equal lengths of sides and equal angles are the rule. It is not unusual for a GIS polygon to have hundreds of sides.

In GIS we take a lot of liberties in the use of the word "polygon." For one thing, a GIS polygon can contain other polygons (which can contain other polygons, which can contain other polygons, and so on). Further, while polygons in shapefiles may only have "sides" that are straight lines, geodatabase polygon "sides" may be parts of circles or ellipses, or may be Bézier curves. Still further, a GIS polygon may be several polygons, as described later in the chapter.

In geodatabase and shapefiles, a single polygon (an entity with a single row in a table) is formed by a collection of one or more rings—as defined previously. If more than one ring is involved, no rings may touch.

Cells—A (usually) square area that (usually) contains a number related to an entity or condition. Used in raster or grid data models.

Zone—A collection of cells that have the same value. Used in raster or grid data models. The cells of a zone may be adjacent or nonadjacent.

Triangles—Plane, three-sided polygons used in the TIN data model, where each triangle has a calculated maximum slope and direction (aspect). Further, each point on or within the triangle has a z value, such as elevation.

Regions—This term is used in different ways depending on which data model is being considered. In a *raster* or *grid system*, a region is a collection of cells, all of which have the same value (i.e., are all of the same zone) and which are connected to at least one other cell in the zone.[8] (So that you aren't confused later: even though we aren't generally concerned with coverages, the term *region* used with the coverage data model specifies something philosophically different than with the raster model. And geodatabases handle this concept with what are called multipart polygons.)

Three-Dimensional Entities in a Three-Dimensional Field: Triangles and Multipatches

Triangles are plane, three-sided polygons used in the TIN data model, where each triangle has a calculated maximum slope and a direction (aspect). Further, each point on or within the triangle has a "z" value such as elevation.

Multipatches are the outer surface, or shell, of features that occupy a discrete area or volume in three-dimensional space. For example, a representation of a building (discussed in Chapter 9).

---

[8]The connection can be specified to be only along edges or to be both edges and corners.

Specific Esri Spatial Vector Data Storage Mechanisms

Let's now move to the specifics of the storage and manipulation of spatial data based on Esri software. Esri is decades old and both its computer programs and storage schema have evolved considerably. As mentioned, the original, elegant data model—the coverage—that is based on using arcs to represent linear and areal features is virtually obsolete. You should know how to convert coverages (which you may come across because a lot of data sets are in this format) to geodatabases, but that's all you will need to know. You may recall that making this conversion was presented in Chapter 1.

Geodatabases are the current "coin of the realm" in Esri software. All the tools being developed deal with geodatabases. A plethora of topology rules and topology fixes accompany geodatbases. Ideally, all coverages and shapefiles would be converted to geodatabase form. However, that is a monumental undertaking, of which, if you continue in the GIS field, you may be a part.

Sets of spatial data in Esri are primarily stored in geodatbases, shapefiles, TINs, (and "super TINS" called Terrains). The interactions among these types are fairly complex. Let's start with the primary ways to store spatial data based on vectors and rasters: geodatabases.

The Geodatabase Data Structure

Esri developed the geodatabase data model for the following reasons:

❏ To take advantage of increased computing power, data storage, and modern relational database management systems (RDBMSs)

❏ Because ideas of how to store geographic data have become more refined

❏ To "umbrella-ize" the different forms of spatial data storage: vector, raster, and, terrains (but not TINs, unless they are converted to terrains),

❏ To permit the use of "objects" that depict real-world entities, in terms of both description (which is still done with attributes) and behavior

Geodatabase Software

Despite the idea of a geodatabase—storing all of geometry, a spatial reference system, attributes, and behavioral rules for data in a single relational database management system—things are still not simple. Esri software has to "partner" with existing RDBMSs, so the conventions for dealing with spatial data must conform to the different general conventions of these RDBMSs. For these and other reasons, there are two Esri geodatabase flavors for single-user ArcGIS: Personal and File. For multiuser Esri software there is the ArcSDE (Spatial Database Engine). (Multiuser ArcGIS has Desktop, Workgroup, and Enterprise versions.)

Personal Geodatabases: They depend on Microsoft Access RDBMS. The datasets are stored within a data file, which is limited in size to 2GB. This was Esri's first version of storing both the geometry and the attributes in a single database system. The extension on the name is .mdb, standing for "Microsoft database." While personal geodatabases suffer from several disadvantages compared to file geodatabases (discussed below), they also have their strengths. If you plan to remain in the Windows operating system environment, and want to search and work with the ArcGIS software that stores attribute tables with Microsoft Access, personal databases can be quite satisfactory.

File Geodatbases: Datasets are stored in folders in a file system. Each dataset is held as a file that can be as large as one TB (Terabyte—roughly a billion bytes, which would be more than 300 million pages of text. Esri recommends that if you are starting from scratch, you should use a file geodatabase, rather than a personal geodatabase. The files are held in a folder; the folder name has extension GDB. It is not limited to the Windows operating system, and you may choose from several RDBMSs to handle the database.

Personal and file geodatabases lack the ability to automatically keep up with versions of the data, and there are restrictions limiting the people who can make changes to the database. Personal geodatabases are usually smaller, run on less powerful machines, and are intended for only a few users in a working group.

ArcSDE Geodatabase Technology

ArcSDE geodatabase technology exists in both single-user file geodatabases and multi-user systems. Such databases are stored in a commercial RDBMS—currently there is a choice of Oracle, Microsoft SQL Server, IBM DB2, IBM Informix, and PostgreSQL. These geodatabases have virtually unlimited size; they support many users and simultaneous editing sessions. ArcSDE is a required piece of software. Because of the requirements to have an additional proprietary RDBMS, we don't discuss ArcSDE Technology (SDE means "Spatial Data Engine") further. But you should know it exists and supports GIS in large and/or complex organizations where several to many people may be viewing and editing the database at the same time.

For comparisons of the various types of geodatabases, you may examine the help files: use the search tab to look for geodatabase types. Be sure to check out the link to "Types of geodatabases."

Polygons within Polygons—Perimeter and Area Calculations

In representing the natural environment or the human-built world, we frequently want to employ plane areas that are included in other plane areas: lakes in a county, for example; islands in a lake; wetlands that are internal to an island; and so on. As mentioned, many polygons are disjoint (that is, if you look at the area covered by a polygon, you see no other polygons), but others are nested (when you look at a polygon you see other polygons within it). The areas of all ArcGIS polygons are mutually exclusive. Each has an identification, an area measurement, and a perimeter measurement. Each has its own set of attribute values. Geographically, however, they may be arranged in two different ways. Consider Figure 4-7 with feature class representations A and B. The small squares are 1 unit on a side.

Feature class A consists of two lines and two polygons (P and Q). Polygon Q is a nested polygon with respect to polygon P.

Feature class B consists of three lines and two polygons (R and S). Polygon S is not a nested polygon. It is simply an area disjoint from polygon R.

The area of polygon P is 8 square units, calculated as $((3*3) - (1*1))$; that is also the area of polygon R. The area of polygon Q is 1 square unit; that is also the area of polygon S.

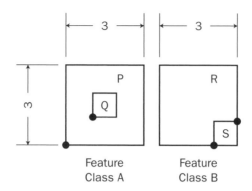

FIGURE 4-7   A nested polygon compared with an adjacent polygon

The perimeter of polygon P is 16 linear units. This is the sum of the length of the line that defines the outside of P (length 12) and the length of the line that segregates the nested polygon Q (length 4). The perimeter of polygon Q is 4 units.

The perimeter of polygon R is 12 linear units, made up of two lines. The perimeter of polygon S is 4 units, made up of two lines.

Again, polygons may be nested to (almost) any depth. Polygon Q could contain three nested polygons, one of which might contain five nested polygons, each of which might contain 22 nested polygons.

In determining areas and perimeters when nested polygons are involved, it might be useful to invoke a rural analogy. Maybe you are raising llamas and want to know how much area would be available for grazing. If the animals are to be confined to a given polygon, say X, the polygon is the area an animal could roam in, which does not include the area of any nested polygons.

Did you ever wonder how the computer "knew" which polygon your mouse cursor was in when you used the identify tool? You, of course, can look at the cursor and the image and tell which polygon the cursor is in. Your head contains a remarkable spatial data processing system. But how does the computer know? If you are interested in an explanation, locate information (on the Internet or elsewhere) on the "point in polygon" problem.

Multipart Polygons

A polygon feature class consists of a set of polygons; each polygon refers to some surface area on the Earth. However, an ArcGIS "polygon" may consist of several geometrical polygons. For example, suppose that you wanted a dataset that depicted of the area in square units of all the states in the United States. You would find yourself delineating two types of areas. First, obviously, there would be those defined by traditional state boundaries, which divided the landscape. But you would also find states like Massachusetts. It consists of a mainland part plus islands. Delineating the surface area of Massachusetts requires several polygons, since the water around those islands could not be considered land area belonging to the state.

Geodatabases—Layout in the Computer

I said earlier that a GIS was the marriage between a (geo)graphical database and an attribute database. The geodatabase still adheres to this in concept. I also said that usually the attribute database was housed in a commercial relational database management system (RDBMS) as mentioned earlier. The new wrinkle is that the entire thing—geographic part and attribute part—is housed in a single RDBMS file. This means, from the point of view of the software, all of the geographic datasets have been rolled up with the attribute data into a single file. For ArcSDE databases this file may be located in one of several commercial relational database systems. To determine which commercial RDBMSs are used by Esri software, consult the help files for the version you are using.

Personal geodatabases are housed using the Microsoft Access database system. With ArcGIS 10. Differences between file or personal geodatabases and Esri's more extensive products include the lack of ability to automatically keep up with versions of the data and restrictions of who can make changes to the database. Also, single user geodatabases are usually smaller and are run on less powerful machines.

With the single-file implementation of a GIS in a geodatabase, there is, therefore, no temptation to go in with the operating system to move, delete, or rename things; the components are somewhat hidden from the user, except through ArcCatalog and ArcMap.

Geodatabases—Logical Construction

Within the single file of a geodatabase, there is the framework for quite a complex hierarchy of elements. You have had some experience with this hierarchy earlier, but here is a summary, with a bit of additional information. The description is based on the file geodatabase, which resides within a folder. ArcSDE geodatabases look somewhat different, but only at the top levels.
The database may consist of the following:

(A) *Freestanding,* and not necessarily related:

❏ Feature classes, resembling the point, line, and polygon classes you have dealt with

❏ Raster datasets, which may represent surfaces (e.g., elevation), areal phenomena (e.g., land cover), or images (e.g., orthophotoquads, scanned maps)

❏ Triangulated irregular network (TIN) datasets[9]

❏ Tables, which are referred to as *object classes*, and which may be imbued with "behavior," as discussed later in the text.

(B) *Feature datasets*, whose constituents share a common geographic reference (datum, projection, units, and so on) and that are composed of the following:

❏ All the elements cited above in (A)

❏ A relationship class that is a set of relationships between the features of two feature classes

❏ A geometric network that consists of

❏ A junction feature class

❏ An edge feature class.

---

[9]As of this writing, TINs are not included with geodatabases, although terrains are.

Geometric networks are useful in a variety of areas, such as routing school buses over a road network and keeping track of electrical or piping systems. While I have avoided trying to divide GIS applications into categories, you could consider that spatial problems that involve flows of entities through conduits to be a major subclass of GIS problems—making GIS of major interest to utility companies. Geometric networks support this sort of activity.

Geodatabases—Feature Shape

The concept that a row in a table contains the attribute values of a single feature remains the same, but geodatabases allow great variety in what constitutes a feature. Specifically, geodatabases allow multipoints, multipart lines, and multipart polygons.

Points

In storing point features, geodatabases allow "multipoints." A *multipoint* is a collection of points associated with only a single row in the database, so all the attribute values in that row apply to all the points. An ecologist may have mapped gopher holes in an area. The only recorded difference between them is location. So, they may be stored together as a single feature. See Figure 4-8 which shows two features—one multipoint feature depicted with dots and another shown with x's.

Two multipoint features

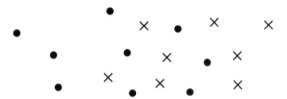

FIGURE 4-8   Two multipoint features: one dots, the other x's

Lines

Linear features are represented by polylines. A *polyline* is composed of one path or several paths. A *path* is composed of sequentially connected segments that may be straight lines, but also geometric curves. You may use a portion of a circle or an ellipse (which are, mathematically, the plots of second-order equations), or you may use a type of spline, called a third-order Bézier curve. The path is a sequence of segments, and it has a left side and a right side. If a polyline is a multipart polyline, then the paths that compose it may be connected, disjoint, or some of each. Look at Figures 4-9, 4-10, and 4-11 for an understanding of segments, paths, and polylines.

Polygons

A single-part geodatabase polygon, without any island polygons within it, is simply enclosed by a single ring. A ring might be thought of as a single path (see Lines above) that starts and ends at the

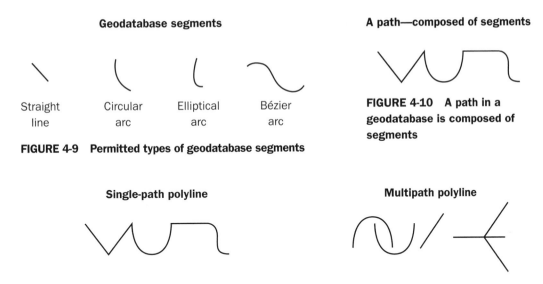

Geodatabase segments

Straight
line

Circular
arc

Elliptical
arc

Bézier
arc

FIGURE 4-9   Permitted types of geodatabase segments

A path—composed of segments

FIGURE 4-10   A path in a geodatabase is composed of segments

Single-path polyline

Multipath polyline

FIGURE 4-11   A polyline may be composed of one or more paths

same point—that is, a ring is a closed figure. Since it is a closed figure, the software knows whether any given arbitrary point is inside the polygon or outside. One effect of using the complete, single entity (ring) to delineate a polygon is that it divorces a given polygon from its neighbors. (It also means that, for traditional disjoint—but adjacent—polygon representation, such as ownership parcels, each vertex and line is stored twice.) So, the traditional coverage topology, which assured that there were no gaps or overlaps between polygons, is no longer present. This sort of topology has been replaced by a much more general set of topological checks which the user can invoke to ensure data integrity.

A geodatabase "polygon" may be what is called a multipart polygon. This may be a set of two or more polygons. Either single-part or multipart polygons may have other polygons nested inside them. So, the term "polygon" encompasses a multitude of conditions. Please look at the illustrations in Figure 4-12.

Nested Polygons in Geodatabases

It is important to look at the calculations of plane area (called Shape_Area) and perimeter (called Shape_Length) for nested geodatabase polygons. Look at Figure 4-13, which has an island polygon D that has dimensions of 2 units by 5 units. Be sure you understand the area and perimeter calculations of polygon A, particularly with regard to the Shape_Area and Shape_Length.

Geodatabases and Attributes

We said that a GIS was the marriage of a geographic database to an attribute database. In a geodatabase, each row in the attribute database refers to an "object" that is a point feature, line feature, or polygon feature.

One of the indications of maturity of the GIS field is the growing emphasis on attribute data correctness and integrity. All large databases, spatial and otherwise, contain errors. With geodatabases, a number

Polygon defined by a single ring

Polygon defined by multiple disjoint rings

Polygon defined by multiple nested rings

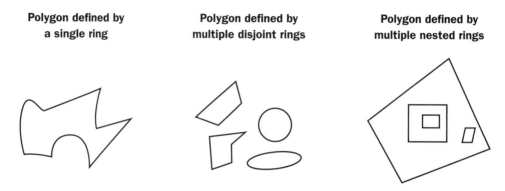

FIGURE 4-12 Different configurations of a "polygon" composed of multiple enclosed areas

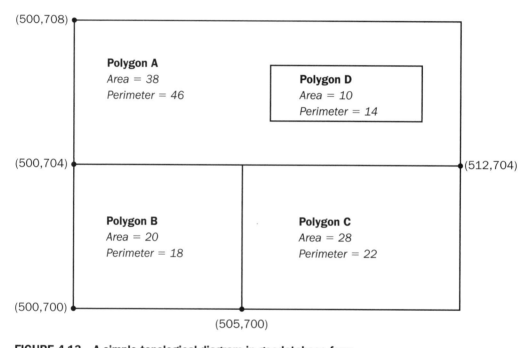

FIGURE 4-13 A simple topological diagram in geodatabase form

of built-in capabilities promote data quality. For example, suppose that you are building a database of the roads in your county. One attribute in the database is the material the road is made of. Perhaps you know that the only allowed materials are concrete, asphalt, macadam, and gravel. With geodatabases you could allow a data entry person to only select among these four. First, this makes data entry faster. Second, it avoids the possibility that someone will type in "asfault" instead of "asphalt." The items concrete, asphalt, macadam, and gravel constitute a domain for the roads feature. You could also set a default value for roads. If no other value is entered, the value of the attribute would automatically be set to "concrete."

Subtypes

Geodatabases go even further in promoting data integrity. Within feature type, say roads, you can define subtypes of roads. Perhaps your planning agency has classed all the roads as freeway, major, or minor. You could set up the database so that "concrete" was the default material for freeways, "asphalt" for major roads, and "macadam" for minor roads. Also you could set the domain for minor roads so that the number of lanes could only be one or two. A subtype is basically an attribute of the feature that gets special attention from the software.

Objects—First Acquaintance

A more profound difference between the depiction of features in the coverage data model and that of the geodatabase is that features are not just geometric entities with attributes, but *objects*, in the computer science sense of the word.

The study of electricity and magnetism is customarily divided into two general areas: fields and circuits. The term *fields* refers to the characteristics of the invisible forces that are caused by magnetic material or by current flowing in a wire. The term *circuits* refers to the study of electricity where electrons are confined to wires and other elements. You might think of a loose analogy between fields and areal features, on one hand, and circuits and linear (network) features, on the other. In networks, entities such as trucks and gas molecules are confined within physical structures, as the electrons are confined within the wires of a circuit. While Esri products have had a network capability for a long time—mainly to deal with transportation systems—the geodatabase takes this capability to new heights. Using the networking features of ArcGIS 10 geodatabases, you can represent and simulate complex and extensive linear, human-built infrastructure—loosely: pipes, wires, and roads.

This new networking capability is facilitated by storing geographic features and their attributes in a database system that is "object-oriented." Each row represents an object. Objects are described by attributes. But objects can also have "behavior." For a human example, you might describe a person's characteristics with attributes, such as weight, hair color, and irritability. However, if in addition, you "allow behavior," then the person might be instructed to drive to the store for a jar of pickles. This process could involve other objects: a particular automobile instructed to allow the person to drive it, a cashier who would accept money for the pickles, and so on. To bring this closer to GIS, a road object might be allowed to connect to another road object, but not to a freeway object. For another example, a high-pressure gas line could be connected to a high-pressure valve, but not to a low-pressure valve. Objects in geodatabases bring us one step closer to integrating the various ideas and components of GIS. A detailed discussion of geodatabase objects is beyond the scope of this text.

The Shapefile Data Structure

Geodatabases are powerful and sophisticated data structures. In addition to topology, you get for free the area and perimeter of delineated areas and the lengths of linear features. However, Esri also supports a much less complex data structure: the shapefile.

A shapefile combines the same two essential major elements that geodatabases do: a (geo)graphic component and an attribute database. The database software is a relational database management system named dBASE.

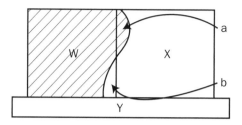

FIGURE 4-14   Shapefile polygons illustrating problems with overlaps and gaps. (Polygon W has a curving right boundary; polygon X has a straight left boundary.)

A particular shapefile is restricted to represent only one of these types: points, multipoints, polylines, or polygons. With points, each individual point has a record in the relational database. If a number of points are considered the same object, then that object has only one record in the attribute table. As with geodatabases, polylines can be composed of one or more paths, connected or disjoint. However, the paths are allowed to be composed only of straight-line segments.

A polygon in a shapefile bears similarity to a polygon in a geodatabase, but no topology is present and none can be created. Each polygon is a stand-alone affair. It is delineated completely by one linear entity: a sequence of segments that starts in one geographic location and returns to that location. There may be adjacent polygons or not. Other polygons may overlap it. See Figure 4-14.

A problem to which shapefiles are particularly susceptible is that there may be slivers of overlap or slivers of vacancy (gaps) between intended polygons, and there is nothing you can do about it if the data remains in shapefile format. In Figure 4-14 the sliver "a" is claimed by both W and X, while sliver "b" is in neither.

A geographic area partitioned into mutually exclusive shapefile polygons will have duplicate information. Since the total boundary of each polygon is defined for that polygon, any common lines are "double digitized." Further, there are two independent boundaries, and you have no assurance that they are congruent. With geodatabases, you can make topological rules to ensure that polygons do not overlap or have gaps, but not with shapefiles.

With shapefile polygons, you do not get the area and perimeter as attributes.

The advantages of shapefile representation are simplicity, processing speed, drawing speed, and, usually, economy of storage. Shapefiles are useful when you do not need sophisticated geoprocessing. Do be aware that a lot of GIS data sets have been put into shapefile format. There may be considerable conversion to geodatabase format in your future if you want to use those data sets in geoprocessing.

Shapefiles—Layout in the Computer

The format for a shapefile on a disk drive is also much simpler than that of a coverage or a geodatabase. Basically, at least three files in a folder are required for a shapefile. If a folder named AMENITIES contains a shapefile named LAWN_SPRINKLERS, then AMENITIES will contain at least these files:

❑   lawn_sprinklers.shp (contains the geographic information)

❑   lawn_sprinklers.shx (contains the spatial index to the geographic information)

❑   lawn_sprinklers.dbf (contains the dBASE table for the attribute information)

Other files (lawn_sprinklers.sbx, lawn_sprinklers.sbn, lawn_sprinklers.prj (containing the projection information) and lawn_sprinklers.shp.xml (containing the metadata) may also be present. So the term "shapefile" is somewhat a misnomer, if one expects a single computer file. ArcCatalog, of course, represents as shapefile as a single entity, but to the operating system, it is several files. You could actually move a shapefile from one folder to another using the operating system. You would simply move all the constituent files, but this is not a recommended practice; use ArcCatalog instead. (Recall that with geodatabases there is no issue with moving or renaming because you can't see the constituent feature classes with the operating system; they are locked away in a single database file.)

When you access a folder containing the files that make up a shapefile in ArcMap or ArcCatalog, you see only the designation—continuing our example—lawn_sprinklers.shp. The other files are hidden, so you may think of the shapefile as just one entity.

Shapefiles are important because they have been around for some time and immense numbers of datasets are in shapefile format. Shapefiles were the data format of choice for Esri's ArcView software (up through version 3).[10]

Summarizing Vector Dataset Features

For various reasons relating to technology and history, Esri now has two major ways of storing vector datasets: geodatabase, and shapefile. Some of the differences relate to functionality and some mostly to terminology. Shapefiles and geodatabases have a lot in common. Geodatabases and shapefiles allow multipoints. They both delineate polygons with rings. They use polylines, composed of segments, with one or more paths. The similarities end there. Geodatabase segments may be circular arcs, elliptical arcs, or Bézier curves. Geodatabases allow for topological relationships; shapefiles do not. A major feature of geodatabases is the ability to use a geometric network (composed of junctions and edges that possess not only attributes, but behavior). Geodatabases calculate the lengths of lines and rings, and the areas of polygons—and shapefiles do none of that. Geodatabases allow subtypes of features, while shapefiles do not. So, the semi-witty comparison that a geodatabase is a "shapefile on steroids" greatly understates the case.

Summary of Logical Structures of Vector-Based GIS Datasets

In terms of logical structure and layout within the computer, shapefiles are at the surface. The files that constitute a shapefile may reside anywhere on a hard disk, provided that they are all in the same folder. Attribute data is always stored in dBASE tables. Only a single feature type—point, multipoint, line, or polygon—is stored in a given shapefile.

A personal geodatabase is a single file in Microsoft Access. A file or enterprise geodatabase may use any of several commercially available database management systems.

---

[10]Esri has used the term "ArcView" in two fundamentally different ways. ArcView versions 1, 2, and up through 3.3 are computer programs. The ArcView associated with ArcGIS (that is, version 8, 9 and 10) describe a level of functionality. The term has been dropped with version 10.1 when ArcGIS Basic describes this level of functionality.

Raster-Based Geographic Data Sets—Logical Construction

In GIS a raster is a set of equally sized squares[11] that cover a rectangular surface. Rasters are used in two basic ways in GIS:

❏ As *rasters* that store attribute information about the area covered by the square (such as elevation, or soil type), in which case individual squares are called cells.

❏ As *images* such as orthophotoquads, where the squares are picture elements (referred to as pixels), containing values that prescribe the intensities of visual colors (e.g., red, green, blue) as well as IR (*infrared*), *UV (ultra violet)*, and thermal spectral elements.

Rasters (Grids)

The *raster* data model, which can be very useful in spatial analysis,[12] comes in two flavors: those in which the cells contain integer numbers and those in which the cells contain floating-point numbers. In each case, each raster cell may contain a single number. If that number is in integer form, then the raster represents categorical or discrete data. Each different integer represents a type of object or a condition. If the numbers in cells are floating point—that is, they contain numbers that may have decimal fractions—then the raster may represent continuous data such as an elevation surface over the area of interest. Integer rasters usually[13] have associated with them a value attribute table (VAT). Floating-point rasters usually do not have a VAT.

Each record in the VAT of an integer raster has a minimum of three fields: an ID field, a Value field, and a Count field that indicates the number of cells in the data set with the given value. For example, Table 4-1 shows a part of a Kentucky land use dataset you examined earlier.

You may recall that value "5" indicated water and "14" indicated transportation, communication, and utilities. A set of cells that contained the same number is called a "zone," whether the cells are adjacent to one another or unconnected. The Count column is useful because, when it is multiplied by the area covered by a single cell, the total area of the zone is obtained.

TABLE 4-1

ObjectID	Value	Count
1	5	2,387,059
5	14	2,606,086

Each cell in the raster can contain a number (value) or can contain an indicator that says no number is assigned to the cell: NODATA.

---

[11]Or almost square—some rasters are based on fractions of a geographic degree.
[12]Use of grid rasters will be discussed in detail in Chapter 8. Here we want only to acquaint you with the data structure and how it is stored in the computer.
[13]If there is an immense number of different integers, a VAT might not be created, due to storage limitations.

A floating-point raster has frequently has no VAT because, usually, almost all the values in the cells are unique. Cell values would be numbers like 45.312 and 46.789. Thus, the number of records would come close to the number of cells and the COUNT for most records would be "1," since it is likely that very few cells would contain identical numbers.

Image Rasters

Raster images come from scanning an area. The two primary types of area scanned are portions of the Earth (scanned by satellites, piloted aircraft, and drones) and maps (scanned by various hardware devices such as flatbed or drum scanners).

You saw the table of an image raster in your examination of the digital raster graphics file COLE_DRG. That table looked partially something like Table 4-2.

The "value" here is simply a code for a color (or grayscale) with the intensity of red, green, and blue shown by the floating-point numbers in the indicated columns.

TABLE 4-2

ObjectID	Value	Red	Green	Blue
0	0	0.00	0.00	0.00
3	3	0.79	0.00	0.08
4	4	0.51	0.25	0.14
12	12	0.80	0.64	0.55

Raster-Based Geographic Data Sets—Layout in the Computer

The simplest storage of a raster band would be something like this:

6666666677777775444449999999

6666667777777777777444444999

66666777777777777744444445555

That is, the integers are stored in sequential locations in the memory of the computer or on a disk. In the preceding case, the number of columns (i.e., the length of a row) is 29 and the number of rows is 3.

The nice thing about storing a raster band is that very little addressing needs to take place to know the location of a raster cell in geographical space. If you know, for example (a) the location of the upper-left corner of the upper-left cell, (b) the cell size, and (c) the orientation of the raster, you can then easily calculate the location of any cell, given its row number and column number. To know what the value of the cell at column 14 and row 3, you need only multiply the row number minus one by 29 and then add the column number. In the example, this gives 2 × 20 + 14 or 72. If you start counting in the upper-left corner of the sequence raster values, wrapping around the end of a row to the beginning of the next row, you will find a "7" when you get to the 72nd value. Though this example is highly simplified, it gives you an idea of an addressing scheme that may be used to store raster bands.

Rasters may be very large. A raster composed of hundreds of millions of cells is not unusual. (The Kentucky-wide land use raster that you looked at has about 300 million cells.) The number of cells is the product of the number of rows and the number of columns. If there are 10,000 rows and 10,000 columns, then there 100 million cells. If you make each cell half the size, then the number of rows and columns is doubled (to 20,000 each), so the number of cells goes up to 400 million.

The sheer size of rasters produces a problem, which in years past limited either the amount of real estate covered or limited the level of detail—that is, the cell size. Great advances in computer storage—both electronic (RAM) and mechanical (hard drives, DVDs)—have solved part of the problem, but huge rasters are also made possible by advances in data compression techniques. One such approach, which has been around for a long time and is simple to implement, is called run-length encoding (RLE). The idea is based on the fact that if you pick any cell in an integer raster, the cell to its right is likely to have the same value. So, we might take advantage of this fact and encode the preceding data as follows:

Row 1: {8:6}, {7:7}, {1:5}, {6:4}, {7:9}

Row 2: {6:6}, {14:7} and so on.

This says that in row 1 there are eight sequential values of 6, seven values of 7, one value of 5, and so on. There are several variations of RLE and, of course, coding economies are used—the braces, colons, and commas don't explicitly appear in the string.

Whether this approach uses less storage depends on the data. For example, if every cell were different from its neighbor, this scheme would be much more costly, but usually the savings in memory are enormous.

As I have said several times, a computer file is composed of 1s and 0s. The idea of a compression scheme is to store the same information in a new file in many fewer 1s and 0s, and then be able to reconstitute the original file exactly, so that no information is lost. This is vital if the file is, say, a computer program where one wrong bit can sink the whole enterprise. Zipping programs do this sort of compression, called *lossless*. However, if one is not picky about being able to exactly reproduce the original file—say, it is a photo in which you will accept a near replication in exchange for a great reduction in file size—then you could use a *lossy* compression method. To know more, type "lossless" and "lossy" into a Web search engine.

Formats accepted by ArcGIS 10 for raster data sets are as follows:

Esri GRID

ERDAS IMAGINE

TIFF (TIF)

MrSID

JFIF (JPEG)

Esri BIL

Esri BIP

Esri BSQ

Windows Bitmap

GIF

ERDAS 7.5 LAN

ERDAS 7.5 GIS

ER Mapper

ERDAS Raw

Esri GRID Stack File

DTED Levels 1 and 2

ADRG PNG NTIF

NTIF

CIB

CADRG

TINs

A surface is a mathematical entity—in particular, it is a function of two variables. That is, if you consider the Cartesian plane, with variables x and y, for every coordinate pair, there is one, and only one, third value z. Visually, you can imagine the flat plane with a cloth billowed above it. The distance, measured perpendicularly from the plane to the cloth is the z value. (Actually, some values of z might be negative, which would indicate that part of the cloth was below the z = zero surface.)

An obvious example of such a surface is elevation[14] above sea level. Another is the daily pollution level of a particular contaminant. A third might be wind speed at a given moment at a given altitude. For every geographic point, there is a z value for these themes. Since there exists an infinite number of points on a plane, there exists an infinite number of z values. Obviously, we could not store an infinite number of values, even if we knew what they were, in a finite computer store. We, therefore, apply the usual GIS techniques, which by this time you are used to, of storing some data and inferring information as we need it. The triangulated irregular network (TIN) is such a device. TINs, which you met in Chapter 1, are described in more detail in Chapter 9, which deals in part with 3-D GIS and is entitled the Third Spatial

---

[14]Elevation is an obvious, but not perfect, illustration. A mathematical surface may have only a single value at a given point and almost everywhere on Earth's surface this is the case. However, in a few places, like where there is an overhanging cliff, the surface of the Earth can have three or more values. For GIS to take these into account would create immense complication, so it gets ignored.

Dimension. The goal just here is to describe the data model that lets the computer tell you about the surface that the TIN represents.

The idea of a TIN is to approximate a surface with a set of planes—triangles, to be specific. A triangle has some nice characteristics: If the computer knows the x-, y-, and z-coordinates of each of its three vertices, it can calculate

❏   The z value (e.g., elevation) of any point on the surface of the triangle

❏   The triangle's slope (the maximum quotient of rise over run)

❏   The triangle's aspect (the direction of the projection on the x-y plane of a line perpendicular to the surface of the triangle–loosely, the direction in which the triangle's face is pointed.)

One could also calculate the area and perimeter of the triangle, but it turns out these aren't usually important. Since the triangle is (usually) at an angle to the x-y plane, the area of the triangle will be greater than its projection onto the plane.

The trick to making a TIN yield useful information is to connect a number of irregularly spaced x-y-z points, of known value, with straight lines so that they form triangles that are not too skinny. The software, based on clever algorithms,[15] takes care of this for you.

TIN-Based Geographic Data Sets—Layout in the Computer

If you look at a TIN in terms of folders and files, you find a single folder, residing within a work space, with a bunch of files within it. As with all spatial data sets, if you want to copy it, move it around, or rename it, you must do so with ArcCatalog, not the operating system.

The primary apparent difference between a TIN data set and other GIS data sets is that there is no attribute table. You can, however, use the Identify tool with TINs. It provides the elevation, slope, and aspect, as well as some additional information tags that the user may add. How does it do this? By calculating information on the fly when you click on a point in the TIN.

For particular problems such as surface analysis, surface display, and hydrological analysis, TINs can be quite useful. You will see more of them in future chapters.

Terrains

An Esri Terrain dataset is a TIN-based surface that may be viewed at several levels of resolution derived from measurements stored in one or more feature classes in a geodatabase. An Esri Terrain is not actually stored in its entirety but is calculated as needed, because of the great size of the datasets that underlie it and the fact that only smaller portions of those datasets are needed at any one time. The principle is sort of similar to a handheld calculator approach to providing functions such as square root or sine. Rather than storing the values in a table they are calculated from a formula as needed.

You met a Terrain in Chapter 2. In Chapter 9 you will build one. The somewhat unusual way in which Esri Terrains are "stored" will be discussed then.

---

[15]Algorithms involving Delaunay triangulation and Thiessen polygons (also known as Voronoi cells and Dirichlet regions).

Spatial Reference

Of course, every instance of every spatial data model discussed previously must have as a foundation a spatial reference to the real world. As discussed, this is not simple. If it were, perhaps Microsoft, Google, or IBM would be the principle vendor of GIS software. And if the world had been created as a big cube, rather than a big sphere, life would be easier for GIS specialists.[16] But we are stuck with the complexities of spherical trigonometry, geodesy, and myriad coordinate systems and datums. You could say that the spatial reference consists of three parts:

❏ A coordinate system, with its associated datum, map projection and parameters, and in some cases an elevation (denoted by z values), and, in some cases, distances along lines or paths (denoted by m values).

❏ A spatial extent (domain) which defines latitude and longitude (or x and y), boundaries.

❏ A scale, when display is involved, that relates units of linear measure on the map to those on the ground.

The method of coordinate system application is handled differently with different Esri products. One way is a projection file, which might look like this:

```
PROJECTION        STATEPLANE
ZONE              3976
DATUM             NAD83
Zunits            NO
Units             FEET
Spheroid          GRS1980
Xshift            0.0000000000
Yshift            0.0000000000
Parameters
```

Of course, this is only part of the story. The lines of text of this file point to additional complexity, which might look like this:

```
PROJCS["NAD_1983_StatePlane_Kentucky_North_FIPS_1601_Feet",
GEOGCS["GCS_North_American_1983",DATUM["D_North_American_1983",
SPHEROID["GRS_1980",6378137.0,298.257222101]],
PRIMEM["Greenwich",0.0],UNIT["Degree",0.0174532925199433]],
PROJECTION["Lambert_Conformal_Conic"],
PARAMETER["False_Easting",1640416.666666667],
PARAMETER["False_Northing",0.0],
PARAMETER["Central_Meridian",-84.25],
PARAMETER["Standard_Parallel_1",37.96666666666667],
PARAMETER["Standard_Parallel_2",38.96666666666667],
PARAMETER["Latitude_Of_Origin",37.5],
UNIT["Foot_US",0.3048006096012192]]
```

And then there is the software that has to interpret all this.

---

[16]Except for those living on the edge. (Pun intended.)

Your job, as a GIS specialist, is not to understand it all, but to be sure that, when you combine datasets, there is agreement in all the specifics. If there isn't complete agreement, then you need to consult an expert in geodesy or do research to determine what errors are introduced by mixing datasets with different parameters, and whether those differences are important to your work. If you absorb only a single from this chapter, it should be the one in this paragraph.

Structures for Storing Geographic Data

____ Open your Fast Facts text or document file.

____ Open the Color Figures file, so you can see the illustrations in more detail.

The exercises in this chapter will acquaint you intimately with both the logical structure of Esri data sets and the methods used for storing those data sets on disk. In the process, you will be introduced to a major component of ArcGIS: ArcToolbox. You will also learn more about ArcCatalog and ArcMap. As before, you should view almost every step as containing a nugget of information that you will find useful in the future. Since the software is complex, it is a good idea to ask yourself at each step if the capability it suggests is one you will be able to remember how to tap, or whether it belongs in your Fast Facts File.

In Exercise 4-1, you are introduced to ArcToolbox. ArcToolbox is available, through an icon on the Standard toolbar, from both ArcCatalog and ArcMap. In the exercises that follow, you look at some trivial Esri datasets, basically to see their underlying structure. Exercise 4-2 looks at point, line, and polygon personal geodatabase feature classes. In each case you will see the minimum in terms of attribute names. You will look at the geographics, then at the attribute tables. You will augment the attribute tables of some of the datasets by adding the x- and y- coordinates of points, using ArcToolbox.

Exercise 4-1 (Warm-Up)

Meet ArcToolbox

____ **1.** Start ArcMap with a Blank Map. Bring up the Catalog tree by clicking on the Catalog icon on the Standard toolbar. Verify that the folder Trivial_GIS_Datasets is in

___IGIS-Arc_*YourInitials*

and that there is folder connection to it.

2. The ArcToolbox icon is a red tool chest on the Standard toolbar.[1] Click it and the ArcToolbox appears. ArcToolbox is a window that can be docked or resized.

3. In ArcToolbox, collapse everything that will collapse (i.e., close any little tool chests that might happen to be expanded). The result should look something like Figure 4-15. (If, when ArcToolbox appears, it covers the T/C, making a tabbed label group, you may want to separate them. You know how to do this, from our discussion in Chapter 1.)

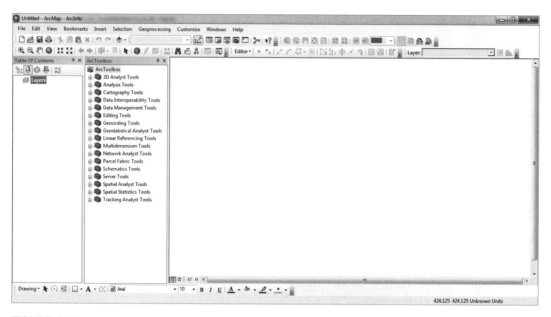

FIGURE 4-15

You will see many basic categories of tools that are available for your use:

❏ 3D Analyst Tools

❏ Analysis Tools

❏ Cartography Tools

❏ Conversion Tools

❏ Data Interoperability Tools

❏ Data Management Tools

❏ Geocoding Tools

❏ Editing Tools

❏ Geostatistical Analysis Tools

---

[1]ArcToolbox is also available as an icon on the Standard toolbar in ArcCatalog.

❏   Linear Referencing Tools

❏   Multidimension Tools

❏   Network Analyst Tools

❏   Parcel Fabric Tools

❏   Schematics Tools

❏   Server Tools

❏   Spatial Analyst Tools

❏   Spatial Statistics Tools

❏   Tracking Analyst Tools

Some of these tool categories we discussed in the Overview of this chapter. However, there are hundreds of tools, so the approach you should use is to believe that, for whatever you want to do with GIS datasets, there is a tool for it and you just have to find out what it is and how to use it.

(In the event that a toolbox you need does not appear in the ArcToolbox pane, you can easily add it. Right-click a blank area in the ArcToolbox pane, click Add Toolbox, navigate to the top of the Catalog, scroll down to Toolboxes, navigate to the toolbox you want in System Toolboxes, and click Open.)

_____   **4.** Leave ArcMap running, with the ArcToolbox window open, for the next exercise. If you have a connection to

   ___IGIS-Arc_*YourInitials*\Trivial_GIS_Datasets

   disconnect that folder, using a right-click on the name.

Exercise 4-2 (Warm-Up)

A Look at Some Trivial Personal Geodatabase Feature Classes

_____   **1.** *In ArcMap, use a different method to make a connection with a folder:* Click on the Add Data icon. In the Add Data window you will see a Connect to Folder icon. Connect to

   ___IGIS-Arc_*YourInitials*\Trivial_GIS_Datasets

   if you don't have this connection already

_____   **2.** From PGDB.mdb\PGDBFD add the personal geodatabase feature class pgdbfc_line_1.[2] As you know, the Identify operation highlights features. Look at the lines. How many lines are there? _____.

---

[2] Ignore any warning about spatial reference. These datasets do not represent the real world.

—— **3.** Open the pgdbfc_line_1 attribute table. Write down the names of the attributes (i.e., columns, fields). _____, _____, _____ What lengths are the lines? _____ Close the table.

—— **4.** Add the pgdbfc_line_2 feature class. Turn off pgdbfc_line_1. Using the Identify tool again, how many lines are there? _____ What are their lengths? _____.

Comparing pgdbfc_line_1 and pgdbfc_line_2: the reason for the difference in the number of lines and different lengths is that, in pgdbfc_lines_1, the lines cross but do not intersect. An example might be that the lines represent dual-lane highways, where one highway passes over the other. In pgdbfc_lines_2 the lines do intersect like the streets on a city grid.

—— **5.** Open the pgdbfc_line_2 attribute table. Verify that the columns are the same as above.

—— **6.** Turn off pgdbfc_line_2. Add the pgdbfc_poly feature class. Open the pgdfc_poly attribute table. Write down the names of the fields. _____, _____, _____, _____.

—— **7.** Turn off the pgdbfc_poly feature class and add the pgdbfc_point feature class. Open the attribute table. Write down the names of the attributes. _____, _____. Dismiss the attribute table. Dismiss the Identify Results window.

In the next steps, you are going to use ArcToolbox to perform an almost trivial operation. As you may recall, the x- and y- coordinates of the fire hydrants were included in the attribute table of the Village Data (from Chapter 1). You will add x- and y-coordinates here, so you get to see a toolbox used in a simple situation.

The tool you want to use is called Add XY Coordinates. How do you find it among all the tool categories and tools available?

—— **8.** On the Standard toolbar find the Search button. The ToolTip says[3] "Opens the Search window so you can search for data, maps, tools, etc.". Click the button. The Search window, which you met in Chapter 1, should appear on the right-hand side. This time, instead of searching for Data we will be searching for Tools, so click the Tools link in the window. Type "Add" into the search text box. Click the search button to see the results. There may be two Add XY Coordinates tools. One (Coverage) works on coverages; the other (Data Management) works on geodatabases. In the three lines of text about each you will see information about the tool, and the path to the tool. You could also search for Coordinates. Again you may see the two Add XY Coordinates tools. Click the green link of the (Data Management) entry. The ArcCatalog sidebar will open and the reference tool will be highlighted. By scrolling up in the ArcCatalog window, you see that it could have been accessed by navigating to ArcToolbox > Data Management Tools > Features > Add XY Coordinates—if you had known it was there.

More Help

—— **9.** Pounce on the tool to start it. In the Add XY Coordinates window—which may need some expanding—notice that at the lower-right corner there is a button that controls whether a help pane appears to the right of the main window. Show the help pane. Read about the tool.

---

[3]I'm describing in this section what you do for version 10.1. Version 10.0 is almost the same.

Click the Input Features text box and the help pane changes to provide information about that field. Press the Tool Help button. That takes you to the appropriate place in the ArcGIS Desktop Help system. Peruse and dismiss.

____ **10.** Back in the Add XY Coordinates window, what you want to do is add the coordinates to the pgdbfc_point feature class attribute table. So browse using the little yellow folder next to the Input Features text box to

___IGIS-Arc_*YourInitials*\Trivial_GIS_Datasets

\PGDB.mdb\PGDBFD\pgdbfc_point.

Highlight pgdbfc_point to bring it into the Name text box. The Input Features window should look like Figure 4-16.

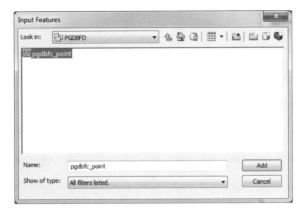

FIGURE 4-16

____ **11.** Press Add. Then press OK on the Add XY Coordinates window. The box disappears and computation proceeds. A scrolling message may appear on the Status bar, next to the coordinates. When the processing is done, a message window will appear in the lower-right corner of the desktop, announcing that the tool has successfully run. You can click on the name of the tool in that message to display a Results window, which shows the input and output. Also, there should be a green check mark next to the operation you just performed. If there is a red circle with a white x, the process failed. Read the reason why the tool failed and, if it reads

"ERROR 000464: Cannot get exclusive schema lock. Either being edited or in use by another application"

close ArcCatalog (or the ArcCatalog sidebar) and try again in ArcMap.

____ **12.** Open again the attribute table of pgdbfc_point. List the names on the column heads. _____, _____, _____, _____. Zoom the map to full extent. Use Select Features to select a point. Use Identify on that point and note the (approximate) correspondence between the location of the cursor on the status bar, the coordinates shown in the location text box of the Identify window, the POINT_X and POINT_Y values in the Identify window, and the values in the attribute table. Dismiss all open windows. Dismiss ArcMap.

Exercise 4-3 (Minor Project)

Adding Tools and Toolboxes to Your Toolset

Although ArcToolbox provides many tool options, you may be facing a task for which a tool isn't provided. It is certainly possible that someone else needed to solve the same problem and, if you're lucky, the other person will have shared their solution on ArcGIS's resource website. So, instead of reinventing the wheel, you should search Esri's sites—specifically, the Geoprocessing Model and Script Tool Gallery—to see what other users have developed. Should a tool you want exist it turns out that it is easy to add to your copy of ArcGIS.

Also, you might accidentally delete a toolbox, or delete a toolbox that you thought you'd never use and then, once you start a new project, you discover that you need that toolbox. For all these reasons, being able to add toolboxes is a skill you should have—or at least put in your Fast Facts File for reference.

Here, you will be adding a toolbox that was supplied by Esri in past versions but then removed. (The elegant term for removing something from the software is "depreciated.") It is called the Samples toolbox. Samples tool box has been provided on the DVD. Currently, it is also on the Internet at the following URL:

```
http://resources.arcgis.com/gallery/file/geoprocessing/
details?entryID=F25C5576-1422-2418-A060-04188EBD33A9_
```

___ **1.** Verify that you know where the toolbox is residing, whether you download the file from the link or plan to use the one provided on the DVD in the Samples_Toolbox folder

\SampleTools.tbx.

___ **2.** Start ArcCatalog. Open the set of Toolboxs by clicking the ArcToolbox button on the standard toolbar.

___ **3.** Right-click in a blank space within the toolbox window and select Add Toolbox. Navigate to the location on the DVD

Samples_Toolbox\SampleTools.tbx

select it, and click Open. You should now see the toolbox, named Samples in your toolbox window. In the next Exercise, you will use tools from this toolbox.

Making a Personal Geodatabase Feature Class Named TextToFeature

___ 1. Use ArcCatalog to make a folder named TextToFeature in

> ___IGIS-Arc_*YourInitials*

In that folder, make a personal geodatabase named TextToFeature_DB.mdb. In that database, make a feature dataset named TextToFeature_DS without specifying a coordinate system.

The plan here is to make an Esri feature class that corresponds to a drawing composed of lines and coordinates. You will do this by the cumbersome method of typing the coordinates of each vertex of each polygon into a text file. After doing this a couple of times, you will really appreciate more efficient methods of data entry. The text file you make will become the input to the Create Features From Text File tool.

Virtually everything in this exercise will be named TextToFeature. It can be confusing. It can also be illustrative of whether or not you know when you are dealing with a folder, a geodatabase, a geodataset, a feature class, a map, or an ArcToolbox tool. Your Fast Facts File should contain references to the file extensions for a personal geodatabase (mdb), file geodatabase (gdb), and map (mxd).

Specifications of your Input Text File for the "Create Features from Text File" Tool

The **symbolic format** of a text file for the lines of the feature class is as follows.

```
<polyline>
<line_id> <part_number>
<vertex_id> <x> <y> <z> <m>
<vertex_id> <x> <y> <z> <m>
<vertex_id> <x> <y> <z> <m>
<line_id> <part_number>
<vertex_id> <x> <y> <z> <m>
<vertex_id> <x> <y> <z> <m>
<vertex_id> <x> <y> <z> <m>
(and so on)
END
```

The symbols < and > are used to enclose characters (letter and numbers) that will appear in the text file. The text file you create will contain only letters and numbers; the text file will **not** contain the symbols < and >.

The first text line should contain the word "polyline" to indicate the geometry type. The second text line contains the number that identifies the first geometric line. On this text line is also a part number; if there is only one part to the line a zero here indicates that. (Subsequent parts would be numbered 1, 2, and so on.) The third text line consists of five numbers, which specify the first vertex:

the vertex ID (which starts at 0 and increases by 1 for each line)

the x-coordinate of the vertex

the y-coordinate of the vertex

the z-coordinate (in our case it will be zero)

the m-coordinate (in our case it will be zero)

Subsequent text lines are of this format until the line is completely specified. Then a second geometric line is begun with.

```
<line_id> <part_number>
```

and the vertices of that line are specified. Study the symbolic format above until you understand this format. (The z-coordinate relates to the vertical height if there is one. The m-coordinate relates to the distance the vertex from the line's beginning to the vertex. (The m-coordinates will be discussed in Chapter 9 when we take up linear referencing.)

Once all geometric lines are specified the word "END" constitutes the last text line of the file.

2. Compare the defining format above with an actual text file below. What are the x- and y-coordinates of the second vertex of the line with the ID of 103? _____, _____ This text file will be used with the Create Features From Text File tool to create a feature class.

```
Polyline
101     0
0       8303.0      7100.0      0.0      0.0
1       8300.0      7100.0      0.0      0.0
2       8300.0      7106.0      0.0      0.0
3       8303.0      7106.0      0.0      0.0
4       8303.0      7100.0      0.0      0.0
102     0
0       8303.0      7100.0      0.0      0.0
1       8309.0      7100.0      0.0      0.0
2       8309.0      7106.0      0.0      0.0
3       8303.0      7106.0      0.0      0.0
4       8303.0      7100.0      0.0      0.0
103     0
0       8304.0      7101.0      0.0      0.0
1       8308.0      7101.0      0.0      0.0
2       8308.0      7105.0      0.0      0.0
3       8304.0      7105.0      0.0      0.0
4       8304.0      7101.0      0.0      0.0
END
```

In the steps that follow, the preceding text file will be used to create a trivial set of three lines. Each line forms a rectangle; since it is a closed figure, the last vertex is the same as the first. Verify that the drawing in Figure 4-17 has indeed been specified by the preceding text file. In Step 3, you will create the text file shown immediately above by typing into a simple text editor, not with a word processor. Note that the text file **does not** contain the characters < and >.

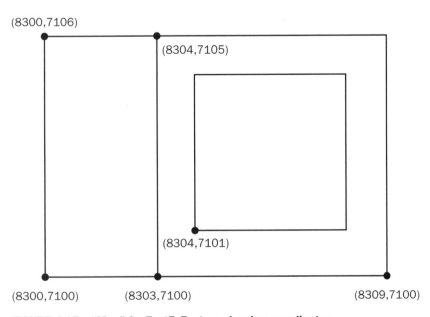

(8300,7106)

(8304,7105)

(8304,7101)

(8300,7100)   (8303,7100)   (8309,7100)

FIGURE 4-17   "Map" for TextToFeature showing coordinates

___ **3.** Using a text editor, create a file from the text above **exactly** as shown.[4] Leave one, and only one, blank space between the various elements in the record. Save this file as

TextToFeature_Lines.txt

in the folder TextToFeature in

___IGIS-Arc_*YourInitials*

Be sure that you save the file with a txt extension. You will create the Feature Class from the Text File.

___ **4.** Show the ArcToolbox pane in ArcCatalog. In ArcToolbox navigate to the tool "Create Features From Text File" in the toolbox Samples > Data Management > Features. Pounce on the tool. For the Input Text File field, in the Create Features From Text File window that appears, browse

---

[4]You could use Notepad or WordPad found in Windows: Start > All Programs >Accessories.

to the text file you just created and add it. For the Input Decimal Separator, type the word "period."[5] In the Output Feature Class field browse to and pounce on:

___IGIS-Arc_*YourInitials*\TextToFeature

\TextToFeature_DB.mdb\TextToFeature_DS

Type the name of the feature class you are going to create:

TextToFeature_Lines

Click Save. Examine the Create Features From Textfile window. When it indicates Completed close it.[6] To gain screen space, make a tabbed label group from ArcToolbox and the Catalog Tree.

_____ **5.** In ArcCatalog, under the File menu, pick Connect Folder (or use the Connect To Folder icon on the Standard toolbar). In the window that appears, navigate (by expansion, not double-clicking) to the folder

___IGIS-Arc_*YourInitials*\TextToFeature

With TextToFeature highlighted, click OK.

_____ **6.** Notice that the folder entry is selected in the catalog tree. As I indicated before, the entry will remain there (on this computer) even when ArcCatalog is stopped and started again. As you remember, this provides a shortcut to this folder to make navigation to it easier. Navigate to the feature class (down a couple of levels in this folder) TextToFeature_Lines that you just made.

_____ **7.** Click the Contents tab and notice your TextToFeature_Lines feature class is named in the right side of the ArcCatalog window. Click the Preview tab. Examine the image for errors. If it is correct, proceed with the step below. If it is incorrect, assess what you did wrong, then select TextToFeature_Lines in the left pane of ArcCatalog, and delete it. Using the text editor, fix up the TextToFeature_Lines.txt text file and then return to Step 4 above and perform the steps to make the TextToFeature_lines feature class again.

_____ **8.** Using Preview, look at the attribute Table for the TextToFeature_Lines feature class. Verify the Shape_Length values for the lines are correct by comparing the table values with the distances between coordinates you specified in the text file.

Labeling Features

_____ **9.** Start ArcMap from ArcCatalog, choosing a Blank Map. Dismiss ArcCatalog. Add the feature class TextToFeature_Lines. (Since we didn't define a spatial reference a warning will appear.

---

[5]In many countries the decimal point is represented as a comma, rather than a period.
[6]Since this is a depreciated tool, it operates with the older style of notification windows. This differs from the tools you saw in exercise 4-2 when you used the Add XY coordinate tools.

Ignore it.) Right-click the feature class name in the T/C and choose Properties.[7] Pick the Labels tab. As the Text string Label Field, choose File_ID. If necessary place a check in the box Label Features in this layer. Your Layer Properties window should look like Figure 4-18. Click Apply, then OK.

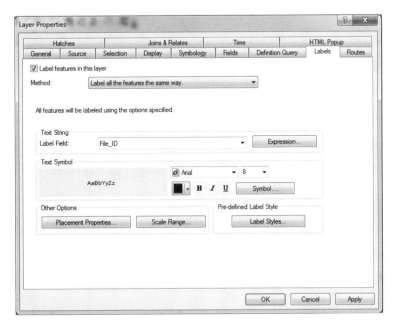

FIGURE 4-18

___ **10.** Your map should now look like Figure 4-19. The lines will have the identification numbers that you provided in the text file—that is, 101, 102, and 103. But you notice something peculiar: labels 101 and 102 seem to be on the same line. Well, they aren't. There are two lines there: one belonging to the leftmost polygon and the other belonging to the adjacent polygon. ArcMap is bright enough not to put one label on top of another, but not bright enough to differentiate one line from the other. Use the Identify tool to determine which line is which. (Hint: Click on a line other than the vertical line. In moving the Identify window around and then zooming to the layer, ArcMap may move the labels so they make more sense—but it also might not.)

Now you might be thinking, "Isn't having two identical lines dividing the polygons a waste of computer storage? Not to mention that I had to type in almost twice as many vertices." Actually, there are good reasons for this duplication, and as we talk later about topology and topological errors, these reasons will become apparent. (There is the older data structure—the coverage, with which you had a passing acquaintance in Chapter 2—that uses single lines to separate polygons. If you are interested in the data structure of coverages, you may take a look at the file Discussion of Coverages from the 1st edition, found on the DVD that accompanies this book.

---

[7]If your T/C disappeared you can restore it with Windows > Table Of Contents.

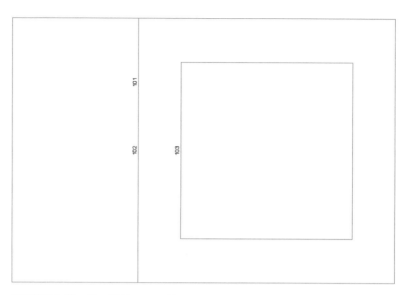

FIGURE 4-19   TextToFeature_Lines with lines identified

_____ **11.** Bring up the attribute table of TextToFeature_Lines. It consists of only four fields. What are they? _____, _____, _____, _____. Dismiss the attribute table. Save the map as TextToFeature.mxd in the TextToFeature folder.

Making Polygons from Lines

Now that you have created a line feature class, you are able to make polygons from it. The process is a little convoluted and certainly deserves a line or two in your Fast Facts File.

_____ **12.** Restart ArcCatalog (not just the Catalog Tree in ArcMap but the entire program). Open ArcToolbox. Expand the following: Data Management Tools > Features, and then pounce on Feature to Polygon. Peruse the Help. For the Input Features text box, using the browse icon, navigate to the TextToFeature_Lines feature class entry and click it so it appears in the Input Features text area. Change the name of the Output feature class to TextToFeature_Polygons in the path

___IGIS-Arc_*YourInitials*\TextToFeature

Click OK. (You may get a rude shock in the form of an admonition. If so, close ArcMap and try again. If not, just proceed with the next step. See the discussion immediately below to understand why ArcGIS sometimes refuses perfectly reasonable requests.)

The need to modify a geographic data set when it is being used by another program crops up frequently and is usually frustrating. An error message you will sometimes encounter is "Cannot acquire a schema lock because of an existing lock . . ." How you correct this depends on circumstances. If you have an extensive map up and you don't want to remove a data set from it, you can easily save the entire map, with all its settings, make the modifications to the data set, and then reload (Open) the original map with the modified data set. However, sometimes you get the "lock" based on having used data sets previously, but that are no longer in use. Esri is working on the problem of data sets that are locked out because they were previously used by other programs. One of their current approaches is the ArcCatalog sidebar in ArcMap.

_____ **13.** In ArcCatalog, refresh the Catalog Tree (F5 will do that if Folder Connections is highlighted) and then alternately select the previews of TextToFeature_Lines and TextToFeature_Polygons. Now look at the attribute table of each feature class. Note that the polygon feature class table indicates area. What is the area of the largest polygon? _____.

Areas and Perimeters Examined

If you have trouble understanding the areas and perimeters, recall our rural analogy: Suppose that each polygon represents a field of a farm in which different animals are kept. The questions to be answered are these: How much area does each animal have to graze in, and how much fence is required to keep each animal where it belongs? In the steps that follow, you will use the labeling capabilities of ArcMap to display the areas and perimeters of each of the three polygons.

_____ **14.** Go back to ArcMap (restart it if necessary and open TextToFeatures.mxd) and add TextToFeature_Polygons. Right-click TextToFeature_Polygons. Click Label Features. What values appear within the polygons? _____, _____, _____. Verify from Figure 4-17 that these are the perimeters of the three polygons. Write the perimeter value on Figure 4-17: P=xx. Be sure you understand where the 40 came from. Going back to our analogy, it would take 40 units of fencing to keep the animals in that polygon: the sum of the perimeters of the two squares.

_____ **15.** Right-click again on the TextToFeature_Polygons entry, then select Properties. Navigate to the Labels tab if necessary. Change the Label Field to Shape_Area. Click Apply and then OK. Now you will see the polygons labeled with the area (instead of perimeter as before) _____, _____, _____. Inside each polygon on the diagram in your textbook, write the area value: A=yy. Verify that each number you have written in a polygon represents the area of the polygon.

Labeling Features with Selected Attributes

_____ **16.** Right-click again on TextToFeature_Polygons. Remove it from the map, then right-click on TextToFeature_Lines and select Properties. Navigate to the Labels tab if necessary. Change the Label Field to Shape_Length. Click Apply and then OK. Now you will see the lines labeled with their lengths. _____, _____, _____. Now label each line with its OBJECTID. _____, _____, _____. Finally label each line again with its File_ID (the values you provided when you created the features). _____, _____, _____. Dismiss ArcMap without saving the map.

Exercise 4-5 (Quick Quiz)

Areas and Perimeters

For each pair of polygons in Figure 4-20, write the Shape_area and Shape_length (perimeter) in the table. The length of each side of the large square is 12. The length of each sided of each small square is 4.

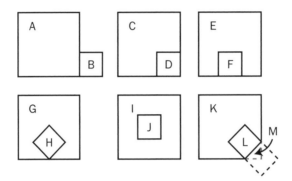

FIGURE 4-20   Area and perimeter quiz

Polygon	Area	Perimeter
A		
B		
C		
D		
E		
F		
G		
H		
I		
J		
K		
L		
M[8]		

---

[8]The dashed lines shown are a hint regarding the area of polygon M, which is a triangle. If a right isosceles triangle has a hypotenuse of length 1, the length of a side is approximately 0.7—or less approximately, 0.70710678118.

Making a File Geodatabase Feature Class for Foozit_Court

Now you know how to make a line feature class with ArcToolbox's Create Features from Text File tool and how to create polygons from the lines. In what follows you will make a more complex feature class and examine it.

A Foozit court is a (hypothetical) game surface, somewhat like a tennis court in that it has painted lines. The diagram in Figure 4-21 shows where the lines are. The units are in feet. The lower-left corner and upper-right corner coordinates are shown in the diagram. Almost all nodes and vertices fall on Cartesian points with *integer* coordinates. If you are familiar with the Cartesian coordinate system, this exercise may seem like so much busy work—but it will go quickly. If you aren't familiar, you will learn a lot from this exercise.

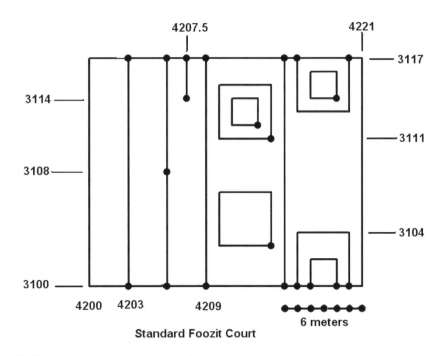

FIGURE 4-21 Foozit court coordinates

—— **1.** In ArcCatalog make a *file* geodatabase named Athletics.gdb in

___IGIS-Arc_*YourInitials*\TextToFeature.

—— **2.** Make a feature dataset named State_Athletics within Athletics.gdb. Make its projected coordinate system State Plane, NAD83 U.S. feet, Idaho Central FIPS 1102. It doesn't need a vertical coordinate system.

271

___ **3.** From [___]IGIS-Arc\Other_Data, use the operating system to copy the file Foozit_Court_Lines.txt to the folder

___IGIS-Arc_*YourInitials*\TextToFeature.

___ **4.** Use the Create Feature From Text File tool (remember: it's in the Samples toolbox) to make a feature class named Foozit_Court_Lines in the feature dataset State_Athletics, using Foozit_Court_Lines.txt. Examine the feature class with ArcCatalog.

___ **5.** Refer to Figure 4-21. How many polygons are there? ____. Compare the Figure with the on-screen representation. Notice that three polygons are missing. Determine what text is missing from Foozit_Court_Lines.txt that is causing the omissions.

___ **6.** Using a text editor (not a word processor), correct Foozit_Court_Lines.txt. Number the new polygons 901, 902, and 903.

Save the file as Foozit_Court_Lines_Fixed.txt.

___ **7.** In ArcCatalog delete the feature class named Foozit_Court_Lines. Then remake it with the repaired Foozit_Court_Lines_Fixed.txt. If the feature class Foozit_Court still isn't right, repeat the above steps until it is satisfactory.

___ **8.** Create a polygon feature class named Foozit_Court_Polygons, using Foozit_Court_Lines as input and the procedure you learned in Exercise 4-4. Check out Foozit_Court_Polygons with ArcCatalog.

___ **9.** Start ArcMap. Add Foozit_Court_Polygons. Right click the T/C entry and click on label features. You get some strange numbers, which I explain later. Instead let's use ArcMap's labeling capabilities to label each polygon with the object identifier (OBJECTID).[9] Use Properties, Labels, then fix up the Label Field.

___ **10.** Print Foozit_Court_Polygons. Now on the paper printout, label each polygon with its Shape_Area. What is the indicated area of the largest polygon(s)? _____ Given the original definition of the Foozit court and the locations of its vertices, what is the actual area of the largest polygon(s)? _____ Depending on the version of the software you are using, the indicated values may not be exactly correct. Where you would expect to see 4.000000 you might instead see something like 3.999984. Why? An explanation will be found in Exercise 4-7.

Exercise 4-7 (Exploration)

Understanding Some Things That Don't Look Right

___ **1.** Start ArcMap with a Blank Map. In

___IGIS-Arc_*YourInitials*\Trivial_GIS_Datasets\FEAT_NUM

---

[9]With a file geodatabase or personal geodatabase feature class the object identifier is OBJECTID; in a shapefile it is FID. Why the differences? Because different pieces of the software were developed at different times and used different, already written computer code.

you will find the following:

A personal geodatabase named Feat_Num.mdb containing three feature classes:

feat_num_pts, feat_num_lns, feat_num_ply

Three coverages:

> feat_num_pts, feat_num_lns, feat_num_ply

Three shapefiles:

> feat_num_pts, feat_num_lns, feat_num_ply

The datasets are identical in what they represent. Each dataset contains three features: one of points, one of lines, and one of polygons. (The personal geodatabase feature classes were converted from coverages, as were the shapefiles.) What we will do is look at the differences in the tables. In particular, we will look for differences in the FEATure NUMbers and at anomalies in calculation.

2. Add the appropriate components of all three of the coverages to ArcMap. You will see three points, three lines, and three polygons. The map should look something like Figure 4-22.

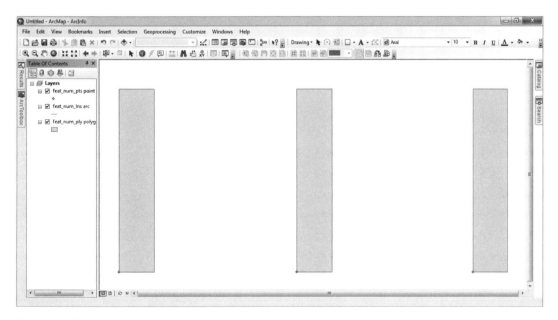

FIGURE 4-22

Open, and leave open, the attribute table of each layer. They may pile up all in the same space, making a tabbed label group, perhaps on top of the T/C. Move the table off the T/C if it is there. You can separate them by clicking on Table Options and choosing Arrange Tables > New Horizontal Tab Group. (You might have to do this twice.) If they are still jammed up together, drag the Table header up toward the top of the page; then drag the very bottom of the set of tables toward the bottom of the page. The points have Feature Identifiers (FID) of 501, 502, and 503. The lines have IDs of 701, 702, and 703. The polygons have IDs of 901, 902, and 903. The lines were created to have length 100. The polygons were created to have area 2000 and perimeter of 240.

The anomaly of interest is that the FIDs of the polygons are 2, 3, and 4, when you might expect 1, 2, and 3, as is the case with points and lines. The reason for this is that the first record (not shown) with FID 1, is the so-called outside or external polygon—as mentioned before, it consists of the entire Earth except for the areas delineated by the other polygons. We will come back to this shortly.

─── **3.** Start a new map. Add all the shapefiles. Open and arrange their attribute tables. (Some of the fields have been carried over because of the conversion from the coverages. Some are useless. Others, like AREA and PERIMETER were just copied, but you should realize that area and perimeter are *not* automatically calculated for shapefiles.) What I want you to see here is that the FID records begin with zero instead of one or two.

─── **4.** Start a new map. Add all the personal geodatabase feature classes. Open and arrange their attribute tables.

Here, in keeping with the idea that features are objects, the identifier is called an OBJECTID instead of FID (feature identifier). In each case the records are numbered starting with 1.

So, there is a lack of consistency with the feature identifiers between geodatabase feature classes, shapefiles, and coverages. We just wanted you to be aware of this so that it won't mystify you in the future when dealing with real data in these formats.

There is another reason to examine the geodatabase feature class tables. Leave these tables up so that you can examine them.

Computers and Inexact Computation

The LENGTH, AREA, and PERIMETER field values in the attribute table were carried over from the coverages. They are, respectively, 100, 2000, and 240, exactly. The Shape_Length of each line, in Attributes of feat_num_lns, however, is calculated in the conversion. It displays as 99.999999. The Shape_Lengths and Shape_Areas in the polygon table are likewise slightly inaccurate. As previously mentioned, computers cannot be counted upon to give exact answers. The difficulty is that, while computers can usually do exact arithmetic with integers, they cannot be exact with floating-point numbers, which may have fractional parts.

These errors occur because humans do arithmetic with decimal (base 10), and computers do arithmetic with binary (base 2). For example, it is not possible to exactly represent the decimal number one-tenth (0.1) in binary. One-tenth in binary is represented (imprecisely) by the sum of some of the fractional powers of two: one-sixteenth, one-thirty-second, one-two-hundred-fifty-sixth, and so on. In binary one tenth looks something like 0.000110011... To be exact, the number would have to have an infinite number of bits.[10] (Some decimal fractions can be represented exactly in binary. For example, decimal 0.5 is 0.1 in binary (that's 2 to the minus one power); 0.25 is 0.01; 0.75 is 0.11; you see the pattern.)

---

[10]Don't look down on binary, however. Every number system has this problem. For example, you cannot represent the number one-third in the base 10 [decimal] system in a finite number of significant figures. An example approximation is 0.3333333333; and you can never do better than an approximation.

The errors that can occur when using floating-point numbers with decimal parts are illustrated by the following steps.

5. In ___IGIS-Arc_*YourInitials*\Other_Data

there is a text file that looks like this:

```
Polyline
5  0
0    1000.00    4000.00    0.0    0.0
1    1002.10    4000.00    0.0    0.0
2    1002.10    4002.10    0.0    0.0
3    1000.00    4002.10    0.0    0.0
4    1000.00    4000.00    0.0    0.0
6  0
0    1000.00    4000.00    0.0    0.0
1    1002.10    4002.10    0.0    0.0
7  0
0    1002.10    4000.00    0.0    0.0
1    1000.00    4002.10    0.0    0.0
END
```

Verify that the text file describes a square whose sides are 2.1 units and which contains two diagonals, which cut the square into four triangles. Start ArcMap with a blank map. From ArcToolbox start the tool "Create Features From Text File" in the toolbox[11] Samples > Data Management > Features. For the input Text File use the file above from Other Data. The decimal separator is a period. Put the output feature class in

___IGIS-Arc_*YourInitials*\TextToFeature

\TextToFeature_DB.mdb\TextToFeature_DS

calling it Four_Triangles_Lines. Make a polygon feature class called Four_Triangles_Polygon by running the Feature to Polygon tool

in Data Management Tools > Features. Input features should be the Four_Triangles_Lines feature class. Make the Output Feature Class Four_Triangles_Polygon in the same location as Four_Triangles_Lines. Click OK. Add both feature classes to the map. Look at the lines feature class attribute table. You can easily calculate that the perimeter of the square as 8.4 (4 x 2.1). What does the lines table show? _____. Look at the polygon feature class attribute table. The area of each triangle is 1.1025. What does the polygon table show? _____.

The Shape_Length of each line and the Shape_Area of each polygon is a calculated value. The fact that they are wrong, however slightly, points up a "problem" for those users who want exact answers: computers cannot be counted upon to give them.

---

[11] You added the Samples toolbox earlier in this chapter, in Exercise 4-3.

Moral: If you tell a computer to ask a simple question of its data, such as "Is A equal to B?," the answer may be "no" even though A and B are meant to be the same and are very close. The computer will report that they are not the same, but the reason for the difference may be that they are calculated in different ways by the computer. A better question would be something like "Is the absolute value[12] of A minus B less than some appropriate very small value (such as 0.0005)."

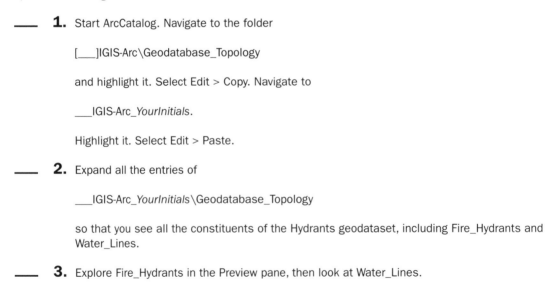

Exercise 4-8 (Project)

Geodatabase Topology

With geodatabases all sorts of conditions of data misbehavior can be tested for. Topology can be constructed between different data sets, *as long as they are in the same personal or file geodatabase feature data set*. (Recall that a geodatabase can contain a feature data set, which can contain feature classes. Those feature classes all have the same extent, projection, and so on. Topological relationships can be formed among those feature classes. A geodatabase can also have free-standing feature classes, which may bear no relationship with each other. Topological relationships may not be formed within or among those feature classes.)

Let's look at a simple example. Recall that the first geodatabase you looked at in this book contained a feature class consisting of fire hydrants in a village. Here you look again at the same data, but it has been converted from a File Geodatabase (Water_Resources.gdb) to the form of a Personal Geodatabase named Water_Resources.mdb. Your first step is to copy the data from this Personal Geodatabase into your personal working area.

_____ **1.** Start ArcCatalog. Navigate to the folder

[____]IGIS-Arc\Geodatabase_Topology

and highlight it. Select Edit > Copy. Navigate to

___IGIS-Arc_*YourInitials*.

Highlight it. Select Edit > Paste.

_____ **2.** Expand all the entries of

___IGIS-Arc_*YourInitials*\Geodatabase_Topology

so that you see all the constituents of the Hydrants geodataset, including Fire_Hydrants and Water_Lines.

_____ **3.** Explore Fire_Hydrants in the Preview pane, then look at Water_Lines.

---

[12]The absolute value of a number is, crudely, the number with the algebraic sign stripped off (so it assumed to be positive). Formally, if "q" is the absolute value of a number "p," then "q is equal to "p" if "p" is positive, and "q" is equal to "negative p" if "p" is negative. (Recall that a negative of a negative number is a positive number.)

The idea is that the Water_Lines (called laterals) are supposed to connect to the Fire_Hydrants. You will use the topology capabilities in geodatabases to see if they do, or if they don't, but are close enough, to move the water lines.

It is rare that two sets of coordinates meant to refer to the same point in space will be identical if the coordinates of the points are created in different ways. (Think back to the previous exercise regarding problems with computer computation.) So, it is wise to ask if two represented points are sufficiently close together to be considered in the same place. You, the user, can indicate what is "sufficiently close" by specifying a "cluster distance" when you develop topology for a feature dataset. You may have noticed this option in several other places, but now instead of ignoring it we will utilize it.

Creating a New Topology

___ **4.** Highlight Hydrants (the feature dataset, not the feature class Fire Hydrants). Select File > New > Topology. Read the New Topology window. Next. Accept the default name Hydrants_Topology. Enter a cluster tolerance of 5.0 Feet. See Figure 4-23. Click Next.

FIGURE 4-23

___ **5.** Put checks in the boxes of Fire_Hydrants and Water_Lines, so they will participate in the topology. Click Next.

Specifying Which Feature Moves When Features Are Adjusted: Rank

The process of generating and validating topology in geodatabases may result in moving some features. You have control over which features move. Each feature class is given a rank. The highest rank is 1; 2 would be a lower rank. When a feature must move to correct a topological error, the feature that moves is the one whose feature class has been assigned the lower rank. Let's presume that the locations of the fire hydrants are highly accurate, but that the locations of the water lines are less well known. Therefore, we will give Fire_Hydrants a higher rank (lower number), so their coordinate representations will not change.

_____ **6.** Leave the number of possible ranks at 5. Change the rank of Water_Lines to 2. Leave the rank of Fire_Hydrants at 1. See Figure 4-24. Click Next.

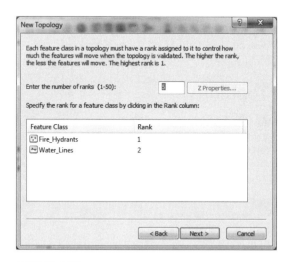

FIGURE 4-24

Topology Rules

Geodatabase topology is based on rules that define the required spatial relationship between two feature classes, or sometimes specify a requirement within one feature class. For example, a rule might be that no two polygons may overlap each other. Currently, there are thirty-some rules defined in the software. Below you will choose a rule that says that the points of one feature class must be covered by (that is, have a point coincident with) a feature in a given line feature class.

_____ **7.** Press Add Rule. In the Add Rule window, select Fire_Hydrants in the Features of feature class drop-down menu. Click the down arrow in the Rule text box to see the rules. There are six rules that relate to point feature classes. List them below:

_____

_____

_____

_____

_____

_____

8. Select the rule Point Must Be Covered By Line, since we want the water lines to go to Fire_Hydrants. The third text field (Feature Class) should indicate Water_Lines. Now give your attention to the right side of the window, where a cartoonish explanation of the rule is displayed. Read what it says. Then, toggle the Show Errors check box to see which points satisfy the rule and which don't. Points that turn red indicate points that don't obey the rule. Click OK.

Validating Topology

9. Since we will only need the one rule, you may press Next after you examine the New Topology window. Read through the Summary and, if correct, click Finish. You get a message that the new topology has been created, and are asked if you want to validate it. Say Yes.

In this case validation means that the computer has looked at every point in the Fire_Hydrants feature class to see if a line from Water_Lines comes to it, crosses it, or comes within 5 feet of it. Those points that do not meet these criteria will be flagged as errors. For those cases where the water line is within 5 feet but does not touch the hydrant, the water line will be moved so that it does touch the hydrant.

10. Once validation is complete, the Hydrants dataset will contain an entry labeled Hydrants_Topology. Click it in the T/C and preview its Geography. You should see two red squares. This means that two fire hydrants were not within the cluster distance of a water line segment.

Let's use ArcMap so we can look at all three layers at once to see what's going on. See Figure 4-25 for a preview of what you will see in ArcMap.

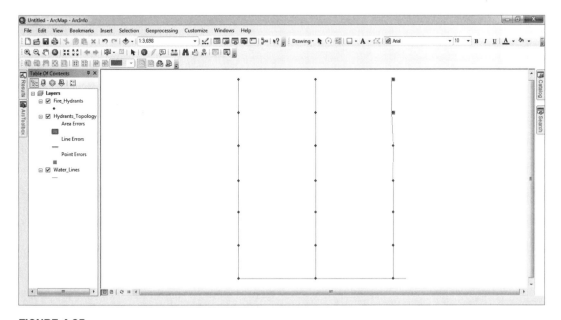

FIGURE 4-25

_____ **11.** Launch ArcMap with a Blank Map. Add Fire_Hydrants to the map. Now add Water_Lines. Observe the map. Finally, add Hydrants_Topology. (You will be asked if you want to add all the feature classes that participate in the topology; since you have already added them, say No.) Making sure that the T/C is in the "list by drawing order" mode, put the Topology layer immediately below the Fire_Hydrants layer. You can see that the two hydrants shown in the red squares are in fact slightly off the east-most lateral. Zoom in and measure the distance (use feet) from the end of the east lateral to the northeast hydrant. See Figure 4-26. What is it? _____ feet. Zoom back to Full Extent.

FIGURE 4-26

_____ **12.** Use the Identify tool to determine which water pipes are not lined up properly. Their OBJECTID numbers are _____ and _____.

_____ **13.** Look further into this matter of misaligned water lines. Add as data WL_copy, which is a copy of the original Water_Lines feature class. Pull it to the bottom of the Table Of Contents. Make its size 4 and its symbol a bright green color. Make the symbol for Water_Lines bright red, size 1.

_____ **14.** Using zooming and panning, look at each of the hydrants on the east lateral, starting with the southernmost one. What you will find is that the original water line diverges away from the hydrants as it goes north. However, the water line that participates in the topology covers the first five hydrants, but not the last two. What has happened here is that the water line came within 5 feet of the first five hydrants, so that, during validation, the lines were snapped over to each hydrant. The other two hydrants were further away than 5 feet, so they were reported as errors, and the water lines were not moved to the hydrants. In fact, moving the lines to five of the hydrants actually caused the lines to diverge from their original path, pushing them further away from the last two hydrants, as indicated by the lack of congruity between Water_Lines and WL_copy.

A Warning: Changes Made through Topology Are Permanent

Please note that the changes made to the Water_Lines feature class during validation cannot be reversed, except by editing. It is important to keep a copy of the original of any feature class involved in a validation as you did with WL_Copy, in case the results are not what you want.

Exercise 4-9 (Review)

Checking, Updating, and Organizing Your Fast Facts File

The Fast Facts File that you are developing should contain references to items in the following checklist. The checklist represents the abilities to use the software you should have upon completing Chapter 4.

- ❏ To make ArcToolbox appear

- ❏ The categories of tools available in ArcToolbox are

- ❏ Two lines may cross in different ways. They are

- ❏ Sometimes you will see two tools that appear to do the same things. That's because they operate on

- ❏ To find a particular tool

- ❏ Help for ArcToolbox can be found in several ways:

- ❏ To label features

- ❏ "Unable to obtain schema lock" means

- ❏ The row identifiers in the attribute tables for polygons in geodatabase feature classes, shapefiles, and coverages are different in these ways:

- ❏ If two numbers are in theory the same the computer may calculate them to be different because

- ❏ To participate in a geodatabase topology the feature classes must be part of a

- ❏ The objects with lower feature class ranks are

- ❏ Topology rules are defined in the software. These rules define

- ❏ Since validation of a topology may move objects, it is a good idea to

281

Geographic and Attribute Data: Selection, Input, and Editing

OVERVIEW

IN WHICH you explore some ideas about collecting and selecting data, including using the Global Positioning System. Also you digitize and edit map data, transform spatial data, and examine combining attribute data with geographic data.

"Garbage in, garbage out."

An often quoted, but seldom heeded, admonition in the computing world.

In the Step-by-Step section of this chapter, you will gain experience working with mostly small datasets. You will examine the "nitty-gritty" of digitizing and manipulating data. However, most GIS projects deal with large amounts of data. Sometimes these datasets are found. Sometimes data are collected from the field. What follows here is advice for getting the right data for the products that might be produced by a medium to large GIS project.

Concerns about Finding and Collecting Data

Datasets form the basis of GIS. These systems are sometimes referred to as being "data driven" to emphasize the importance adequate data plays in their operation. The products of a GIS are the most important contributors to its utility; the data are the chief ingredients of that product. The single message emphasized by this section is that any determination of what data are needed, and what the characteristics of those data should be, comes after a very careful look at what sorts of information products are required for specific decision making. The products, one would hope, are developed to satisfy the needs discussed in Chapter 2 and the requirements for analysis that we will take up in later chapters.

Determining what datasets are needed is not an easy process. It requires the concentrated effort of experts, from the decision makers who will use the ultimately produced information to the scientists who gather and analyze the data. The ideal order of matching data to needs is from the specification of the information product back to the data gathering—that is, in the direction opposite from the one in which the process of producing the information takes place.

The fact that a lot of already collected data exists will influence not only the process of turning those data into information but also the types of information produced. However, to allow the fact that some inappropriate data are at hand to dictate the output of a GIS reminds me of the story of the man who, one evening, lost a ring on the north side of the street but searched for it on the south side because "the light was better."

Looking for Data on the Internet

The fact that suitable spatial datasets are hard to find, and that a major impediment to sharing datasets is simply knowing of their existence, the U.S. government undertook two distinct but related efforts in the 1990s: the Federal Geographic Data Committee (FGDC), set up by an Executive Order (12906) from the President, and the National Spatial Data Infrastructure (NSDI).

The FGDC (www.fgdc.gov) is a governmental interagency committee composed of representatives from the Executive office of the President, Cabinet-level, and independent agencies. The FGDC is developing the NSDI (www.fgdc.gov/nsdi). Cooperating in the effort are state, local, and tribal governments, the academic community, and the private sector. The NSDI encompasses policies, standards, and procedures for organizations to cooperatively produce and share geographic data. The FGDC also sponsors the National Geospatial Data Clearinghouse to provide a mechanism for searches for geographic data.

At the same time, various governmental agencies (federal, state, regional, and local) are developing spatial datasets, which may or may not be accessible, and which may or may not be free to the public. For example, the U.S. Geological Survey (www.usgs.gov) has amassed large and varied datasets, dealing not just with geology, but elevation, hydrography, land cover, and other themes. USGS produced the Digital Raster Graphic (DRG) quadrangles that you saw in Chapter 2. Also, the Census Bureau develops massive spatial datasets.

Commercial datasets are also available—sometimes for considerable amounts of money, sometimes inexpensively, and sometimes for free. The site www.arcgis.com, sponsored by ESRI, is a place to search for datasets. Finding the data you need on the Internet is a matter of, first, casting a wide net, and, second, drilling down to see if you can find data that meets your needs. If you type "Geographic Information Systems" into a search engine you get millions of hits. (You can also get thousands of hits by typing "Geogrpahic Information Systems," which should tell you something about quality and the Internet.)

Steps in Developing the Database

Once a data product has been defined for the spatial area of interest, a few questions should be asked about the data you are looking for in order to ensure you are finding data which will suit your needs.

1. Determine what types of data are needed to produce the product(s).

2. Determine what characteristics of those datasets (accuracy, timeliness, coverage, and so on) are demanded by the product; set priorities.

3. Make some preliminary studies to determine that the data and characteristics specified will produce the information wanted.

4. Begin a data acquisition effort—search for already collected data that meet your specifications, or begin a subarea data collection effort.

5. Put data acquired into proper form (reformat, or encode it) for inclusion in the GIS database.

6. Check accuracy of each step of the collection process; also check the first form of the data against the form in the GIS.

7. Repeat Steps 4, 5, and 6 for the complete set of data for the database.

8. Employ techniques for monitoring and updating of the database.

We now look at these steps one at a time.

1. Determine what types of data are needed for the information products(s).

This assumes that you have determined the needs of the decision-making apparatus and, further, that you can identify the sort of information which will satisfy those needs.

The most important component in this step is to have people involved who understand (a) the information required (and how it is to be ultimately used), (b) the characteristics of data that might be used to generate the information, and (c) the manipulation of the data necessary to produce the information.

At this early stage, it is wise to consider alternative and innovative ways of getting to the same information. For example, if the information required is a delineation of areas of potential high soil erosion, then calculating the potential soil loss might be produced using the universal soil loss formula and data, including rainfall, soil erodibility, slope length and gradient, and vegetative cover. However such data might be produced by interpreting aerial photography for existing erosion conditions. Many projects that get into trouble at this stage do so probably because communications with the analysis and decision-making group ceased after the initial contact. Problems are constantly changing in both importance and type. Thus, to be most responsive, the data support sector of a decision-making process must have its roots not in the data collection area, but rather in the analysis and decision-making area.

2. Once the basic types of data required have been identified, more thought must be given to the characteristics of the data to be acquired.

In determining the characteristics of the data needed, several fundamental questions should be asked.

What geographic area is involved? What geographic identifiers are necessary for use of the data? With what accuracy must the coordinates be known? Are the values of the data "continuous" (like elevations above sea-level) or "discrete" (classifications of land cover)? How frequently do data values change? What causes these changes? Are the most basic data types in use, or can other data be derived from more basic sources? If the latter, what are the advantages and costs in using the most basic information available? What degree of detail is required? When the detail level desired is "multiplied" by the area involved, how large are the datasets? How sensitive to errors in the data is the process that is used to get information products from the data?

3. Make some preliminary studies to determine that the data and characteristics specified will produce the information wanted.

How this task is done depends greatly on what is available. If a GIS exists and new datasets to support a new product are being added, the best course might be to generate some typical subarea of data to try out the process. If the GIS does not exist but is being installed to produce the product, the issue of preliminary testing of the data-product relationship must be approached differently. Perhaps an analysis of your plans could be contracted to a consulting firm for checking. However it is done, someone other than the originators of the techniques for development of information from data should independently examine the projected course of action.

4. Begin the data acquisition effort with a pilot project.

So far, I have described a rather idealized process for the formation of a portion of a database for a GIS. I think the idealized approach is worth sticking to. Too many times the existence of some collected datasets not only dictates the process used to manipulate them but also the kinds of products that get produced. It often turns out that when the cost of conversion of already collected data is counted, the error rate discovered, and the lack of suitability of the data for the task at hand realized, more money and time will have been spent than if an original data collection effort were begun. And yet, it is unreasonable not to make an examination of existing data sources, after you know what you want, to see if, considering the millions spent on data collection in this country, there are some data sets that will meet your needs.

(a) A search for relevant data may not prove easy. Although there are myriad Web sites that allow downloads (paid or free), the questions of appropriateness and quality will keep occurring. It may be the type of frustrating undertaking during which one never really knows when to surrender; how long do you keep looking before you elect another route? A common problem is that very few people seem to have both the depth of understanding required to manipulate data into information in particular areas and the overall view of how to collect the data that could be relevant to the decision-making process.

A real search, therefore, should be undertaken, and it should look widely, not eschewing any possible source of data. Interviewing of individuals in various agencies and companies is probably as profitable as searches through documents on the Web; interviews may produce more up-to-date information about data sources or data collection efforts.

There is no dearth of data, spatial or otherwise, but there are three major problems with existing data:

(1) The data files themselves are spatially distributed, hither and yon, in offices and computing centers, in desk drawers and filing cabinets. Perhaps the first task in developing data for a specific geographic area should be a list that includes data sources, characteristics, and owners.

(2) The data sets are not in a common format. Granted, different sorts of data should be presented in different form because of their inherent characteristics and uses, but the variability vastly exceeds the requirements for different formats.

(3) Already existing datasets are getting older and less correct every day. The accuracy of the data will decay slowly over time.

A method for assessing the correctness of data after a length of time might be borrowed from the concept in physics of a radioactive half-life period. Such a period, measured in time units (from millionths of a second to years to millennia) and different for each radioactive substance, is the length of time required for half the mass of the substance to have decayed into something else. In general form this idea could have a parallel with "correctness of a set of data" as the variable instead of mass of material. The correctness of spatially distributed data of a given type, say, land use, decays at different rates depending, as you might imagine, on its location. A greater rate of decay would be expected adjacent to urban areas than distant from them. In any event, the age of the data used is one of its most important characteristics.

Assuming a set of already collected data has been found that approximately meets the specifications, the data should be carefully analyzed according to the number of characteristics:

❏ Can the data sets be used directly or must they be manipulated before they can be used in analysis?

❏ Are the data the most specific and detailed available? When accuracy is critical, and you have to resort to digitizing maps, can you find a version printed on nonshrinking, nonstretching Mylar?

❏ Is the resolution of the data sufficient to fill information needs?

❏ Are data mapped at an appropriate scale for the resolution required?

❏ In what geographic coordinate system are the data recorded? What complications will occur in converting the data if conversion becomes necessary? Will accuracy or resolution be lost in the process?

❏ What process was used to collect the data? Do statements about the precision and accuracy of the data accompany the dataset? Did the creators of the data sets seem to take the idea of metadata seriously?

❏ Are the data uniform? Is the medium on which they are recorded also uniform and free from the kind of distortion found in some un-rectified aerial photos?

❏ Are the datasets truly available? Who owns them? Are they in the public domain? Can "originals" be obtained, or only copies; what information is lost in the copying process?

❏ What process was used to collect the data? Are they subject to provisions of confidentiality? Are they classified by the military?

❏ How much time will be required to obtain the datasets? How much time must be allowed to reformat or encode them?

❏ Will information updates be available from the same source? If not, will new updates, possibly from different processes, mesh with existing data?

❏ When all considerations are combined, what will the data cost?

(b) If you can't find data sets that meet your specifications—or even if you can—you may embark on a data collection effort. In many ways, if data sets that approximately meet your specifications are available, the issue of whether to use them or collect your own is much like the issue of whether to buy a used or new car. There are advantages, disadvantages, and uncertainties associated with both courses of action. The decision can become very much the classic avoidance-avoidance conflict that college sophomores learn about in psychology courses: the more you look at the other people's data for your requirements, the more you want to collect your own; the more you examine what you have to go through to collect your own data, the more attractive the existing data sets seem.

If you decide to collect data anew, many of the concerns for characteristics still apply, but the question changes, from "Do these data have the properties I want?" to "How do I construct a process to produce the data and characteristics I want?"

Probably the best advice to anyone planning a large data collection effort is to start slowly . . . and carefully. In fact, with all operations involving a data-handling program, one should probably use a "10 percent planning rule." This rule says that if x dollars are to be spent over a period of time, then 10 percent of x dollars should be spent over a previous period on the same subject. For example, if 1 million dollars is to be spent on GIS data development in a year period, $100,000 should have been spent in the months before for planning, analysis, and testing. And, by extension, $10,000 should have been spent before that to determine how to spend the $100,000. The 10 percent planning rule suggests, then, that a small but substantial and representative amount of data be collected, encoded, and validated before the major data collection effort gets under way.

The process of data collection must be carefully planned and executed. It is possible to spend a lot of money at it and wind up with nothing very useful. Among the points to consider are the following:

(1) Some work may well be contracted out. Certainly all understandings with the contracting firm must be written down. As important is that such understandings are completely comprehended by both parties. (It's not hard to mess this up. (Famous example: The Mars Climate Orbiter crashed into the planet instead of cruising around it because NASA and a contractor miscommunicated—using different units (English instead of metric).)

(2) There aren't very many firms that do good intermediate- and high-altitude orthophoto work. Those that exist may be scheduled for months or years in advance.

(3) Rigid timetables for collection of data about the environment cannot be followed. Clouds form, trees get leaves, airplanes malfunction, the ground gets wet. Timetables must be based on probabilities.

(c) Consider datasets that are collected in an ongoing fashion by satellite or aircraft. Datasets showing features in color at resolutions on the order of a meter are now available—available, but not cheap. These datasets primarily depict land cover (from which, in many instances, land use may be determined). The use of data collected in this way has a number of advantages. Among them: (a) the data may be obtained in already digitized form, (b) updating takes place on a periodic basis, and (c) sophisticated computer software is available to manipulate these data.

5. **Put the data acquired into proper form for inclusion in the GIS database.**

Basically, the process of encoding the data means transforming it from the basic form in which it is collected (or acquired if already collected data are used) into the symbolic or graphic form required by the GIS. The process depends on the types of data, the precisions required, the equipment available, the scheme used to represent the data in computer memory (the storage paradigm), and other factors, discussed in Chapter 4.

6. **Check the accuracy of each step of the collecting or reformatting process; also check the first form of the data against the form in the GIS.**

Two elements must be constantly monitored: (a) the process of collecting and reformatting the data, and (b) the quality of the data sets themselves. Perhaps the most important statement that can be made about this "checking" process is that it be accomplished by someone other than the person or group doing the collection.

Such independent checking has many virtues: it provides for a more objective view by those doing the checking; it ensures that the checking activity is a project in itself and not just an adjunct to the data collection effort; and it reduces the temptation to use the same techniques to check the data as are used to develop the data.

There must be more to the checking process than simply ascertaining and reporting error rates. An understanding of why errors occur must be developed. If the encoded value for evaluation at a certain point is 1023 feet and a checker with an altimeter set on the spot finds 999 feet, what happened? Was the problem in measuring altitude? Are the positional coordinates off? Is there some systematic or random error in the encoding process or equipment?

It is vital to understand that all large databases contain errors. If a variable in a database is a continuous quantity, such as elevation, there will be values outside the established accuracy standards. If the base is one of classifications, such as land use activity, some uses will be misclassified. If the base is geographically referenced, there will be disagreements of actual locations between points of the Earth's surface and where the system has them located. The purpose of validating data is to develop an understanding of how great these error rates are and, if they are too great, take steps to reduce them.

7. **Repeat Steps 4, 5, and 6 for the complete set of data for the base.**

Realize, however, that the database will be a growing, evolving entity, supplying useful information to decision makers for years to come. These steps are simply the first ones. By the time the pilot data have been satisfactorily collected or acquired, encoded, and validated, the database developers should have a firm grasp on the associated problems, costs, and techniques. Serious thought should be given to the differences between the pilot project and the major data-gathering effort.

There should be no letup in the testing of data as they come in and encoding. And if you really want to put your data collection techniques on the line, recollect some data from the pilot area and compare it against that originally collected. The results of that may be very instructive, if disheartening.

One of the worst losses at this point can be of key personnel, now much more valuable than when the pilot data collection began. Their importance to the success of the system should be recognized and appropriately compensated, if possible.

8. **Employ techniques for monitoring and updating the database.**

Even as the major database collection effort is going on, the world will be changing, and new problems will be appearing; updating must be a constant activity to keep the database reasonably current.

GPS and GIS

In Chapter 1 you learned a bit about the NAVSTAR Global Positioning System. Let's explore some of the main reasons for making GPS a primary source of data for GIS.

❏ *Availability*—in 1995, the U.S. Department of Defense (DoD) declared NAVSTAR to have "final operational capability." Deciphered, this means that the DoD has committed itself to maintain NAVSTAR's capability for civilians at a level specified by law, for the foreseeable future, at least in times of peace. Therefore, those with GPS receivers may locate their positions anywhere on the Earth.

❏ *Accuracy*—GPS allows the user to know position information easily and with remarkable accuracy. A receiver operating by itself can let you locate yourself within 2 to 4 meters of your true position. (And using two GPS receivers, when one is positioned over a known (accurately surveyed) point, a user can get accuracies of 1 to 3 meters.)[1] At least two factors promote such accuracy:

First, with GPS, we work with primary data sources. Consider two alternatives to using GPS to generate spatial data: the physical digitizer and heads-up digitizing on a computer screen. A digitizer is essentially an electronic drawing table, wherewith an operator traces lines or enters points by "pointing"—with "crosshairs" embedded in a clear plastic "puck"—at features on a map. Or, in heads-up digitizing, the operator manipulates the crosshairs on a computer screen with a mouse.

One could consider that the ground-based portion of a GPS system and a digitizer are analogous: the Earth's surface is the digitizing tablet, and the GPS receiver antenna plays the part of the crosshairs, tracing along, for example, a road. But data generation with GPS takes place by recording the position on the most fundamental entity available: the Earth itself, rather than a map or photograph of a part of the Earth that was derived through a process involving perhaps several transformations.

Second, GPS itself has high inherent accuracy. The precision of a digitizer may be 0.1 millimeters (mm). On a map scale of 1:24,000, this translates into 2.4 meters (m) on the ground. A distance of 2.4 m is comparable to the accuracy one might expect of the properly corrected data from a medium-quality GPS receiver. It would be hard to get this out of the digitizing process. A secondary road on our map might be represented by a line five times as wide as the precision of the digitizer (0.5 mm wide), giving a distance on the ground of 12 m, or about 40 feet.

One larger-scale maps, of course, the precision one might obtain from a digitizer can exceed that obtained from the sort of GPS receiver commonly used to put data into a GIS. On a "200-scale map" (where 1 inch is equivalent to 200 feet on the ground), 0.1 mm would imply a distance of approximately a quarter of a meter, or less than a foot. While this distance is well within the range of GPS capability, the equipment to obtain such accuracy is expensive and is usually used for surveying, rather than for general GIS spatial analysis and mapmaking activities. In summary, if you are willing to pay for it, at the extremes of accuracy, GPS wins over all other methods. Surveyors know that GPS can provide horizontal, real-world accuracies of less than one centimeter.

❑ *Ease of use*—Anyone who can read coordinates and find the corresponding position on a map can use a GPS receiver. A single position so derived is usually accurate within 4 meters or so. Those who want to collect data accurate enough for a GIS must involve themselves in more complex procedures, but the task is no more difficult than many GIS operations.

❑ *GPS data points are inherently three-dimensional*—in addition to providing latitude-longitude (or other "horizontal" information), a GPS receiver may also provide altitude information. In fact, unless it does provide altitude information itself, it must be told its altitude in order to know where it is in the horizontal plane. The accuracy of the third dimension of GPS data is not as great, usually, as the horizontal accuracies. As a rule of thumb, variances in the horizontal accuracy should be multiplied by 1.5 (and perhaps as much as 3.0) to get an estimate of the vertical accuracy.

---

[1]These accuracies pertain to "mapping grade" data collection. By spending more money and much more time one can shift to "survey grade" GPS, with accuracies down to a centimeter.

Anatomy of the Acronym: GPS

Global Anywhere on Earth — Well, almost anywhere, but not (or not as well):

❏ Inside buildings

❏ Underground

❏ In very severe precipitation

❏ Under heavy, wet tree canopy

❏ Around strong radio transmissions

❏ In "urban canyons" among tall buildings

❏ Near powerful radio transmitter antennas

or anywhere else that does not have a direct view of a substantial portion of the sky. The radio waves that GPS satellites transmit have very short lengths—about 20 cm. A wave of this length is good for measuring because it follows a very straight path, unlike its longer cousins, such as AM and FM band radio waves that may bend considerably. Unfortunately, short waves also do not penetrate matter very well, so the transmitter and the receiver must not have much solid matter between them, or the waves are blocked as light waves are easily blocked.

Positioning—Answering brand-new and age-old human questions: Where am I? How fast am I moving and in what direction? What direction should I go to get to some other specific location, and how long would it take at my current speed to get there? *And, most importantly for GIS, where have I been?* To collect GIS data with a GPS, one moves the receiver antenna around areas of the Earth, leaving a tracing of points in the memory of the receiver, which one later transfers to GIS software.

System—A collection of components with connections (links) among them. Components and links have characteristics. GPS might be divided up in the following way.[2]

The Earth

The first major component of GPS is Earth itself: its mass and its surface, and the space immediately above it. The mass of the Earth holds the satellites in orbit. From the point of view of physics, each satellite is trying to fly by the Earth on a horizontal path at 4 kilometers per second. The Earth's gravity pulls on the satellite, so it falls vertically. The trajectory of its horizontal movement and its vertical fall is a track that parallels the curve of the Earth's surface, so it never crashes. All satellites, including our moon, are subject to the same principle.

The surface of the Earth is studded with little "*monuments*"—carefully positioned metal or stone markers—whose coordinates are known quite accurately. These lie in the "numerical *graticule*," which we all agree forms the basis for geographic position. Measurements in the units of the graticule, and based on the positions of the monuments, allow us, through surveying, to determine the position of any object we choose on the surface of the Earth.

---

[2]Officially, the GPS system is divided up into a space segment, a control segment, and a user segment. We will look at it a little differently. One of the many places to see official terminology, at the time this book went to press, is http://tycho.usno.navy.mil/gpsinfo.html#seg.

Earth-Circling Satellites

The United States GPS design calls for a total of at least 24 and up to 32 solar-powered radio transmitters, forming a constellation such that several are "visible" from any point on Earth at any given time. The first one was launched on February 22, 1978. In mid-1994 twenty-four were broadcasting. The minimum "constellation" of 24 includes three "spares." As many as 31 have been up and working at one time.

The GPS satellites are at a "middle altitude" of about 11,000 nautical miles (nm), or roughly 20,400 kilometers (km) or 12,700 statute miles above the Earth's surface. This puts them above the standard orbital height of the erstwhile space shuttle, most other satellites, and the enormous amount of space junk that has accumulated. They are also well above Earth's air, where they are safe from the effects of atmospheric drag. When GPS satellites "die," they are sent to orbits about 600 miles further out, where they will remain virtually forever.

GPS satellites are below the geostationary satellites, usually used for communications and sending TV, telephone, and other signals back to Earth-based fixed antennas. These satellites are 35,763 kilometers (which is 19,299 nautical miles or 22,223 statute miles) above the Earth, where they hang over the equator, relaying signals from and to ground-based stations.

The NAVSTAR satellites are neither polar nor equatorial, but slice the Earth's latitudes at about 55º, executing a single revolution every 12 hours. Further, although each satellite is in a 12-hour orbit, an observer on Earth will see it rise and set about four minutes earlier each day. (For an explanation use your browser to search for solar and sidereal days.) There are four to six satellites in slots in each of six distinct orbital planes (labeled A, B, C, D, E, and F) set 60 degrees apart. The orbits are almost exactly circular.

GPS satellites move at a speed of 3.87 kilometers per second (8,653 miles per hour). Different versions of the satellites have evolved over the years. They weigh 1100 to 2200 kilograms (1 or 2 tons) and have a width of about 11.6 meters (about 38 feet) with the solar panels extended. Those panels generate roughly 1000 watts of power. The radio on board broadcasts with about 40 watts of power. (Compare that with your maximum permitted FM station with 50,000 watts.) The radio frequency used for the civilian GPS signal is called "GPS L1" and is at 1575.42 megahertz (MHz). Each satellite has on board four atomic clocks (either cesium or rubidium) that keep time with 3 billionths of a second or so, allowing users on the ground to determine the current time to within about 40 billionths of a second.

Ground-Based Stations

While the GPS satellites are free from drag by the atmosphere, their tracks are influenced by the gravitational effects of the moon and sun, and by the solar wind. Further, they are crammed with electronics. Thus, both their tracks and their innards require monitoring. This is accomplished by four ground-based stations near the equator, spaced around the world. Each satellite passes over at least one monitoring station twice a day. Information developed by the monitoring station is transmitted back to the satellite, which in turn rebroadcasts it to GPS receivers. Subjects of a satellite's broadcast are the health of the satellite's electronics, how the track of the satellite varies from what is expected, the current *almanac*[3] for all the satellites, and other, more esoteric subjects that need not concern us. Other ground-based stations exist, primarily for uploading information to the satellites. The GPS Master Control Station at Schriever Air Force Base near Colorado Springs, Colorado.

---

[3]An almanac is a description of the predicted positions of heavenly bodies.

Receivers

A GPS receiver usually consists of the following:

- ❏ An antenna (whose position the receiver reports)

- ❏ Electronics to receive the satellite signals

- ❏ A microcomputer to process the data that computes the antenna position and to record position values

- ❏ Controls to allow the user to provide input to the receiver

- ❏ A screen to display information

More elaborate units have computer memory to store position data points and the velocity of the antenna. This information may be uploaded into a personal computer or workstation, and then installed in a GIS software database. Another elaboration on the basic GPS unit is the ability to receive data from and transmit data to other GPS receivers—a technique called "real-time differential GPS" that may be used to considerably increase the accuracy of position finding.

Receiver Manufacturers

In addition to being an engineering marvel and of great benefit to many people concerned with spatial issues as complex as national defense or as mundane as refinding a great fishing spot, GPS is also big business.

Dozens of GPS receiver builders exist—from those who manufacture just the GPS "engine," to those who provide a complete unit for the end user. Prices range from somewhat under $100 USD to several thousands.

The U.S. Department of Defense

The U.S. DoD is charged by law with developing and maintaining NAVSTAR. It was, at first, secret. Five years elapsed from the first satellite launch in 1978 until news of GPS came out in 1983. In the approximately three decades since—despite the fact that parts of the system remain highly classified—citizens have been cashing in on "The Next Utility."

There is little question that the design of GPS would have been different had it been a civilian system from the ground up—but then, GPS might not have been developed at all. Many issues must be resolved in the coming years. A Presidential Directive issued in March of 1996 designated the U.S. Department of Transportation as the lead civilian agency to work with DoD so that nonmilitary uses can bloom. DoD is learning to play nicely with the civilian world. The use of GPS for automobile navigation, location-based services, cell phone location, fleet management, emergency services, and so on has been little short of astounding. And many devices with GPS devices report their positions. The great advantage of GPS in daily use is you know where you are. The scary part is "they" know where you are (and where you have been) as well. George Orwell's 1949 novel *Nineteen Eighty-Four* begins with:

> *"There was of course no way of knowing whether you were being watched at any given time. How often, or on what system, the Thought Police plugged in on any individual wire was guesswork. It was even conceivable that they watched everybody all the time. But at any rate they could plug in your wire whenever they wanted to. You had to live—did live, from habit that became instinct—in the assumption that every sound you made was overheard, and, except in darkness, every movement scrutinized"*

Anyone who isn't worried about what the combination of GPS, warrantless wiretapping, and the capacity of computers to process mammoth amounts of data can do to society and to individual freedoms is naïve. Orwell's only error may well have been that he was off by 30 or 40 years.

Users

Finally, of course, the most important component of the system is *you*. A large and quickly growing population, users come with a wide variety of needs, applications, and ideas. From tracking ice floes near Alaska to digitizing highways in Ohio. From rescuing sailors to pinpointing toxic dump sites. From urban planning to forest management. From improving crop yields to laying pipelines.

What Time Is It?

Although this is a text on GIS—and hence primarily concerned with positional issues—it would not be complete without mentioning what may, for the average person, be the most important facet of GPS: providing human beings with a universal, amazingly precise, and accurate time source. In fact, GPS probably should be called a GPTS (Global Positioning and Timing System). Allowing any person or piece of equipment to know the exact time has tremendous implications for things we depend on every day (like getting information across the Internet, like synchronizing the electric power grid and the telephone network). Further, human knowledge is enhanced by research projects that depend on knowing the exact time in different parts of the world. For example, it is now possible to track seismic waves created by earthquakes, on one side of the Earth, through its center, to the other side since the *exact* time[4] may be known worldwide.[5]

---

[4]Well, okay, there is no such thing as "exact" time. Time is continuous stuff like position and speed and water, not discrete stuff like people, eggs, and integers, so when we say "exact" here we mean within a variation of a few billionths of a second—a few nanoseconds.
[5]The baseball catcher Yogi Berra was once asked "Hey Yogi what time is it?" to which Berra is said to have replied, "You mean right now?" Yes, Yogi, RIGHT NOW!

Geographic and Attribute Data: Selection, Input, and Editing

—— Open your Fast Facts File.

—— Open the Color Figures file.

You will want ArcCatalog to display file extensions and to indicate data set projections. Relevant information on setting this up should be in your Fast Facts File—recorded from the early steps in Chapter 1. Also, for exercises in this chapter you will need 3D Analyst and Spatial Analyst.

Exercise 5-1 (Warm-Up)

Looking at Areal Representations of the Real World

The land of the Earth is partitioned into areas. Nature started this. First take a look at a Globe view of the planet.

—— **1.** Start ArcCatalog. Hide ArcToolbox if it is present. Disconnect from all folders except for IGIS-Arc, your personal workspace folder, and C:\. Click Folder Connections in the Catalog Tree. Choose View > Refresh. You are going to be looking for Esri ArcGlobeData. The first place to look might be

   C:\Program Files\ArcGIS\Desktop10.x\ArcGlobeData.

—— **2.** Expand ArcGlobeData. Click wsiearth.tif. Click the Preview tab. The geography preview doesn't give you any hint that the world is spherical, so click Globe View in the Preview drop-down menu. If necessary, click once on the black background that appears. You get a version of the Earth that can only be described as neat, especially when you:

   ❏ Rotate it by dragging the mouse cursor, and

   ❏ Zoom by holding the right mouse button down and dragging back and forth.[1]

---

[1]If you have a "wheel mouse," you can zoom with the wheel.

___ **3.** Find a view in which the Earth appears to be almost all water. Reflect on how much of earth is covered by water.[2] Now find the continent of Europe.

It's pretty hard to understand, from visual inspection, if the word "continent" has any physical meaning at all, just how Europe and Asia were ever considered separate entities. This may be the first major indication that human beings divide land up in strange ways. But read on. It does get better.

___ **4.** Click country.shp, which you will find in

[___]IGIS-Arc\Other_Data\Countries

And wait. And, depending on the speed of your computer, wait some more. If you get tired of waiting, click the map. When a black screen finally appears, click it once. Look around.

Here you should see the divisions human have recently imposed on the land masses. They have divided up planet Earth into areas for reference and jurisdiction (a generous way of describing the results of, mainly, wars and political strife over millennia). Now, perhaps more than ever, it takes documents and time to cross many of these division lines.

___ **5.** Click countries.lyr. Look at the Globe View of this, then shift to Geography in the Preview drop-down menu. Examining the coordinates in the lower right of the window, what would you say the coordinate system of countries.lyr is? _____. Zoom in on your country. Start the identify tool and click the country to bring up the Identify Results window. Set it so that All Layers will be identified. Determine the populations of your country and two or three around it, according to the data presented. Look at some other countries. What is the monetary currency type used in Mexico? _____. Is Bolivia (latitude 18° south, longitude 65° west) landlocked according to Identify? _____. How many square miles are there in the United States? _____. What population is listed for Antarctica? _____.

___ **6.** Zoom back to the full extent. Note the distortion in areas well south of the equator. For instance, in most of South America the distance covered by a degree of longitude is much less than a degree of latitude. The most dramatic example is Antarctica, where a single point (the South Pole) becomes a line several times the width of the United States, which itself looks pretty squashed.

___ **7.** Look at the table for countries.lyr. How many countries are there, according to this table? _____.

We begin now to look at more rational ways of dividing up the landscape.

Looking at Reference Systems

___ **8.** In the Catalog Tree contract the ArcGlobeData entry. Expand Reference Systems. Here are 10 plus ways of dividing up the planet's surface that are scientifically, if somewhat randomly,

---

[2]Bill Bryson, in *A Short History of Nearly Everything*, comments on how fortunate it is that the earth is bumpy, which allows for land masses. He says that otherwise there might be life, but there wouldn't be baseball.

devised. Click usgs250q.shp. These are polygons that represent the United States Geological Survey 1-to-250,000-scale quadrangle maps. Zoom in on one quadrangle. Using the coordinate values shown in the lower-right corner of the window determine how many degrees it is from west to east? _____. How many degrees from south to north? _____. Zoom back to the full extent. Pick a quadrangle at random with the Identify tool, and look at the attributes. Why do you think there are four possible state names? _____
_____.

____ **9.** Look at the table of usgs250q.shp. Select the QUAD_NAME column. Click Table Options > Find. Find Boston. Parts of what states would you think would appear on that quadrangle, based on the ST_NAMEn columns? _____, _____, _____,
_____. Select the ST_NAME1 column. How many quads are contained completely by Washington State? _____. Just for a diversion, click Globe View. Spin it around a little. Also notice that a quadrangle in the northern United States covers less land area than one in the southern United States, since the "rectangles" are defined in terms of meridians and parallels. Now bring back the Geography.

____ **10.** Look at the USGS 1-to-100,000-scale map divisions. What is the name of this feature class? _____. Zoom in on one quadrangle. How many degrees is it from west to east? _____. How many degrees from south to north? _____. Look at the 1 to 24,000-scale quadrangles. Zoom in on one quadrangle. How many degrees is it from west to east? _____. How many degrees from south to north? _____. Go to full extent, then, if you are familiar with the geography of the United States, zoom up on the southernmost tip of Lake Michigan and determine the name of the quadrangle there. _____. What is the state? _____.

____ **11.** Examine georef15, which divides the Earth up in to 15-degree by 15-degree "squares." Also look at georef1, whose table, together with identifiers from georef15, identifies every one degree by one degree "square" on the planet. How many such "squares" are there? _____.

____ **12.** Click "World Time Zones.shp" and display its geography. Use Identify. What does the ZONE attribute refer to? _____. How many square kilometers are in the zone that contains Greenwich, England?[3] _____. From the table: how many time zones are there? _____. How many would there be if time zones were defined simply by meridians that were one hour apart? _____. For extra credit: What place relates to a time zone that is 5:45 later than GMT? _____. (Hints: find the polygon, zoom in, go back to Countries, or use the Internet.)

____ **13.** FYI: JOG stands for Jet Operation Graphic. ONC means Operational Navigation Chart. Take a quick look at each.

The two major reference systems that are left in this list are the United States State Plane coordinate system of 1983 (ustpln83.shp) and the Universal Transverse Mercator coordinate system (utm.shp). UTM is a very regular, almost worldwide system.[4] The State Plane system is a set of one or more zones defined by each state in the United States, and not intended to be used outside that state. You saw some examples of data sets in these reference systems earlier.

_____

[3]UTC (Universal Time Coordinated), previously designated GMT (Greenwich Mean Time), is the basis for all time zones, so the difference between UTC and time in the city of Greenwich is zero.

[4]The areas around the poles are represented with different projections.

Looking at Coordinate Systems

There is not much to see in the coordinate systems references, in terms of geography. These are the files that store the formulae for converting from one coordinate system to another. What is impressive is how many different ways people have referenced positions on the Earth.

——— **14.** Collapse Reference Systems. If you are using ArcGIS version 10.0 Expand Coordinate Systems > Geographic Coordinate Systems. If you are using ArcGIS 10.1 go to

[___] IGIS-Arc\Other_Data

and expand Coordinate Systems\Geographic Coordinate Systems.

——— **15.** Expand Europe and look at the Catalog Tree. The point here is that many countries and some cities each have (or had) their own coordinate systems—just as a century ago every locale had its own time system. And these are latitude and longitude systems; consider that each of these lat/lon systems may be projected on to the Cartesian plane in dozens of different ways, producing hundreds of different ways of assigning a pair of numbers of a given point on Earth's surface. If you double-click on a coordinate system you can view its Properties.

——— **16.** Collapse the Europe folder (and the subsequent systems after you explore one or two to get an idea of what is included). Check out North America, where there are fewer geographic coordinate systems, but still a lot. If you right-click on a coordinate system you can view its properties. Check out the planets under Solar System. Based on year 2000 data, what is the diameter of Venus? _____ meters.[5]

——— **17.** Check out World. At the bottom of the list is WGS 1984. This is the most current estimation of where the latitude and longitude graticule falls on the Earth's surface. What is the diameter of Earth? _____ kilometers.

——— **18.** Collapse Geographic Coordinate Systems and expand Projected Coordinate Systems. Look at World and you will see names of projections that may be familiar from some past geography class. Expand the World (Sphere-based) folder, and you will see a lot more. The number of approaches taken to representing the quasi-ellipsoidal Earth on a flat plane is mind numbing. Collapse World and World (sphere-based) folders.

——— **19.** You see a folder for UTM)Universal Transverse Mercator), with which you have some familiarity. Open the WGS 1984\Northern Hemisphere folder. Widen the Catalog Tree as necessary to see all the text. There you will find some complex UTM Zones (which you may ignore) and zones numbered from 1 to 60: 1N (meaning zone 1, northern hemisphere), 2N, . . . 60N.[6] To repeat the warning issued earlier: in the United States, the UTM system based on the World Geodetic System of 1984 is just different enough, in terms of geographic space (up to a few hundred meters), from UTM based on the North American Datum of 1927 to cause major problems for those trying to determine accurate locations. Contract the UTM folder.

---

[5]The Semi-major Axis is equivalent to the radius at the equator.
[6]UTM zones customarily have a letter suffix, such as P in 18P. These letter designations are not necessary for most applications; they are a latitude reference, but the northing component of the coordinate system takes care of that. The N and S suffix in the Esri designations are not those, but rather reference the northern or southern hemisphere.

___ **20.** You also see there a folder for State Plane. Expand it. Even if you restrict yourself to NAD 1983 (almost exactly equivalent to WGS 1984 in the United States), you will notice the units of measurement in the state plane system are a hodgepodge. Meters are available for all states (under NAD 1983), and some states use meters as the primary survey unit. However, other states use feet (called, variously and equivalently, feet, survey feet, or US feet), and still others use international feet.[7] Further, it is usual for multiple zones to represent each state. How many zones is Michigan divided into? (Hint: Michigan uses international feet.) _____. What are the FIPS (Federal Information Processing Standard) numbers of the zones? _____. Collapse the State Plane folder.

Using the Reference System to Discover the Boundary Coordinates of a State Plane Zone

We can take advantage of what we saw earlier in "Looking at Reference Systems" to set the extent of a geodatabase. We begin by setting the data frame to the correct coordinate system.

___ **21.** Using ArcCatalog, make a new folder under

___IGIS-Arc_*YourInitials*

Rename the new folder Digitize&Transform. Make a Folder Connection to that folder.

___ **22.** Start ArcMap with a new, empty map. Depending on whether you are using ArcGIS version 10.0 or 10.1, add data from:

C:\Program Files\ArcGIS\Desktop10.0
\Reference Systems\usstpln83.shp

or

C:\Program Files(x86)\ArcGIS\Desktop10.1
\Reference Systems\usstpln83.shp

You will see a somewhat squashed coterminous U.S. map and a really large Alaska. Bring up the Data Frame Properties window (View > Data Frame Properties; or double-click the word Layers in the T/C; or right-click the data frame and choose Data Frame Properties). If necessary, activate the General tab. What do you find for display units? _____. Activate the Coordinate System tab. What is the coordinate system? _____.

___ **23.** Under the Coordinate System tab, navigate: Projected coordinate Systems > State Plane > NAD 1983 (US Feet) > NAD 1983 State Plane Kentucky North FIPS 1601 (US Feet). Highlight it. Under the General tab, the map units should read Feet and you may need to set the display units to Feet. Click Apply, then OK.

---

[7]The meter is the fundamental unit of length measurement in the world. Both the survey foot and the international foot are based on it. The survey foot is defined by 1 meter = 39.37 inches exactly. The international foot is defined by .0254 meters equals exactly 1 inch. These two definitions differ by about 2 parts in 100,000. So, for example, in a distance of 100 miles there would be about an 11 foot difference.

A map appears—distorted in a different way from the previous one. The reason it looks strange is that the latitude and longitude coordinates that are natural to this dataset have been converted "on the fly" (meaning instantaneously, for display purposes) to the Kentucky state plane coordinates. Those coordinates work reasonably well in the middle of America, but not elsewhere, particularly, as you note, in Alaska. In the next step you will find the Kentucky north zone, isolate it, and export it as a shapefile all by itself.

___ **24.** Zoom in on Kentucky (KY), using Identify if necessary. Zoom in on its north zone. Use the Identify tool to make sure you are in the correct place—the ZONENAME83 should be in KY_N. What is its FIPSZONE83 number? _____. To the nearest tenth, how many square miles are in the zone? _____.

___ **25.** From the Tools toolbar, pick Select Features by Rectangle. Click the north zone. It should now have a thick, colored border. Right-click usstpln83 in the T/C. Click Data > Export Data. In the Export Data window, make sure that you are about to export Selected Features. Press the radio button to select "Use the same coordinate system as the data frame" so that the north zone will be changed from geographic coordinates to state plane coordinates. For output, browse to

___IGIS-Arc_*YourInitials*\Digitize&Transform

and make the name KY_N_sp83.shp. Make sure your "Save as type" is set to shapefile. Click Save, then, after checking that it looks like Figure 5-1, OK the Export Data window. When asked if you want to add the data to the map, select Yes. Remove the layer usstpln83.

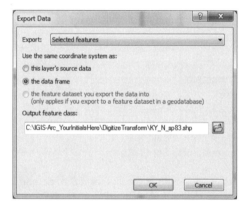

FIGURE 5-1

___ **26.** Using the Select Elements tool as a cursor, slide the cursor around the map. Verify that the map display is in feet – if not, change it. Imagine a rectangle around the entire north zone. Consider the north-south line that is west of the entire zone, and whose easting is a nice round number like 1110000. Write it down below, along KY_N_with three other boundaries (again with round numbers):

Western boundary: _____

Eastern boundary: _____

Northern boundary: _____

Southern boundary: _____

You will use these numbers later when you make a personal geodatabase feature dataset that encompasses the Kentucky north zone.

___ **27.** Dismiss ArcMap but save the map as Temp.mxd in

___IGIS-Arc_*YourInitials*\Digitize&Transform

In ArcCatalog click on Folder connections and press F5 to collapse the Catalog Tree, then dismiss ArcCatalog.

Primary Lesson

If you take only one point away from this exercise, it should be that there are a multitude of ways to reference points on the surface of the Earth, and that if you combine data sets, either (a) they must all agree in all parameters or (b) you must know exactly what you are doing. For the case of (b), in the absence of having access to a geodesist, we recommend two resources to start with. First are the texts *Understanding Map Projections* by Esri and *Lining Up Data* by Margaret M. Maher also published by Esri. The second resource is to navigate to the Esri website, www.esri.com. Then select the support page and type 21327 in the Search Support box. You should see an article entitled "(How To:) Select correct datum transformation when projecting between datums." Reference material, labeled "Related Information," will be found on this Web page.

Exercise 5-2 (Project)

Look at Geographic Data on the web

Starting with some of the Web address in the Overview of this chapter, locate five spatial datasets both in your geographic area and on a topic interesting to you. Make a list of those sources you find.

_____

_____

_____

_____

_____

Exercise 5-3 (Project)

Digitizing and Transforming

"Digitizing" is a process in which a graphic representation (a drawing, a map) is turned into numbers (digits) and characters that give the computer some of the information that is "in" the drawing. Generally, that information makes it possible for the computer to reproduce the drawing on its screen or on paper and to answer questions about it. We, of course, are most interested in digitizing maps.

You have had experience with one method of digitizing: creating a geodatabase feature class by typing in the "digits" of the coordinates of TextToFeature project for the Foozit Court for the elementary school. Doubtless you have had all of that sort of data entry you want. Sometimes, however, it is necessary when the absolute coordinates of some important feature are known precisely.

Another way of digitizing, requiring much less human effort, at least initially, is *scanning*, where a map is placed on a rotating drum, or laid flat on a table, and through a combination of electronics and mechanics, the map's contents are finely gridded into little squares. The pixels that result are stored in a computer file. At this point, all the user has is a picture. However, with some direction from the user, sophisticated computer programs can then interpret the picture according to rules and can create an intelligent GIS map. For example, the program could recognize a contour line—perhaps even to the extent of reading its height off the map and placing it as an attribute in a table. High-precision scanners are expensive, and the work to make programs that interpret map image files is ongoing.

A third technique of digitizing is to mount the paper map[8] on an electronic drawing board, called (surprise) a digitizer, and trace over the lines of the map with a stylus or puck; this process transmits x-y coordinates to the computer. The user can add attributes to the lines or areas that are digitized.

A Plan for Digitizing and Transforming

Here we will look at a fourth way of getting graphic information into the computer; it is called "heads-up" digitizing—so named because you look up at the monitor rather than down at a paper map, as in traditional digitizing. In heads-up digitizing, an electronic image of a paper map is made by *scanning*. In this case the map is the Foozit Court. You will be able to see this image on the screen of your computer. By tracing around the lines of the map with crosshairs (a cursor or pointer, controlled by the mouse), you will be able to automatically supply coordinates to ArcMap.

This process is a little like learning to ride a bicycle: not really hard, but difficult to describe in all its detail (e.g., if the bicycle begins to tip over to the left, turn the handlebars to the left—just a little now—and . . .), so I will simply provide general directions and let you figure out how to digitize mostly on your own.

A caveat: In the process of scanning and then displaying a map systematic error may creep in—since scanners, and particularly computer screens do not maintain exactly the same aspect ratio(height to width) as the paper or Mylar map. It's something to think about, measure, and correct on the final product if necessary.

The product of your digitizing will be an ArcGIS *shapefile*. As you know, since we have discussed the nature and characteristics of these elements in the last chapter, a line shapefile is made up of *beginning points*, *ending points*, and *vertices*, which define *polylines*.

---

[8]Or Mylar map—Mylar is used because it is a "stable" medium, unlike paper, which stretches and shrinks with environmental factors like humidity.

Our ultimate goal is to produce a personal geodatabase polygon feature class and have it appear at a particular location on a real-world map. For purposes of illustration, you will do this in two different ways. Here is an overview of Exercise 5-3:

_____ **1.** Make a blank shapefile with ArcCatalog.

_____ **2.** Add the image of a scanned "map" as data in ArcMap.

_____ **3.** Add the blank shapefile as data in ArcMap.

_____ **4.** Use the Editor to digitize (trace over) the lines of the map.

_____ **5.** Save the shapefile with the digitized lines in non-real-world coordinates.

The features in the shapefile will then be converted to a geodatabase feature class, with real-world coordinates. The steps for this will be as follows:

_____ **6.** Make a blank personal geodatabase feature class (PGDBFC) in the proper projection and with the proper extent.

_____ **7.** Convert the shapefile to a geodatabase feature class using ArcToolbox.

_____ **8.** Move the Foozit_Court feature class into the real world with the Spatial Adjustment tool.

Getting Started

_____ **1.** Make sure your Fast Facts File is open; you'll need to enter information into it. Your FFF should contain at least the name of each ArcGIS command, module, feature, tool, or wizard that you use.

_____ **2.** With ArcCatalog running, highlight your Digitize&Transform folder in the Catalog Tree. Choose File > New > Shapefile to bring up the Create New Shapefile window. Create a shapefile named Dig_Lines_shape. Make the feature type Polyline. Click the Edit button. Choose Projected Coordinate Systems > State Plane > NAD 1983 (US Feet) > NAD 1983 StatePlane Kentucky North FIPS 1601 (US Feet), then OK. In the Create New Shapefile window, click the Show Details box and read over the properties of the coordinate system you have selected. Click OK. Dig_Lines_shape.shp will appear in the Contents area of ArcCatalog.

Loading an Image File as a Layer in ArcMap

_____ **3.** Launch ArcMap using a blank map. Dismiss ArcCatalog.[9] From

[____]IGIS-Arc\Image_Data

_____

[9]Throughout this project you will be asked to close ArcMap and ArcCatalog and then reopen them. This may not be necessary. The problem that sometimes occurs is that you are "locked out" of performing operations on a feature class by "program A" if "program A" believes (rightly or wrongly) that "program B" is using that feature class. The solution that always works is to close "program B," but other solutions may exist. Symptoms of the problem are messages like "cannot acquire schema lock" or simply grayed-out fields or buttons.

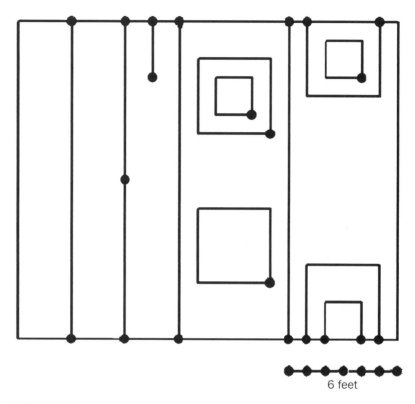

FIGURE 5-2

add the image file Foozit_Court.TIF as a layer in ArcMap. If the Unknown Spatial Reference message appears, ignore it. If asked do not build "pyramids." An image of the court should appear. See Figure 5-2. No reference numbers are provided, because, unlike when you typed in the values, the node positions and line lengths will be determined by where you click when you trace the image with the cursor.

____ **4.** Make the ArcMap window occupy the full screen. Zoom in so the line drawing occupies the full map display area. Slide the cursor over the map area and notice that the coordinates are (probably) in Decimal Degrees. This is because, if the coordinate numbers are small, ArcMap assumes a geographic "projection." Obviously, you want to fix this: With the Select Elements pointer, right-click the map, pick Data Frame Properties > General, and change the map and display units to Feet. OK. Use Measure, with its units set to Feet, to determine the dimensions of the court to two decimal places. The result: _____ by _____. Dismiss Measure.

____ **5.** Click the pointer (Select Elements) icon. Slide the cursor around the image, noticing the coordinates. Zoom in on the southwest corner. By placing the + part of the cursor on the corner, write the coordinates in the table that follows with precision of two decimal places. Zoom back. Find the coordinates for the northwest corner. Repeat this procedure for each of the other two corners of the overall rectangle; write the coordinates of each corner of the court to two decimal places:

Corner	X-Coordinate	Y-Coordinate
Southwest		
Northwest		
Northeast		
Southeast		

These local coordinates will be used later to provide the reference locations you will need to move the court to real-world locations. Zoom back to the full extent.

Loading the New, Blank Shapefile into ArcMap

___ **6.** Press Add Data. Navigate to the empty shapefile:

   ___IGIS-Arc_*YourInitials*\Digitize&Transform
   \Dig_Lines_shape.shp

   and add it as a layer.

Adding Line Features to a Shapefile by Using the Editing Facility in ArcMap

ArcGIS Desktop contains a rather extensive editing facility. To use it, you must add the Editor toolbar to ArcMap.

___ **7.** Use Customize > Toolbars to get the Editor toolbar onto the ArcMap window if it isn't there already. Park it where you want it—but be sure it is horizontal. It consists of two main portions: (A) tools along the toolbar itself, and (B) a drop-down menu which contains commands to the Editor or options for the user. The section (b) becomes quite extensive. For example, Snapping leads to (1) a snapping toolbar and Options, which leads to a Snapping Options window which leads to a Text symbol window which leads to a Symbol Selector window and a Style references window which leads to a Create New Style window – I could go on but I'm sure you will be happier if I don't. The point is that the editing facility is quite extensive and its abilities are found in a hierarchical tree. Staying with just the top levels of "A" above, list in your Fast Facts File the menu items you see on the drop-down menu.

   (A) Done? Yes ___ No ___

___ **8.** Click Editor > Start Editing. Ignore the warning about different coordinate systems. Continue. List in your Fast Facts file the tools that are revealed when you when you mouse over the icons on the Editor toolbar.

   (B) Done? Yes ___ No ___

A Create Features window may appear on the right side of your ArcMap window. If it does, dismiss it. Then restore it by clicking on the Editor drop-down window and select Editing Windows > Create Features. Dismiss the window again. Restore it with the Create Features icon on the Editor toolbar. The window is divided into two sections. The top section contains a list of layers available to be edited. You should only see Dig_Lines_shape (twice, actually). Click the name. The bottom section then lists different input methods, called Construction Tools. Depending on which type of feature you are editing, you will see different options. What construction Tools appear for editing Dig_Lines_shape?

_____, _____, _____, _____, _____.

Select Line. (If the Create Feature window covers up part of the image click Full Extent, followed by the Go Back To Previous Extent arrow.) (If you don't have the correct Construction Tools, you should go back to Step 2 and recreate the empty shapefile, paying close attention to the file type.)

___ 9. In the Editor toolbar drop-down menu, click Options. Select the General tab. Display measurement using two (2) decimal places. Click Apply, then OK.

___ 10. In the Editor toolbar drop-down menu, click Snapping > Options. Set or confirm the snapping tolerance set to 10 pixels. Make sure Show Tips is checked as well as Layer Name and Snap Type. Click OK.

Snapping is a process that helps you connect parts of features (e.g., ends of lines) that need to be connected. The word probably comes from the positive effect that occurs when the two parts of a snap on a piece of clothing are pressed together correctly. In the case of putting the ends of lines together, it means that when the cursor that is making a line comes within a preset distance of the end of another line, it moves over exactly to the end of that line, allowing no gap or overlap. That preset distance is specified as the snapping tolerance. You have set it to 10 pixels because that will provide for ease of editing lines but will prevent ends being snapped together that shouldn't be.

___ 11. In the Editor menu, click Snapping > Snapping Toolbar. Find the Snapping toolbar on the screen. Using ToolTips, determine the effects of the four icons on the toolbar: _____, _____, _____, _____ Activate the End Snapping mode, which will cause the *ends* of the polylines of Dig_Lines_shape to snap together when you are editing. Also, click the Snapping Toolbar's Snapping drop-down menu and activate (if it isn't already) Snap To Sketch.[10] This will ensure that, as you draw a "sketch" consisting of lines, you can snap the end of a polyline back to the beginning of that line, as when you have a single polyline around a polygon. There is no Apply or OK here – just click a blank area in the T/C to lose the menu. You can double check your settings by clicking the Snapping drop-down menu and making sure "Snap To Sketch" has a check mark beside it. See Figure 5-3. (In the event that red or green squares appear on the map it means that digitizing has started prematurely. To erase the effects right-click on the Data Frame and choose Delete Sketch.)

___ 12. Move the cursor into the map window—it should appear as crosshairs. (If not, inspect the Create Features window and make sure that Dig_Lines_shape is activated and Line is selected under Construction Tools.) With the crosshairs cursor, you can begin to create polylines by

---

[10]The term "Edit sketch" is a noun phrase, not an adjective followed by a noun, as you might suspect. An "Edit sketch" is a graphic entity that you will work with.

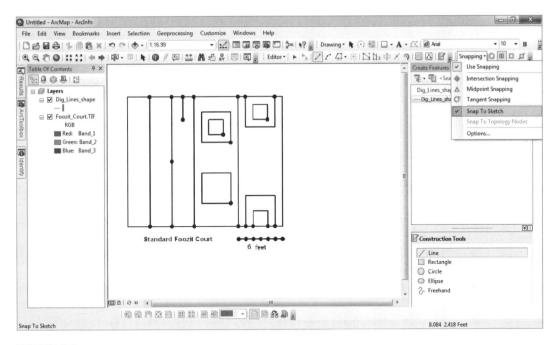

FIGURE 5-3

FIGURE 5-4

digitizing. First you will form the three of the segments of the leftmost rectangle. Put the cursor in the center of the black disk (as shown in Figure 5-4). Now click once and you will create a starting vertex in Dig_Lines_shape. Move the cursor to the left to the place where the line changes direction (at the corner of the court), and click again to create a second vertex. Create the next vertex at the northwest corner of the court. Create the ending vertex (at the northwest dot), again by clicking once, then pressing F2 to complete the sketch of the polyline. Notice the bright cyan line that indicates the polyline that you have digitized.

_____ **13.** Begin the second polyline at the southern endpoint of the first polyline. Notice when you move the cursor near the already existing endpoint, the cursor displays the four-square grid, indicating it will snap to an endpoint. This is the effect of snapping. Notice also the Snap Tip. What does it say? _____. Since you set the snapping tolerance to 10 pixels, this effect occurs when you get within 10 screen pixels of an existing endpoint. When you click you get a little red circle at the endpoint and you may begin the new polyline. Move the cursor to the northern end of the line. Instead of using a single click followed by F2 to end the sketch, use a double-click on the northern point.

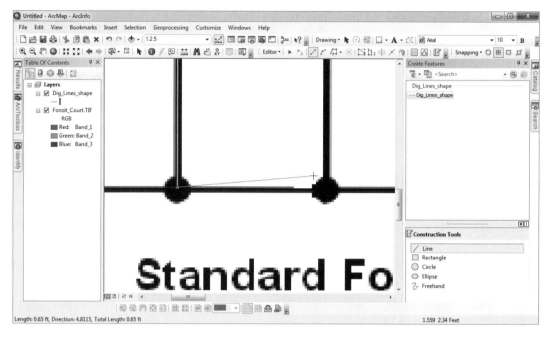

FIGURE 5-5

___ **14.** Zoom in on the portion of the drawing shown in Figure 5-5. Click the Line Construction Tool in the Create Features window again. You see that you can be as precise as you like in beginning a line on a disk on the graphic image. However, you are stuck (for the moment) with the locations of the endpoints you have already digitized. Digitize the line in the figure. Return to the previous zoom extent.

If you wanted to be really precise, you could zoom in on every vertex. Don't bother on this project, since you are simply learning the digitizing process, so accuracy is not a major consideration.

___ **15.** You may have trouble distinguishing the lines of the shapefile from the lines of the image. So change the symbology of the shapefile polylines to a red line with width of 2. Now you can clearly see where you have digitized the elements.

___ **16.** Continue to make polylines until you have created each polyline that corresponds to a line in Foozit_Court.TIF. Be careful to make polylines **between** the black disks shown on the image. Where there are no black disks but the line changes direction, make vertices. Do not ever make a digitized line go through a black disk!

In case of "whoops": If you start to make a line where there shouldn't be one, you can press Ctrl-Delete to do away with it. If you complete making a line where there shouldn't be one, you can click the Edit tool (next to the Editor drop-down menu), select the offending line by clicking it, and press the Delete key. It wouldn't hurt to try both of these procedures out. And write them down in your Fast Facts File.

The order in which you digitize the polylines is unimportant. The direction in which you digitize the polylines is unimportant. It's a good idea to save edits (in the Editor drop-down menu) along the way.

___ **17.** Check carefully that all the black lines of the image have red polylines over them. Turn off Foozit_Court.TIF. Check again. Everything there? If not, turn Foozit_Court.TIF on again and continue digitizing. When finished, click Editor. Click Save Edits. Click Editor again. Click Stop Editing.

___ **18.** Close ArcMap without saving the map. The empty shapefile, Dig_Lines_shape, that you started now contains the polylines you digitized, because you saved the edits. Its coordinates are not real-world coordinates, but you will be able to move it, intact, to where you want it. Further you will be able to rotate it and change its size.

___ **19.** Start ArcCatalog (the complete version—not just the sidebar). Navigate to

___IGIS-Arc_*YourInitials*\Digitize&Transform
\Dig_Lines_shape.shp

and click it. Preview it (it might be kind of ugly—jagged lines may occur because of the screen resolution—these will be much reduced if the Foozit Court is plotted or printed out). Again highlight the name. Press Ctrl-C to copy its contents onto the ArcGIS clipboard. Click Digitize&Transform. Press Ctrl-V to paste Dig_Lines_shape.shp into the folder. The name will appear as Dig_Lines_shapeCopy.shp. You do this in case something goes wrong in the upcoming transformation to real-world coordinates and you need to start again. Backups are almost always a good idea.

Converting a Shapefile to a Geodatabse Feature Class and Giving It Real-World Coordinates

The first step in getting the contents of the Foozit court shapefile into a geodatabase feature class is to make the personal geodatabase, and within it, a feature data set that will contain the spatial parameters of datum, projection, X/Y domain, coordinate system, and units. Finally, you will create the feature class.

___ **20.** In ArcCatalog click

___IGIS-Arc_*YourInitials*\Digitize&Transform

to highlight it. Choose File > New > Personal Geodatabase. In the Catalog Tree, change the name of the PGDB to Recreational_Facilities.mdb. Make sure the new entry is highlighted in the Catalog Tree. Select File > New > Feature Dataset. For Name type North_ Zone. Click the Next button. Starting (with a vertical slider all the way at the top, if necessary) navigate to

Projected Coordinate Systems > State Plane > NAD 1983 (US Feet) > NAD 1983 StatePlane Kentucky North FIPS 1601 (US Feet)

and then click Next. Click Next again, since we aren't concerned with Vertical Coordinate Systems. Click Finish.

Converting the Shapefile to a Geodatabase Feature Class

____ **21.** Open ArcToolbox. Expand Conversion Tools[11] > To Geodatabase > Feature Class To Feature Class. Pounce. Read the Help in the right pane (Use Show Help if necessary). Click the Input Features text box and read the right pane, then browse with the Input Features open folder icon. You want to navigate to the shapefile you created earlier: Dig_Lines_Shape.shp. Add the shapefile.

____ **22.** Click the Output Location text box. Browse to the North_Zone data set (make sure the name appears in the Name box of the Output Location window), and click Add. You can use the arrow keys, and Home and End keys to see any hidden contents of the Output Location box in the Feature Class To Feature Class window.

____ **23.** For the Output Feature Class name, type School_Court_A. Expand the window vertically or use the slider bar to see the rest of the window. Under Field Map do away with the unneeded shapefile identifier ID by clicking it, then clicking the Delete button (an X) in the window. Click OK. Once processing completes, close the ArcToolbox window. Preview the Geography and the Table of School_Court_A.

Moving the Foozit Court Feature Class into the Real World

____ **24.** Start ArcMap with a new, blank map. Add School_Court_A. Change the lines in School_Court_A to Mars Red, with a width of 2.

____ **25.** Right-click the Data Frame. In the Data Frame Properties window note that the map units are Feet, as you defined them to be. Make sure the Display units are set to Feet as well. You will notice, as you slide the cursor around the Foozit court, that the coordinates on the screen show in feet. These will be similar to the coordinates you wrote down previously. If you measure the width of the court, you get something on the order of 5 feet—clearly not correct. It's nothing to worry about. When you transform the court into real-world coordinates, the feature class dimensions will change and it will be rotated to fit the spot. All that is important here is that the right lines are connected and that the image is not distorted.

____ **26.** Add the image file six.tif from [___]IGIS-Arc\Image_Data. (Do not build pyramids if asked. Ignore the warning about the unknown spatial reference.) Zoom to the layer. In the southwest corner of the image is a school with a running track. Zoom to the school, including the track. Bookmark this image, calling it School. In the south-central part of this image, you should be able to discern most of a large parking lot.

It has been decided to make a full-size Foozit Court in the parking lot of the school. A surveyor was hired to specify the corners of the court in the Kentucky state plane coordinate system you have been working with. So, you will place the Foozit court in the parking lot of the school, at the coordinates shown in Table 5-1, which have been provided to you by a surveyor.

_____

[11]You may need to add Conversion Tools. If so, refer to Exercise 4-3. The toolbox is located in Toolboxes > System Toolboxes.

TABLE 5-1

Corner Number	Easting	Northing
1	1573792.8	176684.5
2	1573800.0	176700.0
3	1573819.0	176691.1
4	1573811.8	176675.7

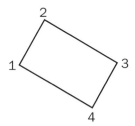

FIGURE 5-6

The corner numbers are given by Figure 5-6. Corner 1 is where the original southwest corner of the digitized image will go. Corner 2 is the northwest corner. Corner 3 is the northeast corner.

In order to move the court to the correct coordinates, you will need to create a text file that specifies the "links" from the existing corners of the digitized feature class to the new corners in the parking lot. Each line of the file consists of three elements (a) a point number (the corner number), (b) the digitizer coordinates you wrote down earlier in Step 5, and (c) the real-world coordinates provided by the surveyor— so five numbers will be placed in each row of the text file.

Where you see the designation "corner1x-digcoord" (see below), you need to substitute the coordinates of the corners from the digitizing process that you wrote down in the table in Step 5 of this Exercise. For example, corner1xdigcoord might be something like 0.21. Be sure to get the right pairs of digitizer coordinates associated with the right pairs of surveyor's coordinates. For example, the southwest corner is represented in the preceding table by line 1. Also note that the x-coordinate (easting) precedes the y-coordinate (northing). That is, the text file looks like this:

```
1   corner1x-digcoord   corner1y-digcoord   1573792.8   176684.5

2   corner2x-digcoord   corner2y-digcoord   1573800.0   176700.0

3   corner3x-digcoord   corner3y-digcoord   1573819.0   176691.1

4   corner4x-digcoord   corner4y-digcoord   1573811.8   176675.7
```

You will put in actual numbers for strings like corner1x-digcoord.

Use tabs to separate the values! The first line of your file will look something like this:

```
1   0.21   2.28   1573792.8     176684.5
```

except, of course, you would use the numbers you found when digitizing instead of 0.21 and 2.28.

_____ **27.** Make this four-line file with a text editor, calling it Foozit_Parklot.txt and save it in

_____IGIS-Arc_*YourInitials*\Digitize&Transform

_____ **28.** Back in ArcMap, zoom to the full extent of both datasets. You may (or may not) see a tiny red dot on the left of the map window at around (0,0), which is School_Court_A, and a little black smudge at the right at around (1575000,180000), which is six.tif. (To see the coordinates, look at the status bar at the bottom right of the window.) Start editing School_Court_A from the Editor toolbar. Turn on the Spatial Adjustment toolbar, under More Editing Tools in the Editor toolbar drop-down menu.

_____ **29.** In the Spatial Adjustment toolbar drop-down menu, mouse over Set Adjust Data and read its ToolTip. Click. Click the radio button to pick All features in these layers. Make sure School_Court_A is checked as the data to adjust (i.e., transform). Click OK. Again, in the drop-down menu, make sure that Adjustment Methods > Transformation - Affine is checked for the adjustment method. Under Links pick Open Links File, specifying Foozit_Parklot.txt in Digitize&Transform (looking at All Files). Click Open.

If you have done everything right, you will now note lines running from the tiny red dot (the digitized school court) in an east-northeast direction toward the black smudge (the School_Court_A orthophotoquad).[12]

_____ **30.** Zoom in on the digitized court. You should see lines running from each corner, in accordance with the text file table you made. See Figure 5-7.

_____ **31.** For the following operations refer to Figures 5-8 and 5-9 as necessary; but mostly pay attention to what is happening on the computer monitor. Zoom back to the bookmark School. Here you will see the easternmost ends of these "transformation lines." Zoom in on the parking lot where the Foozit Court is to be placed. From the arrowheads note where the corners of the Foozit Court will be. From Spatial Adjustment > Links > View Link Table, make the Link Table appear. One at a time, click the ID value in the link table and note the flashing link to the parking lot.

You can also see, as the last column in the link table, the residual error for each link, which is the number of feet by which the corners of the court don't exactly fit the digitized image. It should be 0.05 or less. That is, the rectangle specified by the corners of the digitized feature cannot be stretched **_linearly_** to exactly fit the surveyor's corners, but the small numbers tell you that the transformation is going to go very well. If these numbers are large, or if the arrowheads don't appear to look right, something has gone wrong. If there is a problem, select Delete Links and go back and fix things—which probably means you need to revise the text file.

---

[12]The lines may appear to be running in the wrong direction or may not appear at all, due to a bug in the program. If they do, just keep zooming in. My experience is that at a zoom level of 1:9 (a "9" typed into the text box on the Standard toolbar) will produce the correct image.

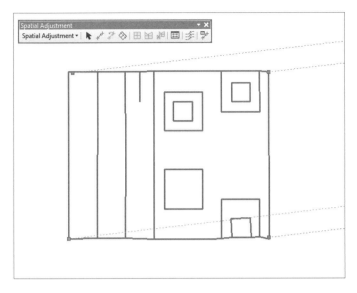

FIGURE 5-7

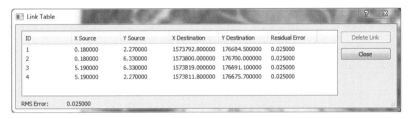

ID	X Source	Y Source	X Destination	Y Destination	Residual Error
1	0.180000	2.270000	1573792.800000	176684.500000	0.025000
2	0.180000	6.330000	1573800.000000	176700.000000	0.025000
3	5.190000	6.330000	1573819.000000	176691.100000	0.025000
4	5.190000	2.270000	1573811.800000	176675.700000	0.025000

RMS Error: 0.025000

FIGURE 5-8

FIGURE 5-9

___ **32.** Close the link table. From the Spatial Adjustment toolbar pick Adjust. Bingo. The Foozit Court appears. Zoom in further to check that it arrived intact. Use the measure command (set so it displays feet) to check that the dimensions are correct. What are they (to the nearest foot)? _____ feet by _____ feet. Check the diagram from which you digitized. Are these dimensions correct? _____. Zoom out to see the court in the parking lot.

___ **33.** If everything looks okay, from the Editor drop-down menu, click Save Edits. Click Stop Editing. If things do not look right, from the Editor drop-down menu, click Stop Editing. DO NOT SAVE EDITS. Go back to Step 27.

___ **34.** Save the map as School_map1.mxd in

   ___IGIS-Arc_*YourInitials*\Digitize&Transform.

Dismiss the Spatial Adjustment toolbar and then dismiss ArcMap.

Exercise 5-4 (Project)

Digitizing Directly into a Real-World Coordinate System in a Geodatabase

In this exercise, you will again do heads-up digitizing, but instead of transforming the product, you will instead move the image file into the appropriate coordinate system and digitize directly. This project focuses on five fictional islands at about 37 degrees north latitude and 171 degrees west longitude, which puts them in UTM Zone 2 North. An ancient, sketchy map of the islands (one of which is artificially square for purposes of illustrating computation) may be found in

   [___]IGIS-Arc\Image_Data

Some UTM coordinates are written on the sketch for reference. See Figure 5-10.

Preliminaries

You will use ArcCatalog to copy Five_Islands.TIF from

   [___]IGIS-Arc\Image_Data

to

   ___IGIS-Arc_*YourInitials*\Digitize&Transform

—actually making a total of three images of the TIF file.

___ **1.** In ArcCatalog make a copy of the image data file Five_Islands.TIF by highlighting the name, pressing Ctrl-C, highlighting Digitize&Transform, and pressing Ctrl-V. Again highlight Digitize&Transform and press Ctrl-V. Do this once more. In the Digitize&Transform folder, you should now have Five_Islands.TIF, Five_IslandsCopy.TIF, and Five_IslandsCopy2.TIF.

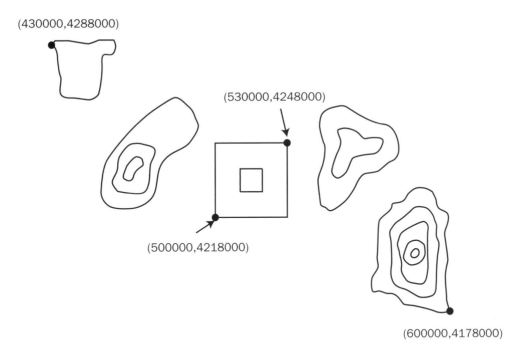

(430000,4288000)

(530000,4248000)

(500000,4218000)

(600000,4178000)

FIGURE 5-10   Five islands to be digitized

Making the Feature Class That Will Be the Object of the Digitization

___ **2.** Using ArcCatalog in

     ___IGIS-Arc_*YourInitials*\Digitize&Transform

make a personal geodatabase named Islands.mdb. Inside that, make a feature dataset named UTM_Zone_2 with the following specifications: WGS 1984, UTM Zone 2 (North) and Vertical Coordinate System > World > WGS 1984 Geoid. Click Next. Make the XY Tolerance 1 meter and the Z Tolerance 1 meter. Finish. Within the UTM_Zone_2 feature dataset, make a feature class named North_Island_Lines. Specify its Type as Line Features. (Don't add any new fields.)

___ **3.** Start ArcMap with a blank map. First Add North_Islands_Lines to the map (which will, of course, show up as nothing since it's an empty feature class). Then add Five_Islands.TIF to the map. Zoom to the TIF layer.

Georeferencing

___ **4.** Choose Customize > Toolbars > Georeferencing. Place the Georeferencing toolbar where you want it.

Georeferencing is a method of moving images or grids to real-world coordinates. Basically, you identify control points on the source layer and then specify where in a real-world coordinate system these points are to be placed. It is, like spatial adjustment, a procedure that results in transformation. Spatial adjustment moves points, and associated features, in feature classes. In contrast, georeferencing moves images and grids.

In what follow you are going to set up the image of Five_Islands with UTM coordinates, so you can digitize directly in the UTM coordinate system. First, however, you will play with the image and the transformation process, both to see what it can do and how it can get you into trouble. We'll start by moving an image a very short distance.

_____ **5.** The Georeferencing toolbar should be active. Mouse over the Add Control Points button to gather information about it. In version 10.0 this information will come from the Status Bar; in 10.1 it will come from the ToolTip. Click the button.

_____ **6.** Move the cursor on the image to the northeast point on the square island. Observe the y-coordinate on the Status bar of that point on the image. What is it? _____. Click this point. Now move the cursor vertically about 2 inches up. Click again. Notice the image moves up one inch. So far, so good.[13] The entire image maintains its orientation and is adjusted by the amount you specified. After reading about the View Link Table button with the ToolTip or Status Bar, click the button. A link table window appears, showing the link you just made. The Y source should be about what you wrote down for the position of the point you moved. The Y map location should be somewhat greater. Close the window.

_____ **7.** Now repeat the operation (two inches vertically up), but use the southwest corner of the square island instead. If you did this the way we intended, the entire image rotated clockwise. Did you expect this? Probably not. So the lesson here is that, unless you change some options, each adjustment of the image is based on the preceding control point(s). In this case, the second adjustment resulted in a rotation around the point that was placed by the first adjustment.

_____ **8.** On the Georeferencing menu, click Reset Transformation. You see the original image plus what sort of amounts to the history of the transformations you made. Click Delete Control Points in the Georeferencing drop-down menu so you can start over.

_____ **9.** Click the drop-down menu in the Georeferencing toolbar and *remove the check in front of Auto Adjust*. Repeat the procedure in the preceding steps where you made two vertical two-inch links. *Turn Auto Adjust back on*. The image will shift.

The result probably showed some rotation, because you don't get the two vertical lines precisely the same length, but nothing like the distortion you saw before. The reason is that the transforming process considered both links at once.

_____ **10.** Make a random 1-inch line from one place on the image to another. Probably things go all weird. Make another such link. More distortion. Since Auto Adjust is back on the links are again considered sequentially (each individually).

The moral of this story is that, if you want to make positional transformations, you have to keep Auto Adjust off until you have defined all the links. When you turn Auto Adjust on all the links will be considered at once, moving the figure where you want it.

---

[13]If at any time you get in trouble, click the Georeferencing drop-down menu and choose Delete Control Points.

___ **11.** View the Link table. Click the last link (#4) in the Link Table to highlight it. Press the Delete key on the keyboard. Note that the image reverts to the form it had before you added that link. Delete link #3 and note the results. Once more, with link #2. On the Georeferencing menu, click Reset Transformation. Click Delete Control Points. Close the Link table.

Moving the Sketch to UTM Zone 2

You have seen how you can change the coordinates of the elements of the sketch. Now you change the coordinates so that they conform to locations in UTM Zone 2. These coordinates represent locations much further away so you can't just click-click. Also, you want to place the control points in precise locations.

___ **12.** Make sure Auto Adjust is set to off. Make the Add Control Points tool active. Click the most western point on the image that has UTM coordinates identified. Right-click and select Input X and Y from the resulting menu. Put in 430000 and 4288000 for X (Easting) and Y (Northing), respectively, and click OK. Notice a line extending from the designated point approximately north-northeast by north. Look at the Link Table to be sure everything is going according to plan. Click each of the other specified points and make links from them as well. The values of the coordinates, from left to right, are as follows:

```
(430000, 4288000)  (extreme northwest)
(500000, 4218000)  (southwest of square island)
(530000, 4248000)  (northeast of square island)
(600000, 4178000)  (extreme southeast)
```

The result should look something like Figure 5-11.

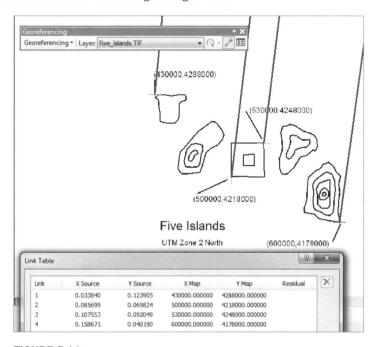

FIGURE 5-11

___ **13.** Turn on Auto Adjust. Poof. The image disappears. Dismiss the Link table. Select Zoom To Layer on the TIF file. If you have done everything right, you will see the image, undistorted. By sliding the cursor around, you can assure yourself that the new coordinates are now being used. Change the display units to kilometers. Use the Measure tool (also set to kilometers) to determine the dimensions of the square island in kilometers. _____ by _____. Does this agree with the UTM coordinates? _____.

Digitizing the Line Boundaries of the Islands

___ **14.** View the Link table. Note that the Link Table shows residuals—probably of a few hundred meters. Like in other transformations this is a measure of how much "less than perfectly linear" the transformation was. Dismiss the Link Table. Turn off the Georeferencing toolbar.

___ **15.** Click Editor. Click Start Editing. Click Create Features. You will create new features with the Line Construction tool. Bring up the Snapping tool bar. Set Snapping to Snap To Sketch (on the drop-down menu). Make sure Ends will be snapped to. Under Options, set the snapping tolerance to 5 pixels.

___ **16.** Digitize the square island and the square lake inside it. Open the attribute table of North_Islands_Lines. The Shape_Length of the outer line should be about 120,000 meters. What is it? _____. What is the length of the inner line? _____. Dismiss the attribute table.

___ **17.** Digitize the other islands. This will be easier and more accurate if you zoom in on each island to do the digitizing. The island in the southeast has a lake within it, an island in the lake, a volcano on that island, and a cauldron defined within the volcano. Each time you zoom or pan you will have to choose the Construction Tool again.

___ **18.** Click Save Edits. Make the feature file line symbol a bright color of width 2, zoom to the layer's extent, and check that everything has been digitized. If not, complete the editing. Click Save Edits again. Stop editing. Turn off the TIF file; it has served its purpose. Save the map under Digitize&Transform as North_Islands_map1. Dismiss ArcMap.

Making Polygons of the Digitized Lines

The next task is to make polygons based on the lines you have digitized.

___ **19.** In ArcCatalog, be sure you can see the contents of the feature dataset UTM_Zone_2. (This may require that you highlight the entry and press F5 to Refresh the Catalog Tree.) Open ArcToolbox. Expand Data Management Tools > Features and then pounce on Feature to Polygon. From the T/C, drag North_Islands_Lines into the input area. Name your feature class North_Islands_ Polygons and make sure it is being saved to the UTM_Zone_2 Feature dataset within the Islands geodatabase. Click OK twice and, when the process is complete, examine the results in the Preview tab. (If not all the polygons show up you may have failed to close one or more lines in your digitizing. Lines to polygon only works on completely closed figures. You will need to go back and fix the problem. Zoom up on each island to see if there are any gaps in the lines surrounding the features. Snapping to the sketch plays an important role in making sure that lines close completely.)

_____ **20.** Restart ArcMap. Add North_Island_Polygons. Use the Identify tool to make sure you have proper GIS polygons. Open the attribute table and inspect it. How many entries are there? _____. Dismiss the attribute table and the identify window.

Making Multipart Polygons

It turns out that several jurisdictions are involved in the chain of islands. As shown in Figure 5-12, the land area is divided into a West County and an East County, by a north-south line that runs from 515000, 4300000) south to (5151000, 4160000). This line splits the square island. However, all inland lakes are controlled by the Region Aqua Board. The volcano is responsibility of the Safety Board, except for its cauldron, under the jurisdiction of the Seismic Group.

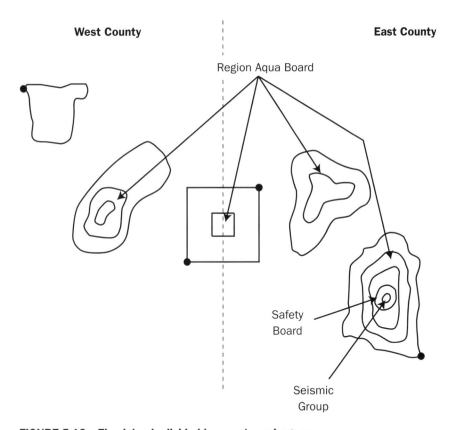

FIGURE 5-12   Five islands divided by county and agency

Five islands divided by county and agency

As a first step, you will split the square island, except for its lake, into two polygons.

_____ **21.** Start Editing. Use the Select features tool on the Tools toolbar to select the land part of the square island. You should see the outer border of the island and its inner border (separating it from the lake) highlighted. Select the Cut Polygons tool from the Editor toolbar. Your mouse cursor will become crosshairs. Right-click the data frame and select Absolute X, Y. Type 515000 for X and 4300000 for Y. Press Enter. Press F6 to get the Absolute X, Y window back, and enter the southern coordinates: 515000 and 4160000. Press Enter. Right-click and select Finish Sketch. You should see two polygons where one was before. The lake will not be split, since it was not selected. Click Save Edits, then Stop Editing.

Merging Multipart Polygons

Now we have separate polygons for all the relevant areas. What we want are all of the polygons of a given jurisdiction as a multi-polygon feature. We do this by merging the polygons that fall under each jurisdiction.

____ **22.** ***First select all the land polygons in the West County:*** Use the Select Features (by Rectangle—somewhat misnamed) tool. Hold down Shift and click each land area in the western part of the island chain. (This area includes an island that is ***in*** a lake that is ***on*** an island—look back at the image for reference to this.) You should have four polygons selected. Right-click North Islands_Polygons and pick Properties > Labels. Label all polygons with OBJECTID. Verify with the attribute table that the correct polygons are selected. Click Editor, then Start Editing. Click Editor again, then Merge. Check the information in the Merge window, click the text lines and observe the drawing. End up highlighting the polygon with the lowest number label. This will make this polygon the one with which the other three will be merged. Click OK. Clear selections with Selection > Clear Selected Features.

____ **23.** In the same way, select all the lakes and merge their polygons.

____ **24.** Merge the polygons of the land areas in the East County except for the volcano and its cauldron. Check your work with the Identify tool by flashing the polygons that belong to each jurisdiction. If you find an error, stop editing *without* saving the edits, and start again.

____ **25.** The attribute table should now show you five features, one for each jurisdiction. Click Save Edits, then Stop Editing. Dismiss the attribute table. Make sure ArcCatalog is not running. Save the map as North_Islands_map2 in Digitize&Transform.

____ **26.** Examine the North_Islands_Polygons attribute table. Click Table Options > Add Field. Name the new field Jurisdiction. Make it a text field of length 20.

____ **27.** Start Editing. Select a record. In the Jurisdiction field of that record, by noting what is selected on the map, place one of the following, as appropriate: East County, West County, Aqua Board, Safety Board, or Seismic Group.

____ **28.** Fill in the other four table values. Click Save Edits, then Stop Editing. Close the attribute table.

____ **29.** Use the Identify tool to make sure that the jurisdiction of each polygon is correct. Clear all selections. Save the map in Digitize&Transform as North_Islands_map3. Dismiss ArcMap.

Digitizing Geodatabase Polygons and Exploring Topology

Previously, in Exercise 5-4, you digitized lines and then converted them into polygons. The conversion process took care of the topology issues of overlaps and gaps. You can also digitize polygons directly, as you will see in this exercise, but then you have to cope with topology more directly. We look at some examples first, examining two different ways of dealing with the topological problems that occur. Then, in Exercise 5-7, you will digitize the polygon features of a newly discovered sixth island.

___ **1.** Start ArcCatalog. In

___IGIS-Arc_*YourInitials*\Digitize&Transform

highlight the Catalog entry UTM_Zone_2, in Islands.mdb. Choose File > New > Feature Class. Type Small_Squares for the name of the new Feature Class. Set the Type of features stored in this feature class drop-down to Polygon Features. Click Next and then click Finish.

You are going to make a square 3 meters on a side as one polygon. Within it you will make a square 1 meter on a side.

___ **2.** Launch ArcMap with a Blank Map. Add Small_Squares as data. Click Start Editing. Within the Create Features window, the template should be Small Squares. Select Polygon from the Construction Tools. Move the crosshairs onto the map. Right-click. Pick Absolute X,Y from the menu. Type 500000 for X and 6200000 for Y to make the beginning vertex. Press Enter.

___ **3.** Right-click and pick Delta X, Y, which will let you provide relative coordinates for the next vertex. Type 3 for X and 0 for Y. Press Enter.

___ **4.** At this point all you will see is a red dot. The problem is one of extent. Choose the Zoom In tool on the Tools toolbar. Repeatedly drag a small box around the red square until you see a line of reasonable length. Click again on Small_Squares in the Create Features window to continue editing.

___ **5.** Make the third vertex: this time, use Ctrl-D to bring up the Delta X,Y window, type 0 for X and 3 for Y. Press Enter.

In this digitizing process, the software understands that you are making polygons (because that is the type of feature you set up when you made the feature class), so the lines will automatically close to form a polygon. Thus, you will see a polygon outlined in the sketch.

(Problems? If things are going wrong during the construction of a sketch, you can press Ctrl-Delete on the keyboard to delete the entire sketch. To delete a particular vertex, click the Edit tool (leftmost button on the Editor toolbar), place it over the vertex, right-click, and choose Delete Vertex.)

___ **6.** Make the last vertex: -3 (that's negative 3) for X and 0 for Y. Right-click and select Finish Sketch, then Zoom To Layer. Save your edits. Open and resize the Small_Squares attribute table and considering docking it.

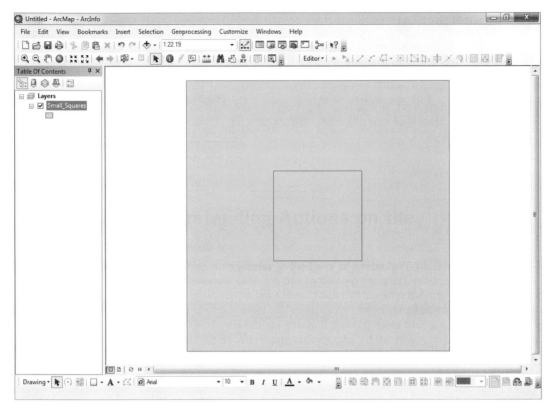

FIGURE 5-13

(If you have problems: If an unwanted polygon is already completed, you can delete it by clicking the Edit tool, clicking the offending polygon, and pressing Delete (either on the keyboard or on the Standard toolbar).)

——— **7.** Click the Sketch tool again. Make a square 1 meter on a side whose southwest corner is at 500001 and 6200001, which places it inside the square you just made. Don't forget to finish the sketch. Save your edits and stop editing. The image should look like Figure 5-13. Dismiss ArcMap.

Making Copies of the Feature Class

——— **8.** Using ArcCatalog, in the Catalog Tree, highlight Small_Squares. Press Ctrl-C. Highlight UTM_ Zone_2. Press Ctrl-V. Okay the Data Transfer Window. Check that Small_Squares_1 has been added to the Catalog. Also make Small_Squares_2.

——— **9.** Examine the geography of Small_Squares. What you are seeing is a larger square with a small square *laid on top of it*. Esri refers to these features as non-planar. By this it is meant that these features do not tessellate a plane. That is, here it is not the case that the larger square has a hole in it in which the smaller square fits. The larger square is a complete square; the smaller square is also a complete square. Verify this by looking at the table and examining areas and perimeters. Fill in the following.

Large polygon perimeter: _____   Large polygon area: _____

Small polygon perimeter: _____   Small polygon area: _____

So you can see by the areas and perimeters that each of these squares is independent of the other, although they, in part, cover the same area of the Earth. We couldn't, for example, grow corn on one and wheat on the other. In general, in a given feature class we want disjoint polygons, so this is a bit of a challenge. Not having disjoint polygons is a problem that occurs whenever we digitize polygons inside polygons.

Using "Clip" to Remove Overlaps from the Feature Class

There are at least two different ways to solve the problem of the non-planar features. The first one is to use the small square as a "cookie cutter" to clip away the duplicated areas of the large square that is under it.

____ **10.** Start ArcMap with a blank map. Add as data Small_Squares_1. Open its attribute table and dock it or reduce its size so that you can see the map. Make the color of the polygons Hollow or No Color, so you can see the boundaries when you are editing. Sometimes the large square blocks out the small square.

____ **11.** Start editing. Click the Edit tool. Click the map somewhere away from either square. Click the center of the small square, and notice one square or the other is selected in the table. Now press the N key on the keyboard. This selects the Next feature, namely, the other square. Repeated pressings of the N key alternates the selections. If the N key doesn't seem to work well for you, use the attribute table and select the features with it. Or use the tiny toolbar that may show up, whose drop-down menu lets you select the square you want. However you do it, *end up with the small square selected*.

____ **12.** Drop the Editor menu down and click Clip.

This tool lets you create a new area, derived from the large square. What will remain from the large square is either (a) that area that intersects the small square or (b) what is left of the large square after an area equivalent to the small square is taken out. To return to our cookie-cutter analogy, we can either keep the cookie (the small square) or keep the dough around the cookie.

____ **13.** Click Preserve the area that intersects and click OK. Notice that the large square disappears and, as you can verify from the attribute table, you are left with two small squares. Not what you want. You are looking for two non-overlapping polygons. Choose (from the Main Menu) Edit > Undo Clip to return to the previous configuration. Again make sure the small square is selected. Clip again, but this time select Discard the area that intersects—that is, throw away the "cookie" part of the big square that is under the selected small square. Now look at the attribute table. Record the values in the following.

Large polygon perimeter: _____   Large polygon area: _____

Small polygon perimeter: _____   Small polygon area: _____

Contrast these values with those you wrote previously, and you will see that you have genuine areal polygons that don't overlap.

—— **14.** Click Save Edits, then Stop Editing. Dismiss ArcMap.

Using Topology to Remove Overlaps from the Feature Class

Here we explore a second way of disentangling these two polygons. We make use of the topology rules that accompany geodatabases.

—— **15.** Back in ArcCatalog, highlight UTM_Zone_2 in the Catalog Tree. Select File > New > Topology. Read the discussion in the New Topology window. Click Next. Name the new topology Squares_Topology. Click Next again. For the feature classes that will participate in the topology, select (with a check box) only Small_Squares_2. Click Next. Since only one feature class will be involved, there is no need to change the rank values. Click Next. Add the rule Must Not Overlap. The New Topology window should look like Figure 5-14. Click Next. Read the summary. Click Finish After a little computation you will be asked if you want to validate the New Topology. Click Yes.

FIGURE 5-14

—— **16.** Launch ArcMap with a Blank Map. Add as data Squares_Topology. When asked if you also want to add the feature classes that participate in the topology, select Yes.

—— **17.** Polygons with errors are shown with red borders. Areas of overlap are shown with pink. Turn the two entries in Layers off and on so you can get an idea of what is being shown. (If both are off and there is still an image in the map pane, press the Refresh View button at the lower left of the map window.) Turn both on.

___ **18.** Open the attribute table of Small_Squares_2. (The topology has no attribute table.) Dock or reduce its size so that you can see most of the rest of the screen easily. Experiment with selecting the two features. It should be clear that these are still non-planar (independent – not nested) polygons.

___ **19.** Use Customize > Toolbars > Topology to bring the Topology toolbar onto the screen. It is inactive because you are not yet editing. Click Start Editing and notice that the toolbar becomes active. Find the Fix Topology Error Tool and press the button. Click the pink area. Right-click and pick Show Rule Description to see what rule is being violated. Click OK.

___ **20.** Right-click again and notice that you have several choices. The ones of main interest are as follows:

- ❏ Subtract

- ❏ Merge

- ❏ Create Feature

To find out what these do, press F1 to get the ArcGIS Desktop Help. Make it full screen. Navigate in the Contents to Professional Library > Data Management > Editing Data > Editing topology > Editing geodatabase topology > Geodatabase topology rules and topology error fixes. Like the descriptions of topological errors that you saw in Chapter 4, there is a lot of material here. Under Polygon Error Fixes, read the potential fixes for the Topology rule: Must Not Overlap. Dismiss Help.

Use a second way to get help for the three possibilities. Right-click the pink square. Highlight the word Subtract by placing the cursor over it. Press Shift-F1. (In 10.1 mouse over the word Subtract.) Read the information box that appears. Press Esc to close the box. Repeat for Merge. Repeat for Create Feature.

It turns out that you want Create Feature. What will happen is that the original small square will be discarded, as will the part of the large square that it overlapped. In their places, a new polygon will be generated that is congruent to the area of overlap (i.e., the equivalent of the small square).

___ **21.** Right-click again on the pink square and choose Create Feature. Note the change in the display. Press Ctrl-Z to undo the change you just made. Use Create Feature again and this time observe the change in the attribute table. Write the results in the following:

Large polygon perimeter: ____ Large polygon area: ____

Small polygon perimeter: ____ Small polygon area: ____

Note the difference between the large polygon now and what you wrote earlier. Clearly you now have two nested, planar, polygons.

___ **22.** Click Save Edits and click Stop Editing. Dismiss the Topology toolbar. Dismiss ArcMap.

Learning all about editing in ArcMap is a course by itself. The Editor has many tools, particularly computer-aided design (CAD) tools, which are quite sophisticated. The Esri manual on the subject is a tome of almost 500 pages. So the best we can do here is give you a insight into a few operations that you will surely need if you do spatial analysis, and a taste of what editing in ArcMap is like. In the next few steps, you will simply splash around in the Editor. Actually learning to swim with the editor is something you will have to do on your own or in another course.

Exercise 5-6 (Project)

Learning Some Editor Fundamentals

Now that you have seen something of the Editor, let's practice using some of its capabilities, and hint at some others. As I indicated, one could have a whole course on just how to use the Editor. So what you learn here will necessarily be incomplete. You will be saving very little of your work and nothing that is significant. This exercise is simply to acquaint you with some of the editing facilities.

The Concept of the Edit Sketch

The Edit Sketch (or Sketch, as we will usually refer to it) is a companion to every line feature or polygon feature (or feature to be created) in a geodatabase feature class or shapefile.[14] When you want to create a feature with the Editor, you create a sketch. Once you save your edits, the sketch becomes a feature. If you are modifying an existing feature, when you make it a target of editing, it is the sketch of the feature that you are editing. Not until you stop editing the sketch do the modifications take effect on the feature.

Making Sketches with Snapping

—— **1.** Start ArcCatalog. By now you know the drill for making feature classes. In

    ___IGIS-Arc_*YourInitials*\Digitize&Transform
    \Islands.mdb\UTM_Zone_2

    make Edit_Play_Lines and Edit_Play_Polys, each with appropriate Type.

—— **2.** Start ArcMap with a Blank Map. Add as data Edit_Play_Lines. Click Start Editing. Let's start by making a 100 meter long east-west line. A Create Features window may appear on the right side of your ArcMap window. If it doesn't appear, click on the Editor drop-down window and select Editing Windows > Create Features. The template should be Edit_Play_Lines. The Construction Tool should be Line. Move the crosshairs onto the map. Press F6 to bring up the Absolute X,Y window. Type 500000 for X and 6200000 for Y to make the first vertex. Press Enter.

—— **3.** Right-click the data frame and pick Delta X,Y, which will let you give relative coordinates for the next vertex. Type 100 for X and 0 for Y. Press Enter.

—— **4.** At this point all you will see is a red dot. The problem again is one of extent. Instead of zooming in with the zoom tool, which requires leaving the sketch tool, you can just press and hold the Z key, which puts the Editor in zoom-in mode as long as the key is held down.[15] Repeatedly drag a

---

[14]The ArcMap Editor cannot be used to edit coverages. You have to use a command-line-driven package called ARCEDIT, which you find by choosing Start > (All) Programs > ArcGIS > ArcInfo Workstation and typing ARCEDIT at the arc prompt. Basically, the authors recommend that you convert coverages to geodatabases. Coverages may be valuable for the data they contain but are fundamentally useless these days as a data structure.

[15]There are lots of keyboard shortcuts like this. If you do much editing you will want to know some of them. In the help interface, you can search for "keyboard shortcuts" or navigate to Professional Library > Data Management > Editing data > Fundamentals of Editing > Keyboard shortcuts.

small box around the red square until you see a line of reasonable length. Release the Z key to return to the sketch tool.

___ **5.** Make a few random vertices by clicking. Click the Undo arrow icon in the Standard toolbar. Click again. What is its effect? _____. Make some more random vertices, ending with a double-click to finish the sketch. Again click Undo. What is the effect? _____.

___ **6.** Use a different method to again make an east-west line of about 100 meters: this time click (in about the center of the lower-left quadrant of the window) to make the first vertex, then observe the Status bar to decide where to place the next vertex. The distance should, of course, be about 100. The direction may be approximately 0 or approximately 90 (depending on whether ArcMap is thinking mathematically or geographically). Click.

This is one of those instances in which two standards come into play. Both geographers (e.g., cartographers, navigators) and mathematicians commonly use 360 degrees to describe a complete revolution. There the similarity ends. Northern hemisphere geographers assign zero degrees to north and the number of degrees increases clockwise; mathematicians (using the polar coordinate systems) assign zero degrees to the positive Cartesian x-direction (east) and the number of degrees increases counterclockwise. GIS relies on mathematical depictions of geography. So we have to deal with this inconsistency. ArcMap allows us to choose which system we want.

___ **7.** Choose Editor > Options > Units. Change the Direction Type to South Azimuth. Click Apply, then OK, and look at the Status bar as you move the cursor. Experiment with the other possibilities under Units, ending with North Azimuth, Decimal Degrees, and two decimal places.

___ **8.** Press F2 to finish the sketch.

___ **9.** Open the attribute table of Edit_Play_Lines. Notice that a polyline of about 100 meters is the only feature referenced by the table. Dock or reduce the size of the table to get it mostly out of the way.

Next you will make a multipart polyline.

___ **10.** Make three more lines of about 100 meters parallel to the first (each 20 or so meters north of the one before it—look at the y-coordinate on the status bar), but at the end of each, click just once, then right-click, and choose Finish Part. When all three lines are done, press F2 to finish the sketch. All three lines will turn cyan, indicating that they are all part of a single selected feature. Now observe the attribute table. You should see the additional feature as a single feature of length about 300.

Once you make a multipart feature (as you just did, and as you did earlier by merging features of the islands), you may disaggregate them if you wish. Here's how.

___ **11.** Choose the Edit tool. Click the original horizontal line to highlight it. Then change the selected features by clicking the three-line feature you just made so it is selected. Chose Customize > Toolbars > Advanced Editing. On that toolbar, using the Status Bar (in Version 10.0) or ToolTips (version 10.1), find the tool that "explodes" a multipart feature. Click it and the three lines become separate features, as you will be able to see from the attribute table.

____ 12. Pick two of the parallel lines (using Shift-click) and merge them (Editor dropdown menu) back into a single feature, using the editor menu. Save your edits.

____ 13. While you have the Advanced Editing toolbar up, run the cursor over the tools and read about each. Write the ToolTips descriptions in your Fast Facts File.

Done? Yes ___. No ___.

____ 14. Get more information about each Advanced Editing icon by clicking the What's This? Icon on the Standard toolbar (version 10.0) or (in version 10.1) by mousing over the icon and, if availability of a help file is indicated, pressing F1. Making reference to this Help information, experiment with these tools as your interest (or instructor) dictates. When you are through, stop editing but don't save edits, so you can continue below with the parallel lines. Dismiss the Advanced Editing toolbar by right-clicking one of its tools and clicking Advanced Editor to turn off its check mark.

____ 15. Start editing again. Choose Editor > Snapping > Snapping Toolbar. Be sure snapping is set so that lines in a developing sketch will snap to the Ends of Features in Edit_Play_Lines. Leave the snapping toolbar visible. Choose Snapping > Options and set the Snapping Tolerance to 5 Pixels. Turn on Show Snap Tips and include Layer Name and Snap type. Click OK.

____ 16. Select the Edit_Play_Lines as the template and Line as the Construction Tool. Starting in the southwest corner, move the cursor to the west end of the line. You have used snapping before, so you are not surprised when the crosshairs changes to a square and jumps to the exact end of the line. Note the information provided by the Snap Tip when the cursor locks onto the end of a line feature. Click. Now make a vertical line from there to the west end of the parallel line just above, and click. Continue to make vertical line segments to each of the two remaining horizontal lines. Double-click the last vertex (or press F2).

____ 17. In the snapping toolbar activate Edge snapping. (A tool on the toolbar is activated when it is shown within a blue square.) Now move the cursor around; when you get close to any part of a line, an edge snapping symbol and ToolTip jumps to the line.

____ 18. Select the Edit tool. Draw a box that contains or crosses all the lines you have made *except* the original 100-meter line. Everything should turn cyan except that original line. Press the Delete button on the Standard toolbar or on the keyboard. The features will disappear, as will the associated rows in the attribute table.

There is even another toolbar to assist with editing. Open the Editing Options window (Editor > Options > General), and make sure the check box next to Show mini toolbar (version 10.0) or Show feature construction toolbar (version 10.1) is checked. Now, when you start a sketch you should see the Feature Construction toolbar. Mouse over the tool icons to get an idea of each tool's purpose. You can guess what Undo and Finish Sketch do. Constrain Perpendicular, for example, is less obvious, until you try it.

____ 19. With the line construction tool activated, click somewhere to start a sketch. Click the Constrain Perpendicular, then click the 100 meter line you made earlier to indicate it is the line you want the line segment you are creating to be perpendicular to. Move your mouse and you will discover that any line you make with the next vertex will be perpendicular to that line – whether it goes to that line or not. Click to make a vertex and your first line segment, and then click the Constrain Perpendicular again. This time, click the line segment you just made. Move your cursor and click again. You should have a 90-degree corner. Make a number of segments, each at right angles to the one before. Right-click and select Delete Sketch. If your map appears to be a mess you can stop editing, without saving edits, and start with just your original four parallel lines. Start editing Edit_Play_Lines again.

____ 20. Sketch a line with several segments. Finish the sketch. Click the Edit tool. Double-click the sketch you just made. The Feature Construction toolbox becomes the Edit Vertices toolbox. Put the cursor over a vertex. When the cursor changes to a square and four triangles, drag that vertex to a new location.

____ 21. Right-click the map and select Edit Vertices to make the old vertex go away. Stop editing without saving edits. Start editing again.

____ 22. Click the Edit Tool. Single-click the original horizontal line. The cursor changes to include a four-headed arrow. Drag the entire line a few meters away from its current position. Click Save Edits, then click Stop Editing.

Now you at least have some ideas of the editing capabilities of ArcGIS with regard to lines. While all this seems pretty cumbersome you can imagine, with all the tools available, that someone past the learning stage could become really proficient at editing feature classes. Rather than trying to make expert graphic editors out of you we look at editing polygons next.

Experimenting with Editing Polygons

____ 23. Remove Edit_Play_Lines from the Table Of Contents. Add as data the blank feature class Edit_Play_Polys. Click Start Editing. Using the Create Features interface, the Polygon construction tool, and zooming, make a triangle with approximately 1000 meter sides and the northern-most vertex at (500000, 6300000). Save Edits. Stop Editing. Make the layer color No Color or Hollow. Open the attribute table.

Experimenting with Editor's Union

Basically, a *union* of two polygons, say A and B, is a polygon whose area includes both the area of A, and the area of B (which, of course, includes any area that is common to both A and B).

____ 24. Make a second feature, say a rectangle, that partly overlaps the triangle.

____ 25. Select both polygons. (Use either the attribute table with Ctrl-click or the Edit tool with Shift-click.) Using Union in the Editor menu, make a third polygon.

____ 26. Prove to yourself that a new polygon has been created by (a) looking at the attribute table, and (b) dragging it with the Edit tool. You could, of course, make this third polygon effectively replace the first two, by deleting them, but *don't*. Instead, select Stop Editing *without* saving edits.

____ 27. Start editing. Make a second polygon that lies outside the triangle. Again perform the union operations. Observe the result. Stop editing without saving edits.

____ 28. Start editing. Make a second polygon that lies inside the triangle. Again perform the union operations. Observe the result. Stop editing without saving edits. Then start editing again.

Experimenting with the Editor's Intersect[16]

Basically, an *intersection* of two polygons, say A and B, is a polygon whose area includes **only** the areas of A and B that are common to both A and B.

___ 29. Look at the Editor drop-down menu. It may include the Intersect command. If not, customize the Editor menu to include Intersect by Customize > Toolbars > Customize (which is the last item in the toolbar list). Click Commands > Editor > Intersect. Drag Intersect to the Editor menu, which will open automatically when you move the cursor over it, and drop the Intersect tool between Buffer and Union. Close the Customize window.

___ 30. Repeat steps 24 through 28, but use the Intersect command from the Editor drop-down menu.

Experimenting with the Editor's Buffer Capabilities

___ 31. Stop editing without saving edits. Then start editing the feature class with the single triangle again. This time select the Buffer command on the Editor drop-down menu. When asked for a distance, use 100 meters. Note the figure that is created, particularly its rounded corners. Every point on the line of the new polygon is 100 meters from the original triangle, measured along a line perpendicular to a side of the original triangle, or is on a circle with radius 100 meters with its center at a corner of the triangle. In the table, clear selections. Now use Buffer on the original triangle again (you have to select it), specifying a distance of −100 meters. Observe. Stop editing without saving edits.

Using Undo, Redo, Copy, and Cut

___ 32. Start editing. Add a second polygon that overlaps the triangle. Union the two. Move the new polygon. Now place the cursor on the Undo tool icon on the Standard menu. What does it say? _____. Press the Undo tool icon. Place the cursor again on the Undo tool icon. What does it say? _____. Press it. Place the cursor on the Redo tool. What does it say? _____. Press it. Place the cursor again on the Redo tool. What does it say? _____. Press it. Observe. Recognize that you can undo and redo back and forth on the chain of actions. Stop editing *without* saving edits.

___ 33. Start editing. Select the triangle. Press Ctrl-C. Press Ctrl-V. Select Edit_Play_Polys as the target layer to create feature(s) in. Verify, with the attribute table, that an identical polygon has been created. Move one of them. Select Stop Editing *without* saving edits.

___ 34. Start editing. Select the triangle. On the Editor toolbar, select Cut Polygons Tool. Sketch a line that starts outside the triangle, goes inside (make a vertex), changes direction, and returns outside. Note what happens as this polygon is completed. Cut the triangle again, making four polygons. Now use the Edit tool (after selecting two records in the attribute table) to move those two polygons. Again use the Edit tool (after selecting the remaining two polygons graphically) to move those polygons. Stop editing *without* saving edits. Dismiss the attribute table.

---

[16]You may have to add the Intersect command to the Editor drop-down menu.

Working with Line Editing Again

_____ **35.** Add back Edit_Play_Lines. Zoom to that layer. Start editing. The lines feature class may not show up in the Create Features template selection list. To add it back, enter the Organize Feature Template window by pressing the Organize Templates button on the Create Features window toolbar. Within the Organize Feature Template window, press the New Template button to enter the Create New Templates Wizard. Make sure Edit_Play_Lines is selected with a check beside its name and click Finish. You should see that Edit_Play_Lines is now a selectable template within the Create Features window. Close the Organize Feature Template window.

_____ **36.** Select Edit_Play_Lines as the editable template. After selecting elements on the map with the Edit tool, experiment with Move, Split, Buffer, and Copy Parallel (found on the Editor menu). Stop editing *without* saving edits.

_____ **37.** Close ArcMap without saving changes.

In the previous exercises, you learned how to digitize lines and polygons directly into a personal geodatabase feature class, and also learned two different ways to cope with the topological problems that result.

In the next exercise you use some of these methods to add a sixth island to the UTM_Zone_2 feature class North_Island_Polygons.

Exercise 5-7 (Follow-on)

Adding the Sixth Island

It turns out that the map you were working with (Figure 5-10 in Exercise 5-4) was very old. A volcano has erupted since, making a new island. A digital orthophoto, converted into a grid named MSH, is available. (Preliminaries: In the first step you will copy a line feature class, and you will make a second, empty line feature class, both of which you will use later.)

_____ **1.** Using ArcCatalog, make sure the 3D Analyst extension is turned on. *Begin* to make a geodatabase Feature Class named Trail in

___IGIS-Arc_*YourInitials*\Digitize&Transform
\Islands.mdb

When you come to the New Feature Class window that lets you define the Data Type and the Field Properties, make the Type of features "Line Features" and place a check in the box beside Coordinates include Z values. Click Next. Click Finish. From

[___]IGIS-Arc_AUX\D&T\Islands.mdb\UTM_Zone_2

Copy the PGDBFC named Trail_horizontal to

___IGIS-Arc_*YourInitials*\Digitize&Transform
\Islands.mdb

——— **2.** In ArcMap open the map named North_Islands_map3. Add as data the geotif file

IGIS-Arc\Image_Data\MSH

This is a pretty big file, so build pyramids if asked, which takes a bit of time now but results in faster access in the future. Zoom to the MSH image.

——— **3.** As you observe the image, you can imagine that you see two triangular harbors along the south coast. Using the Georeferencing toolbar (don't forget to turn Auto Adjust off), move the apex of the southwest harbor to (Easting 630000, Northing 4100000) and the southeast harbor apex to (Easting 660000, Northing 4100000). View the Link Table to be sure you have the right coordinates in. Turn on Auto Adjust so the move happens. When the move is complete, zoom to the image. Then zoom to full extent to see where this new island is with respect to the others.

Recall that you previously digitize the five islands as lines and then converted the lines to polygons. When you did this, the polygons that were created obeyed rules that resulted in no overlaps and no gaps. There are several ways to add the new island to North_Island_Polygons, but let's use one in which you digitize *polygons* directly into the geodatabase feature class.

——— **4.** Start editing. The target should be North_Island_Polygons. You want to create new polygons. Zoom to MSH. Digitize first the outline of the island. When digitizing polygons with curved lines, use more points along the lines where the curvature is greatest. See Figure 5-15.

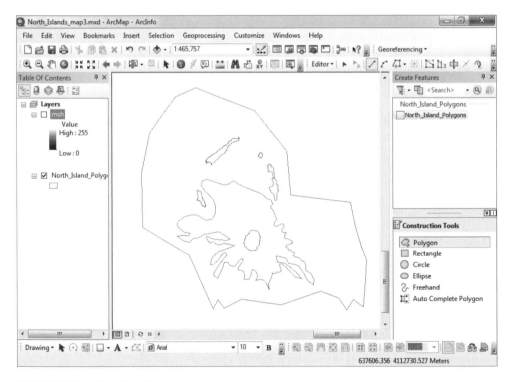

FIGURE 5-15   The sixth island

When you create polygons, you see a closed figure after you have put in the first three vertices. To complete the figure, double-click the last vertex, or single-click the last vertex and press F2 to complete the sketch. The polygon will automatically close.

5. If necessary, rearrange the entries in the T/C so that MSH is again at the top of the list and the image is visible. Digitize the cauldron of the volcano. There appear to be several lakes (they appear black on the orthophoto) on the island; zoom in and digitize them. Digitize what appears to be new lava flow that covers about a third of the island, and in some cases, runs down to the ocean. Click Save Edits. Click Stop Editing. Turn off the image layer MSH. Make the North_Islands_Polygons layer hollow so you will be sure to see the boundaries of all the polygons. The map of the newly digitized island should roughly resemble Figure 5-15. Save edits. Stop editing. Save the map as North_Islands_map4 in your Digitize&Transform folder. Launch the ArcCatalog sidebar.

From this point on in the exercise, the instructions will become very general. You should use the tools you learned in the previous exercise to accomplish the tasks.

6. Using ArcCatalog, make two copies of North_Islands_Polygons in the same dataset that you have been using.

7. In ArcMap, use clip operations on one copy to make separate polygons of the lava flow, the lakes, and the cauldron. Since polygons are nested within polygons that are themselves nested this can be a little tricky. Things will probably go better if you so two things: (1) start with the most deeply nested (smallest) polygon first, and (2) open the attribute table and use it to select the polygons, rather than relying on selecting them graphically.

8. Merge the cauldron into the Seismic Group.

9. Merge the lava flow area with the Safety Board.

10. The lakes go to the Aqua Board.

11. The remaining land area goes to the East County.

12. Save edits and remove the first copy of North_Islands_Polygons from ArcMap. Add the second copy of North_Islands_Polygons to ArcMap. Close ArcMap so you can make topology with ArcCatalog.

13. Use topology on the second copy. Perform the same merging operations as immediately above.

14. Not that you ever make errors, but if you had merged the polygons that should not be merged, what toolbar and tool would you use to separate them again? _____, _____.

Creating a 3-D Feature

15. If you happen to have dismissed ArcMap, start it again with North_Islands_map4. Turn on the image layer MSH. Zoom in on the sixth island. You should see the image of the volcano. (If not, you will have to Georeference the image again.)

__ **16.** In ArcMap add the Feature Class named Trail_horizontal from

    ___IGIS-Arc_*YourInitials*\Digitize&Transform
\Islands.mdb\UTM_Zone_2.

    Make it a red line of width 3.

Trail_horizontal is a trail from the water's edge to the rim of the volcano, but with all segments in the horizontal, sea level plane. What we will do is make a three-dimensional trail, named Trail, with the Editor, using the pattern of Trail_horizontal.

__ **17.** In ArcMap add the empty Feature Class named Trail (that you made earlier in this exercise), from

    ___IGIS-Arc_*YourInitials*\Digtize_Transform
\Islands.mdb\UTM_Zone_2

__ **18.** Zoom to Trail_horizontal. Start Editing Trail–making sure that the target is Trail. Starting at the ocean make a polyline with five segments that overlay the segments of Trail_horizontal. End each segment, ***including the last***, with a ***single click***.

__ **19.** Find Sketch Properties on the Editor toolbar. Click it. You will see the three dimensional coordinates of the six vertices you have made. All the z-coordinates are presently zero. Starting with the vertex on the ocean, click each z-coordinate to change the values to:

    `0, 400, 900, 1500, 3000, 4000`

__ **20.** Click Finish Sketch (or press F2). Dismiss the Edit Sketch Properties window. Click Save Edits, then click Stop Editing. Dismiss Sketch Properties. Dismiss Create Features. ____21. Start ArcScene by selecting Start > (All) Programs > ArcGIS > ArcScene. Add Trail and North_Islands_ Polygons from

    ___IGIS-Arc_*YourInitials*\Digitize&Transform
\Islands.mdb\UTM_Zone_2. Zoom to Trail.

    Make ArcScene full screen. Change the Trail symbol to a black line of width 2.

__ **22.** Add Trail_horizontal from

    [___]IGIS-Arc\Digitize&Transform\Islands.mdb\UTM_Zone_2.

__ **23.** Explore, using the Navigate cursor (with the left mouse button) and the zoom control (use the right mouse button). Convince yourself that the polygon of Trail does have a three-dimensional component. Using Identify to determine the Shape_Length of Trail. _____kilometers. Close ArcMap, ArcScene, and ArcCatalog.

Obtaining Field Data and Joining Tables[17]

A Discussion of the Project:

This is a cooperative class exercise with several objectives:

❏ To collect "field data" in the classroom by making measurements

❏ To understand some coordination issues associated with collecting data

❏ To gain an appreciation of how errors can occur in GIS data collection

❏ To enable each student to work in a small team of classmates

❏ To learn to enter non-spatial data into ArcGIS

❏ To learn to join tables associated with a spatial geodataset with non-spatial data

❏ To discover problems that can occur with data collection

❏ To do something different (and fun) in lab

The process will go like this:

❏ Each student will be assigned to one of four teams (A, B, C, or D).

❏ Each student will be assigned a three digit identifying number, (e.g., 612).

❏ Each student will write the ID number on a sticky note, along with an "X" (See Figure 5-16) and will place that on the top of his or her computer terminal or desk.

❏ Each team will perform a measurement or recording task.

❏ Each team will perform a second measurement or recording task.

❏ The data from the teams will be shared by all students.

❏ Each student will enter data in

___IGIS-Arc_*YourInitials*\Student_Data_Collection.

❏ The spatial data will go into a text file table and the non-spatial data into a database table.

❏ Each student will make a map showing the locations of the computer terminals or desks.

❏ The tables of spatial and non-spatial data will be joined so that the map can provide all recorded information about each student.

---

[17]This is an exercise that requires about an hour of class time in which students in teams collect data in the classroom. It works best if there are between 12 and 20 students participating. Instructors should see the Wiley web site for advice on how to conduct the exercise. If you decide not to collect the data but just do the software part of the exercise you may use some sample data in [___]IGIS-Arc_AUX. The two required datasets are in an Excel spreadsheet named Exercise_5-8_Student_Computer_Data.xls. They are saved as Sheet1 and Sheet2. However, you should read through the entire exercise so you will understand how these datasets are used.

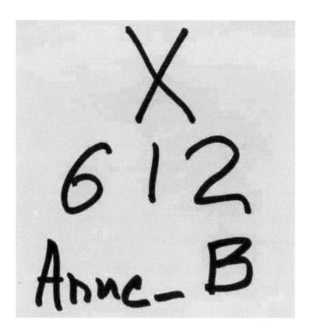

FIGURE 5-16 Sticky note identifying computer location

Organization

❑ Students: Assign yourselves sequential numbers, beginning with number 601. Remember your number, which will serve as both the Computer (or desk) ID number and your Student ID number. Write the number on a sticky note, along with an "X." Stick it on top of your computer terminal or desk.

❑ Students: Assign yourselves to teams (A, B, C, or D) with the same number of students in each team, plus or minus one.

❑ Students: Get together with your teammates in a vacant corner of the room and decide on a plan for accomplishing the tasks.

Environment and Measurement (Spatial Data)

❑ Use the back wall of the lab as southern latitude.

❑ Use a Cartesian coordinate system.

❑ Consider the intersection of the back wall and the "west" wall as the Cartesian point (1000,2000).

❑ For each location's easting, measure perpendicularly from the west wall to the center of the cross on the sticky note.

❑ For each location's northing, measure perpendicularly from the south wall to the center of the cross on the sticky note.

❑ Measure to the nearest 0.01 (one hundredth) of a foot.

Measurements (Non-spatial Data)

- ❑ Record the name of each student in the class in this form: *FirstName_LastInitial* (e.g., Anne_B).

- ❑ Measure the *armspread* of each student to the nearest 0.1 (one-tenth) foot. (Armspread is the distance measured **across the person's back**, from fingertip to fingertip, with arms outstretched.)

- ❑ Measure the *height* of each student to the nearest 0.1 (one-tenth) foot.

- ❑ Assess the eye color of each student.

- ❑ Record the Student ID Number (also that is the Computer (or desk) Number)

Recording Data

Record the data on the black board or white board in two tables. The first table should have the headings:

> Computer Number, Northing Easting.

The second table should contain

> Student Name, Armspread, Height, Eye Color, Student Number

Each student should make a paper copy, or computer text file copy that can be printed, of these datasets.

Also record the data in two separate computer text files. (Two people on the recording team could work on the boards, and two others could type into the text files.) The text files will be mailed to the instructor for distribution to the class.

Team Assignments

All measurements will be taken twice and recorded twice. The first time around:

- ❑ Team A will measure the x-coordinates (the eastings) of the set of computers (i.e., the distances from the west wall to the "X" on the sticky notes).

- ❑ Team B will measure the y-coordinates (the northings) of the set of computers (i.e., the distances from the south wall to the "X" on the sticky notes).

- ❑ Team C will measure the armspreads and heights of the people in the class.

- ❑ Team D will record the data from the teams A, B, and C on the board and in computer files.

When this process is complete, the boards will be erased and the computer files sent to the instructor. Then:

- ❑ Team A will measure the armspreads and heights of the people in the class.

- ❑ Team B will record the data from the other teams on the board and in computer files.

❏ Team C will measure the x-coordinates (the eastings) of the set of computers and will measure the width of the room.

❏ Team D will measure the y-coordinates (the northings) of the set of computers and will measure the length of the room.

These data sets will also be e-mailed to the instructor.

The instructor will put the four data files in a shared area on the class computer disk. Students will examine the pairs of data files for errors and inconsistencies.

Undertaking the Data Entry Process

If you did the fieldwork data collection (detailed in the preceding discussion of this exercise) skip ahead to Step 1 below. If you chose not to do the fieldwork and want only to do the software-based part of the exercise, proceed as follows:

Find the Excel spreadsheet in [__]IGIS-Arc_AUX named

Exercise_5-8_Student_Computer_Data.xls

There are two sheets there, one named Computers and the other named People (accessed with tabs at the bottom of the window). The data in Computers should be typed into the text file Coordinates.txt (described below). The data in People is to be placed in the dBase table named Student_Info.dbf (also described below).

—— **1.** Using MS Windows, in ___IGIS-Arc_*YourInitials*, make a new folder called Student_Data_Collection.

Making a Table That Contains the Coordinate Data

—— **2.** Using Notepad, or another text processor, make a text file that contains the coordinate data. Call it Coordinates.txt and put it in the Student_Data_Collection folder. Each line should contain three values, using CSV (comma separated values) format:[18]

Each student's three-digit ID
The easting value (x-coordinate) of the computer terminal
The northing value (y-coordinate) of the computer terminal
For example, lines 2, 3, and so on should look like:

```
612, 10.28, 20.05
606, 10.99, 21.22
```

etc.

---

[18]If you have not done the data collection part of this exercise go to the spreadsheet in [__]IGIS-Arc_AUX named Exercise_5-8_Student_Computer_Data.xls and use the data there.

As the first line of the text file, use the following text **exactly** as shown including the quotation marks:

"Comp_ID","Easting","Northing"

_____ **3.** Save the file as Coordinates.txt in

   ___IGIS-Arc_*YourInitials*\Student_Data_Collection.

_____ **4.** Start ArcMap with a Blank map. Make a folder connection to Student_Data_Collection. Add Coordinates.txt as data. The T/C will show the path to the file. (Notice that List By Source is active in the T/C. If you click the List By Drawing Order button, the T/C will show nothing, since this is not—yet—a feature class. Make sure the List By Source button is active.)

_____ **5.** Right-click Coordinates.txt and choose Display XY Data to bring up a window of that name. In the X Field drop-down menu choose, of course, Easting. Make the correct choice for the Y Field. When you OK the window, you should see point symbols for each computer or desk (and maybe a warning message; disregard). The T/C will show an entry named

   Coordinates.txt Events.[19]

Open the attribute table of Coordinates.txt Events. You should see the text file you created, but here in attribute table format. Examine it; dismiss it.

Making a Table That Contains the Student Data

_____ **6.** Open the ArcCatalog sidebar. Navigate to and highlight the folder

   ___IGIS-Arc_*YourInitials*\Student_Data_Collection.

   Right-click. Select New > dBase Table. In the Catalog Tree, change the name of the table to Student_Info.dbf. (If New_dBase_Table appears in the T/C of ArcMap, remove it.) Add Student_Info.dbf to the map as data.

_____ **7.** Open the Student_Info.dbf table. You will see only two column heads: OID and Field1. Click Table Options and pick Add Field. In the window, type St_Name in the Name field. For the Type, choose Text. For Length, type 20. Click OK.

_____ **8.** Add another field called Armspread. Make it a floating-point number (Float) that can contain two digits overall (Precision value: 2) with one decimal place (Scale value: 1). Add another field, Height, with the same characteristics. Add Eye_Color as a Text Field of length 8. Finally, add St_ID as a Short Integer.

_____ **9.** Right-click the column Field1. Delete the field.

---

[19]Event has a specific meaning in ArcGIS. It refers to a geographic location that is stored in a table.

Populating the Student_Info Table with Data

____ **10.** Start editing. In the Start Editing window, select Student_Info. Click OK. The table should now allow you to enter data.[20] Start with the first student name, in the form of Anne_B. Fill in the rest of the data for this student. Fill in data for the rest of the students.

____ **11.** Click Save Edits, then Stop Editing. Dismiss the table.

Joining the Two Tables to Make a Single Table

What you have now are two datasets. The first, the Events layer, has the geographic location of the computer terminals. Use the Identify Results cursor to look at one point. In addition to the Easting and Northing, you have a key value, the Computer ID that also references the student using the computer. The second dataset is not a spatial one but contains information about students. It also has a key field, St_ID. What you want to do now is to create a single table with both the geographic and the non-spatial information together. There are several ways to do this. We will pick the most straightforward one: the join.

____ **12.** Right-click Coordinates.txt Events. Choose Joins and Relates > Join to bring up a Join Data window. You want to Join attributes from a table. The ***key field*** in the layer is Comp_ID. The table to join to the layer is Student_Info.dbf. The ***key field*** in that table is St_ID. Continue through any warnings. Click OK.

Seeing the Results of the Join

____ **13.** Using the Identify cursor, click a point in the map display. Notice that you get information not only about that point but the associated student information as well.

____ **14.** From the T/C, open the attribute table of Coordinates.txt Events. Notice that all the information has been put together. Notice that, for a given record, Comp_ID and St_ID are the same.

If you now removed Coordinates.txt Events from ArcMap, or closed ArcMap (don't do either!), the joined data would no longer be part of the table, should you ask for it again. There are two ways to make the join permanent. The first is to save it as a layer file. The second is to export it as feature class or shapefile). Do both in the steps that follow.

____ **15.** Right-click Coordinates.txt Events and choose Save As Layer File. Navigate to the Student_Data_Collection folder. See Figure 5-17. In the Save Layer window, click Save.

____ **16.** Right-click Coordinates.txt Events and choose Data > Export Data, which you should put, as a Shapefile, into the Student_Data_Collection folder. See Figure 5-18. Click OK. Do ***not*** add the exported data to the map.

____ **17.** Ask for a new blank map. To see the first method of saving the joined table, add as data

       ___IGIS-Arc_*YourInitials*\Student_Data_Collection
       \Coordinates.txt Events.lyr

---

[20]If you have not done the data collection part of this exercise go to the spreadsheet in [__]IGIS-Arc_AUX named Exercise_5-8_Student_Computer_Data.xls and use the data there.

FIGURE 5-17 Save Layer window

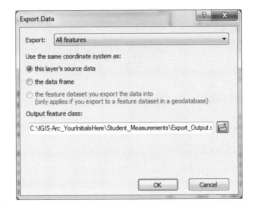

FIGURE 5-18 Export Data window

and check to see that the geographics and the table are present and contain all the information. Verify that each student's information is correctly hooked up with her or his computer location.

_____ **18.** To see the second method of saving the joined table, add as data

___IGIS-Arc_*YourInitials*
\Student_Data_Collection\Export_Output.shp

and check to see that the geographics and the table are present and contain all the information.

_____ **19.** Make a layout of your own design showing the information you have generated with this exercise. The map should label each point with the student's name. Print the map showing the Lab.

Checking, Updating, and Organizing Your Fast Facts File

The Fast Facts File that you are developing should contain references items in the following checklist. The checklist represents the abilities to use the software you should have upon completing Chapter 5.

❏   ArcGlobe data is located in

❏   You can see Globe View, along with Geography and Table Views, in

❏   Reference system information is located in

❏   Geographic coordinate systems information may be found in

❏   A problem will occur if you attempt to combine data from UTM based on NAD 27 with data based on NAD 83 or WGS 84 because

❏   Several difficulties in using state plane coordinates occur because

❏   Four ways of getting (geo)graphic map data into digital form are

❏   To do heads-up digitizing

❏   To make a blank shapefile

❏   To load an image file into ArcMap

❏   The toolbar used in making lines in the shapefile is

❏   The purpose of setting the snapping distance is

❏   To use the Editor, you need to set a number of parameters:

❏   Two ways to finish a sketch are

❏   To convert a shapefile into a personal geodatabase feature class

❏   Getting the extent of a personal geodatabase feature dataset correct to begin with is important because

❏   To move a personal geodatabase feature class to a new location

❏   To convert a shapefile into a personal geodatabase

❏   The toolbar to use when moving an image file is

❏   Before you attempt to move an image, it is important to turn Auto Adjust off because

❏   To make polygons of digitized lines

❏   To make multipart polygons from single part polygons

❏   To merge multipart polygons

❏   Two ways of providing coordinates by typing in the editor are

❏   In the Editor, to delete a sketch under construction

- ❏ In the Editor, to delete a vertex
- ❏ In the editor, to delete an already formed feature
- ❏ To copy a feature class
- ❏ Two ways of removing polygon overlaps are
- ❏ An "Edit Sketch" is
- ❏ Finish Part makes
- ❏ Finish Sketch makes
- ❏ Snapping can be used to precisely locate coordinates in a variety of ways:
- ❏ The snapping interface options
- ❏ The snapping toolbar
- ❏ Snap tips help by
- ❏ A vertex can be moved by
- ❏ A feature can be moved by
- ❏ A union of two polygons includes
- ❏ An intersection of two polygons includes
- ❏ A buffer around a polygon includes
- ❏ The cut polygon feature tool allows
- ❏ To make a LINE feature class, when the assumption is a polygon feature class:
- ❏ To make a LINE feature class that can have a third coordinate, indicating vertical displacement:
- ❏ To put 3-D Coordinates
- ❏ To join two tables
- ❏ To make the joined table permanent

PART II

Spatial Analysis and Synthesis with GIS

Analysis of GIS Data by Simple Examination

OVERVIEW

IN WHICH, by inspecting or simply viewing data in a variety of ways, you take steps toward analysis.

Information

There are two simple ways of looking at a packet of information:

❑ A small amount of matter-energy with the potential to produce large effects

❑ A sequence of 1s and 0s—for example, 10100011

The poem "Paul Revere's Ride" (excerpted) by Henry Wadsworth Longfellow, illustrates both:

> LISTEN, my children, and you shall hear
> Of the midnight ride of Paul Revere,
> On the eighteenth of April, in Seventy-Five;
> Hardly a man is now alive
> Who remembers that famous day and year.
> He said to his friend, "If the British march
> By land or sea from the town to-night,
> Hang a lantern aloft in the belfry arch
> Of the North Church tower, as a signal light, —
> One, if by land, and two, if by sea;
> And I on the opposite shore will be,
> Ready to ride and spread the alarm
> Through every Middlesex village and farm,
> For the country-folk to be up and to arm."
>
> [. . .][1]
>
> Meanwhile, impatient to mount and ride,
> Booted and spurred, with a heavy stride
> On the opposite shore walked Paul Revere.
> Now he patted his horse's side,

---

[1]Indicates omission.

Now gazed on the landscape far and near,
Then, impetuous, stamped the earth,
And turned and tightened his saddle-girth;
But mostly he watched with eager search
The belfry-tower of the Old North Church,
As it rose above the graves on the hill,
Lonely and spectral and somber and still.
And lo! as he looks, on the belfry's height
A glimmer, and then a gleam of light!
He springs to the saddle, the bridle he turns,
But lingers and gazes, till full on his sight
A second lamp in the belfry burns!

A hurry of hoofs in a village street,
A shape in the moonlight, a bulk in the dark,
And beneath, from the pebbles, in passing, a spark
Struck out by a steed flying fearless and fleet:
That was all! And yet, through the gloom and the light,
The fate of a nation was riding that night;
And the spark struck out by that steed, in his flight,
Kindled the land into flame with its heat.

[. . .]

You know the rest. In the books you have read,
How the British regulars fired and fled, —
How the farmers gave them ball for ball,
From behind each fence and farm-yard wall,
Chasing the red-coats down the lane,
Then crossing the fields to emerge again
Under the trees at the turn of the road,
And only pausing to fire and load.
So through the night rode Paul Revere;

And so through the night went his cry of alarm
To every Middlesex village and farm, —
A cry of defiance and not of fear,
A voice in the darkness, a knock at the door,
And a word that shall echo forevermore!
For, borne on the night-wind of the Past,
Through all our history, to the last,
In the hour of darkness and peril and need,
The people will waken and listen to hear
The hurrying hoofbeat of that steed,
And the midnight-message of Paul Revere.

Excerpted from Paul Revere's Ride by
Henry Wadsworth Longfellow.
Publicly appeared first in the
Atlantic Monthly, January 1861

Certainly in this age—early in the twenty-first century, when what computers do seems indistinguishable from magic—it is difficult to remember or believe that a computer is, at its heart, a conceptually simple device. The way in which it is simple has a profound impact on its use for GIS. Further, the simplicity imposes constraints on what we can do.

The elementary nature of a computer is based the idea that just the existence or nonexistence of a single thing, when put together with other such things, can be a powerful mechanism for representing meaning—a code, if you will, consisting of 1s and 0s. In the case of Paul Revere's Ride, the things were lit lanterns—two of them hanging in a belfry arch. If a lamp burned we could call that a "1." If not, a "0." Therefore, the code was as follows:

00:	No British
10 (or 01):	British by Land
11:	British by Sea

As for the other definition of information—a small amount of matter-energy with the potential to produce large effects—it's pretty clear, if the poem can be believed, that in some sense the United States was saved that night. The beginning of the battle has been described as "the shot heard round the world."

Computer Hardware—What a Computer Does

The information in this section is intended to demystify computers for you. As computer power increases—with abilities like speech and facial recognition and championship chess playing—the tendency is to regard a computer as basically incomprehensible. Since the fundamental logical structure behind the operation of a computer is quite simple, I believe such understanding should be part of a student's knowledge base. It's not that hard, and you may find it interesting. And the knowledge may help you to figure out why things go wrong sometime when you are using a computer to work on a GIS project.

A refresher: All the operations of a computer are based on the concept that combinations of binary states can represent information. Binary states can be: on or off; A or B; tied or untied; yes or no; exists or does not exist. Customarily, we represent those two states by BInary digiTS (BITS). With these a computer does three things:

❑ Strings of 0s and 1s are fed in (input) to the machine and placed electronically into a "store" or "memory."

❑ Bits from the store are manipulated according to a number of exact rules (operations of arithmetic making up a good-sized subset of those rules), and the results of those manipulations are placed back into the store.

❑ Strings of 0s and 1s are sent out from the store to output devices.

That's it! Everything else is simply elaboration on this basic theme.

Why binary states? Because, in the physical world, it is easier to identify the existence (1) or nonexistence (0) of something than it is to identify the degree to which something is present. For example, it is easy to tell by looking at a light bulb whether the switch that feeds the light bulb is on or off (i.e., whether the switch that controls it is set to 1 or 0). It is not so easy to tell, by looking at the brightness of a bulb, the position (say, 1, 2, 3, or 4) to which a light bulb dimmer (rheostat) is set.

Input

When you press and hold the Shift key and then press the K key on a computer keyboard, a string of bits[2] is sent to the computer. When you move the mouse pointing device, sequences of bits are sent to the computer. When you speak into a microphone attached to your computer's sound card, a string of bits is generated by the sound card (also a computer, by the way) and winds up in the computer's store.

Representation

For a computer to do computation the way humans like it done, it has to represent decimal (base ten) numbers in binary (base two). Here are some decimal numbers and their binary equivalents:

Base Ten	Base Two
0	00000
1	00001
2	00010
3	00011
4	00100
5	00101
6	00110
7	00111
8	01000
9	01001
10	01010
11	01011
12	01100
13	01101
14	01110
15	01111
16	10000

Given a number in base ten, you may interpret it in this way: Starting from the right and summing the values, the first position represents the number of 1s, the second position represents the number of 10s,

---

[2]Assuming that the computer uses the most commonly accepted convention of the relationship of characters, such as "K," to bits—namely the American Standard Code for Information Interchange (ASCII)—the string of bits sent from the keyboard to the computer would be 11010010. Such a string of 8 bits is called a byte. The size of a computer's memory and disk storage is commonly expressed in terms of bytes—always a whole lot of them. A kilobyte is 1024 bytes. A megabyte is 1024 kilobytes. A gigabyte is 1024 megabytes. A terabyte is 1024 gigabytes. A petabyte is 1024 terabytes. An exabyte is 1024 petabytes. Some hard drive manufacturers cheat and call a kilobyte 1000 bytes, a megabyte 1000 kilobytes, and so on, thus inflating the advertised hard drive capacity. Check the Internet for more information on this scuffle.

the third position the number of 100s, the fourth the number of 1000s, and so on. So the number 342 means the sum: 2 times 1, plus 4 times 10, plus 3 times 100, which is 342.

The binary number system is essentially the same, but much simpler. When you have a number in base two, it may be interpreted in this way: Starting from the right, the first bit position represents the number of 1s, the second bit position represents the number of 2s, the third position the number of 4s, the fourth the number of 8s, and so on and so on, doubling each time you move a position to the left. The binary number 01110 represents the decimal number fourteen, calculated as follows: the sum of zero times 1, plus one times 2, plus one times 4, plus one times 8, plus zero times 16, which is 14.

Computation

For a simple example of computer computation, consider addition. The table to give the results of addition of two binary numbers is quite simple, compared to the addition table for decimal numbers. It is as follows:

$0 + 0 = 0$
$0 + 1 = 1$
$1 + 0 = 1$
$1 + 1 = 0$ (with a carryover to the next column to the left)

For the computer to add two numbers together, say, the integers 3 and 11, the computer has them represented in its store as 00011 and 01011. So:

```
  00011
+ 01011
  01110
```

Most fundamentally, the central processing unit (CPU) of a computer can

1. Add, subtract, multiply, divide, and so on, two sequences of bits

2. Compare one sequence of bits with another to determine if they are the same, or if one is numerically greater than the other. Based on the result of the comparison, the computer can begin executing one set of instructions, or another set of instructions.

The CPU has other capabilities, but these are the principal ones. (Much of the progress, and complication, in computing—and why it seems to be so magical) comes from the ability to transfer bits from machine to machine wirelessly, from graphical user interfaces, from parallel computing, and saving partial results (caching). And then there is the state of the art of putting transistors on a chip: 4000 of them in a space the width of a human hair.

Output

When you see a color image on a computer monitor screen, it is composed of, approximately, a million little dots—called pixels (picture elements)—the color of each being controlled by, say, 32 bits. Thirty-two bits allows a large number of combinations, so that many shades of color can be presented. When you hear music coming from your computer's speakers, the sound is generated by strings of bits sending impulses to a speaker cone at varying frequencies—just as sound from a CD is generated by bits: A hole in the surface of the CD is a "1" bit, while no hole is a "0" bit. Whether or not a hole exists in the CD is determined by a laser beam in the CD reader.

(You can contrast this technology, called "digital," with another technology, called "analog," in which the elements are not limited to two states, 0 and 1, but, as discussed previously, may have many more. An example of an analog system is a music recording on a vinyl disk, where a wiggly grove in the disk creates an equivalent wiggle in the speaker cone.)

Why do you need to know this rather arcane stuff? ("Hey, I just want the car to go when I push down on the accelerator and stop when I press the brake. I don't care what makes it happen.") You should know because the "real world" is a mixture of continuous phenomena (water flowing in a river) and discrete phenomena (number of people living in a school district). A computer can only store discrete values. Therefore, results from analysis of GIS data that come from computer storage and manipulation may be in error—by amounts that may make an important difference.

Continuous and Discrete Phenomena

For this subject you have to think somewhat abstractly. Consider any system you like.[3] At any distinct moment in time, the elements of the system may be characterized by an exact condition, or "state." The state may be viewed as the values of a set of variables, which could (practically or theoretically) be measured. We can talk about those variables as being independent or dependent. For example, if the subject under consideration is elevations of the Earth's surface, the elevation of a point might be described by the dependent variable "height above sea level" based on the independent variables "position coordinates" (e.g., latitude and longitude) and "time." The terms independent and dependent are a little misleading, because cause and effect is suggested, where none exists. The idea is, rather, that we (independently) specify place and time, which corresponds to a particular elevation. The answer depends on the specifications, but the place and time do not cause the elevation.

(I should say here that the difference between discrete and continuous phenomena has been the source for much debate in philosophy, mathematics, and physics for centuries. Look, for instance, at Zeno's Paradox (check the Internet). At very small sizes, a mixture of the continuous and the discrete apply to many phenomena. Electromagnetic radiation (e.g., light) comes both in packets (discrete objects) and waves (continuous phenomena). In theory, every moving object has associated with it a wave, called the de Broglie wave. It applies primarily to subatomic particles (e.g., electrons), but one could calculate the frequency of a de Broglie wave for an SUV moving at 30 miles per hour. We won't. Instead, my intention here is to illustrate the difference between continuous and discrete in the practical, human-size world.)

So, a system (phenomenon) may be thought of as consisting of (or being in) a given state at a given moment in time. Following are some examples.

Discrete: A chess game has each player's pieces on particular squares of the board after a particular move. It does not matter, in terms of the game, where in a square a piece is. When the last piece was moved, the path it took or the length of time required for the move is of no consequence.

---

[3]System: a collection of elements (things: objects, ideas, organs, equations, banks, planets, and so on) with connections of some sort between (proximity, gravity, pedagogy, electrical, chemical).

Continuous: A billiard ball on a table has, at a particular moment in time, a certain position. When it moves, its velocity, acceleration, jerk, direction, and so on, are vital to the outcome of its ending position, and the ending positions of other balls. The smallest difference in position, velocity, and spin can make a major difference in whether a ball drops into a pocket or not.

Continuous phenomena are characterized by the following:

1. The existence of an infinite number of states over independent variables, for instance, time.

2. When there is a finite but extremely small (infinitesimal) difference between values of an independent variable, there is at most an infinitesimal difference in one or more dependent variables. (A served tennis ball will change position very slightly in a fraction of a second.)

3. No matter how carefully measurements are undertaken, the state of the system can never be determined exactly.

Discrete phenomena are characterized by the following:

❑ A finite number of states (There are only so many combinations of pieces in positions on a chess board.)

❑ The smallest possible difference in an independent variable may result in a significant finite difference between states. (If "move number" increases by one, the positions of pieces on the chessboard will be in a distinctly different state.)

❑ The exact state of the system can be determined.

To further illustrate the difference between continuous and discrete phenomena, let's look at graphs of each. Figure 6-1 might illustrate the amount of water in a stoppered sink as it is being filled from a faucet, using time as the independent variable. Note the connectedness of the line.

Figure 6-2 shows the number of students "in" a classroom as they enter in the minutes before the class starts. Since students come in packages of one, you see jumps in the graph.

With discrete phenomena, exactness is possible. With continuous phenomena, it is not. A basket may contain exactly eight eggs. A bathtub cannot contain exactly eight liters of water.

Table 6-1 shows some examples of continuous and discrete phenomena. We might argue over some of the categorizations. Further, discrete phenomena may have continuous parts and vice versa. But you will probably get the idea from the list.

With our languages, we respect the difference between continuous and discrete. "How much" applies to continuous things. "How many" applies to things of a discrete nature. We wouldn't speak about an amount of votes—we would speak of a number of votes.

If we can use computers with enough precision to represent the world, does it matter that they are discrete machines and that that world is a mixture of continuous and discrete? (Can you tell if recorded music comes from a CD or an LP? Actually, some people can.) Is there an impact in using discrete machines to represent continuous phenomena? Usually only a little—but sometimes a lot. The idea of this section is to be sure that you understand that the potential exists for inaccuracies and errors when we use completely discrete machines to represent continuous phenomena. Punch line: computers are completely discrete machines!

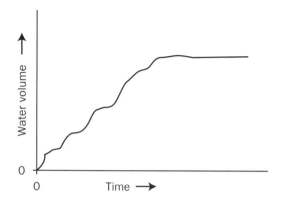

FIGURE 6-1　Continuous phenomena

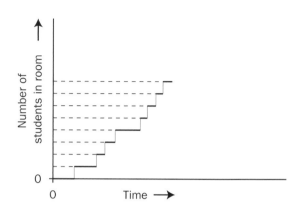

FIGURE 6-2　Discrete phenomena

TABLE 6-1

Some Continuous Phenomena:	Some Discrete Phenomena
Ramp	Steps
Water (Liquid)	Ice cubes
Time	Days
Real numbers	Integers, rational numbers
Musical representation—LP vinyl disk	Musical representation—compact disc
Musical performance	Musical score
Facial expressions	Wink
Sloping land	Terraces
Scrambled eggs	Soft-boiled eggs
Ping–Pong	Checkers
Meaning	Words
Analog (rotary) watch	Digital watch
Pond scum	People
Poison gas	Bullets
Belt-type people mover	Automobile
Light dimmer (rheostat)	Toggle switch
Steering (automobile)	Turn signal (automobile)
Brake (automobile)	Horn (automobile)
Pile of sulfur	Load of bricks
Handwriting	Typing
Slide rule (see Figure 6-3)	Abacus (see Figure 6-3)
Analog computer	Digital computer

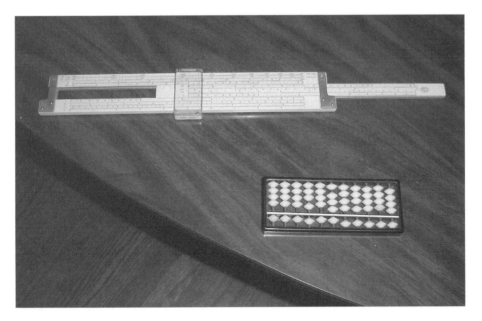

FIGURE 6-3   Early computers: abacus and slide rule

Some Implications of Discrete Representation for GIS

Because, with GIS, the world is represented in a discrete machine, everything a GIS tells you is an approximation—sometimes a poor approximation—of reality. For example, a GIS may represent a curving road by a series of straight-line segments, which are themselves represented by numbers. Sets of numerical coordinates (x and y pairs) placed along the centerline of the roadway define the beginnings and endings of the segments. The sum of the lengths of the segments will, unfortunately, underrepresent the true length of the curving road. The degree of error may be reduced by using shorter, and therefore more, segments, but the fundamental problem remains the same.

Further, the "real world" is virtually infinite in the level of detail that exists (look through a microscope if you doubt this), while a computer store is finite, and in many ways, quite small—compared, say, with what is in your own head.

How does a digital machine store information about a continuous environment? By digitizing—using this term in the most general way. We describe the world with numbers—integers (such as 7 and 2383) and "floating-point" numbers (such as 1.618034 and 6.626 times 10 to the negative 34th power). We also use strings of letters and other symbols: A, a, B, #, %.

To even have a chance at being sufficiently accurate when we specify a location on the Earth's surface, we need to use a lot of digits. For example, the location of a particular fire hydrant near Vancouver, British Columbia, Canada, was reported by a GPS receiver to be 49.2773361 north latitude and 122.8793473 west longitude. Since each decimal digit requires some two and a half bits, the resultant binary number can be quite lengthy, especially when you must add more bits for the exponent needed for floating-point numbers. The capability to represent very precise numbers is sometimes called double precision or extended precision, and is frequently required in GIS. With regular "real" numbers, called floating-point, you can count on six significant digits. With double-precision numbers you can count on 16 significant digits.

Scientific Notation, Numerical Significance, Accuracy, and Precision

How Old Is the Dinosaur?

The curator of a natural history museum had a habit, from time to time, of walking around and listening to the guides give their lectures at the various exhibits. One day he arrived just in time to hear that the museum's Tyrannosaurus rex was sixty-five million and three years old. He went back to his office and told his secretary: "Gladys, tell George I want to see him as soon as he's done with his tour." When George appeared, the curator exasperatedly asked him, "What do you mean telling people that the Tyrannosaurus rex fossil is sixty-five million and three years old?" George looked abashed, but said confidentially, "Well, you hired me three years ago and you told me then that the skeleton was sixty-five million years old, so . . ."

A GIS will deal with very precise numbers—that is, numbers that contain many digits. For example, locating the longitude of a point on Earth's surface within a centimeter (not at all an unrealistic expectation nowadays) requires a number with 10 significant digits—for example, 123.4567890. Because of the number of bits devoted to "ordinary" numbers by most computers, and because fractional decimal numbers may not be represented exactly by fractional binary numbers, GIS frequently use "double-precision" numbers to represent positions.

Also, a GIS may deal with very big and very small numbers. Very large and very small numbers are stored in the computer in a fashion similar to scientific notation: The number is represented as a mantissa (which contains the significant digits of the number) and an exponent (which tells how many digits and in which direction to move the decimal point). For example, the number in scientific notation:

2.0×10^{-7} (which is two times ten raised to the negative seventh power) represents the number 0.0000002

To use this mantissa-exponent notation to represent 7,009,181,222 you would write the following:[4]

7.009181222×10^9

Precision vs. Accuracy

The terms "accuracy" and "precision" do not refer to the same idea. Both are important to GIS. A classic example of the difference follows:

Weather forecaster A indicates that it will be between 40 and 50 degrees tomorrow at 4 p.m. The actual reading turns out to be 43. Thus, the forecast was accurate, but not very precise. Forecaster A provided a true statement but without much detail. Forecaster B states that it will be 52.47 degrees at 4 p.m. tomorrow. The temperature turns out to be 43 degrees. Forecaster B was very precise, but not accurate.

---

[4]The estimated world's human population as of Wednesday 25 April 2012 at 00:23 UTC (Greenwich Mean Time (GMT)).

The terms accuracy and precision can also be applied to a *set* of readings or measurements. Again, a classic example: Darts are thrown at targets. The target of person "R" looks like Figure 6-4.

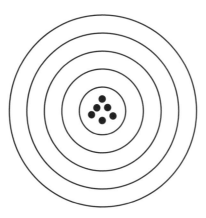

FIGURE 6-4   Good accuracy and good precision

Her dart throwing was both accurate and precise. The darts are where they are intended to be (accurate), and very close together (precise).

Person S's target looks like Figure 6-5.

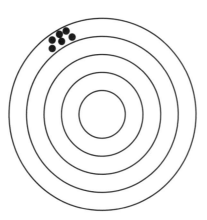

FIGURE 6-5   Poor accuracy but good precision

The darts are tightly clustered (they hit a spot almost precisely) but lack accuracy, as they missed the intended spot. (Perhaps some sort of systematic error is present—such as a breeze blowing up and to the left.)

Person T's target is as shown in Figure 6-6.

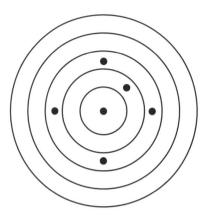

FIGURE 6-6   Good accuracy but poor precision

Figure 6-6 indicates a lack of precision, but when the dart positions are averaged out, they are quite accurate—they come quite close to the intended spot. (Note that this doesn't count for much in the game of darts, but would be useful in finding an intended spot if one had a number of readings, say, latitude and longitude, from a process or piece of equipment, such as a GPS receiver, known to be accurate.)

Well, what can we say about the target of Person U?

The darts are not clustered (not precise). They do not average out to the intended spot (inaccurate). See Figure 6-7.

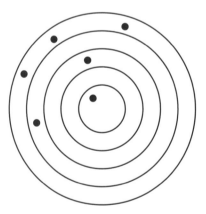

FIGURE 6-7   Poor accuracy and poor precision

To summarize: Precision refers to the degree of specified detail. A number with many significant digits is a precise number. (Unfortunately, the "significant digits" may not truly be significant.) Accuracy refers to the correctness of the specified number. What counts is its truth. It may be precise or imprecise.

For information to be useful it must have substantial degrees of both precision and accuracy. "How much" of each? Obviously the issue is complex.

Basic Statistics

Statistics may be defined in very erudite ways, but fundamentally, it is the art and science of making a big bunch of numbers into a little bunch of numbers so you can understand stuff about the big bunch of numbers. A GIS can usually do statistics without even breathing hard. Here are some basic statistical measures, most of which you probably know:

Average

The mean or average: add up all the numbers in a set and divide by the number of numbers. Expressed as a formula the average, called x-bar here, because of the x with the bar over it on the left-hand side, looks as shown in Figure 6-8.

$$\overline{x} = \frac{\sum_{i=1}^{n} x_i}{n}$$

$$\sum_{i=1}^{n}$$

FIGURE 6-8   Formula for average of a set of numbers

FIGURE 6-9   Sigma: symbol that means to sum a set of numbers

This formula scares some people. It shouldn't. It simply says what was said in the preceding text. More specifically: Assign integers to each number, like 1to the first number, 2 to the second, and so on until you reach the last number, which we refer to as n. (If there were 12 numbers, n would be 12.) The numbers then are $x1, x2, x3, \ldots x12$. So when we say xi and i = 7, we are referring to the seventh number, named $x7$. The symbol shown in Figure 6-9 means "take the sum of" whatever follows it. So in the case of the symbol in Figure 6-9, what follows it is x_i. What values of i do you use? Those that run from 1 through n, namely, 1, 2, 3, . . . 12. Finally, the formula in Figure 6-8 says divide the sum by the number of numbers, which is n, or, in this case, 12.

Median

The median number is simply the middle number (or the average of the two middle numbers) in a set of numbers that has been sorted from smallest to largest (or largest to smallest—it makes no difference). The median is not sensitive to very large or very small outlying values, as is the mean.

Mode

The mode of a set of numbers is the most frequently occurring number. A set of numbers could have several modes or none at all.

Range

The range of a set of numbers is the largest number minus the smallest. This gives you an idea of the spread of the numbers. A better idea of the spread is perhaps the standard deviation, indicated next.

Standard Deviation

The formula for standard deviation is shown in Figure 6-10.

$$\sqrt{\frac{\sum (x - \bar{x})^2}{n - 1}}$$

FIGURE 6-10   Formula for standard deviation

The formula says this: (1) Find the average (that's x-bar, from Figure 6-8), (2) make a new set of numbers by subtracting that average from each of the numbers in the original set, (3) make a third set of numbers by squaring each of those in the second set, (4) add up this third set, (5) divide the sum by n – 1, and (6) take the square root.

If the data are normally distributed—as, for example, the heights of adult men in a town might be—then the standard deviation has a graphical meaning as well as a numerical one. Suppose Figure 6-11 is the curve for the heights, with each height rounded off to the nearest inch.

The x-axis direction represents the possible heights and the y-axis direction represents the number of individuals who have each given height. The curve is "bell shaped about the mean"; the mean is 65 inches. Graphically, the standard deviation is defined by the *x* values on the curve at its points of inflection—where the curve goes from being cupped down to being cupped up. The standard deviation is 4 inches. This means that approximately 68 percent of the individuals are between 61 inches tall (65 – 4) and 69 inches tall (65 + 4) The idea behind the standard deviation is to give you an idea of how spread out your data values are. The standard deviation is meaningful, but only if the data are close to a normal distribution are you able to make statements like: 47 percent (34 percent + 13 percent) of the individuals are between 65 inches and 73 inches tall (65+4+4).

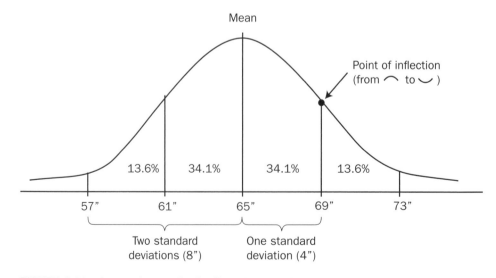

FIGURE 6-11   Curve of normally distributed data values

Putting Values into Classes

Another way of reducing a large group of numbers to a more manageable size is to put similar values into classes or categories. For example, we might have the data on real estate parcel sizes. We could put each parcel into one of several categories: (1) greater than 10 acres, (2) 10 acres to 5 acres, (3) 5 acres to 1 acre, and less than 1 acre. In the Step-by-Step section of this chapter, you will do an exercise on classification of values.

Measurement Scales

When you see a number, its context determines how you interpret that number and how you may use it in computation and comparison. A number that represents some description or measurement of something in the real world will fall into one of four categories:

❏ Nominal—A nominal data set is just a set of names, except the names take the form of numbers. If in a teaching laboratory you randomly assign numbers to test tubes (say, to prevent student experimenters from knowing what is in them), you are forming a set of nominal data. There is no numeric relationship between the numbers. Each is merely a name. The only operators you may use on the numbers are equal (=) and not-equal (< >). For example, suppose the soil type found in area A is stored as x and the soil type found in area B is stored as y, you can ask: "Does x = y ?" (assuming x and y are integers or text strings)?

❏ Ordinal—The numbers in an ordinal data set indicate an order among the entities they represent. Perhaps 1, 2, and 3 indicate first-born, second-born, and third-born children. You know that, in a given family, a child numbered 1 was born before a child numbered 3. But you don't know how many months or years before. In addition to using the equal and not-equal operators on ordinal data, you may use less than (<), less than or equal (<=), greater than or equal (>=), and greater than (>). For example, house numbers on a given side of a street in the United States are generally in order. Suppose house numbers increased along the right side of a north-south road as you proceeded south. You could determine which of two houses was the southmost house by asking: is house number of X greater than house number of Y?

❏ Interval—An interval data set consists of numbers that come from a measurement scale where a unit amount is established and values along the scale are linear multiples of that amount. Examples are the common, nonscientific temperature scales: Celsius and Fahrenheit. With interval data, you may use all the operators described previously and also the arithmetic operators addition (plus, +) and subtraction (minus, –) as well. For example, if it were 20°F in the morning and it became 60°F in the afternoon, you may say that it had become 40°F warmer $(60 - 20) = 40$.

You have to be very careful when using multiplication and division operators on interval data. For example, you may not say, given the preceding numbers, that it became three times as warm in the afternoon. The reason is that a set of interval data does not necessarily have a meaningful zero point from which to measure.[5] For example, if the morning temperature were negative 20°F and the afternoon positive 20°F, the difference would still be 40°F, but you can see the difficulty in saying that one

---

[5]The zero in the Fahrenheit scale was originally determined by the temperature of a mixture of ice and salt. Check the Internet for the bizarre determination of 100 in that scale. The zero in a Celsius (centigrade—changed in 1948 to honor the inventor of the Celsius thermometer, Anders Celsius) scale is arbitrarily defined to be the freezing point of water and 100 is its boiling point.

is a multiplicative factor warmer than the other. (Three times –20 is –60—a clearly wrong result.) The use of multiplication (*) and division (/) with interval data is not completely ruled out, however. As long as the answers are understood to be in the context of the arbitrary zero, you can make limited use of * and /. For example, if the morning temperature were 20°F and the afternoon 60°F, you could say that the mean temperature was 40°F ((20 + 60)/2) = 40.

❑ Ratio—A set of ratio-scale data is like an interval-scale data set, with the additional proviso of a non-arbitrary zero point. The simplest example might be measured distances. A ruler has an absolute zero and a unit value of 1 inch. I can say both that, if desk A is 30 inches long and desk B is 45 inches long, desk B is 15 inches longer than desk A, and also that desk B is 1.5 times as long as desk A. With ratio data you may use any comparison or arithmetic operators (=, < >, <, <=, >=, >, +, –, *, /, and exponential (^ or **)) plus a whole batch of functions, such as square root. Of course, you must be careful to know the limits and characteristics of your data. You may not, for example, take the logarithm of a negative number. And when taking the cosine of 450° you must understand what that means.

Analysis of GIS Data by Simple Examination

— Open your Fast Facts File.

— Open the Color Figures file.

Reviewing and Learning More of ArcMap

The goal of this exercise is to review ArcMap and to become familiar with more of its features and complexities. You may have the thought: "Hey, I've already learned this." By going over things you already know and integrating them with new material, you will discover a way toward using GIS for analysis of spatial data. These steps involve basic statistics, thematic mapping, changing symbology for emphasis, and classification. You can probably breeze through the steps very quickly, but you shouldn't. Each step is one that you will have to know how to do in the future, without much explanation. After doing each step, you should ask yourself: Will I know how to do this in a month, if I don't use it until then and simply rely on my memory? If not, write it in your Fast Facts File. Later you can reorganize the information you record during these exercises.

To make sure that everyone is working from the same geodatabase, you will copy a pristine version from the folder IGIS-Arc_AUX. You will then load the personal geodatabase featureclass Landcover onto a new empty map:

— **1.** If they are running, close ArcCatalog and ArcMap. Use the operating system of the computer to delete the folder

   ___IGIS-Arc_*YourInitials*\Wildcat_Boat_Data.

   Launch ArcCatalog. Find the folder

   [___] IGIS-Arc_AUX\Wildcat_Boat_Data

   noting carefully that it is the AUX folder you want, not the one you have been working with.

   Copy the folder Wildcat_Boat_Data from

[___] IGIS-Arc_AUX

to ___IGIS-Arc_*YourInitials*.

_____ **2.** Launch ArcMap with a Blank Map, and make the window occupy the full screen. Open the dockable ArcCatalog window, navigate to

___IGIS-Arc_*YourInitials*\Wildcat_Boat_Data\Wildcat_Boat.mdb\Area_Features \Landcover

and drag the feature class into the map. Landcover will show up as a homogeneous color background with lines dividing polygons. If the cursor has somehow slipped into "Pan" mode, click on the Select Elements icon.

_____ **3.** In Data Frame Properties > General, make sure Display Units is set to Meters. Map Units is already set to meters because of the coordinate system of the feature class.

Examining the Toolbars

_____ **4.** If you right-click anywhere in the areas containing toolbars (or use Customize > Toolbars), you get a list of all the available toolbars, with an indication of which ones are displayed. To get an idea of the extensive capability of ArcMap (and for future reference), type the list into your FastFactsFile. Give a moment of thought to each toolbar, regarding what sort of capability it might provide as you type it.

Done? Yes ___ No ___

At the bottom of the list notice that there is a toolbar named Customize. ArcMap is software that you can "adjust" to your own specifications. You can customize toolbars and commands. Even when you find out how (I touch only lightly on this subject in this book), customize carefully. The more you customize an interface, the more difficulty you will have when confronted with the standard interface, say, on another computer. Further, if you share your computer with another person, he or she may have trouble operating the interface.

_____ **5.** Turn off all toolbars except Standard, Draw, and Tools.

_____ **6.** Click the Identify button. Point to a polygon with the mouse cursor and click. Examine the Fields and Values that appear in the Identify window. LC_CODE[1] is the number of the land cover type (see table below); COST_HA is the cost in U.S. dollars per hectare.[2] Dismiss the Identify window.

---

[1]The land cover code explanation may be found in Exercise 1-1
[2]A hectare is an area equivalent to a square 100 meters on a side—about two and a half acres.

LC-CODE	LC-TYPE
100	Urban
200	Agriculture
300	Brushland
400	Forest
500	Water
600	Wetlands
700	Barren

____ **7.** Open the Landcover attribute table. How many records are in it? _____ So, how many land cover polygons are there? _____ Sort the table by polygon area by right-clicking the heading Shape_Area and choosing a sort command. To one decimal place, what is the area of the largest polygon that is *not* land cover code 500 (water)? _____ (The units are square meters.) What is the land cover code of the smallest polygon? _____ What sort of land cover is it? _____ If any records or polygons have been selected, press Clear Selection.

____ **8.** Calculate statistics on areas: Again right-click the heading Shape_Area, and click Statistics to get a Statistics of Landcover window. What is the average of the areas of the polygons (to the nearest tenth of a square meter)? _____ While you have the statistics window open, choose Shape_Length in its Field text box. Shape_Length gives the perimeters of the polygons. If you added the perimeters of all the polygons together what would be the sum? _____ meters. What is the approximate cost per hectare of the most expensive land? $_____ Dismiss the statistics window.

____ **9.** Sort the records back into their original order based on the Object Identifier (OBJECTID). Clear any selections.

Pointing at Records

____ **10.** Experiment with various ways of pointing at records: Click any cell of any record. Note that, in the gray box to the left of the record, a triangular marker appears. The Record indicator at the bottom indicates the position of that record in the table. By pressing the buttons on either side of the Record indicator, you can control which record is pointed to. Only one record may be pointed to at a time. Pointing to records has nothing to do with selecting records.

Two Windows Are Available for Selecting

As with many operations with this software, there are multiple ways of accomplishing a task. Using the Selecting by Attributes option is no exception. It's worth taking a look at the two windows you can use for this. You get to one window through the Options button on an attribute table. You get to the other one through the Selection menu from the Main menu. As a side trip here, look at both.

_____ **11.** On the attribute table, select Table Options > Select By Attributes. Move the window down a bit and to the right. From the Main menu, choose Selection > Select By Attributes. Move this window off to the left and down. Compare the two windows. See Figure 6-12. If you use the window from Options, then the Layer is obviously set for you, since you picked Options off an attribute table of a layer. In the other window you must pick the layer. Close both selection windows.

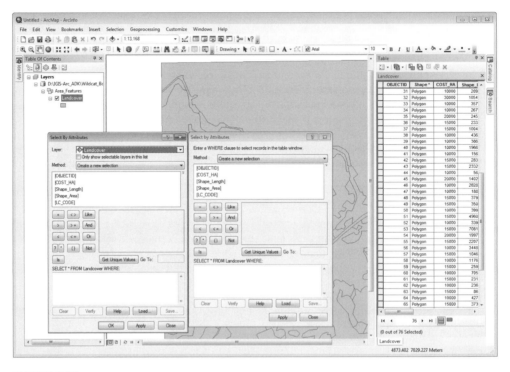

FIGURE 6-12

Selecting Records (and, Thereby, Features)

(A lot of what follows is review and practice. It provides the sort of information that needs to become second nature to you as you operate the software.)

A record is selected if it is highlighted in cyan. Any number of records may be selected—from none to all—or any combination in between.

_____ **12.** Experiment with various ways of selecting records: In the Table, click the gray box at the left of any record. The entire record is selected. Click another box of another record, and that record is selected instead. If you hold down the Ctrl key and click the box of an unselected record, it is added to those already selected. If you hold down the Ctrl key and click the box of a selected record, it is deselected. If you hold down the Ctrl key and drag the cursor over consecutive boxes, the selected records are deselected and the unselected records are selected. The number of selected records is shown at the bottom of the table. Also at the bottom of the table, you can choose whether the window displays all the records or only those selected. Choose records with OBJECTIDs 2, 3, 5, 7, 11, 13, 17, 19, and 23. Show only those selected records. Then show all records.

13. Use simple logic to select records: Press the Table Options button on the table window. Press Switch Selection. Note that all records except 2, 3, 5, 7, 11, 13, 17, 19, and 23 are selected. Now switch the selections in a different way: press the Switch Selection button on the table menu bar. Choose Clear Selection, then Switch Selection, then Clear Selection. Find a way to Select All, and finally, Clear Selection from the Table Options menu.

14. *Select records by attributes:* Click Table Options > Select By Attributes to bring up a window of (almost) the same name. Type the following expression (observing capitalization) in the large blank in this window:

[Shape_Length] <= 201

and click Apply. This selects all those records that have a Shape_Length value less than or equal to 201. Dismiss the Select by Attributes window. Examine the table. Show only selected records. How many selected records are there? _____ out of _____. Run Statistics on the Shape_Length field. What is the average? _____. Now dismiss the Selection Statistics of Landcover window, select Show All Records, and run Statistics again. What is the average? _____. Choose Clear Selection. Run Statistics again. What is the average? _____. Make a note to yourself in your Fast Facts File that if any records are selected, no matter what records are shown, statistics are calculated *only on the selected records*. Forgetting this fact is a good way to get wrong answers. Dismiss the statistics window.

15. Again click Table Options > Select By Attributes. Erase the expression (there is a Clear button). Experiment by building the partial expression

[Shape_Area] <=

by clicking or double-clicking various buttons and values in the window. This query-building method is a quick way to create an expression while avoiding the possibilities of mistyping. (Even when you use the query builder, you still can type parts of an expression, and at times you may have to.)

Surrounding feature class names with square brackets (i.e., [and]) works with personal geodatabases but not with file geodatabases, where double quotation marks are used instead. (There are several other differences in syntax between the types of geodatabases. When you want to see these differences, you can go to HELP, click the Search tab, type "Migrating to the file geodatabase" into the text box, press the "ask" button, and then display Migrating to the file geodatabase. For now, just put this in your Fast Facts File.)

Press Get Unique Values to bring up all the possible Shape_Area values in the Unique Values pane. Scroll down to the value that begins 7710.087 and double-click. This will complete the expression:

[Shape_Area] <= 7710.08730661013

Click Verify to determine that the expression is valid. Click OK. Click Apply. Dismiss the Select by Attributes window. Note what happens when you show all records and sort them in descending order by Shape_Area to determine that the query worked properly.

___ **16.** Make the table narrower and move it out of the way so you can see the map. You will notice that many of the small polygons on the map have a border around them. They are the polygons that correspond to the selected records—those with area less than or equal to 7710.08730661013. The lines that define a selected polygon are shown in cyan (it's cyan, by default—you can change the color if you wish).

___ **17.** Using Selection on the Main menu, clear all selected features. Sort the table (ascending) according to OBJECTID. Select the record with OBJECTID number 15. Note that on the map you see two cyan rings. Why? Use Identify or your Fast Facts File if you need help.

_____

Looking at the Other Capabilities of the Options Menu

___ **18.** Open the attribute table again if you closed it. Besides the several ways of selecting records that you just explored, the Options button leads to additional capabilities. List those entries in your Fast Facts File. Done? Yes ___ No ___

Selecting Features (and, Thereby, Records)

Selecting works the other way too. If you select features on the map the corresponding records are selected. Many ways of selecting geographic features are supplied by the software.

Quick Selection of Features

___ **19.** Set the attribute table so that it shows all records. Clear all selections. Set the table so it shows only Selected Records (which will make the window empty). Find and mouse over the Select icon ("Select Features by" icon in version 10.0) on the Tools toolbar. Note what the ToolTip or Status bar has to say about the operation of the tool. Explore the dropdown menu. Press the button. Click a polygon and notice that its record is selected. Click another. To select several polygons, hold down the Shift key (rather than Ctrl, as you do when selecting multiple records in a table) and click the intended polygons.

___ **20.** Drag a box inside a polygon. Notice that that polygon is selected. Draw a box that is encompasses parts of several polygons. Note that they are all selected. Clear selections. Select three or four adjacent polygons (use Shift+Click).

___ **21.** Under Selection in the Main menu, pick Zoom To Selected Features. So here you have a technique of zooming in on exactly what you selected (whether selected by attribute or location). Zoom back to Full Extent.

___ **22.** To clear all selections, click somewhere off the map with the Select Features tool. Close the attribute table.

Selecting by Location

In what follows you will examine some of the interesting and useful ways ArcMap has of selecting features by their locations.

___ **23.** From

    ___IGIS-Arc_*YourInitials*\Wildcat_Boat_Data
    \Wildcat_Boat.mdb\Area_Features

add the feature class Soils to the map. Make the symbol color of soils Light Apple green. Make sure the Soils entry appears above the Landcover entry in the T/C.

From

    ___IGIS-Arc_*YourInitials*\Wildcat_Boat_Data
    \Wildcat_Boat.mdb\Line_Features

add the feature class Roads. Using the Symbol Selector window, represent the roads as Highway under Symbology, reached by double clicking the entry name. The Highway symbol will show up as red, but change the width to 2. Make sure it appears at the top of the T/C.

___ **24.** Choose Selection > Select By Location. Read the text at the top of the box. Under Selection method you have four choices. List the contents of the drop-down text box to fix the possibilities in your mind. _____, _____, _____, _____

For the Selection method, choose "select features from". In the text box "Target layer(s)" put a check by Roads. In the Source layer text box choose Soils. For the Spatial selection method, choose from the drop-down menu "Target layer(s) features are completely within the source layer feature". If you were expressing this in English you would say: "I want to select all features from the layer Roads that are completely within the boundaries of any polygon of the features of Soils." If a feature of Roads crosses a polygon boundary it will not be selected.

___ **25.** Click Apply, then Close. From the map, you can see that the roads that are completely contained within any single polygon are highlighted in cyan. Those segments that cross polygon boundaries are not highlighted.

___ **26.** Open the Roads attribute table and note, using Identify on a few of the selected Road features, that the corresponding records (based on OBJECTID) are selected. Close the attribute table.

___ **27.** If necessary, drag the layer Landcover to the bottom of the T/C. Clear all selected features. Turn off Roads, but leave both Landcover and Soils on. What layer appears on the map?
_____

____ **28.** Again choose Select By Location. This time select features from Landcover that are completely within the features in Soils. Click Apply. Click Close. Notice that although the Soils layer obscures the Landcover layer, the selected Landcover polygons are shown in cyan. Click Soils off and on to assure that this is the case. Open the attribute table of Landcover. How many polygons are selected? ____ out of ____.

____ **29.** Open the Select By Location window one more time and look at the possibilities in the drop-down menu near the bottom of the window. Type them into your Fast Facts File. Think what each means as you enter it.

Done? Yes ___. No ___.

Reviewing and Understanding Actions on the Table of Contents

____ **30.** *Look at getting information from entries in the table of contents (T/C):* Drag Soils to the bottom of the Table of Contents. Click the box containing the check mark next to Landcover. Note the result. Click it again so the layer is on. Click the box with the little minus sign in front of Landcover, and note that the symbol that represents Landcover goes away. Click the little plus sign. Click the List By Source tab at the top of the T/C. Note that you are given the path, in a hierarchical form, to each data set. Experiment by clicking the boxes that contain either plus (+) or minus (–) in the T/C. Understand these actions. Leave all entries fully expanded. Clear all selections.

____ **31.** Press the List by Selection tab at the top of the Table of Contents. The check boxes here let you set the layers that are selectable. That is, if a layer's box isn't checked, the Select Features tool cannot select features from that layer. Prove this to yourself by clicking the "Click to toggle selectable" box that is associated with Landcover to make Landcover un-selectable. It will appear in a different part of the T/C. Now click the map with the Select Features tool. Features will be selected, but they will be from the soils layer underneath. Turn off the display of the soils layer by clicking "Click to toggle visibility". Now notice that Select Features has no effect. The lesson: For Select Features to work, a layer must be both displayed and selectable.[3] Make all layers selectable. Clear any selections.

Layers and the Data Frames

____ **32.** Understanding layer properties versus data frame properties: In the Layers pane, click the List By Drawing Order tab. In the T/C make all the +/– boxes say minus, so that nothing is hidden. Right-click the text Landcover. Click Properties. A Layer Properties window, with a number of tabs, appears. The Layer Properties window tells you (or lets you specify) the properties of the particular layer that you right-clicked on, such as how it is symbolized, what the fields of the table are, what its extent is, how selected features are to appear, aspects of feature labeling, the coordinates and characteristics of its data source, and several other parameters.

_____

[3]However, features from un-selectable layers can still be selected by selecting their records from the attribute table.

Write the names of the tabs in your Fast Facts File.

Done? Yes ___ No ___

Dismiss the Layer Properties window for Landcover.

___ **33.** Now right-click in the right pane of the overall window—where the map is shown. Click Data Frame Properties. The Data Frame Properties window tells you (or lets you specify) the properties of the display of the map, such as the scale, the map units, the coordinate system, whether a reference grid appears and so on. Write the names of the tabs in your Fast Facts File. Dismiss the Data Frame Properties window.

Done? Yes ___ No ___

It is important to be cognizant of the difference between layer properties and dataframe properties.

Changing Layer Properties

As noted, Landcover is displayed using a homogeneous color background with lines dividing polygons. The polygons represent different land uses. We can make the map display these different uses in a variety of ways.

___ **34.** Bring up the Landcover Properties window again—this time by double-clicking the Landcover layer name. Choose the Symbology tab. You have your choice of five major possibilities of what to show. List them:

_____, _____, _____

_____, _____

Note that features are displayed with a Single symbol. (If you don't see this, click the word Features.)

___ **35.** Instead, let's display the layer using Categories > Unique Values. For the Value field, pick LC_CODE, which, as you may recall, is an integer (100, 200, and so on). Click Add All Values to bring up the list of values and a random list of colors with which those values will be symbolized. Click Apply and OK in the Layer Properties window. A garish-looking map appears. Observe the map. Rather than being displayed with a single color, different polygons are shown with different colors, based on the value in the LC-CODE field.

___ **36.** *Label the polygons with text:* Bring up the Landcover Properties window again. Click the Labels tab. Check the box that says Label features in this layer. Make the Label field read LC_CODE. Click Apply. Click OK. The land cover code appears in each polygon. Right-click on the Landcover name in the T/C. Remove the check beside Label Features. The labels go away. Restore them by turning Label Features back on.

___ **37.** Suppose that you want the cost per hectare shown instead. Bring up the Layer Properties window, and click the Symbology tab. In the Value Field drop-down menu, pick COST_HA. Add All Values. Click Apply, then OK. The map colors have changed, but the labels remain the same. So, go into properties again and select the Labels tab. In the Text String Label Field box, select COST_HA. Click Apply, then OK. The cost figures appear.

___ **38.** The numbers are a bit too big for the map. You don't need all those zeros. In the Layer Properties window, Labels tab, click Expression. In the Label Expression window, add / 1000 to [COST_HA]. The window should look like Figure 6-13. When executed, this expression will cause each label value on the map to be divided by one thousand.

FIGURE 6-13

Click OK, Apply, and OK. Now only two-digit numbers should appear in the polygons.

___ **39.** *Change the color symbology for a subset of features:* The Land cover code 500 is for water. Water in large quantities is usually blue. Change the field on which to base the Symbology of the Landcover layer back to LC-CODE. (Note that to change the label field back to LC-CODE you have to put [LC-CODE] by itself in the expression text box.)

Display the map with random colors as before. In the T/C, right-click the color rectangle next to 500. Pick a blue from the color pallet that shows up. Note the effect on the map.

___ 40. Experiment with other layer properties and data frame properties. These windows allow you a great deal of control over what is displayed. You should become familiar with them. A large part of successfully operating ArcMap is knowing where to find and use the various controls.

___ 41. Right-click the Landcover layer and remove it from the map. Remove Roads as well. Make sure Soils is turned on. The map now appears consisting of a homogeneous color background with lines dividing polygons of Soils.

As before, you could identify individual polygons with the Identify tool. You could place text in each polygon with the labeling facility. You could open the attribute table and, by selecting polygons, see the corresponding rows in the table, and vice versa. All of these techniques are useful, but none gives you a very good picture of what features are where, and it doesn't let you compare various features. In the steps that follow, you change this by presenting the polygons as a thematic map, with different colors for each type of soil. First you label the polygons with SOIL-CODE.

___ 42. Bring up the layer's properties window. (What are two ways to do this? _____, _____.) Click the Labels tab. Check the box that says Label features in this layer. Use SOIL-CODE in the Label field. Click Apply and on OK. Check out the resulting display.

___ 43. Bring up the layer's properties window again. Click the Symbology tab. In the Show box you will probably see Single Symbol highlighted under Features. Click Categories and make sure Unique values is selected. Make the Value Field SOIL-CODE. Click Add All Values. Click Apply, then OK. Notice that all polygons labeled Sg are the same color. Change the symbol used to depict the water to Atlantic Blue or Pacific Blue. If there are other polygons with a color close to the blue you picked, change them to some other color.

___ 44. Now you can clearly get an impression of the locations and amounts of the various soil types. What (commonly named) color is associated with Ko? _____

___ 45. Change the labeling so that soil suitability (SUIT) is shown. By inspecting the map and the legend, and using the Identify tool if necessary, determine the soil codes of the soils that have suitability 1. _____.

Suitability 2? _____.

Suitability 3? _____.

What is the suitability value shown for water? _____.

___ 46. Now switch the labeling and the color representation, that is, label the polygons again with the soil code and show the suitability with different colors. Check your answers from the last step.

___ 47. There are practical and visual limits to the numbers of color categories you can add to a map. So ArcGIS has the capability to display only the categories you choose. To see how this works, under Symbology in the Layer Properties window, click Remove All, then click Add Values. Now you have the option of adding only some of the values. Since soils suitable for the Wildcat Boat facility are only 2 and 3, add only those values (use Ctrl-click). Click OK, Apply, then OK. Note that the rest of the map is colored with the <all other values> symbol.

___ **48.** Under the Symbology tab in the Layer Properties window, click Remove All. Now, in Show > Categories, click Unique Values, Many Fields. Note that there are now three possible Value Fields instead of just one. Put SUIT in the first of these text boxes and SOIL-CODE in the second. Now click Add All Values. Click Apply, then OK. Different colors for all the unique combinations of suitability and soil codes that exist in the table will be shown in the Table of Contents and on the map.

___ **49.** You can also fix it so that the labels reflect both suitability and soil code. Under the Labels tab in the Layers Property window, click the Expression button. Arrange it so that the Expression text box says, exactly:

[SUIT] & "-" & [SOIL_CODE]

by typing and/or clicking. This expression says concatenate three values, the middle one of which is a hyphen. The double quotation marks let you place any text string in the expression. Click OK, then Apply, then OK. The result should look something like Figure 6-14. Close ArcMap without saving.

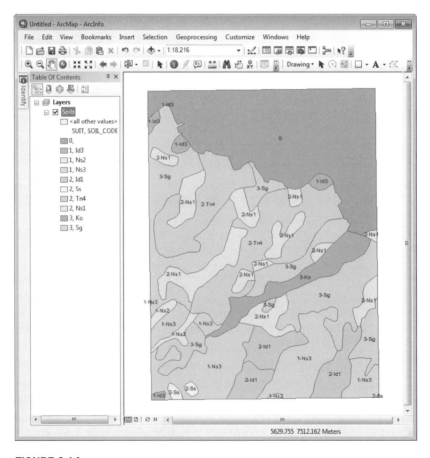

FIGURE 6-14

Categorization and Symbolization
Thinking about Maps Again

Now that the connection between the table and the map is firmly re-established in your mind, we turn back to the map itself. You could broadly classify maps into two categories: general-reference maps and thematic maps. Many maps, like road maps or topographical maps are used for general information, navigation, topology, and to show locations of features. Thematic maps (sometimes called statistical maps), on the other hand, show polygons that divide up the landscape into distinct areas; the areas are coded—with color or other symbolization—to indicate the value of a characteristic or attribute about each individual area. So, in a sense, attribute information is reflected directly, visually, on the map; in GIS you find this information mostly in tables. (Description of areas somewhat true for general-reference maps as well, but in thematic maps this information is the primary reason for the existence of the map.)

Classification (or Categorization) and Symbolization

Basic statistics is one way of understanding bunches of numbers. Another way is to place things in categories, based on some pertinent number associated with each thing. For example, we could categorize (i.e., place into separate groups) the students in a class according to their heights. Put everybody less than 4 feet tall in group A. Those 4 feet or more but less than 4 feet, 3 inches belong in category B. Those who are 4 feet, 3 inches or more but less than 4 feet, 6 inches go in category C, and so on. So here's the general concept: We have k objects and we place each of them in one, and one only, of n categories, where n is less than (or, in a trivial case, equal to) k. Almost without saying, the first category consists of a set of smallest numbers, the next category consists of the set of next smallest of numbers, and so on.

Consider another example: Suppose that we had a set of numbers (which I have put in order to make things simpler).

1 1 2 3 3 4 5 7 8 10 10 11 12 15 19 19 22 23 25

That is, we could place all the numbers from low to high, say, in a text string from left to right.

If we wanted three categories we might partition them as follows by assigning obvious breaks between categories:

1 1 2 3 3 4 5	7 8 10 10 11 12 15	19 19 22 23 25
Category 1	Category 2	Category 3

Here the simplicity ends. The goal is to arrange things so that humans can best understand the nature of whatever is being studied. To do this, three fundamental questions must be addressed: (1) How many categories are appropriate for a given set of data? (2) What approach should be used to partition the objects into categories? (3) How might the results be effectively presented or displayed.

Looking first at the issues raised by question 2, ArcMap allows you to use several techniques to plunk objects into categories, or classes, based on the numbers associated with the objects:

❏ Manual

❏ Equal interval

❏ Defined interval

❏ Quantile

❏ Standard deviation

❏ Natural breaks

We look at classification not only as putting numbers in categories but also as classifying the geographic areas to which they relate. For example, in what follows I've made a cartoon grid of equally sized areas which has different populations of females, males, and wombats in each area.

___ **1.** Start ArcMap with a Blank map and the dockable Catalog Tree. From

___IGIS-Arc_*YourInitials*\Trivial_GIS_Datasets

add as data a shapefile named Classify_This.shp. Ignore a warning about spatial reference. Open the attribute table. Sort the field FEMALES (which indicates the number of females in each rectangle) so that it is in ascending order. How many records are there? _____ How many attribute values are in the vicinity of each of the following numbers? (Hint: Select records by dragging; read the bottom of the window.)

1,000 _____
5,000 _____
10,000 _____
15,000 _____
25,000 _____

Close the table. Clear any selections.

___ **2.** Label each polygon with the field FEMALES. Examine the map. Try to get an idea of where the clusters, and highest and lowest, female population values are. You may find this difficult.

___ **3.** Bring up the Layer Properties of Classify_This. Click Symbology > Quantities > Graduated Colors. Choose the Fields Value FEMALES from the drop-down menu. Find a color ramp that goes from light green to dark green, left to right.[4]

___ **4.** Press the Classify button. In the Classification window, set the Classification method to Equal Interval. Choose seven classes. Observe the Classification Statistics. What's the minimum? _____ Maximum? _____. See Figure 6-15.

---

[4]If you are more comfortable using a name for the color ramp, right-click on the ramp itself, uncheck Graphic View, and pick "Green Bright."

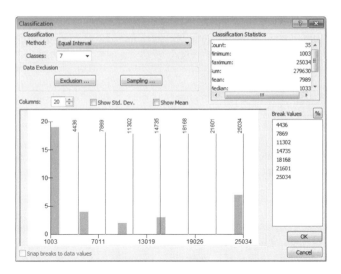

FIGURE 6-15

_____ **5.** Look at the histogram (the graph with the gray columns). These columns represent the number of values that are in the vicinity of abscissa (x-coordinate) values. Change the number of columns to 20. By looking at the vertical axis, indicate how many values (around 1003) are represented by the first column on the left. _____ How many by rightmost column? _____ (Notice the tick marks and reference numbers at the bottom of the histogram. The range of the data set is 24031 (that's 25034 minus 1003). (Regardless of how many classes you asked for the numbers and tic marks divide the range of the data set up into intervals of fourths.) Concentrate on the numbers at the top of the histogram; those are the numbers where the class on the left ends and the next one begins.

_____ **6.** Examine the Break Values box. These numbers—4436, 7869, . . ., 25034—represent the top end of each category. They divide the range of the data set (that's 24031) up into seven equal subranges. If you click one of these numbers, the break line separating its class and the class above it, turns red. At the bottom of the window, you can read the number of numbers (elements, as they are called) in that class. How many numbers are in the class whose top value is 7869? _____

By clicking the % button (next to Break Values), determine the percentage of elements that fall in the first four categories. _____. Press % again to return to the numerical break values.

_____ **7.** Make sure the Classification Method is still Equal Interval. Change the number of classes to 5.

User Selection of Classes

_____ **8.** Notice each vertical blue line. The value at its top end indicates the break value as well. So, the first category consists of 23 values, 19 of which are at the lower end (look at the height of the column), and 4 of which are toward the upper end of the category. The range of this category is from 1003 to 5809.

____ **9.** Use the mouse cursor to drag one of the lines. Notice that the Classification Method has changed to Manual—the method selected when the software realizes that you are going to be in charge of where the class breaks are. By moving the lines around, you can choose the Break Values (look in the pane to the right). Also observe, as you move a break line, that the number of values in the class to the left of the break line is dynamically displayed at the bottom of the window. Set the break values—the value at the high end of each category—between classes to approximately 2000, 6000, 11000, 16000, 26000. By doing this, you see that you are putting each natural group of values in a class by itself. Click OK. In the Layer Properties window, click Apply, then OK.

____ **10.** Examine the T/C. The five categories into which you have put the populations of FEMALES are represented by increasingly darker colors of green. Note that the range of each class is shown next to its symbol. Look at the map. Notice how much easier it is to see the clusters of similar values.

____ **11.** In Layer Properties, click Classify again. By highlighting and typing in the Break Values pane, make the break values exactly 2000, 6000, 11000, 16000, and 26000. Click OK, click Apply, then Click OK again. Observe now the Table of Contents and the map. You see that you can easily make divisions between sets of values more comprehensible. The T/C shows the low and high values in each category.

____ **12.** Using Layer Properties > Symbology > Classify bring up the Classification window again. Click the box Snap breaks to data values. Each break line moves to the closest data value. Click OK, then Apply. Now the Table of Contents indicates the upper bound of each class. In the Layer Properties window, turn on Show class ranges using feature values (version 10.1) or Snap breaks to data values (version 10.0). Click OK. Now the T/C shows the lower and upper bounds of each class, based on actual values in the class. See Figure 6-16.

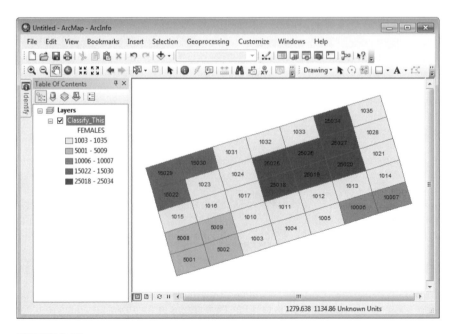

FIGURE 6-16

A More Careful Look at Equal Intervals

___ 13. Classify the values again. Choose the Equal Interval method again. Notice that the break lines move back to the positions they had before. These positions are calculated so that each class occupies the same distance along the horizontal axis. What is that distance? _____. The calculation takes the form of "greatest value in the data set minus the least value, divided by the number of classes." Click OK, Apply, OK. Notice that the resulting map has five categories in the T/C, but only four colors show up in the map, because one class is empty. What is the range of the class that has no values in it? _____ – _____ Notice that the values around 1000 and those around 5000 are lumped together. Look at the Classification window again to notice that one class contains two clusters of values that beg to be separated, while another class is empty. This is a difficulty with equal intervals.

Defined Interval

___ 14. Choose the method Defined Interval. This is an equal interval method, but instead of the interval size being chosen by the formula (High – Low)/N, it is chosen by the user and adjusted so that the numbers look nice. Type 4000 in the Interval Size box and click the histogram. Note that defining the interval determines the number of classes. How many? _____. Click OK. Uncheck Show class ranges using feature values. Click Apply, then OK. Note the results on the map and in the T/C. Again there are empty classes.

Quantiles

The problem of empty classes may be solved by the approach of quantiles. Quantiles places approximately the same number of values in each category.

___ 15. Classify and select the Quantile method. Change the number of classes back to 5 (this gives you quintiles; other popular classifications are quartiles, deciles, and percentiles). Click OK, Apply, and OK. Notice the map and the Table of Contents. Again, the pattern looks somewhat similar to previous ones, but here the values around 1000 are broken into two groups. A further complication: two of the values around 5000 are lumped in with some 1000s, while the other two are in with the 10000s and the 15000s. Again, the divisions seem inappropriate.

Standard Deviation

The standard deviation method is not really appropriate for this set of data, because the values are not distributed "normally"—that is, in a bell-shaped curve. But just because a method is inappropriate doesn't mean we can't have the software apply it. It just means that we get bad results. (Garbage in, garbage out, as the saying goes.)

___ 16. Select the Standard Deviation method, and make the Interval Size one half of a standard deviation. Find and click boxes to show both the Standard Deviation and the Mean on the histogram (with vertical dashed lines). Click OK. Notice that the color ramp changes so that the categories reflect a change at the mean (-0.25 to +0.25), rather than gradual change from one end to the other. Click OK, then Apply, then OK. Observe. You might find the display in the T/C confusing. That's because the "–" sign is used both as a minus and as a separator.

Natural Breaks[5]

To understand "natural breaks" (also known as the Jenks method), you need to comprehend the idea of "variance from a mean." If you have a cluster of values, they have a mean, or average, value. If you take each value and subtract the mean from it, you get another set of numbers. The variance used by the Jenks approach is calculated by squaring each of these new numbers, taking their sum, and dividing by the number of numbers. You can see that the variance would be larger the further the original values were from the mean. So the variance is an indication of the dispersal from the mean of the set of values. If we had, say, five clusters of numbers, each with its own variance, we could add those five variances together to get a total variance. The idea behind the Jenks approach is to minimize the total variance by moving values around from cluster to cluster. At the end of the process, it should be the case that no value can be moved from one class to another without raising the total variance. This method works so well that it is the default that ArcMap uses. It is illustrated by using a different data set.

_____ **17.** In Layer Properties > Labels, change Label Field to WOMBATS (click Apply) and, under the Symbology tab change Value Field to WOMBATS. Change the Show box to Categories–Unique Values. Click Add All Values. (Make sure WOMBATS is chosen in the Value Field) Click Apply, then click OK. Look at the quilt. Open the attribute table and sort the WOMBATS field into ascending order. Note that there are seven groups of values (100s, 200s, and so on). Each group has five values, except for the first group, which has six, and the second group, which has four.

_____ **18.** Go to the Layer Properties window and pick Quantities > Graduated Colors. Now classify areas according to the number of WOMBATS contained in each, using seven categories and the equal interval method. Looking at the Classification window, you can note that in two categories two groups are placed together. Also note that there are two empty classes. Display the labeled map.

_____ **19.** Classify according to septiles (quantiles, seven classes), showing class ranges using feature values. Show the labeled map. Look at the ranges shown in the Table of Contents. What is the problem with this method? _____

_____

_____ **20.** Classify according to Natural Breaks, using seven classes. Problem of inappropriate class ranges solved? _____

---

[5]A method first published by Professor George Jenks, Department of Geography, University of Kansas. If you use the Natural Breaks method in ArcGIS check your answers carefully. There are differences between the results produced by Natural Breaks and other Jenks programs.

Normalization

Sometimes we are interested in seeing values that have been divided by other values. The first set of values is said to be "normalized" by the second set of values. For example, if we were interested in population density of counties, we would be interested in the value of population divided by the area in which it resides. Or, as in the next step, we might want to display the ratio of females to males, so we would apply normalization.

____ **21.** Normalize the number of females by the number of males, using the Layer Properties window, Symbology tab, and the Fields area. (Females goes into Value and Males goes into Normalization) Classify into seven classes using Jenks. With the labels tab, label the areas using the expression [FEMALES]/[MALES]. Look at the map. Notice that the labeling makes an unreadable mess of things, with ratios carried out to 14 decimal places.

____ **22.** Go back to the Labels tab and use the expression

Round ([FEMALES] / [MALES], 2)

which has the effect of rounding the ratios off to two decimal places, and should give you the result shown in Figure 6-17.

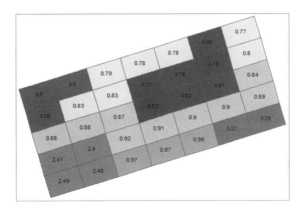

FIGURE 6-17

Using Charts and Graphs

____ **23.** Examine another way of comparing values in an area: In Layer Properties > Symbology > Charts, click Pie. Highlight FEMALES and move it to the right pane with the > button. Do the same with MALES. Make the symbol colors distinct: very light for females, very dark for males. Click Apply, then OK. Look at the map. You see a disk for each area that indicates the ratio. If you think the labels now just get in the way, right click Classify This and turn off Label Features.

____ **24.** Change the chart type to Bar/Column. Notice that you get a sense of the relative numbers of females to males, and also a sense of the total population of the area.

____ **25.** Change the chart type to Stacked and observe.

Making a Layout

___ **26.** Switch to Layout View. Make a layout with the pie charts, a title, the legend, your name, and the top portion of the attribute table. You can reposition the charts with the pan tool on the Tools toolbar. The Insert menu works to get you the legend, the title, and text for your name. To get the table onto the layout: On the Attributes Of Classify_This window, select Options > Add Table to Layout. Print the layout. Close ArcMap.

Exercise 6-3 (Short Project)

Comparing Data Sets: Medically Underserved Areas (MUAs) and Health Professional Shortage Areas (HPSAs)

The U.S. Government's Health Resources and Services Administration (HRSA) has two (of several) databases that relate to areas that do not have sufficient medical resources to meet the needs of the people in those areas. The databases relate to Medically Underserved Areas (MUAs) and Health Professional Shortage Areas (HPSAs). Information about them, and a lot of other interesting data, may be found at the U.S. Health Resources and Services Administration (www.hrsa.gov).

Since each of these two databases deals with areas that have something in common, it would be interesting to contrast them.

___ **1.** Start ArcCatalog. Type Ctrl-F to bring up the Search window. Select Local Search. In the Search Option drop-down menu select Search Options. Under the Index tab press Add and Register the folders:

___IGIS-Arc

___IGIS-Arc_AUX

Press Re-index from Scratch. This process will take a minute or two. Follow the progress (Indexing Status) in the lower pane of the window.

Once indexing is complete (Status Active), search for HRSA_Geography, which is a Personal Geodatabase. When it is found click the third line (it is green) to display the name in the Catalog Tree. In what folder is the geodatabase? _____.

In the geodatabase you will find four PGDB feature classes. (By the way, note that this is a case in which feature classes exist directly within a geodatabase (called "freestanding"), without an intervening feature dataset.) Name them:

_____

_____

_____

_____

____ **2.** Click through each of the four, looking at the geography. Form an impression of the amount of area in each, compared to the others. Then look at each attribute table, writing the number of records for each data set next to the preceding blanks. Does the number of records seem to correlate with the areas involved? _____. What is the coordinate system used? _____. (Hints: Notice the squashed appearance of the country. Run the cursor over the map and observe the coordinates. Check the Description > Spatial Reference. (If asked if you want to update metadata, say No.))

____ By the way these feature classes display a good example of fairly complete, well-done medatata. Spend a bit of time looking at both the Esri metadata and the FGDC metadata.

____ **3.** Copy the folder

[___] IGIS-Arc\Health_Areas_Data

to

___IGIS-Arc_*YourInitials*

Make a folder connection to it. Work with this folder for the remainder of the exercise.

Suppose that you are interested in the total of the medically underserved areas in the United States. You might imagine you could run statistics on the Shape_Area column.

____ **4.** Display the MUAs' attribute table. What is the smallest area (according to the values in Shape_Area—just show three significant digits)? _____. The largest? _____. What are the units of these areas? _____. Not much help here.

A problem is that the "areas" are in decimal degree angular units—which are useless. To determine actual areas, you have to convert the MUAs feature class to a projection that preserves area. Recall that no projection preserves all of the four measures of major interest: area, shape, distance, and direction. One projection that works well for the United States for area is the Albers Equal-Area Projection.

____ **5.** In ArcToolbox invoke the Project tool by selecting Data Management Tools > Projections and Transformations > Feature. Make the Input US_20021105_MUAs. Make the output Name MUAs_Albers. For the Output Coordinate System, navigate to North America Albers Equal Area Conic. Select Projected Coordinate Systems > Continental > North America.). Run the tool. When it finishes you will see an indication in the lower right of the window. Close ArcToolbox.

____ **6.** Start ArcMap with a Blank map. Add MUAs_Albers. Examine the data frame properties. What are the map units? _____. Make the display units Miles. What is the central meridian on which the projection is based? _____. What are the two standard parallels for the projection?

_____, _____.

___ **7.** Just to reinforce the idea that ArcMap does on-the-fly projections, add the US_20021105_MUAs feature class. Make its color symbol distinctly different from MUAs_Albers. Notice that it lies precisely on top of the Albers projection. Click it off and on. Flip back to ArcCatalog to see it both in its decimal-degree form and in its Albers projected form. Switch back and forth between the two. Close ArcCatalog and go back to ArcMap. Remove the US_20021105_MUAs feature class.

Examine the attribute table of MUAs_Albers. As you determined previously, the units are meters. So, the calculated areas are in square meters—not very useful numbers because the areas are so large. Suppose we prefer square miles. A square mile contains about 2,589,988 square meters.

___ **8.** Under Table Options, click Add Field. Call it Area_SQMI. Make it a floating-point number. Click OK. Right-click the column and activate the Field Calculator. Ignore the warning—if you make a mistake, you can just do the calculation over. What expression should you use in the Field Calculator? _____ Apply it. Run Statistics. Is the largest area about 13,678 square miles? (If not, look at your expression and correct it.)

___ **9.** How many square miles are there of Medically Underserved Areas? _____ If the area of the coterminous ("lower 48") United States is about 3 million square miles, what percentage of the country fits the "medically underserved" designation, according to these data? _____

Geographically Comparing Two Datasets

___ **10.** Create (by using the Project tool) HPSAPrimaryCare_Albers from US_200301_HPSAPrimaryCare. Add it to ArcMap if that doesn't happen automatically.

Since the two data sets (MUAs_Albers and HPSA_PrimaryCare_Albers) tend to get at the same concept—places where there isn't sufficient medical care—it might be interesting to see how they compare.

___ **11.** Start by examining the layers visually. Make MUAs_Albers red and HPSA_PrimaryCare_Albers green. Then observe what happens when you turn them off and on. Finally, leave them both on.

One measure of how areas compare is whether or not there is overlap between them—that is, do they intersect or not?

___ **12.** Under Selection, clear any selected features. Use Selection > Select By Location. Select features from MUAs_Albers (make it the target layer) that intersect HPSA_PrimaryCare_Albers (which will be the source layer). See Figure 6-18. Click OK.

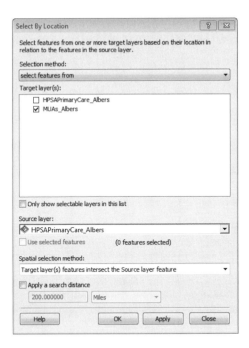

FIGURE 6-18

____ **13.** Examine the resulting MUAs_Albers map. Verify that there seems to be a sizable amount of overlap. Zoom in on some areas to get an idea of what the selected areas look like, compared with the Primary Care areas. Again, turn the layers off and on, finally leaving them on. From the attribute table of MUAs_Albers, determine how many records are selected. _____ out of _____.

____ **14.** Determine how many MUAs_Albers polygons (target layer) relate to the HPSA_PrimaryCare_Albers polygons in the following ways—being careful to clear selections before each test.[6]

Intersect (same as previous step) _____

Completely contain _____

Are completely within _____

Are identical to _____

Contain _____

---

[6]For a thorough discussion, with graphic portrayals, of all the relationships, see the Esri website resources.esri.com/ help. If you can't get to it directly use Google, searching for Clementini ArcGIS. (Eliseo Clementini was an author of a paper on spatial databases.)

___ **15.** Reverse the questions posed previously. That is, determine how many polygons of HPSA_
PrimaryCare_Albers (target layer) relate to the MUAs_Albers polygons:

Intersect _____

Completely contain _____

Are completely within _____

Are identical to _____

Contain _____

Close ArcMap.

Exercise 6-4 (Major Project)

Important note on running this exercise: The initial stages of this exercise ask you to download data. If you have access to the Internet, you may be able obtain census data from the U.S. Bureau of the Census. This process sometimes succeeds and sometimes fails— perhaps because of a firewall at your institution or changes in the website at the Census Bureau. However these early steps can be quite useful to those of you who plan to use ArcGIS to work with census data. If you don't have the access or tools to do some of these steps (or lack the inclination to do them), you can still do the bulk of the exercise. I have indicated where you can skip over some steps, using the data I supply.

Combining Demographic and Geographic Data

Suppose that we have decided to open a day-care center on Galbraith School Road in Knox County, Tennessee. We have a banker who tells us that a loan can be obtained if we can show that 25,000 people reside within 3 miles of the road on which we plan to build the center. We make a plan as to how to proceed: We will combine data obtained over the Internet, firstly from the U.S. Bureau of the Census and secondly from Esri. The census dataset is in the form of a spreadsheet that contains information about the population in census blocks in our area of interest. It is strictly tabular data, with no geographic component.

The data we will get from Esri consists of a modified TIGER street file and a modified TIGER census block file. (Actually, this information came originally from the Bureau of the Census but has been converted by Esri to shapefile format for geographic display and analysis.) The project will involve combining these data sets so that we can determine how many people reside within 3 miles of Galbraith School Road.

This project has several stages. The initial ones (which are optional) require access to the Internet, figuring out how to get data from the www.census.gov website and saving a table in Microsoft Excel spreadsheet format. Then you will get Tiger-based data from the Esri website, www.esri.com. There are a lot of steps, but the whole process is instructive and can be useful, given that a lot of spatial data is available from the Bureau of the Census and Esri. Here are the steps:

❑ Download data from the Bureau of the Census (optional for the exercise).

❑ Manipulate the Census data (or data I supply) in a Microsoft Excel spreadsheet.

❏   Convert the spreadsheet data so it becomes an ArcGIS table.

❏   Download data from Esri (optional for the exercise).

❏   Combine Census data and Esri Data. (using data provided on the DVD)

❏   Display the data.

❏   Analyze the data.

A warning: The downloading data part of this exercise requires a lot of pointing, clicking, and waiting. Its annoyance component is high and there are lots of places to go off the rails. As such, it is emblematic of many of the activities you will undertake when doing real GIS projects. However, as previously indicated, sample data is provided so you can skip the more annoying parts if you run into trouble with them.

_____   **1.** In ArcCatalog make a new folder named Day_Care_Data in

___IGIS-Arc_*YourInitials*

by right-clicking that entry and picking New > Folder. Make a folder connection to Day_Care_Data.

Obtaining Data from the U.S. Bureau of the Census

_____   **2.** On the Internet, browse to www.census.gov. What time is it? _____ What is the U.S. population? _____. Determine what the world's population is. _____.

Previous editions of this textbook provided instructions for finding the data for Knox County, Tennessee, only to (1) have students find that the website had changed and it wasn't obvious how to find the data, and/or (2) discover that their institutions had firewalls erected so that the extensive data files couldn't get through. So, I have abandoned the idea of trying to tell you how to get the files; rather I will provide them to you. If you plan to do a lot of work with census data, you might want to determine how to get these files on your own. Otherwise, just take a pass, skip to Step 4, and use the data I have provided.

_____   **3.** You ultimately want to find and obtain the following files on www.census.gov:

dt_dec_2000_sf1_u_data1.xls
dt_dec_2000_sf1_u_geo.xls
dt_readme.txt
readme_dec_2000_sf1.tx

The one you must have is:

dt_dec_2000_sf1_u_data1.xls

Good luck to you in finding and downloading it. Look around the census site. You may be able to get to it with American Fact Finder.

_____ **4.** If you don't find and download the files, examine the folder IGIS-Arc_AUX/DCD. There you will find Extracted_Census_Files > output. In that folder note the existence of the above named files. You will need these files in a later step if you were not able to (or inclined to) download the files from the census bureau.

_____ **5.** Before leaving the Census site, look at the population clocks again. (You might have to refresh the Internet page so that the time changes: the F5 key usually works.) What time is it now? _____ What is the U.S. population? _____. What is the world population? _____. How many people per hour are added to the U.S. population? _____. How many people per hour are added to the world population? _____. In 1950 the world population was about two billion. Now it is about seven billion. We probably need to do something about the dramatically increasing population—maybe using our GIS skills. Leave the Census site.

Converting the Census Data Spreadsheet to dBASEIV Format

_____ **6.** Whether or not you were able to download the census data, you can do the conversion to ArcGIS formats in the following steps. Use the computer's operating system to find the file

dt_dec_2000_sf1_u_data1.xls

in

[____] IGIS-Arc_AUX\DCD\Extracted_Census_Files\output

and copy it to

___IGIS-Arc_YourInitials\Day_Care_Data

_____ **7.** Still in the computer's operating system, pounce on the dt_dec_2000_sf1_u_data1.xls file. An Excel spreadsheet should open. If not, start Microsoft Excel and go File > Open and choose

dt_dec_2000_sf1_u_data1.xls

Make the spreadsheet full screen. How many rows (expect thousands) are there in the spreadsheet that contain population data?__ _____. That's how many census blocks there are in the Knoxville CCD. (A CCD is a Census Collection District. In this case, it is a subset of the Knoxville metropolitan area.) What are the five headings of the spreadsheet columns (row 1)? _____, _____, _____, _____, _____.

You are mainly interested in the GEO_ID and the P001001 columns. The Geography Identifier is (a part of) the Block Number. The population column cells indicate the number of people living in the block.

_____ **8.** Click the A1 cell, to make sure the spreadsheet is selected. Choose: File > Save As. Save it in Day_Care_Data, but use the File name Census1.xls. Now Click Save As again, this time saving with the File name Census2.

Our next steps will move us towards converting the spreadsheet into a dBASE4 table. When we make this conversion, the text in the top row becomes the headings and the remainder of the table goes into the cells of the attribute table. Therefore, row number 2 of the spreadsheet is in the way. So we will delete it.

_____ **9.** Click on the "2" of row 2. Right-click. Pick Delete.

The next step is to get a key field—a column we can use to join the spreadsheet information with the Esri blocks shapefile table that we will obtain shortly.

_____ **10.** Working with Census2.xls, put your cursor on the divider that separates the A column heading from the B column heading and drag the separator to the right. (Clever of them to hide all those other digits in GEO_ID!) What we want are just the last four digits (L4D) of GEO_ID. So we will make "column F", currently blank, our working GEO_ID and put the desired digits there. Click in cell F1. Type the name GEO_ID_L4D and press Enter. Widen the column as necessary to get the heading in the box.

_____ **11.** Click in cell F2 and type

=RIGHT(A2,4)

which will invoke a spreadsheet function that grabs off the last four characters of the A2 cell and puts them in cell F2. Press Enter. The number 1000 should appear in cell F2. The software has evaluated the formula and placed the result in cell F2. Check that cell F2 now contains "1000". Click on F2 again and observe that the formula bar at the top of the window contains "=RIGHT(A2,4)".

_____ **12.** Press Ctrl-C to copy the formula that underlies the number in F2 to the clipboard. Cell F2 should be activated. Click cell F3. With the vertical slider, scroll down the rows, so you see the last one (6022). Hold down Shift and click cell F6022, so the entire column is highlighted. Press Ctrl-V (paste) to evaluate the formula "=RIGHT(cell number,4)" for each of the selected cells of column F. Check the results. Each cell F value in a given row should equal the last four characters of the GEO_ID in the A column of that row. Press Esc to clear the activation of cell F2.

_____ **13.** Click in the A1 cell. Save the file with File > Save As, this time using the name Census3.xls.

Our spreadsheet is now ready to be converted to an ArcGIS attribute table. We do this with an ArcGIS tool.

_____ **14.** Close the spreadsheet. Start ArcCatalog. Start ArcToolbox. Start the tool "Table to Table" found in

ArcToolbox > Conversion Tools > To Geodatabase

For Input Rows navigate to Day_Care_Data. Click on Census3. Click Add. Instead of what you might expect, Sheet0$ appears. Click on that to bring it into the Name field. Click Add. For output location you want Day_Care_Data in the Name field. Name the Output Table Census4. (The six column headings will appear in the Field Map.) Click OK. After a bit, Census4.dbf will appear in the Catalog Tree. Preview it to see that it is all there. See Figure 6-19. (If it is not there, or not correct, you can try again. We kept Census1 as a pristine version of the spreadsheet. That is especially important for those who downloaded the census data. I'm sure you don't want to go through that again!)

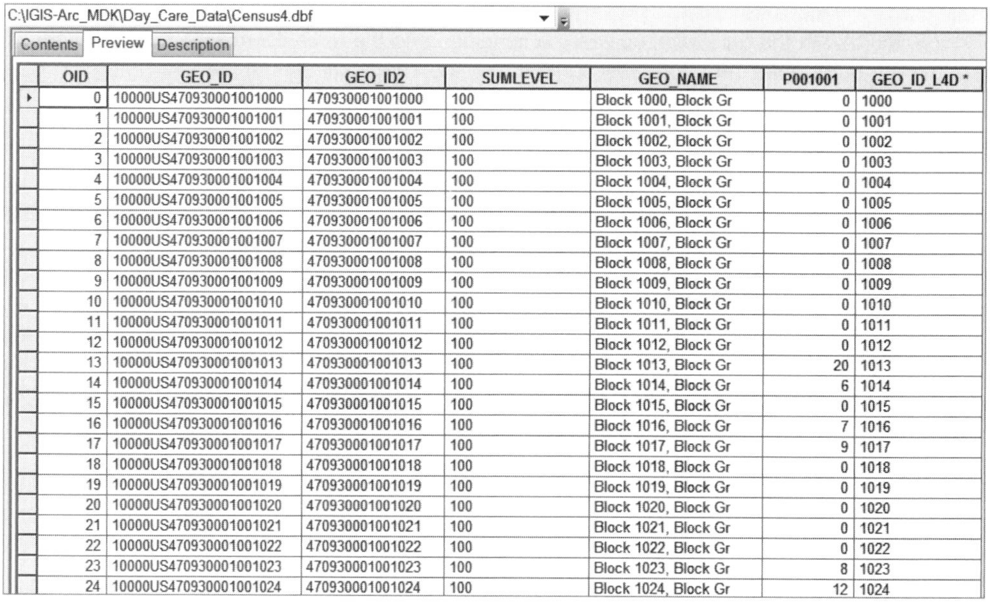

C:\IGIS-Arc_MDK\Day_Care_Data\Census4.dbf

Contents | Preview | Description

OID	GEO_ID	GEO_ID2	SUMLEVEL	GEO_NAME	P001001	GEO_ID_L4D *
0	10000US470930001001000	470930001001000	100	Block 1000, Block Gr	0	1000
1	10000US470930001001001	470930001001001	100	Block 1001, Block Gr	0	1001
2	10000US470930001001002	470930001001002	100	Block 1002, Block Gr	0	1002
3	10000US470930001001003	470930001001003	100	Block 1003, Block Gr	0	1003
4	10000US470930001001004	470930001001004	100	Block 1004, Block Gr	0	1004
5	10000US470930001001005	470930001001005	100	Block 1005, Block Gr	0	1005
6	10000US470930001001006	470930001001006	100	Block 1006, Block Gr	0	1006
7	10000US470930001001007	470930001001007	100	Block 1007, Block Gr	0	1007
8	10000US470930001001008	470930001001008	100	Block 1008, Block Gr	0	1008
9	10000US470930001001009	470930001001009	100	Block 1009, Block Gr	0	1009
10	10000US470930001001010	470930001001010	100	Block 1010, Block Gr	0	1010
11	10000US470930001001011	470930001001011	100	Block 1011, Block Gr	0	1011
12	10000US470930001001012	470930001001012	100	Block 1012, Block Gr	0	1012
13	10000US470930001001013	470930001001013	100	Block 1013, Block Gr	20	1013
14	10000US470930001001014	470930001001014	100	Block 1014, Block Gr	6	1014
15	10000US470930001001015	470930001001015	100	Block 1015, Block Gr	0	1015
16	10000US470930001001016	470930001001016	100	Block 1016, Block Gr	7	1016
17	10000US470930001001017	470930001001017	100	Block 1017, Block Gr	9	1017
18	10000US470930001001018	470930001001018	100	Block 1018, Block Gr	0	1018
19	10000US470930001001019	470930001001019	100	Block 1019, Block Gr	0	1019
20	10000US470930001001020	470930001001020	100	Block 1020, Block Gr	0	1020
21	10000US470930001001021	470930001001021	100	Block 1021, Block Gr	0	1021
22	10000US470930001001022	470930001001022	100	Block 1022, Block Gr	0	1022
23	10000US470930001001023	470930001001023	100	Block 1023, Block Gr	8	1023
24	10000US470930001001024	470930001001024	100	Block 1024, Block Gr	12	1024

FIGURE 6-19

Using TIGER-Based Street and Block Shapefiles from Esri

Steps 15 through 17 will acquaint you with procedures for obtaining TIGER-like files from Esri. If you choose to skip the data-downloading part of the Exercise and only do the analysis part, simply start with Step 18, where I have provided you with data guaranteed to work. In either event, you should read the material between here to Step 18.

The TIGER/Line feature files, as they come from the U.S. Bureau of the Census, are what are called "flat files." That is, they have no hierarchy. Each file consists of a large number of records, each consisting entirely of text or numbers. Included in each record is a pair of coordinates, latitude and longitude, in decimal-degree format. Esri has taken these records and used that position information to create shapefiles that users can download for free. As you might guess the files are large, so they are zipped up. The steps that follow indicate how you can obtain them. If you are unable to download and unzip the files you can use a workaround that will be prescribed shortly.

—— **15.** Point your Internet browser at

http://www.esri.com/data/download/census2000_tigerline/index.html

Under Free Download click Preview and Download. (You may have to register.) You should see a window with a graphic of the United States. (Alaska and Hawaii will be somewhat out of place.) Select Tennessee (TN). Under Select by County pick Knox. Submit selection. Put checks by

Census Blocks 2000 (1.2 MB) and Line Features—Roads (2.0 MB). At the bottom of the page, click Proceed to Download.

After a brief wait you will be told that your data file is ready. Click Download File.[7]

Here is a procedure that worked in Internet Explorer:

Click Save As. In the Save As window navigate to

Day_Care_Data in

___IGIS-Arc_*YourInitials*

Don't change the filename. Press Save. When the download is complete, close all browser functions.

Using Firefox the procedure is similar. Firefox may not be configured to allow you to specify the download the file location. You may have to find it and copy it to the Day_Care_Data folder.

___ **16.** With the operating system, examine the Day_Care_Data folder. You should find a ZIP file there with a name something like

at_tigeresri1234567890.zip

Unzip the file, using either a commercially available unzipping program (like WinZip or PKZIP) or the unzipping facility in Microsoft Windows (right-click the zip folder, then select Extract All). Unfortunately the structure of downloads from Esri is somewhat like Russian nested dolls: You now have two additional zip files to contend with. Unzip the folder blk0047093 (the census block file), and you should get the three basic components of a shape file:

tgr47093blk00.dbf
tgr47093blk00.shp
tgr47093blk00.shx

(What is 47093? Tennessee is state number 47, and Knox County is county 93 in the state, according to the government's FIPS classification).

Also unzip the roads folder lkA47093.zip to produce tgr47093lkA.dbf, tgr47093lkA.shp, and tgr47093lkA.shx.

___ **17.** Back in ArcCatalog, click the Folder Connections entry and refresh the Catalog Tree. Find the two shapefiles somewhere in the tree of Day_Care_Data. If necessary, move them directly into

---

[7]You should probably find out where your browser is going to put the files.

the Day_Care_Data folder. Change the name of tgr47093blk00.shp to Blocks.shp. Change the name of tgr47093lkA.shp to Roads.shp. Look at both the tables and geography. Dismiss ArcCatalog.

Now you have everything you need to determine how many people live within 3 miles of the road of the proposed day care site. But since Web sites change and datasets get updated things may not have gone smoothly up to this point. Again, that's the real practice of GIS. Frustration and workarounds are part of the job. But to ensure that the rest of the project goes smoothly and that we all get the same answers, I'm going to ask you to delete some of the work you have done and substitute some data files that I know will work.

_____ **18.** Start ArcCatalog. From

___IGIS-Arc_*YourInitials*\Day_Care_Data

delete Roads.shp, Blocks.shp, and Census4.dbf if they exist there.

From

[___] IGIS-Arc_AUX\DCD

copy Roads.shp, Blocks.shp, and Census4.dbf to

___IGIS-Arc_*YourInitials*\Day_Care_Data.

_____ **19.** Launch ArcMap with a Blank Map. Dismiss ArcCatalog. From

___IGIS-Arc_*YourInitials*\Day_Care_Data

Add (on at a time, in this order) Roads.shp, Blocks.shp, and Census4.dbf. If you slide the cursor around the map, you notice that the coordinates are in decimal degrees. The display would look more like the real world (what you would see if you could look straight down from an airplane or satellite) if we used a Cartesian projection.

_____ **20.** Right-click the data frame and pick Data Frame Properties. Change the Coordinate System to Projected Coordinate Systems > State Plane > NAD 1983 (US Feet) > Tennessee FIPS 4100. Ignore any warnings. Under the General tab of Data Frame Properties change the display units to Feet. Notice now the coordinates shown are in Feet. Recall that you have only changed the display, not the underlying data.

_____ **21.** Using Selection > Select By Attributes, look for an FENAME of 'Galbraith School' in Layer Roads. (Be sure you use Galbraith School as the FENAME. Don't include the word Road. Don't leave off the word School (there is a Galbraith road—it's not the one you want). (If your computer is slow, it would probably best not to ask for Unique Values, since there are a ton of them. Waiting for them to be enumerated and then finding the right one is more work than typing. If you type, be careful of the quotation marks (single).) See Figure 6-20.

_____ **22.** Click Apply, then Close. You should see a little cyan patch showing the selected road on the map. Choose Selection > Zoom To Selected Features, and then zoom out some. Have ArcMap label the roads with FENAME. See Figure 6-21.

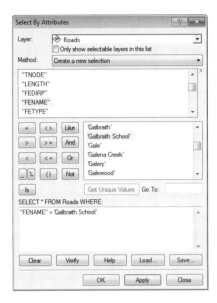

FIGURE 6-20

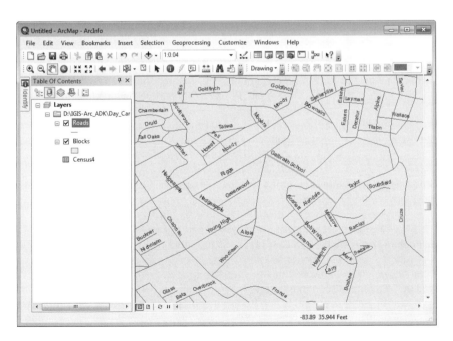

FIGURE 6-21

Assessing What We Have and What We Need to Solve the Problem

What we are going to do is find the census Blocks polygons that lie within three miles of Galbraith School Road. But since our objective is to find population we have to first join the census table to the Blocks table, so that the number of people in each census block will be available to us. To join these two tables, we need to find a key field in each table whose values serve as block identifiers. In the census table that key field is GEO_ID_L4D. In the Blocks table the key field is BLOCK2000.

___ **23.** Look at the attribute table of Blocks. In the column BLOCK2000 we have the four-digit block number. In the Layer Properties window check out Fields. Note that BLOCK2000 is a String field of Length 4.

___ **24.** Look at the table Census4. In the GEO_ID_L4D column you see four-digit block numbers. In the Layer Properties window check out Fields. Note that GEO_ID_L4D is a String field of Length _____. (We will hope that it will match up with BLOCK2000 string field of length 4 when we join the tables.) Also note that we have P001001 which is the population of each block. Dismiss both tables.

Next we need to join the Blocks table and the Census4 table so we can combine the geographic information of the shapefile with the tabular information in the census file.

___ **25.** Right-click Blocks and choose Joins and Relates > Join. Under "What do you want to join to this layer?" choose "Join attributes from a table." For Step 1 choose BLOCK2000. For Step 2 choose Census4. What's Step 3? _____. Click Validate Join for reassurance. Click OK. (Read the Create Index box if it shows up. Since we are dealing with large tables, let's take ArcMap's word for it that indexing the join field will improve performance. Click Yes.)

___ **26.** Right-click Blocks, open, and examine the table. The important thing to note is that now the population of each census block (field: P001001) is part of the shapefile table. Dismiss the table.

___ **27.** Launch the Identify tool, click on the map, set the Identify from field to Blocks, and check out the two blocks northeast of Galbraith School Road. What are their populations? _____, _____. Dismiss the Identify Results Window.

___ **28.** Start the Select Features tool with its icon and, using Shift-Click, select the same two blocks. In the Blocks table, choose Show selected records. Verify the populations.

___ **29.** Clear all selections. (If you don't, when you export the table, as you will you later, only the selected records will be exported!) Dismiss the table.

You might recall that, even though you have the shapefile together with the table, the connection between the original shapefile table and the census data is pretty tenuous. All you would have to do is leave ArcMap and—poof—no join. So set this shapefile and its expanded table in concrete using the following step.

___ **30.** Right-click Blocks in the T/C and choose Data > Export Data. From the Export Data window browse to

___IGIS-Arc_*YourInitials*\Day_Care_Data

In the Saving Data window change the Name from Export_Output.shp to Blocks&Pop. Save as type Shapefile. Click Save. Click OK. Add the exported data to the map as a layer. Remove Blocks. Remove Census4. Check out the new table of Blocks&Pop. How many census blocks are there in the Knoxville CCD? _____ Close the table.

Converting the Relevant Files to Cartesian Coordinates

You are going to be working with distances, and that suggests that you should probably convert our files, which are stored in decimal-degree format, to a coordinate system where you can be surer that distance calculations are done simply and properly. You are displaying in the State Plane system, but for the sake of neat and tidy, you should convert the basic data and remap it.

___ **31.** Ask for a new blank map in ArcMap, using the New Map icon. Don't save the previous map.

___ **32.** In ArcToolbox invoke the Project tool in Data Management Tools > Projections and Transformations > Feature. Make the input Blocks&Pop.shp. Call the output Blocks&Pop_TN_SP.shp, because we plan to put the data in the Tennessee State Plane Coordinate System. For the Output Coordinate System navigate to the

NAD 1983 StatePlane Tennessee FIPS 4100 (Feet)

coordinate system. Make sure that you are placing the output in

IGIS-Arc_*YourInitials*\Day_Care_Data

(In version 10.0 you may have to select a Transformation from the drop-down window: WGS 84 to NAD 83.) Click Add. OK the Project window. After the tool executes (look at the lower right of the window to see the action taking place—it may take some time), the new layer is automatically added to the map.

___ **33.** Use this same procedure and projection to Project the Roads shapefile to Roads_TN_SP.shp and notice that this layer is added to the map. Hide ArcToolbox.

Finally

With all this preparation, you may have forgotten the goal of this exercise: to determine the population that resides within 3 miles of Galbraith School Road. Now that everything is in place, we can use Select by Attribute to select 'Galbraith School' road. Then, we can use Select by Location to find the census blocks that are within 3 miles.

___ **34.** To simplify things, remove Roads and Blocks&Pop if you added them (***not*** the newly created TN_SP layers) from the T/C. From the Selection menu pick Select By Attributes. As you did before, select Galbraith School. Label the roads features. Zoom in and use the measure tool to determine how long the road is, to two decimal places. _____miles.

___ **35.** From the Selection menu pick Select By Location. We want to ***select features*** from Blocks&Pop_TN_SP (the Target layer) that are within a distance of the selected features of Roads_TN_SP, applying a buffer of 3.0 Miles to the selected roads features. Click Apply. Close the Select By Location window and zoom to selected features in the Blocks&Pop_TN_SP layer.

____ **36.** Open the Blocks&Pop_TN_SP table. How many records were selected? _____ out of
_____. Run Statistics on the P001001 column. Double-check that the number of records
you are calculating from (Count) is the number selected. What is the population in the selected
census blocks? _____ Close the table.

Since the population is well over the 25,000 that the banker required, we are elated. So we invite her
over to see the results. After being suitably impressed with the operation and the software, she says,
"Gee, knowing that area as I do, the spread of census blocks looks like more than 3 miles." With some
trepidation, immediately confirmed, we take a measurement across the area.

____ **37.** What is the greatest distance across the selected blocks? _____ feet. How many miles is
that? _____

The problem here appears to be that any part of any census block that was within the 3-mile radius
causes that entire block to be selected—thus, probably overestimating the population that lies within the
buffer.

As with most operations with this software, there is more than one way to achieve a result. We can make
a graphic circle with the Drawing toolbar and then use Select By Graphics.

____ **38.** Look again at the Blocks&Pop_TN_SP table. With Table Options clear the selections. Show all
records. Dismiss the table. In Roads_TN_SP make sure the four segments of Galbraith School
Road are selected. Select them again if necessary. Change the Display Units of the data frame
to Miles. Zoom to the approximate area of about 10 miles across in the area around the road.

____ **39.** Make sure the Drawing toolbar is available. On the Drawing toolbar, find the first icon that has a
triangle next to it indicating a drop-down menu. (The ToolTip probably says Rectangle.) Choose
the circle. Place the cursor crosshairs in about the middle of the selected line of Galbraith
School Road. Click and drag a circle of Radius 3.00 miles, plus or minus 0.05 miles. (Hint:
Look at the left end of the Status bar). Shortly a large yellow circle will show up.

____ **40.** Under Selection > Selection Options > Interactive Selection, pick "Select features completely
within the box or graphic(s)". Click OK. Under the Selection menu pick Select By Graphics, click
OK, and wait. The boundaries of the selected blocks should appear shortly.

This moves us closer to satisfying the banker's objection that areas of census blocks that don't lie within
the 3-mile radius are being selected.

____ **41.** Zoom in on an area in the northwest where the circle meets the rest of the map. Assure
yourself that no selected census block has any part outside the circle. Go Back To Previous
Extent.

____ **42.** Now open the Blocks&Pop_TN_SP attribute table. How many records are selected? _____.
Run statistics on the populations of the selected records. What is the population within the
blocks that fall completely inside the 3-mile circle. _____.

Uh-oh. Now we don't have the 25,000 people we need. After a brief period of despair we re-examine the
requirement. It says "a loan may be obtained if it can be shown that 25,000 people reside within 3 miles
of the road on which the facility is to be built." Since we centered our circle around the midpoint of the

road, maybe we can get a few more census blocks by looking at two circles, centered at opposite ends of the road. (If this doesn't work, we might be able to argue that while our first method overestimated the population, the current one probably underestimates it.)

____ **43.** Click the circle with the Select Elements tool. Press the Delete key to make it go away. Under Selection click Clear Selected Features. If you made the roads layer un-selectable make it selectable again. Select the Galbraith School Road segments again (the formula may still be in the Select box) and zoom to leave about 5 miles all the way around them. Click the circle icon on the Drawing toolbar and again draw a 3.00 mile radius circle, but center it on the southeast end of the road. Then, click the circle icon again and drag a second 3.00 mile circle centered on the northwest end. Pick the Select Elements tool. Hold down Ctrl and click within each circle so that both are selected. Under the Selection menu pick Select By Graphics, click OK, and wait. The boundaries of the selected blocks should appear shortly.

____ **44.** Now open the Blocks&Pop_TN_SP attribute table. How many records are selected? _____. Run statistics on P001001 in the selected records. What is the population within the blocks that fall completely inside the pair of 3-mile circles? _____. Call the banker back!

____ **45.** Close ArcMap, saving the map as Day_Care_YES.mxd in Day_Care_Data.

Exercise 6-5

Determining Proximity of Points to Lines and Other Points

For this exercise, we return to our fictional islands of Chapter 5. You may recall that you digitized them in

> ___IGIS-Arc_*YourInitials*\Digitize&Transform
> \Islands.mdb\UTM_Zone_2

The data set became North_Islands_Polys. You stored a map named North_Islands_map3. Houses are being built on the square island. A power generation plant is being constructed on the northwest corner of the lake and a power line run from the plant around the island. Also, wells are being dug to supply water to the houses. These feature classes are called Houses, Power_Lines, and Wells. In order to bring the appropriate lengths of electrical wire and water pipe to the island, we want to know the distances from the houses to the power lines and the distance from the houses to the wells. This information will be stored in the attribute table of the Houses feature class.

____ **1.** Start ArcCatalog. From

> [__] IGIS-Arc_AUX\D&T\Islands.mdb\UTM_Zone_2

> copy the PGDBFC named Power_Lines to

> ___IGIS-Arc_*YourInitials*\Digitize&Transform

> \Islands.mdb

> Now also copy the feature classes Houses and Wells. Close ArcCatalog.

_____ **2.** Navigate to the map North_Islands_map3.mxd in

___IGIS-Arc_*YourInitials*\Digitize&Transform.

Double-click the name to start ArcMap. Zoom in on the square island.

_____ **3.** Because of your previous copying actions Houses, Power_lines, and Wells are feature classes in

___IGIS-Arc_*YourInitials*\Digitize&Transform
\Islands.mdb\UTM_Zone_2

Add them as data. Show the wells as green circles of size five. Accept ArcMap's choice of symbol for the power line. Change the color of the houses to black. Looking at the attribute tables: How many houses are there? _____. How long is the power line? _____. How many wells are there? _____.

_____ **4.** Examine the attribute tables of houses and wells. Note that both are dull, as befits basic point feature class tables.

_____ **5.** Measure the diagonal distance across the island. Obviously, the distance from any feature to any other feature cannot be greater than this number. Write it here. _____ meters.

_____ **6.** Use the search facility button to locate the Near (Analysis) tool. Type "near tool" (no quotes) into the search box. You will get several groups, each of which consists of three lines of text, which are links. You are interested in the group that starts off with Near (Analysis) (Tool).[8]

First line: Near (Analysis) (Tool). If you hover the mouse cursor over the word Near you may get a brief description of the tool. If you click the word Near the tool itself will start, displaying a Near window. Click Near. Observe the window, then Cancel it.

Second line: Determines the distance from each feature . . . If you click this line you may get a detailed description of the tool—basically a help file with an extensive discussion of the tool of the tool.

Third line: toolboxes\system toolboxes\analysis tools.tbx . . . If you click this line you may get response from the Catalog Tree. It will show you the path to the tool. Try clicking this line. What is the path to the Near tool?

Toolboxes > _____.

_____ **7.** Right-click the tool and click Open. Show Help for the Near tool, if it is not already there. Click Help within the Help panel. Expand and read the Usage Tips. Dismiss the ArcGIS Desktop Help window.

_____ **8.** For Input Features, click the drop-down menu arrow (since the layers are already represented in ArcMap, they will appear) and choose Houses. For Near Features (which means those features

---

[8]The effect of clicking on a link and hovering over a link seems to vary from one version 10 of the software to another. Experiment to find out what your version does.

that are **Nearest** to the Input Features), pick Wells. For the search radius, type the number you found previously for the diagonal distance across the island. Check to see that the units are meters. Click the boxes that will specify the location of the well (called Function Name) and the angle from the house to the well. Run the tool by clicking OK. Close it when the Close button appears. Dismiss ArcToolbox to get more room on the screen. Make sure ArcCatalog is closed.

____ **9.** Again examine the attribute table of Houses. What are the field names?

_____ , _____ , _____ , _____ , _____ , _____ .

____ **10.** Run statistics on NEAR_DIST. What is the total number of meters of pipe that would be required to connect each house to its nearest well? _____ .

____ **11.** Label each well with its OBJECTID. Label each house with its near feature ID (NEAR_FID) attribute value. "Eyeball" the results to verify that correct houses were assigned to wells. Label each house with its NEAR_DIST. Use the measure tool on a couple of house-well pairs to verify that the Near command got it right in the attribute table.

____ **12.** Label the houses with NEAR_ANGLE. Check to see that things look right. Realize that the angle given from the house to the well is the "math" version (0 degrees is the positive x-axis, the angle increases counterclockwise), rather than the compass version (0 degrees is the positive y-axis, the angle increases clockwise). Check the Help file for details. This is just one of the wrinkles one encounters in ArcGIS—the kind of thing that gets you paid the big bucks when you graduate—unless you go into teaching, that is.

____ **13.** Extra credit problem: Could you reduce the cost of putting in water lines by using mains (say using larger pipes to some intermediate junction) and then smaller pipes from there? You could augment the wells feature class by putting in pseudo-wells (junctions) and experiment with those. What is the shortest length of pipe you can find that would do the job? _____ . Of course, this makes the problem more complicated, since you have to figure in the greater cost of the larger pipe, the additional connections, and so on.

____ **14.** Run the Near tool again, this time determining connections between the houses and the power line. (Re-running the tool will overwrite the previously stored attribute values NEAR_DIST, NEAR_X, and NEAR_Y, and so on; there is no warning to this effect.) What length of electric cable would be required? _____ . Why are so many of the angles 90 degrees? _____

_____

Exercise 6-6 (Review)

Checking, Updating, and Organizing Your Fast Facts File

The Fast Facts File that you are developing should contain references to items in the following checklist. The checklist represents the abilities to use the software you should have upon completing Chapter 6.

____ Double-clicking a toolbar's left end

____ Double-clicking a toolbar's title

____ To get a list of all toolbars

____ Shape_Length is

____ Shape_Area is

____ Pointing at records

____ Two windows for selecting records are available at

____ When a record is selected

____ When a feature is selected

____ Once some records are selected, other records, which are not selected, may be added to the selected set by

____ In selecting records, logical and arithmetic expressions

____ In the Options window, the Statistics button produces

____ When some records are selected, Statistics are calculated only on the

____ It is not usually necessary to type all of an expression because

____ When a user is selecting records, all the values of a variable may be seen by

____ When a polygon feature is selected, it is shown on the map by

____ When a polygon feature has islands within it, it is displayed

____ The Options button on a table also provides these capabilities

____ The difference between the Select Elements button and the Select Features button is

____ Selection of features based on two different layers is accomplished by

____ A graphic representation of Select by Location is available

____ Plus and minus signs in the table of contents

____ A data frame is

____ A layer file

____ To get layer properties

____ To get data frame properties

____ Under the Symbology tab, the user the following options exist

____ To change the field used for labeling

____ To symbolize features based on values in an attribute field

____ To label features with values from more than one attribute

____ ArcGIS has a number of ways to put data into classes:

____ The windows involved in the classification process are

____ A histogram shows

____ When selecting a ramp a user may see either

____ The Break Values box shows

____ On-the-fly projecting means

____ Select by Location, used with intersect, produces

____ TIGER/Line street files

____ To convert a spreadsheet to database format

____ To make a join permanent

____ TIGER/Line Files usually have ____ coordinates

____ Graphics may be used to select features by

____ The Jenks method

____ Normalization means

____ To round off numbers

____ Three sorts of graphs that ArcMap provides are

____ To add a column to a table

____ To calculate values in a column

____ The Internet site of the Bureau of the Census is

____ To determine distances between various features, use

____ The syntax differences between personal geodatabases and file geodatabases are

Creating Spatial Feature Classes Based on Proximity, Overlay, and Attributes

Generating Features Based on Proximity: Buffering

OVERVIEW

IN WHICH you learn three of the fundamental tools of GIS analysis and use them to solve problems. Also, you are introduced to the Esri model building software

Proximity is a word that implies nearness in terms of physical distance. It is not quite the inverse of distance, but greater proximity implies smaller distance. Proximity is a concept that strongly affects our lives and activities. We, along with most of the animals of higher intelligence, innately understand the idea. We want to be close to pleasant and useful things, or to those things we must access on a regular basis, such as friends, places of employment, and shopping. We want to be far from those things that are unpleasant or noxious, like smelly dumps or plants, irritating people, or dangerous environmental conditions. Much of the law regulating land use is written with the concept of proximity in the background—it sets limits of acceptable proximity, usually as a threshold distance. An example would be a law that says a liquor store may not exist within 500 feet of school grounds.

The primary analytical tools of GIS that implement the concepts of proximity create what are called buffers. To create a buffer, a set of features, say X, is specified. A distance, say Y, is specified. The software then generates a buffer: a set of areal features that have the property that every point on or within their borders lies within the specified distance of the original features. Formally, the buffer consists of the locus of points whose distance from each of the features in X is less than or equal to the specified distance Y.

Because we have such an intuitive idea of proximity, understanding what buffers are is not nearly as fearsome as the definition suggests. You encountered the buffer concept in the first exercise of this book, when you solved the Wildcat Boat problem by manual means. Recall that the testing facility had to be within 300 meters of a sewer line and could not be less

FIGURE 7-1   Points to be buffered

FIGURE 7-2   Buffers around points

than 20 meters away from streams. That is, the facility had to be within a buffer of the sewers and outside a buffer of the streams.

Buffering Points

Figure 7-2 shows what a buffer around the set of point features in Figure 7-1 would look like.

You don't see the point features in Figure 7-2 because the *point features are not part of the buffer*.

In addition to the (geo)graphic portrayal of the buffered areas, an attribute table is created. Since the production of areas inside the buffer can also create polygons that are outside the buffer, the attribute table should contain a field that indicates the status of each polygon: inside or outside. For buffering with geodatabase feature classes, a field is generated that contains the distance used in creating the particular polygon. If that distance is zero, the area is not within the buffer.

Buffering Lines and Polygons

Line features may be buffered as well. The area of the polygons generated is that covered by the locus of points that lie within the threshold distance of any point on the line. Examine Figure 7-3 and then Figure 7-4.

You could now see that buffering polygons is merely an extension of the preceding ideas. Again, the polygons that are buffered, although they are areal features, are not present in the buffer. Examine Figure 7-5 and then Figure 7-6.

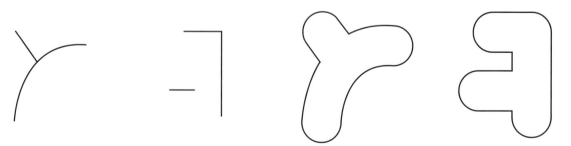

FIGURE 7-3   Lines to be buffered

FIGURE 7-4   Buffers around lines

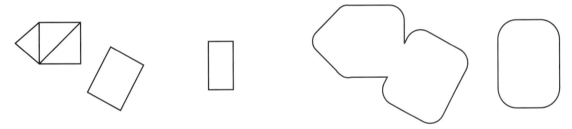

FIGURE 7-5   **Polygons to be buffered**   **FIGURE 7-6**   **Buffers around polygons**

There are many variations on the buffering theme. You can buffer only one side of a line. You can ask that line buffers have flat ends, in which case areas past the ends of the line are not included in the buffer. Very importantly, you can buffer features by different thresholds. The thresholds are taken from information in the attribute table of the feature being buffered. For example, suppose that you wanted to establish "no smoking" zones around wells. Perhaps your features consist of water wells, oil wells, and natural-gas wells. You could set things up so that water wells were buffered by 30 meters, oil wells by 100 meters, and gas wells by 400 meters.

Generating Features by Overlaying

Pick any point of Earth's land area. A large number of attributes might describe that point. It may be within the area owned by a person. The point will be in some country or other. It might be part of a floodplain. It will have associated with it a certain soil type. The representation of the point in a GIS might indicate that it is part of a roadway or stream. Or its GIS representation might show it to be an oil well.

In doing GIS analysis, we are frequently interested in combinations of attributes that relate to points, lines, or areas. For example, we might be interested in those areas that have a certain land use zoning and that are also for sale. Suppose that we have one GIS layer whose polygons show zoning and another, different layer that shows properties for sale. By a process called overlaying, we can create a third layer from which we can identify those properties with the zoning we want that are also for sale.

The term "overlay" comes from the physical process of laying one map of transparent material on top of another—say, on a light table—and examining the effect of the combination. To continue the preceding example, suppose that you wanted to buy some property to start a business. The zoning had to be "B1" and, of course, the property had to be for sale. You might take a clear Mylar map of the area and use a marker to blacken all of the zones that were not B1. On a second such map, you might blacken all of the properties that were not for sale. If you then laid both maps on a light table (assuming that they covered the same area, were at the same scale, were based on the same geographic datum, had the same projection, and so on—not a trivial assumption as you know by now), light would shine through the areas that might be suitable for your intended business. With GIS software we can simulate this activity, with much greater efficiency and garnering much more information in the process. You encountered the essence of overlay, working the Wildcat Boat problem, when you found combinations of suitable soils and land use.

Let's look first at the idea of overlaying a polygon feature class with another polygon feature class. We get two types of results: one (geo)graphic and the other tabular.

Figure 7-7 and Table 7-1 depict a feature class named A, consisting of four polygons. In feature class A there is an attribute named Zone, which has values p, q, r, and s, as shown on the diagram and in the table.

p	q
r	s

**FIGURE 7-7   Feature class "A"
with ZONE attribute**

TABLE 7-1

Shape_area	Shape_length	Zone
		p
		q
		r
		s

(In Table 7-1 and subsequent tables, values of columns not germane to the discussion have been omitted. These include feature identifier numbers, which would be unique. Also, no values have been given for Shape_Length and Shape_Area, since these are not important to our discussion.)

Figure 7-8 and Table 7-2 depict feature class B, of five polygons. Feature class B has an attribute named "Jurisdiction." Values of Jurisdiction are u, v, w, and x. Note that two different polygons are characterized by u.

Suppose feature classes A and B are overlaid to produce polygon feature class C. Then C would consist of ten polygons, as shown in Figure 7-9. Each of those polygons would be homogeneous in pairs of attribute values, as illustrated. The attribute table, Table 7-3, of C would, of course, consist of 10 records. It would contain the attributes, Zone and Jurisdiction, *from each of the constituent feature classes*.

This is the fundamental essence of polygon overlay. You generally get more polygons and more attributes than in either of the input feature classes. You can identify each polygon by the attributes in each of the input feature classes.

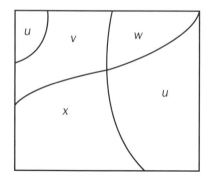

**FIGURE 7-8   Feature class "B"
with JURISDICTION attribute**

TABLE 7-2

Shape_area	Shape_length	Jurisdiction
		u
		v
		w
		x
		u

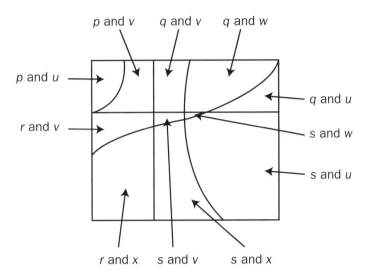

p and v   q and v   q and w

p and u

r and v

q and u

s and w

s and u

r and x   s and v   s and x

FIGURE 7-9   Overlay of "A" and "B"

TABLE 7-3

Shape_area	Shape_length	Zone	Jurisdiction
		p	u
		p	v
		q	v
		q	w
		q	u
		r	v
		r	x
		s	u
		s	v
		s	w
		s	x

Overlaying with Line and Point Feature Classes

When you do an overlay, one of the two feature classes is frequently of the polygon type. The other may be point, line, or, as you saw above, polygon. What does it mean to overlay a polygon feature class and a point feature class? Basically, the change is only the tabular output (although some points may be omitted if they lie outside the scope of all polygons, depending on which command you use).

12 ●
● 14
● 45

88 ● 23 ●

● 77
15 ●

**FIGURE 7-10   Points
indicating parking meters**

TABLE 7-4

Meter_number
12
45
77
88
23
15
14

Overlaying Point Features

Let's suppose that some point features are parking meters, each of which has a number. See Figure 7-10. In the attribute table, the field Meter_number records this number. See Table 7-4. Also assume that the polygon feature class "A" you examined previously, depicted zones of the city, and it was desired to know which parking meters fell into which zones, so that the income from each meter would go to the correct budget.

Now if the point feature class (see Figure 7-10) is overlaid on the polygon feature class A (see Figure 7-7), the graphical result of the new point feature class will be the same as Figure 7-10. However, the attribute table of the new point feature class will look like Table 7-5. Figure 7-11 indicates why, showing as it does with the dashed lines, the geographic location "A" of Figure 7-7.

Overlaying Line Features

A similar effect occurs if lines are overlaid by polygons. The records associated with the lines in the output table acquire the attributes of the polygons in which the lines lie. The output line feature class usually will consist of more lines than the input because when a line from the input crosses a polygon boundary it is cut, forming a line on either side.

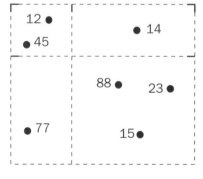

**FIGURE 7-11   Points overlaid with
feature class "A"**

TABLE 7-5

Meter_number	Zone
12	p
45	p
77	r
88	s
23	s
15	s
14	q

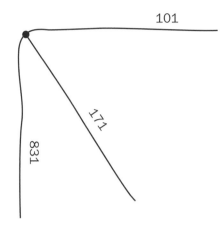

TABLE 7-6

Shape_length	Highway_number	Lanes
	831	4
	101	4
	171	2

FIGURE 7-12   Lines indicating roads

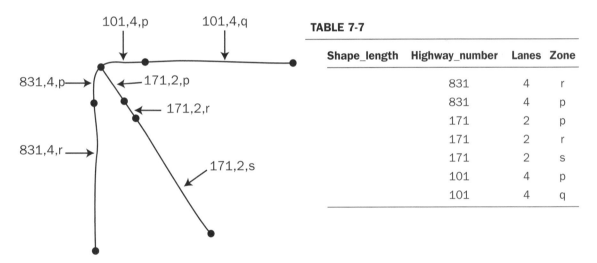

TABLE 7-7

Shape_length	Highway_number	Lanes	Zone
	831	4	r
	831	4	p
	171	2	p
	171	2	r
	171	2	s
	101	4	p
	101	4	q

FIGURE 7-13   Lines overlaid with feature class "A"

Suppose that we have three roads in the area of feature class "A." See Figures 7-12 and Table 7-6. For purposes of maintenance, we would like to know what zone each stretch of road falls into.

The overlay would produce a line feature class with seven lines as illustrated in Figures 7-13 and Table 7-7. The Shape_lengths would be the lengths of the newly created lines.

Spatial Joins in General

The term "spatial join" has been coined to describe overlay and similar processes. Basically, spatial joins share the idea of taking two or more spatial datasets, each with associated attribute tables, and creating another dataset. Both the (geo)graphics and the attribute tables in the new set differ from those of the originating datasets.

You find the tools to perform these actions in ArcToolbox: Analysis Tools (for work with feature classes in general).

Union combines feature classes in such a way that the extent of the derived dataset includes the extents of all the constituent datasets. The idea is that, in the union of P and Q to make R, say, the territory covered by R is the territory covered by P and also by Q, including any overlapping territory. In logic, the union of two sets (A and B) is the set of all objects in A or B or both. If A is all red squirrels and B is all squirrels with fuzzy tails, then A union B is the set of all squirrels that are either red or have fuzzy tails or both. The union is equivalent to the logical *inclusive* OR. (The logical *exclusive* OR—XOR—would consist of those squirrels that were red or had fuzzy tails, but not those that were red and also had fuzzy tails.)

Intersect combines feature classes in such a way that the extent of the derived dataset includes only the overlapping area of the constituent datasets. The idea here is that, in the intersection of S and T to make U, say, the territory covered by U is only the territory covered by S that is also covered by T. In logic, the intersection of two sets (A and B) is the set of all objects that are in both A and B. If A is all red squirrels and B is all squirrels with fuzzy tails, then A intersect B is the set of all squirrels that are red and have fuzzy tails as well. The intersection is equivalent to the logical AND.

A number of other spatial join tools are available in the tool chests:

❏   Identity

❏   Erase

❏   Update

❏   Clip

❏   Split

❏   Symmetrical Difference (the union with the intersection removed)

Further, some of these tools are not confined to polygons. For example, you can use Intersect to determine, from two feature classes consisting of points, which points are coincident (or lie within a certain distance of each other). The same can be true with two sets of lines. Or to find the points that intersect lines.

You will work with some of these in the Step-by-Step section of this chapter. Now we turn to what you can do once you have applied buffers and spatial joins.

Deriving Feature Classes by Selecting Attributes: Extraction

You have perhaps buffered some points, lines, and polygons, and have the results as feature classes. You have overlaid these buffers on some other polygon feature classes. What you now have is a feature class that has dozens, hundreds, maybe even thousands of polygons. That feature class also has an attribute table with a list of fields as long as your arm, since it consists of almost all the attributes of the constituents tables. Now it is time to make a feature class that consists only of those polygons that meet your requirements. You do this by writing a query that will be processed by the Select tool, within the

Extract tool chest. Selection By Attributes is nothing new to you. The difference here is that you will make an entirely new geodatabase feature class or shapefile based wholly on the selected features and records.

Let's assume that you are trying to find the appropriate polygons for the Wildcat Boat problem. Without being rigorous, consider a query like the one that follows, made to the composite table that includes land use, soil suitability, sewer buffers, and stream buffers. You might use Extract to make a feature class of the polygons where:

```
LANDUSE = "Brushland"

AND

SOIL_SUITABILITY = "Fair" OR SOIL_SUITABILITY = "Good"

AND

DISTANCE_TO_SEWERS <= 300

AND

DISTANCE_FROM_STREAMS > 20
```

In summary, then, the EXTRACT Wizard produces a feature class Y from a feature class X, by extracting features based on their *attribute* values. EXTRACT works very much like building a query to select records in a table (e.g., AREA >= 500 AND COLOR = 'Green').

But instead of simply highlighting records in a table (and the corresponding features on the map), EXTRACT creates an entirely new dataset (Y) consisting of selected records.

The way to use EXTRACT is, once you have constructed (by repeated buffering and overlaying, using ArcToolbox), a feature class (A) that contains all the features that those processes produce, you then extract just those features that meet your criteria, to make feature class B.

The concepts in this Overview pretty much cover the sort of geographic data analysis that is created by deriving feature classes from other feature classes. There are, of course, a lot of data management commands that go along with this form of analysis. Also, there are several other types of GIS analysis—geostatistical for one, network for another—covered in Chapter 9.

In the following Step-by-Step section, you will apply the buffer, overlay, and extraction principles and tools.

Finally, in the Step-by-Step section of this chapter, I introduce the Model Builder, which is one way to link and automate the processes in GIS analysis. By building a model:

❏ You can document the flow through the processes of a project.

❏ You can modify those processes easily, change processes, change the inputs to the processes, and so on, in a friendly graphical environment.

❏ You can have the computer repeat some or all of the processes with trivial effort on your part.

Creating Spatial Data Sets Based on Proximity, Overlay, and Attributes

—— Open your Fast Facts File.

—— Open the Color Figures file.

ArcGIS Desktop has three programs that create buffers. While they are similar in their intent, they have different capabilities and allow different inputs. The fact that there are three programs, instead of one that has all capabilities, is indicative of how the computer code for ArcGIS has evolved, rather than having been written from the ground up. The three are:

❑ Buffer Wizard, an older, but still supported, program that has a lot of capability but has to be added to the menu of tools that you can use. In the author's experience Buffer Wizard doesn't always work.[1]

❑ Buffer, in ArcToolbox, creates buffers, but, depending on the final output you want, have to be tweaked with other ArcGIS tools, and which doesn't supply an important column of information in the buffer feature class attribute table with some inputs.

❑ Multiple Ring Buffer, which allows several buffers to be created at the same time, but lacks the ability to make variable-width buffers based on the attribute table of the feature class being buffered.

You will use all three to gain an understanding of their capabilities.

Exercise 7-1 (Warm-Up)

Making Trivial Buffers around a Trivial Feature Class

—— **1.** In ArcCatalog make or confirm a Folder Connection to

___IGIS-Arc_*YourInitials*\Trivial_GIS_Datasets.

---

[1] The Buffer Wizard didn't work for the author in version 10 with the Roads feature class. It produced an error that required closing ArcMap. This problem may have been fixed by Esri by the time you are learning the buffer programs. Or maybe not. In any event, I call your attention to two ideas: (1) All major software programs have bugs, which should be reported to the vendor (I have), and (2) with ArcGIS, there is generally some sort of workaround.

___ **2.** Start ArcMap with a new Blank map. In the folder

___IGIS-Arc_*YourInitials*\Trivial_GIS_Datasets

you will find a file geodatabase named Buffer_TicTacToe.gdb. Find the feature class named TicTacToe (in TicTacToe_FDS). Display it in ArcMap. Set the map units and display units to Meters. After setting Distance to Meters in the Measure window, determine the width of TicTacToe: _____ meters. Open the attribute table from the T/C. How many lines are there? _____ Identify some of them. What is the length of each? _____ meters.

The concept of buffering is pervasive in GIS. Over the years, ESRI has developed several tools to do the operation. For geodatabases, we will explore two of them. The one we will use first is called the Buffer Wizard. It is not automatically available in ArcMap or ArcToolbox, so you will have to add it. In addition to getting the wizard, you will also learn something of the way the software can be customized. The Buffer Wizard will be added to the Tools menu in the following step.

___ **3.** In ArcMap select: Customize > Customize Mode > Commands > tools (from the Categories list). Click Buffer Wizard in the Commands list. Drag it to the Geoprocessing menu on the Main menu, but don't let go of it yet. The Geoprocessing menu will open. Drag it down to just barely below Buffer and release the mouse button. The wizard is now one of the geoprocessing tools you can use. Close the Customize dialog box.

___ **4.** Click on the Buffer Wizard tool. Use it to buffer the lines of TicTacToe, making a File Geodatabase Feature Class named TicTacToe_BUF with the following characteristics:

❏ A specified distance of 5 meters (you have to set the buffer distance units)

❏ Barriers not dissolved between buffers

❏ Saved in ___IGIS-Arc_*YourInitials*\Trivial_GIS_Datasets \Buffer_TicTacToe.gdb\ TicTacToe_FDS

with the name TicTacToe_BUF (Careful: it's not a shapefile.)

___ **5.** TicTacToe_BUF should appear on your ArcMap display, probably in brown. If necessary, drag the T/C entry TicTacToe_Buf below TicTacToe. Make sure you can see both feature data classes, changing the line color and/or width of TicTacToe as necessary. This image should give you a good idea of what buffering is all about. Note that the ends of some buffers cover the ends of others, and that the buffer distance is applied around the endpoints of the lines. Turn off TicTacToe. Note that each of the 12 lines is buffered. Open the polygon Attribute Table of TicTacToe_BUF. What are the names of the fields? _____, _____, _____, _____, _____. How many polygons are there in TicTacToe_BUF? _____. Identify records to determine which polygon is which. What is the area of each polygon, to one decimal place? _____. Why isn't the buffer 1000 square meters (100 meters long times 10 meters wide)? _____
_____

Some points are important here:

❑ The lines of TicTacToe are not part of the feature class TicTacToe_BUF, which is strictly a polygon feature class.

❑ The buffering process created polygons with a BufferDist of 5, indicating that the area of that polygon is within 5 meters of the lines that were buffered. The BufferDist field is the mechanism that allows you to distinguish areas of the spatial field that are within the buffer distance.

_____ **6.** Repeat the buffering process, but this time dissolve the barriers between the features, calling the output feature class TicTacToe_BUF2. When it appears turn off TicTacToe_BUF. Open the TicTacToe_BUF2 polygon attribute table. How many polygons are there in TicTacToe_BUF2? _____. What is the area? _____. You might expect the area to be more than 12,000 square meters (four 300 meter lines times 10 meters wide) plus the rounded ends. Instead it is less. Why? _____.

Exercise 7-2 (Project)

Exploring FEATURE CLASS Buffers with the Wildcat Boat Data

_____ **1.** In ArcCatalog, make or confirm a folder connection with

___IGIS-Arc_*YourInitials*\Wildcat_Boat_Data

Close ArcCatalog.

_____ **2.** In ArcMap with a new Blank map (don't save changes), use Add Data to display the Personal Geodatabase Feature Classes that depict Sewers and Roads, which you will find in the tree of:

___IGIS-Arc_*YourInitials*\Wildcat_Boat_Data

You may bring both feature classes in with a single step by selecting both entry names (click with Ctrl) and pressing Add. Also add the Soils geodatabase feature class. Zoom to the full extent of the map.

_____ **3.** A review: Rearrange the order of the entries in the Table of Contents so that both the layer Roads and the layer Sewers are below the layer Soils. (Drag each entry by its name.) What does this say about the order in which layers are drawn? _____. Note the effect of drawing a polygon layer after line layers. Experiment with turning layers off and on. Click the List By Source tab at the top of the T/C. Note that you cannot now rearrange the order. Click the List By Drawing Order tab again. Then rearrange the layers so that the Sewers entry is at the top and the Soils entry is at the bottom.

_____ **4.** Open the attribute table for sewers. How many segments of sewer pipe are there? _____. What are their diameters? _____, _____. Close the table.

_____ **5.** Make Roads bright yellow with width 3. Make Sewers red with width 1. Arrange things so that where the roads and sewers are coincident you can see both.

_____ **6.** It may be that when you slide the cursor around the map the coordinates are given in unknown units. If so, go into the data frame properties window and, using the General tab, make the Map Units and the Display Units "meters." (ArcMap should have picked up the coordinate system, and the units, from the metadata, but sometimes this doesn't happen.)

Using ArcToolbox to Make Buffer Zones around the Roads

_____ **7.** Start ArcToolbox. Navigate to Analysis Tools > Proximity > Multiple Ring Buffer. Start the tool. You will use Buffer to make a buffer around the roads. For Input Features browse to

___IGIS-Arc_*YourInitials*\Wildcat_Boat_Data\Wildcat_Boat.mdb\Line_Features\Roads

click it and click Add. (Alternatively select Roads from the dropdown menu.)

For Output Feature Class browse to

___IGIS-Arc_*YourInitials*\Wildcat_Boat_Data\Wildcat_Boat.mdb\Area_Features

and supply the Name Buffer_of_Roads. Click Save. For Distances type in 44 and then press the plus icon. Make the Buffer Unit Meters. For Field Name, type Radius. Make the Dissolve Type ALL. OK. Be patient. When computation ceases the buffer will be added to the map. Zoom to full extent.

You noticed that you got the input from Line_Features and put the output in Area_Features. It's appropriate to save it there because, as you will remember, buffers are always polygons regardless of the type of feature being buffered.

_____ **8.** In the T/C move the name Buffer_of_Roads below Roads. Turn the Sewers layer off. Note that you see three distinct buffered areas.

_____ **9.** Display the attribute table for the Buffer_of_Roads layer. Because you dissolved the barriers between the buffers and the tool generates multipart polygons, you will see only one record. The shape length and shape area wili subsume all three polygons. Note that an attribute called Radius was created; its value is 44, which will be useful when you want to identify the buffers. Notice that the tool does not make explicit, separate polygons for the islands that are created.

_____ **10.** Zoom in on a relatively straight part of the road buffer. Use the Measure tool to determine the entire width of the buffer to two decimal places. What value do you get? _____ meters. What would it be if you could measure exactly? _____. Zoom to full extent of Buffer of Roads.

Suppose you want attribute table records for each of the three polygon buffers you see on the map. You can convert the multipart polygons to singlepart polygons.

_____ **11.** In ArcToolbox navigate to Data Management Tools > Features > Multipart To Singlepart. Start the tool. For Input Features use Buffer_of_Roads; for Output Feature Class use Buffer_of_Roads_spp (spp for singlepart polygons). Run the tool.

___ **12.** Look at the attribute table of Buffer_of_Roads_spp. You should see three records. Select each record and note the corresponding polygon.

___ **13.** Use the Multiple Ring Buffer to make another buffer around the roads—this time with a distance of 111 meters. Call it Buffer_111. Except for the buffering distance, use the same parameters as before.

___ **14.** Arrange the T/C so that Buffer_of_Roads is at the top, with Buffer_111 immediately below it. Turn off any other entries in the T/C. The feature class Buffer_111 contains only one polygon instead of three. Why? _____. (*Hint:* Flip the 111-meter buffer off and on.)

___ **15.** Remove the Soils layer. Display the 44-meter buffer with a light yellow. Display the 111-meter buffer with a dark yellow. Using the Identify feature tool (with the Layer set to Top-most layer), click the various portions of the buffers, paying attention to what flashes.

___ **16.** Use a single application of Multiple Ring Buffer to make a 50 meter buffer and a 100 meter buffer around Roads, naming the buffer whatever name you want: _____. As you provide input to the tool (it's a Python Script, actually – a term to be defined later) click on the Dissolve Option and read about the two different options. What are they? _____. Use whichever you want and explore the results.

___ **17.** Dismiss ArcMap. Dismiss ArcCatalog if it is running.

Variable-Width Buffers

As you know, a single layer may contain many features. The features usually have different attributes that distinguish features from each other. For example, in Roads there are some graphic features that are associated with RD_CODE "1" and others with RD_CODE "2." Suppose that you wanted to buffer the "1" roads with a distance "x" and the "2" roads with a distance "y." You can do the following:

___ **18.** Start ArcMap. Add Roads from the Wildcat_Boat geodatabase in ___IGIS-Arc_*YourInitials*. Represent the features by a red line of width 2.

___ **19.** Open the attribute table. Under Options add a field named Buffer_by. Make it a Short Integer field.

___ **20.** Start editing with the Editor toolbar. Note that a tiny pencil shows up at the bottom of the table. Suppose that you want to buffer the RD_CODE = 1 roads by 20 meters and the RD_CODE = 2 roads by 40 meters. Notice that the first records have a 1 in RD-CODE, so we would want to put 20 in the Buffer_by attribute of those records. Click the Buffer_by field cell in the first record and type 20. Now do the same with the second record. And the third. This looks like a long haul. We'll use another way instead.

___ **21.** On the Main menu, choose Selection. Select By Attributes. Construct the expression [RD-CODE] = 1. See Figure 7-14. Click Verify. If the expression is OK, click Apply, then OK to close the Select By Attributes window. Now you can see that all the records with 1 in the RD_CODE are selected.

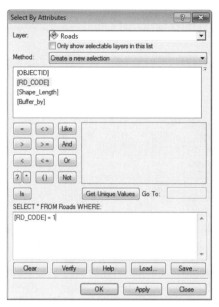

FIGURE 7-14   Select by attributes window Roads

_____ **22.** Right-click the column name Buffer-by and pick Field Calculator. Type 20 in the "Buffer_by =" text box that appears and click OK. Now every RD_CODE = 1 road should have a 20 in the Buffer-by field. Use Options in the attribute table window to switch selection, which will select every RD_CODE = 2 road. Use the Field Calculator to put a 40 in the Buffer_by cells of those records. Clear all selections. Save edits, then stop editing.

Now that you have the proper distances in the attribute table, you can proceed to make the actual buffers.

_____ **23.** Start the Buffer tool (found under Analysis Tools > Proximity). The input, of course, is Roads. Browse (using the little yellow file folder) to save the buffer in

___IGIS-Arc_YourInitials\Wildcat_Boat_Data\Wildcat_Boat.mdb\Area_Features

For the name use Roads_dual_buffers. Press the Field radio button. Choose Buffer_by from the drop-down menu which will provide the correct distance (20 or 40) to buffer each given road by. For variation, choose FLAT for the End Type. The Dissolve Type should be NONE. Click OK.

_____ **24.** The feature class Roads_dual_buffers will be added to the map. Pull its name below Roads, if necessary. In the open attribute table of Roads, select the RD_CODE = 1 records (hold down Ctrl and drag the cursor down the boxes at the left of the records) so that the line features show up selected in the map. Measure the width of some RD_CODE = 1 roads buffers. _____ Measure some RD_CODE = 2 roads buffers. _____.

_____ **25.** Notice that there are as many buffer polygons as there are road lines. You may simply want to represent the two types of buffers as fewer polygons. Use the Search (Ctrl+F) icon to find

"Dissolve" from among the Tools. You want the "data management" one. For Input Features specify Roads_dual_buffers. For the Output Feature Class put Roads_dual_buffers_reduced (into the Area_Features dataset, of course). Check the Dissolve Field(s) box RD-CODE. (What other box could you have checked instead? _____.) Uncheck "Create multipart features". Click OK. When Roads_dual_buffers_reduced is added to the map explore it with the attribute table and the Identify tool. Close ArcMap.

Exercise 7-3 (Project)

Manipulate Polygon Feature Classes with Union and Extract

____ **1.** Start ArcCatalog. Make or confirm a Folder Connection to the

___IGIS-Arc_*YourInitials*\Trivial_GIS_Datasets

folder. In the personal geodatabase Overlay_Exercise.mdb, you will find the dataset Trivial_Feature_Classes. Examine the feature class TwoStalks. Look at both its graphic representation and its table. What are the values of the LOCATION attribute? _____, _____. Using the Identify tool, determine the location, to the nearest integer, of the southwest corner of the left polygon: X = _____. Y = _____. Using the Description tab of the Metadata (under Extents > Extent in the item's coordinate system) determine the location of the northeast corner of the feature class. X = _____. Y = _____.

____ **2.** Again using the Metadata (look at Fields), write down the names of the attributes of TwoStalks:_____, _____, _____,_____, _____

____ **3.** Examine the feature class ThreeBars. Look at its Geography. What are the PLACEMENT item values of the three polygons?

_____, _____, _____.

From the ThreeBars polygon Feature Class Properties window, determine the Data Type and Length (width)[2] of the PLACEMENT attribute: _____, _____.

____ **4.** Examine the attribute Table of ThreeBars. Write down the names of the attributes of ThreeBars: _____, _____, _____, _____, _____. What are the values of the areas of ThreeBars polygons? _____, _____, _____.

____ **5.** Launch ArcMap (and close ArcCatalog). Add both ThreeBars and TwoStalks to the map. Zoom to Full Extent. Notice their positions relative to each other. Slide the Select Elements cursor around the map to verify the locations of the images.

____ **6.** In ArcMap use the Search icon (or Ctrl-F) to find Union(analysis). Click on "toolboxes\system . . ." What is the path to the tool?

---

[2] Sometimes the number of characters in a string is called "width" and sometimes "length." Go figure.

Toolboxs > _____. Click on the path text. Toolboxes will open. Besides Union, what other tools do you find under Overlay?

_____, _____, _____, _____, _____, _____.

Start the Union tool by double-clicking the hammer icon. Read the Help panel for Union. Click the Input Features text box. Read the Help for that. For the first Input Feature class you can either click the down arrow at the end of the Input Features text box and choose Two_Stalks, or you can browse to

___IGIS-Arc_*YourInitials*\Trivial_GIS_Datasets\Overlay_Exercise.mdb\Trivial_Feature_Classes\ TwoStalks

Add Two_Stalks. Specify a second input as the ThreeBars feature class from Trivial_Feature_ Classes. Put the output feature class in Trivial_Feature_Classes (being careful about the path— you will have to use the yellow "browse" icon), calling it Stalks_Bars_mpp. For Gaps Allowed make sure the box not checked. (By not allowing gaps you instruct the software to create polygons in areas completely surrounded by the polygons of the union; the polygons formed, where there would otherwise be gaps, are not, in fact, part of the union.) Click OK and watch the messages go by. Close the Union window. When the resulting map appears spend some time with the Identify tool looking at the resulting polygons and their attributes.

The Union tool, by default, creates multipart polygons (hence we added "mpp"). That is, all the polygons that have the same values for the attributes (like LOCATION and PLACEMENT) become a single, multipart polygon. Sometimes this is useful and sometimes not. You can make the multipart polygons into singlepart polygons easily. You exploded multipart features before with the Advanced Editing toolbar of the Editor, but let's look for a different way.

7. Type Explode into the search box. Explode doesn't come up directly, but you see Multipart to Singlepart (Data Management), which is what you want. If you click on the middle of the three lines, you find a Description of the tool. In that Description are tags, one of which is explode, which is how Search found the tool. In the ArcCatalog window collapse everything related to Toolboxes. Click the third line of the Multipart to Singlepart tool item to bring up ArcCatalog with the Toolboxes expanded to show the Multipart to Singlepart tool. What is the path to the tool?

Toolboxes > _____

Make sure no features are selected. Run the tool on Stalks_Bars_mpp to make Stalks_Bars_ spp (for single-part polygons). Make sure it goes into Trivial_Feature_Classes.

8. When the computation is finished hide ArcCatalog, if that is where the tool came from. In the ArcMap T/C turn off all other layers. Stalks_Bars_spp was probably added to the map automatically. If not, add it. Make the layer color light green.

9. In the Layer Properties window for Stalks_Bars_spp, click Labels. Check Label features in this layer. Click Expression in the Text String pane. Place the following expression in the box:

[LOCATION] & vbNewLine & [PLACEMENT]

Click OK. Change the font size to 7. Click Apply. Click OK.

This last step has the effect of labeling each polygon with the LOCATION attribute and then, starting on a new line (hence the vbNewLine), the PLACEMENT attribute. Make sure ArcMap occupies the full screen. Even so, you may have to zoom in on the parts of the window to read the labels well.

___ 10. Open the Stalks_Bars_spp attribute table. Below list **_all the combinations_** of LOCATION and PLACEMENT (e.g., one combination is LEFT and TOP). Also list the **_number of polygons of each combination_**. (Two are done for you as examples.) Use both the Identify tool and sorting of the Stalks_Bars.spp attribute table to help you.

As you use the Identify tool, notice that the original OBJECTIDs of the input feature classes are present in the Stalks_Bars_spp attribute table as FID_TwoStalks and FID_ThreeBars. (Also present is the ORIG_FID from the multipart polygon feature class.) You will also notice values of negative one ("'1) for some FIDs. These occur in polygons for any feature classes that did not contribute to the union. For example, threebars had no presence in the topmost part of the left stalk, so its FID is shown as -1. Further, one polygon has a negative one in both the FID_TwoStalks and FID_ThreeBars fields. Which polygon would that be?

_____

❏ LOCATION is LEFT, PLACEMENT is blank. There are three such polygons.

❏ LOCATION is LEFT, PLACEMENT is TOP. There is one such polygon.
(Feel free to abbreviate.)

❏ _____

❏ _____

❏ _____

❏ _____

❏ _____

❏ _____

❏ _____

❏ _____

❏ _____

❏ _____

❏ _____

How many _polygons_ are there in total, according to what you wrote above? _____. How many polygons do you get by counting the image? _____. Save the map as Stalks_Bars_no_gaps. mxd in the folder Trivial_GIS_Datasets.

I realize that was not a fun exercise, but the main points to be made here are essential. The overlay process usually creates a large number of polygons and attributes. The attributes of the input polygons are carried over to the output. You should understand where each of the resulting polygons comes from and why those polygons have the attribute values that they do.

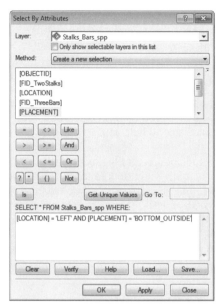

FIGURE 7-15   Select by attributes window Stalks_Bars; different from

_____ **11.** Use Selection > Select By Attributes to highlight the two polygons specified as LOCATION equals LEFT and PLACEMENT equals BOTTOM_OUTSIDE. See Figure 7-15 as a guide. Zoom in on the selected polygons to check their labels. Zoom back. Look at the attribute table.

_____ **12.** Use Selection again> Select By Attributes to _**additionally**_ highlight any polygons specified as RIGHT and TOP. Hints: Clear the previous expression. For Method, use Add To Current Selection. Once the selection is done, close the window and examine the result on the map. How many polygons did this operation add? _____. Dismiss the Select By Attributes window.

_____ **13.** Turn off Stalks_Bars_spp display and turn on the Stalks_Bars_mpp. Open its attribute table. Arrange things so that you can see both the table and the map. Display Stalks_Bars_mpp with a light yellow color. Write down all the attributes of Stalks_Bars_mpp. (Make sure the fields are wide enough so that you see all the characters of the attribute names.)

_____, _____, _____, _____,

_____, _____, _____, _____,

_____ **14.** Select the two columns entitled LOCATION and PLACEMENT. Right-click and sort them in ascending order. By selecting records in the table, you should now be able to easily verify the correctness of the blanks you filled out in Step 10 above, simply by counting the number of polygons that light up with each selection. When you are done, clear all selections. Then use the Select Features tool to look at the multipart polygons and their representation in the attribute table. When you are finished, clear all selections.

When you overlay polygons sometimes new polygons are created where there were none before. Below you will do a Union that illustrates this fact.

___ **15.** Create a Union of TwoStalks and ThreeBars as before, but do it with the box titled Gaps Allowed checked. Call the new feature class Union_with_gaps. Explode the feature class into

Union_with_gaps_spp. How many polygons are created? _____.
How do you explain the different number from the "no gaps allowed"
version?_____. Use Identify on the added
"polygon". What happens. Save the map as Stalks_Bars_gaps_allowed.mxd in the folder Trivial_
GIS_Datasets. Close ArcMap.

You should come away from this exercise understanding of the results of combining two polygon feature classes with Union and what makes up the attribute table. Frankly, it takes some thinking and maybe additional experimenting to understand the concepts involved. As I mentioned earlier, you usually get many more polygons and lots more records in the result of a Union than were in the constituent feature classes. Thinking this all out, throwing in also the ideas of single-part and multipart polygons, is not conceptually trivial.

Make a New Feature Class from a Subset of Polygons: Extract

As you saw, the Union of the feature classes produced many output polygons from a few input polygons. As you also saw, those polygons could be identified by their attributes. Frequently what you want to do after an overlay is to select a subset of the output polygons and make a new feature class. For example, to solve the Wildcat Boat problem, you would want to extract the polygons that met the requirements of soil suitability, land cover, and so on. To do this, you build a query in what is called the Structured Query Language (SQL). There is a lot of similarity between using SQL and what you did above in selecting polygons. The difference is that **the result of the SQL process is a new feature class** that contains the selected features.

Suppose now that we want to make a **new feature class** named BOTH_TOP_and_LEFT, consisting of those polygons on the top plus those polygons at the left.

___ **16.** Start ArcMap by opening Stalks_Bars_no_gaps.mxd. Make the only entry in the T/C Stalks_ Bars_spp by removing any others. From ArcToolbox: Analysis Tools > Extract > Select. Start the Select tool. Read the Help panel. Then read the several different Help discussions, obtained by clicking various places on the Select window—especially Expression. Look at the Tool Help.

When you begin to "build a query," it is assumed that all features (and therefore records) are selected. This is different from the previous assumption you have been using that none are selected. You begin by selecting a subset—that is, reducing the number of records selected, by reselecting some of them.

___ **17.** Browse to the Input Feature class Stalks_Bars_spp in

___IGIS-Arc_*YourInitials*\Trivial_GIS_Datasets\Overlay_Exercise\Trivial_Feature_Classes

Call the Output Feature class

ALL_TOP_POLYS_plus_ALL_LEFT_POLYS

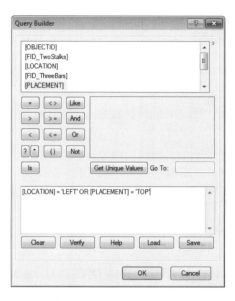

FIGURE 7-16   Query builder

Put it into the Trivial_Feature_Classes folder. The feature class to be produced will contain the polygons with the attribute value TOP plus those with the attribute value LEFT.

_____ **18.** Press the SQL button to develop the SQL Expression in a Query Builder window. Build the Query:

[LOCATION] = 'LEFT' OR [PLACEMENT] = 'TOP'

Once the expression is built (see Figure 7-16), Verify it and press OK to return to the Select window, where your expression will be in place. Press OK. When the computation stops the new feature class will appear in the T/C.

_____ **19.** Verify that the feature class

ALL_TOP_POLYS_plus_ALL_LEFT_POLYS

is added to the map. Make the layer color light blue. Turn off any other layers on the map. Verify that those polygons that were LEFT and those polygons that were TOP are both represented. In the Layer Properties window of

ALL_TOP_POLYS_plus_ALL_LEFT_POLYS,

check the box to label features and make the Label Field LOCATION. Click Apply, then OK. Observe. Change the Label Field to PLACEMENT. Observe. Turn off

ALL_TOP_POLYS_plus_ALL_LEFT_POLYS.

Recognize that you have created a new feature class by selecting polygons from the Union of two other feature classes.

___ **20.** ***Make another feature class by extracting from Stalks_Bars_spp:*** If necessary, add back Stalks_ Bars_spp to the map. This time we will be more restrictive. For a polygon to qualify it has to be both on the TOP ***and*** at the LEFT. For output use the same Trivial Feature Classes dataset and use the name

ONLY_TOP_POLYS_that_are_ALSO_LEFT_POLYS.

Use the Query Builder, putting in the expression

[LOCATION] = 'LEFT' AND [PLACEMENT] = 'TOP'

Display the results, showing the layer in red.

___ **21.** Notice that you have only one polygon. Using Layer Properties, label it, first with LOCATION, then again with PLACEMENT.

More Complex Queries—And's and Or's

Perhaps you are confused about AND's and OR's—like why you get both the left polygons ***and*** the top polygons when you use the expression

[LOCATION] = 'LEFT' OR [PLACEMENT] = 'TOP'

Well, it's good that you are confused. It will make you think carefully about the Boolean expressions you put in the Query Builder. It is amazingly easy to make a mistake!

For further information click on Help on the Main menu, enter "building SQL expressions", and press Ask.

Other Polygon Spatial Joins: Intersect and Identity

___ **22.** To a new map add TwoStalks and ThreeBars.

___ **23.** Again overlay TwoStalks with ThreeBars, but use the Intersect tool rather than Union. Name the Output feature class INTSCT_S_B_mpp, and put it into

___IGIS-Arc_*YourInitials*\Trivial_GIS_Datasets\ Overlay_Exercise.mdb\Trivial_Feature_Classes

Leave Output Type as INPUT, which will give you polygons.Make sure that INTSCT_S_B_mpp is added to the map. Verify that Intersect computed the geometric intersection of the feature classes, producing new polygons only in the areas common to both feature classes. Use Identify to look at each polygon. Examine the attribute table. Notice that the attributes are the same as for Stalks_Bars_mpp, but of course there are fewer records. How many? _____. Create INTSCT_S_B_spp from INTSCT_S_B_mpp. Now how many records? _____.

___ **24.** Perform Overlay with TwoStalks and ThreeBars, but use Identity rather than Intersect. (Read the Help panel.) Use TwoStalks as the Input feature class and ThreeBars as the Identity feature class. Call the Output feature class IDNTY_S_B_mpp. Display on a new map. How many polygons are in the Output feature class? _____. The answer is not 14 (use the Identify tool). Note that only polygons in the areas of the TwoStalks polygons are present in IDNTY_S_B.

When using polygon-on-polygon overlay with the Identity option, it is the areas covered by the Input feature class that determine the area covered by the Output feature class.

_____ **25.** Perform Overlay with Identity again, but this time use ThreeBars as the Input feature class and TwoStalks as the Identity feature class. Call the Output feature class IDNTY_B_S_mpp. How many polygons are in the Output feature class? _____ Note that only polygons in the areas of the ThreeBars polygons are present in IDNTY_B_S_mpp.

_____ **26.** Create a feature class named IDNTY_B_S_spp, made by exploding IDNTY_B_S_mpp. How many polygons are in this feature class? _____.

Exercise 7-4 (Project)

Use Overlay and Extract with Trivial Point and Line Feature Classes

As discussed in the Overview section of this chapter, Intersect and Identity operations may be used with point feature classes and line feature classes. Basically, overlays can take place between datasets of the same dimensionality, or between datasets in which one dataset has a lesser dimensionality than another. To enumerate:

❏ Polygons and polygons

❏ Polygons and lines

❏ Polygons and points

❏ Lines and lines

❏ Lines and points

❏ Points and points

When you use Overlay with a point feature class, the point feature class is always the Input feature class and the Output feature class is always a point feature class. The Overlay feature class is frequently a polygon feature class. With Identity, all of the points in the Input feature class are present in the output; with Intersect only those points that are in the areas covered by the overlay feature class are present.

The main thing that happens when you use Overlay with a point feature class and a polygon feature class is that each point picks up the attributes of the polygon into which it falls. For example, if you had points representing parking meters and a polygon feature class representing enforcement areas, an overlay of the two could tell you into which enforcement area each particular parking meter fell. The steps below illustrate this situation.

Creating Spatial Data Sets Based on Proximity, Overlay, and Attributes

___ **1.** Start a fresh map in ArcMap. Add TrivialPoints and TwoStalks from

   ___IGIS-Arc_*YourInitials*\Trivial_GIS_Datasets\Overlay_Exercise.mdb\Trivial_Feature_Classes

   Zoom to full extent. Examine the geography and table of each layer.

___ **2.** Use ArcToolbox > Analysis Tools > Overlay > Intersect to perform an overlay of TrivialPoints and TwoStalks. Name the Output feature class

   Points_with_polygon_information

and put it in

   ___IGIS-Arc_*YourInitials*\Trivial_GIS_Datasets\Overlay_Exercise.mdb\Trivial_Feature_Classes.

___ **3.** Make sure that Points_with_polygon_information is added to your map. Represent TrivialPoints with a yellow dot of size 5; represent Points_with_polygon_information with a red dot of size 2. Arrange it so the newly created feature class is at the top of the T/C and TwoStalks is at the bottom. Use the Identify tool to explore Points_with_polygon_information (which you have to set in the Identify from text box because, as you see, it has some coincident points with TrivialPoints). The thing to note is that the points have picked up information from the polygons into which they fell. How many points are there in TrivialPoints? _____ How many points are there in Points_with_polygon_information? _____ Describe the differences in *geography* the operation produced, comparing the geography of TrivialPoints and the geography of Points_with_ polygon_information. _____

___ **4.** Describe the differences in *tables* that the operation produced, comparing the table of TrivialPoints with the table of Points_with_polygon_information. _____
   _____

___ **5.** Use ArcToolbox to perform an Overlay of TrivialPoints and TwoStalks, but this time use the Identity option. The input feature should be TrivialPoints. The identity feature should be TwoStalks. Accept the default name for the Output feature class (write its name here _____) and put it in Trivial_Feature_Classes.

___ **6.** Remove the Points_with_polygon_information feature class from the map. Make sure the new feature class is on the map. Open its attribute table. How many points are there in the new feature class? _____ Note that some of those points have picked up information from Two_Stalks. How many points did not pick up new information. _____. Why not? _____
   _____.

   The lesson here is that, with Identity, the input feature defines the extent of the output feature, while the identity feature adds information where appropriate.

Also, Intersect and Identity operations may be used with line feature classes. When you use Overlay with a line feature class, the line feature class is always the Input feature class and the Output feature class is always a line feature class. The Overlay feature class is always a polygon feature class. With Identity, all of the lines in the Input feature class are present in the output; with Intersect only those lines that are in the areas covered by the Overlay feature class polygons are present.

Two things happen when you use Overlay with a line feature class. First, the lines are divided where the polygon boundaries cut across them. Second, each line takes on the attributes of the polygon into which it falls. For example, if you had lines representing roads and a polygon feature class representing highway districts, an overlay of the two could tell you into which district each particular highway segment fell.

_____ **7.** Start a fresh map in ArcMap, add TrivialLine and TwoStalks from

   ___IGIS-Arc_*YourInitials*\Trivial_GIS_Datasets\Overlay_Exercise.mdb\Trivial_Feature_Classes

   Examine the geography and table of each layer.

_____ **8.** Use ArcToolbox to perform an Overlay of TrivialLine and TwoStalks, using the Identity option. The line is the input feature; TwoStalks is the Identity Feature. Call the new feature class LinesThruPolygons and put it in Trivial_Feature_Classes.

   Arrange the T/C so LinesThruPolygons is at the top. Represent TrivialLine with a yellow line of size 6; represent LinesThruPolygons with a red line of size 2. Open its attribute table. How many records are there in the LinesThruPolygons feature class? _____. Select each record to highlight the corresponding line. Clear selections. Describe the differences in geography that the operation produced, comparing the geography of TrivialLine and the geography of LinesThruPolygons feature class. _____

_____ **9.** Use Extract > Select to make a feature class named PieceOfTrivialLine consisting only of the part of LineThruPolygon that lies within the **right** stalk. Display it on the map in black with a width of 4. From its attribute table, determine to two decimal places how long the line in this new feature class is: _____. Close ArcMap.

In summary, with point and line feature classes, the first input feature class is always a point or line feature class; the second input is a feature class of the same or higher dimensionality. The output feature class is always the same type (point or line) as the first input feature class. With the Identity tool, all of the geography of the first input feature class is represented in the output feature class. With the Intersect tool, only the geography of the first input feature class that lies within the polygon feature class boundaries is represented in the output feature class. In all cases, unless you specify otherwise, the elements of the output feature class pick up the attributes and appropriate values of the polygon feature class.

There are other variations on spatial joins or overlays. Two point feature classes could be intersected; the result would be some or all of the points that were coincident (or nearly so, depending on the tolerance being used). A point feature class could be intersected with a line feature class; the result would be (a) points that were on the line, and (b) that the point attribute table would pick up information from the line attribute table. Just about anything you might imagine wanting to do with respect to overlay can be done with the Analysis commands. That is to say it will probably take some effort to determine the workflow and some head scratching to comprehend what commands might be used, but it's unlikely that you will run across a problem involving points, lines, and/or polygons that cannot be solved using the Analysis commands.

Using Buffer and Overlay Together with Geodatabases

Now let's use the principles you just learned regarding buffering and overlaying to solve a somewhat more realistic problem. In this exercise, you use feature classes that come from a feature data set in a personal geodatabase.

The Getrich Saga

You have been presented with some potentially valuable information! In a western state, in a county named Getrich_county, a miner long ago buried gold now worth *at least* 4 million dollars but was never able to come back to dig it up. He buried it within 1900 meters of the *edge* of one of the wagon trails that run throughout the county, but at least 300 meters away from the *edge* of the trail. He put it into sandy soil—more than 2500 meters away from any of the old oil wells. It will cost two cents ($0.02) per square meter to search for the treasure—using metal detectors and Global Positioning System (GPS) receivers. You may assume that if the gold is in the area you search, you will find it.

This information comes to you by way of a friend/client who knows you are taking an excellent GIS course, have access to an Esri ArcGIS software system, and have some data available. Your friend certainly doesn't have money to launch the search, but he has some friends who have money to invest in the search if the cost is low enough and the maximum cost is well known in advance. There is not nearly enough money to search the entire county (besides, that would cost more than the gold is worth). For a modest cut in the profits, you agree to create a map showing the areas where the gold might be buried and to compute the total cost of searching those areas. You will do this by making a *personal geodatabase feature class (PGDBFC)* named Look_Here that contains only the polygons representing the areas to be searched.

The data sets for the problem are all contained in a personal geodatabase named PGDB_gold.mdb, located in a folder named

> [___] IGIS-Arc\Gold_Data

You have four feature classes to work with:

❏ Getrich_County—A very simple POLYGON feature class of the county. The treasure is within the outline of the county.

❏ Oilw—A POINT feature class of all the sites of the abandoned oil derricks and wells.

❏ Wagt—A LINE feature class of the centerlines of the wagon trails. (Each trail is assumed to be 10 meters wide. Recall that the treasure is buried a certain distance from the *edges* of the trails.)

❏ Soils—A POLYGON feature class of the soils in the area. There are four types of soil: clay, dirt, sand, and rock. You also have a separate table that indicates the soil types. In this table is a column with the heading of Characteristic. The feature class contains a key that will let you join this table with the feature class.

___ **1.** Start ArcCatalog. Copy the Gold_Data folder from

[___] IGIS-Arc to your

___IGIS-Arc_*YourInitials* folder.

___ **2.** Make a folder connection to the

___IGIS-Arc_*YourInitials*\Gold_Data folder.

The personal geodatabase PGDB_gold.mdb contains a personal geodatabase feature data set named PGDBFD_gold, which contains the personal geodatabase feature classes Getrich_county, Soils, Oilw, and Wagt. PGDBFD_gold.mdb also contains a table named soil_type.

___ **3.** *Carefully* look at the *geographics* and *attribute tables* of each feature class, so you have a good idea of what you have to work with. Identify the three named wagon trails. What are they? _____, _____, _____ How many oil wells are there? _____ How many different soils polygons? _____.

___ **4.** So that we know precisely where the oil wells are, add the x- and y-coordinates to the attribute table of Oilw. (If you don't remember how to perform the ADD XY operation, check your Fast Facts File or the Help files.)

___ **5.** Look at the attribute table for Oilw, to be sure POINT_X and POINT_Y fields were added. Start ArcMap with a blank map. Close ArcCatalog. Add Oilw to the map. Check that the map units and the display units are set to meters.

The first step in solving this problem will be to make buffers. You will use the Multiple Ring Buffer tool because the wagon trail specifications require two buffers. The first buffer will be around the oil wells. Even though we need only one buffer, we will use the Multiple Ring Buffer tool, if only to get used to using it.

___ **6.** Start ArcToolbox. In Analysis Tools > Proximity, launch Multiple Ring Buffer. Check the tool out again with the Help panel.

___ **7.** For Input Features, browse to Oilw in

___IGIS-Arc_*YourInitials*\Gold_Data\PGDB_gold.mdb\PGDBFD_gold

and add it. (Or get it from the dropdown menu in Input Features.) Put the Output feature class in PGDBFD_gold, calling it Oilw_buf. For Distances put in the number of meters that is appropriate (check the problem statement) and press the + sign. Scroll down and make the Buffer Unit Meters. For Field Name type OilwNoGold. Press OK. When the tool stops close the window. This tool automatically adds the result to the map, so the polygons representing the oil well buffers should appear as a layer in ArcMap. (If not, add the layer.) Examine it.

___ **8.** Open the attribute table of Oilw_buf. Notice that the OilwNoGold field has the value 2500 in it. You will use this when you extract features from the overlay. Since the gold is not within this buffer, you will want to find those polygons where OilwNoGold is **not** 2500. Also notice the table has only one record. Use the Identity tool to click one circle and notice that they all light up.

The Multiple Ring Buffer tool makes a multipart polygon. That is, all the separate circles are together considered as just one polygon. This one polygon might work okay for our analysis, but let's separate them just for the sake of neat and tidy.

____ **9.** Start the Multipart To Singlepart tool. Browse for Input Features: Oilw_buf—or, better yet, since it is on the map, click the drop-down menu arrow and find it there. Browse to name the Output feature class Oilw_bufs (note plural: bufs) and put it in the same location as Oilw_buf. Click OK. When Oilw_bufs appears in the Table of Contents, drag it below Oilw if necessary. Remove Oilw_buf (note singular: buf) from the map. Now check out a few wells with Identify. Also look at the attribute table. Note that there is a record for each oil well. Add the feature class Wagt to the map. Zoom to full extent.

____ **10.** *Use Multiple Ring Buffer to make the two buffers along the wagon trails:* What would the appropriate distances be, considering the original problem statement concerning the width of the wagon trails? _____, _____.[3] Start the tool. Call the Output feature class Wagt_bufs. Put in the appropriate distances, smaller first, clicking the + after each. Units are again meters. For the Field Name use Wagt_areas. Click OK.

____ **11.** Drag Wagt_bufs to the bottom of the table of contents and zoom to the layer. Carefully examine Wagt_bufs with both the attribute table and the Identify tool, zooming up as necessary. Using the Layer Properties of Wagt_bufs, display the 305-meter buffer with light green and the 1905 meter buffer with some color close to gold (Hint: Use Symbology > Categories > Unique values). You can see that, when the Multiple Ring Buffer makes multiple buffers, those buffers are independent of each other. That is, for example, the outer buffer does not include the inner buffer. Look at Figure 7-17. The amount you put in for the outside distance becomes the field value for that buffer. So what is the value of Wagt_areas that you will want in the attribute table of the final feature class that you are creating, in order to indicate only the area that might contain the gold? _____.

Deriving Information by Combining Tables

____ **12.** Add Soils to the map. Drag it to the bottom of the Table of Contents. Look at the Soils attribute table. Write the names of the fields in Attributes of soils.

_____, _____, _____,

_____, _____

Note that, just from this table, you have no way of knowing which of the 20 polygon areas consist of sand. However, each polygon does have a value in a SOILS_ID field. This field will allow you to key into a second table that matches up the values of SOIL_ID with a similar field in another table.

____ **13.** In the PGDB_GOLD.mdb database, you will find a table named soil_type. Add that to the Table of Contents and open it. What fields do you find?

_____, _____, _____

Close all attribute tables.

---

[3] If you didn't write 305 and 1905, check the problem statement again.

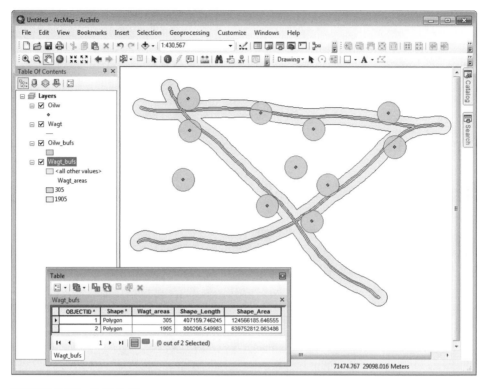

FIGURE 7-17   Gold Buffers

Here are some salient facts:

❏   The values in the field SOILS_ID (in soil_type) correspond to the values in SOILS_ID (in the Soils feature class). Each is a key field in its table.

❏   A key field contains, for each record, a unique value that identifies that record.

❏   A given value in the SOILS_ID field in the soil_type table refers to a polygon with the same number in the SOILS_ID in the attribute table.

❏   When a table B is joined to a table A, the records of table B are concatenated with the records of table A based on the identical values in the key fields.

❏   Joining the soil_type table to the Soils feature class table, using the keys, will result in associating the correct polygon with its CHARACTERISTIC ("sand," "rock," and so on).

❏   The values in CHARACTERISTIC are the soil types you will need to identify the sandy soils.

Putting the table together with the feature class is a bit convoluted, so you may want to be especially detailed in your Fast Facts File.

_____ **14.** With the List By Drawing Order tab active in the T/C, right-click the Soils entry. Under Joins And Relates, pick Join. In the Join Data window, you want to select Join attributes from a table.

___ **15.** In Step One choose SOILS_ID from the drop-down menu. In Step Two, browse to the table soil_type in PGDB_Gold.mdb or pick it off the drop-down menu. In Step Three, pick SOILS_ID from the drop-down menu. Click Validate Join. Read the messages in the two windows; close the validation windows. Click OK. Decline any offer to index the join table. (This is a time-saver for really big tables. It basically puts the keys of the join table in order so that they can be located quickly during the joining operation. Our tables are so small that it doesn't matter.)

___ **16.** Open the attribute table for the Soils feature class, which now is combined with the soil_type table. List the field names.

_____, _____, _____, _____, _____,

_____, _____, _____

___ **17.** Sort the CHARACTERISTIC field. How many "sand" polygons are there? _____ What are the values of the SOILS_ID field of those "sand" polygons? _____, _____, _____, _____, _____.

___ **18.** Zoom to the Soils layer. Turn off all other layers. Dismiss the attribute table. Label the polygons in Soils with SOILS_ID. Observe. Now label the polygons with CHARACTERISTIC. Observe.

You are going to use the information in the new Soils attribute table in an overlay command. Because the table is somewhat "ethereal" at the moment—it exists only in the fast memory of the computer—you need to put it onto disk, where it won't go away if you leave ArcMap.

___ **19.** Right-click Soils in the T/C. Slide the cursor to Data, then click Export Data. In the Export Data window, select Export All Features. For the Output feature class click browse to open a Saving Data window. In this window, set Save as type to File and Personal Geodatbase feature classes. In the Look in field navigate to

IGIS-Arc_YourInitials\Gold_Data\PGDB_Gold.mdb\pgdbfd_gold

and use the name Soils_with_Characteristics instead of Export_Output. Click Save, then OK. When asked, add the exported data to the map. Remove the layer based on the feature class Soils. Soils_with_Characteristics is now the feature class you will use henceforth.

___ **20.** Open the attribute table of Soils_with_Characteristics. Notice that you have (almost) the same fields. What is different from the table you exported? _____.

Overlaying the Feature Classes

___ **21.** Start the Union tool in ArcToolbox by selecting Analysis Tools > Overlay. For input, use Soils_with_Characteristics, Oilw_bufs, and Wagt_bufs. Call the new feature class Union_So_Ow_Wt_mpp, and, with an OK, put it in the feature data set pgdbfd_gold. After it is added to the map, examine its attribute table. How many polygons are represented? _____. What are the names of the fields you will use to extract the polygons where the gold might be? _____, _____, _____. (Hint: There would be one field for each of the three feature classes in the Union.)

____ **22.** Explode the multipart polygons that Union makes, using the Multipart To Singlepart tool. Call the resulting feature class Union_So_Ow_Wt. Look at its attribute table. Now how many polygons are there? _____

____ **23.** Add Getrich_County to the map. Since the gold lies within the county boundaries, you can limit yourself to the part of Union_So_Ow_Wt that lies within Getrich County. Fortunately there is a tool that will trim away the area we don't need to look at. Use ArcToolbox > Analysis Tools > Extract > Clip. Determine by reading the sidebar Help what needs to go in each field. Give the resulting feature class the name GR_Cnty_only. Use the T/C to display that feature class by itself on the map. Use Identify to check out a number of polygons, paying attention to whether the gold might be there, based on CHARACTERISTIC, Wagt_areas, and OilwNoGold.

____ **24.** As the final analysis step, you will use ArcToolbox > Analysis Tools > Extract > Select. Start the Select tool. For input, use GR_Cnty_only. Name the output Look_Here. Before you run the tool, write in the space that follows the expression you will enter in the SQL section in order to extract the polygons that represent the areas that must be searched. (By the way, you may want to use the operator <>, which means "not-equal". Strictly speaking, A <> B means either A is less than B or A is greater than B—which is the same as saying A cannot be equal to B.)

_____

____ **25.** Enter the expression, making good use of the Get Unique Values button. (To check the expression you just wrote – hey, now, try to do this on your own first – look at Figure 7-18.) Click Verify. Click OK. Click OK.

____ **26.** Make the polygons of Look_Here bright yellow. Make the polygons of GR_Cnty_only a light pink. The map should look like Figure 7-19. Verify, from both the Identify tool and the attribute table, that they satisfy the original problem requirements. How many areas are there? _____.

____ **27.** From the attribute table create Statistics on the areas of Look_Here. In the worst case, where you have to look at all possible areas, how many areas would have to be searched? _____. How much area would have to be searched? _____ How much would it cost to search it, at $0.02 per square meter? _____ Would you invest in the search? _____.

Suppose it cost $0.03 per square meter. Would you invest in the search? _____. Why or why not? (Think about probabilities.)

_____

```
[CHARACTERISTIC] = 'sand' AND [Wagt_areas] = 1905 AND
[OilwNoGold] <> 2500


  Clear      Verify      Help      Load...      Save...
```

FIGURE 7-18   Select string for gold

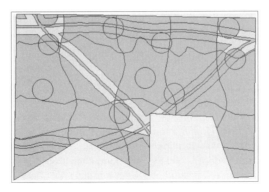

FIGURE 7-19   Final GetRich map

_____ **28.** On the map turn on the following layers, adding feature classes as necessary: Soils, Wagt, Wagt_ bufs, Oilw, Oilw_bufs, Getrich_County, and Look_Here. Turn all other layers off. For point and line layers, pick distinctive colors. For all polygon layers except Look_Here, make the polygon color No Color or Hollow. For Look_Here use Mars Red. Make a Layout of the map. (You will need the Drawing toolbar.) Enhance the Layout to make a map that satisfies you. Print the map.

Exercise 7-6 (Project)

Building a Model of the Getrich Project Solution

Suppose that you have worked the Getrich County problem and given the answer to your friend. He comes back with new information: Actually, instead of 1900 meters for the extent of the buffer for the wagon trails, the number should be 2000. So could you please redo the process. Well, okay, you agree. But are there likely to be more changes? The friend doesn't think so but some better maps may soon be available, so maybe. At this point you begin to wish for a more automated way to run this problem. And there is: the Esri MODEL.

A model will let you make a diagram of the steps to solve the problem and then execute (run) those steps at, literally, the click of a single button. In what follows you will make a model of the Getrich problem.

_____ **1.** Start ArcCatalog. Hide ArcToolbox if it is open. Highlight the folder ___ IGIS-Arc_*YourInitials*\ Gold_Data. Right-click and make a new Toolbox. Right-click the Toolbox and make a new Model.

A Model window opens.[4] Dismiss the Model window. In the Catalog Tree, expand the Toolbox. Change the name Model to Gold_Model. Right-click the new name. In the menu that appears, click Edit. A Gold_Model window appears.

At this point, you will begin building the model in the window. What goes into the model are, primarily, data sets (called parameters) and tools. The combination of a tool, with its input and output data sets, is called

---

[4] If you need to bring up a model window later, you can expose the Model under the Toolbox, right-click it, and select Edit from the drop-down menu.

an operation. The model will consist of operations in series and parallel configurations—where the output of one operation may serve as the input of the next operation.

You may recall that you began the gold-finding project by buffering the oil wells with the Multiple Ring Buffer tool.

——— **2.** Open ArcToolbox from the Standard Menu. (Along the way in this project, you may have to move and resize some windows to make the screen comfortable to work with.) Find the Multiple Ring Buffer tool and highlight it. Drag it into the Model window. Notice that boxes representing the tool and its output, with a directed link between them, come as a package; the handles show that the package is selected. By clicking at a point away from the package, you can unselect it. By clicking one box or the other, you can select it individually. By dragging a rectangle around them, you can select both boxes together. (Actually if any part of the rectangle touches a box it is selected.) You can move whatever is selected by dragging it. The connection arrow (link) stays with the pair.

——— **3.** *Provide the Multiple Ring Buffer tool with input and output:* Right-click the box containing the tool and select Open. You will see a familiar window. For Input Features browse to

____Arc-IGIS_*YourInitials*\Gold_Data\pgdb_gold.mdb\pdgbfc_gold\Oilw.

and Add. In the same location on disk, specify Oilw_buf_M for the Output Feature class. The Distance again is 2500. Use meters for the Buffer Unit. Type OilwNG for the Field Name. The Apply button has no effect when building models, so just press OK.

There is a dramatic change in the model diagram. (Move the elements around as needed to see everything. Expand the Gold_Model window as needed.) An input bubble has been added. And the elements of the model are colored in: blue for initial data, yellow for the tool, and green for output, which will also serve as data (called derived data) for the next operation. This is, by itself, a complete model.

——— **4.** From the Gold_Model Menu bar, click Model and select Run. The tool turns red while it is running. Another Gold_Model window appears, giving details of the process. Wait until the process has completed.

——— **5.** Close both Gold_Model windows, saving the changes. In the feature data set pgdbfd_gold, a new feature class should have appeared: Oilw_buf_M. Check out its geography and table.

——— **6.** Bring up Gold_Model again by right-clicking the name and choosing Edit. Recall that the Wagon Trails required two buffers. Again drag the Multiple Ring Buffer tool into the Gold_Model window. To explore a different way of adding data, drag the feature class Wagt into the Gold_Model window. On the Gold_Model window toolbar, click the Connect icon (a line connecting two little boxes). With the wandlike cursor, click Wagt and then Multiple Ring Buffer(2) to forge a link between them. Click Input Features in the subwindow that appears. With a right-click Open the tool and notice that Wagt has already been filled in for you. For an output feature class, put in Wagt_bufs_M. The distances to put in are 305 and the revised distance of 2005. Pick Meters for the Buffer Unit. Put in Wagt_areas for the field name. Click OK.

——— **7.** Run the revised model. You will get messages from the second Gold_Model window as to the particular process occurring. When Completed Close the window with the details. Minimize

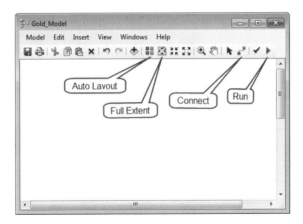

FIGURE 7-20   Model toolbar

the Gold_Model window. Note Wagt_bufs_M in the Catalog Tree. Check out Wagt_bufs_M Geography to be sure that the proper action was taken. Bring the Gold_Model window back. Note the shadows made by the tools that have been run, and by the feature class bubbles, which have had data created for them.

8. At this point, the diagram of the model may be getting a bit unwieldy. There are tools, accessible by buttons below the menu bar, that let you rearrange and resize the elements of the model. See Figure 7-20 for the less familiar ones. Make the model window bigger. Slide all its elements around with the Pan tool. Experiment with the Full Extent tool, the Zoom tools, and the Auto Arrange tool. For more information consult the Help file.

The model-making procedure involves mainly bringing in tools and then defining their input and output. Here is a process for doing that that works well: Drag the tool in. If it has inputs that will have been previously computed in the model, make links from those inputs. Then open the tool and specify the outputs.

9. *Make a union of the two buffer feature classes with the soils feature class:* Click the Select icon from the Gold_Model window menu. Drag in the Union tool from ArcToolbox > Analysis Tools > Overlay. Drag in the feature class Soils_with_Characteristics from the Catalog Tree. Use the Connect icon and Input Features to make connections from all three feature classes to Union. See Figure 7-21.

10. Right-click the Union tool in the Gold_Model window and choose Open. Accept the Input Features. For the Output feature class, specify Union_WT_OW_S_M. Click OK. Click the Run icon on the model toolbar. After some computation Union_WT_OW_S_M appears in the Catalog Tree.

11. Let's take a little side trip. From the Model drop-down menu click Delete Intermediate Data. Notice that the shadows have disappeared and that Oilw_buf_M and Wagt_bufs_M have disappeared from the Catalog Tree. Now find the Run icon on the model toolbar. Click it to rerun the model from the beginning. When it finishes, you can see the buffers are there again. Is this not painless flexibility?

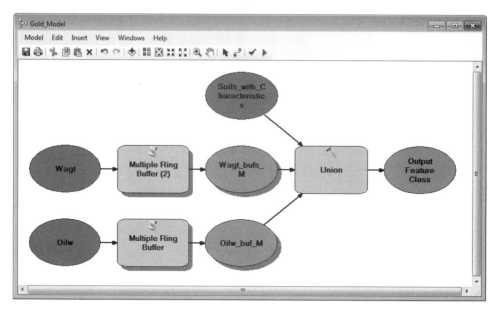

FIGURE 7-21   Partial gold model

_____ **12.** Using Pan if necessary, make room on the right for a new tool. In order to clip the union with the county, add the Clip tool from ArcToolbox > Analysis Tools > Extract. Connect Union_WT_OW_S_M, identifying it as an input feature in the drop-down menu. Drag in the Getrich_county feature class. Connect it as the clip feature.

_____ **13.** Open the Clip box. Check that the Input and Clip Features are correct. For the Output feature class, specify Just_Getrich_M. Click OK. Without deleting intermediate data, run the model. Notice from the dialog box that only the Clip tool is run, since the data needed for it is already been derived. Close the dialog box. Save the model: Model > Save. Slide the Gold_Model window out of the way or minimize it. Check out Just_Getrich_M from the Catalog Tree.

_____ **14.** *Insert the final tool into the model:* Use Select (from ArcToolbox > Analysis Tools > Extract). Connect Just_Getrich_M to it as the input feature. Right-click the Output feature class, click Open, and make the name Look_Here_M. Click OK. Open the Select tool, check the input and output fields, click SQL and create the Expression:

[CHARACTERISTIC] = 'sand' AND [Wagt_areas] = 2005 AND [OilwNG] <> 2500

_____ **15.** Verify the expression. Click OK. Click OK. Run the model. Check out Look_Here_M. Congratulate yourself. Rearrange the model elements so that they look like Figure 7-22. Save the model. Close the Model. Call your friend.

The friend arrives, looking sheepish. It turns out that, of course, when the miner buried the gold, the units of measurement were yards, not meters. One yard is 0.9144 meters. So, the distance from the oil wells is not 2500 but 2286 meters. And the buffers for the wagon trails are at 279 and 1741. Knowing that you have an easily modifiable model available, you simply smile.

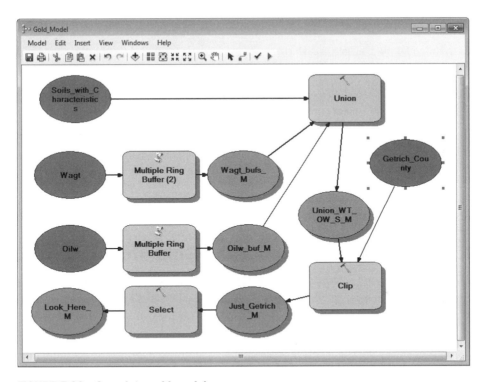

FIGURE 7-22   Complete gold model

____ **16.** ***Bring up the Gold Model again:*** Right-click the name in the Catalog Tree, pick Edit. Open the Multiple Ring Buffer box for the oil wells. In the Distances area delete the 2500, with the little cross under the plus icon, and put in 2286. Click OK. Fix up the distances for the tool dealing with Wagt so they read 279 and 1834. Click OK.

____ **17.** Open the Select box and modify its SQL expression with the new distances. Run the model. Print the Gold_Model from the Model menu. In ArcMap, print the final map for the client. Save the map as Gold_Model_Map. Close ArcCatalog. Close ArcMap.

There is a lot more modeling capability than you have seen here. And more complexity. To learn more, consult the ESRI manual *Geoprocessing in ArcGIS*.

Create a Python Script from the Gold Model

Another way to automate geoprocessing activities is with scripts. These are programming instructions in languages that can be used to direct GIS processes. Writing scripts are beyond the scope of this book, but you should at least know of their existence and be able to execute one. You can use a model to make a Python script.

____ **18.** ***Make a Python script of Gold_Model:*** Start ArcCatalog. Expand: ___ IGIS-Arc_*YourInitials* > Gold_Data > Toolbox.tbx > Gold_Model. Right click Gold_Model and choose Edit. Choose Model > Export > To Python Script. In the Save As window, navigate to

___ IGIS-Arc_*YourInitials*\Gold_Data

and name the file Gold_Model.py. (The extension "py" will probably be the only one available.) Click Save.

____ **19.** Using a text processor (e.g., Notepad), open Gold_Model.py. As you scan the text, you can probably identify the processes that were used to solve the Getrich problem. Comments in the file (which are there for the human reading the code, not for execution by the computer) begin with a # character. Other lines are instructions to the Python interpreter. When Python executes this script, it will produce the same results as the model. There is one problem, however: If Python tries to write a file to a folder, and there is already a file by the same name in the folder, the program will bail out, giving you only an error message. Therefore, we need to slightly modify the program so that it will overwrite existing files that were generated by the Gold_Model.

Modify the Python Script from the Gold Model

____ **20.** *Modify the Python Script:* In the text of the script you will find comments consisting of dashed lines. After the second set of such lines type the command

print "Beginning execution of Gold_Model script"

This command causes the text within quotation marks to appear on the screen when the program begins execution—just to reassure you that something is happening. (This script can take two or three minutes to run; all you will see is a blank screen unless you mark the beginning in some way.) You might also put a similar print command at the end of the program.

____ **21.** The more important modification is to tell Python to overwrite existing files. You will find a command early in the script that says "import arcpy", which brings the ArcGIS version of Python into play. On the next line after this, you should enter, **exactly as shown:**

arcpy.env.overwriteOutput = True

Save the Python script (Ctrl+S). Close ArcCatalog and ArcMap if either is or both are open.

Execute the Python Script

____ **22.** Depending on how experienced you are in dealing with outmoded computer techniques, you may need some help with the following. You are going back to the days when computers were directed by typed commands rather than "point-and-click" actions.

Bring up a Command Prompt window, which has a black background and a flashing cursor. (Probably available in Windows with something like Start > Programs > Accessories > Command Prompt. The first chore is to point the computer to the correct directory. (Folders used to be called Directories.) You do this with the Change Directory (cd) command. Determine the exact path to the folder that contains the textfile, Gold_Model.py, that you just modified and type cd, then type that path into the command prompt window. For example,

cd C:\IGIS-Arc_*YourInitials*\Gold_Data

If you have done this correctly, the computer will respond with the name of the folder, followed by the flashing cursor. (If you've done it wrong, try again.) Now type the command to list the contents of the folder:

DIR

You should see Gold_Model.py. (If not, something hasn't gone right.)

_____ **23.** ***Run the Python script:*** Type the name of the script:

Gold_Model.py and click Enter. The screen should tell you that execution has begun. Be prepared to wait a minute or three for it to finish. When it does the prompt (the name of the folder) will appear. You can then start ArcCatalog and look at the results. They will, of course, be the same results as were there before—from when you ran the model earlier. If you doubt that the results were generated by the Python script, you can look at the time the Geodatabase was last updated.

_____ **24.** Dismiss the Command Prompt window (or type EXIT as a command). Print the Python script from the text processor.

Examining such scripts, while knowing what they do, is a good way to begin learning how to write scripts. Of course, scripts can be, and frequently are, written directly by a human being using a text processor, rather than generated from a model.

Exercise 7-7 (Minor Project)

Making Buffers for Solving the Wildcat Boat Problem

Do your work for this exercise in

IGIS-Arc_*YourInitials*\Wildcat_Boat_Data.

If you have any doubts about the correctness of the Wildcat_Boat_Data folder, delete it from your working folder and re-copy it from IGIS-Arc_AUX. You will use the data sets you create in this exercise in the next exercise.

_____ **1.** Start ArcCatalog. Using the Multiple Ring Buffer tool, make a 20-meter buffer around the Streams feature class. Call it Streams_buf_mpp (for multipart polygon) and put it in the ***Area_Features*** feature data set. Make the field name NoBuild.

_____ **2.** Using the ArcToolbox > Data Management Tools > Features > Multipart To Singlepart tool, explode Streams_buf_mpp into Streams_buf.

_____ **3.** Using the Multiple Ring Buffer tool, make a 300-meter buffer around the Sewers feature class. Call it Sewers_buf_mpp and put it in the ***Area_Features*** Feature Dataset. Make the field name Build.

_____ **4.** Using the Multipart To Singlepart tool, explode Sewers_buf_mpp into Sewers_buf.

_____ **5.** Delete Streams_buf_mpp and Sewers_buf_mpp. Close ArcCatalog.

Exercise 7-8 (Project)

Finding a Site for the Wildcat Boat Facility

You will recognize this problem as the one you solved by manual means at the beginning of Chapter 1. Throughout the text, you have experimented with the data. In Exercise 7-7, you made the buffers around the streams and sewers that were required. In this exercise, you finish the work of finding suitable sites. To restate the problem:

Wildcat Boat Company is planning to construct a small testing facility and office building to evaluate new designs. They've narrowed the possibilities down to a farming area near a large lake. The company now needs to select a specific site that meets the following requirements:

❑　It must reside on soils suitable for construction of buildings. (The value of the soil suitability must be 2 or 3.)

❑　The site should not have trees (to reduce costs of clearing land and preserve an important natural resource). A regional agricultural preservation plan prohibits conversion of farmland. The other land uses (urban, barren, and wetlands) are also out. So, the land cover must be "brush land" (which has Landcover code 300).

❑　A local ordinance designed to prevent rampant development allows new construction only within 300 meters of existing sewer lines.

❑　A recent national water quality act requires that no construction occur within 20 meters of streams.

❑　To provide space for building and grounds the site must be at least 4000 square meters in size.

The data sets for this exercise, including the buffers you made in the preceding exercise, are in

_____IGIS-Arc*YourInitials*\Wildcat_Boat_Data.

This folder contains the personal geodatabase feature classes that describe the area from which the site must be chosen. Your assignment is to present a map showing **all** *the areas* that meet the requirements stated in the preceding list. Those areas should be contained in a single feature class named Final_Sites in your feature dataset:

_____IGIS-Arc*YourInitials*\Wildcat_Boat_Data\Wildcat_Boat.mdb\Area_Features

1. Start ArcCatalog. Show ArcToolbox. Make or confirm a folder connection with _____IGIS-Arc_*YourInitials*\Wildcat_Boat_Data

2. In _____IGIS-Arc*YourInitials*\Wildcat_Boat_Data\Wildcat_Boat.mdb\Area_Features

you should find the following feature classes:

❑　Landcover

❑　Sewers_buf

❏ Soils

❏ Streams_buf

Using the four preceding data sets and ArcToolbox ≥ Analysis Tools ≥ Overlay ≥ Union, make a feature class named Union_mpp in

___IGIS-Arc*YourInitials*\\Wildcat_Boat_Data\\Wildcat_Boat.mdb\\Area_Features.

Union_mpp contains multipart polygons. Look at Union_mpp, clicking a few polygons with the Identify tool. As you can see, you need to make sure each set of boundaries defines an un-subdivided area.

____ **3.** Use the Arctoolbox > Data Management Tools > Features > Multipart To Singlepart tool to explode Union_mpp into Union_Lc_Se_So_St (for union of landcover, sewer (buffers), soils, and stream (buffers)). Now note that Identify lights up only single polygons.

____ **4.** Carefully, very carefully, examine the table of Union_Lc_Se_So_St. How many polygons are there? _____. Ignoring, for the moment, the size criterion of 4000 square meters, write the relevant fields, values, and operators (like =, <>, >=) in the table that follows. These are the values you must use to extract the sites meeting the criteria. (Some values have been filled in for you.) Sorting some columns might aid you in determining the values you need.)

Fields	Operators	Values	(source feature class)
		300	Landcover
			Sewers_buf
			Soils
NoBuild	<>	20	Streams_buf

____ **5.** Write the expression that would extract the suitable polygons from Union_Lc_Se_So_St:

_____

_____

(*After* writing your expression, check it against Figure 7-23.)

____ **6.** Using ArcToolbox > Analysis Tools > Extract > Select with the above expression (be sure to Verify that it will work), make a feature class from Union_Lc_Se_So_St, which you will name Sites_A. Look at the attribute table of Sites_A to be sure that the values you specified for the various fields are proper. How many sites are there? _____

____ **7.** Display Sites_A, checking to be sure that it makes sense. Use Identify to look at the areas to see if they meet the 4000 square meter requirement. You will see some small areas that

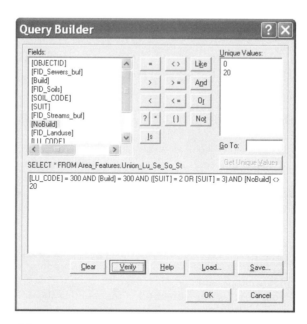

FIGURE 7-23   Wildcat query builder

obviously will not be candidates. But in the center note the two polygons that share a common boundary. Neither by itself is large enough to meet the size criterion, but together they do. Use Identify to understand why they are separated, and explain. (Hint: Look at Soil_Code.)

_____ .

8. You need to dissolve the boundary between those adjacent areas and make them a single polygon. When faced with a geographic problem like this, you can be pretty sure that, among the hundreds of tools supplied by ArcGIS, there will be one to solve it. Use the ArcToolbox Index to find a Dissolve tool that works on geodatabases. Locate it. What is the path?

_____

9. Start the Dissolve (management (not arc)) tool. Explore the Help files of the tool to understand what it does.

In this case, since all the polygons of Sites_A meet our criteria for the Wildcat Boat facility, you can just use dissolve without specifying any fields, to make Sites_C. (I will explain the role of Fields in a later exercise.)

10. Using the Dissolve (management) tool, make the input features Sites_A and the output Sites_C. To have the tool produce singlepart features, uncheck the box in front of Create multipart features. Compare the Geography of Sites_A with that of Sites_C. Note that the lines that separated some of the polygons are gone. How many polygons are there in Sites_C? _____

At this point each of your polygons should have satisfied all criteria except that the area must be greater than 4000.

____ **11.** Use the Select tool on polygons in Sites_C to make Final_Sites, such that Final_Sites contains only those polygons whose Shape_Area is equal to or greater than 4000. Look at the Geographics. How many polygons are left now? _____. Look at the attribute table. How many fields (columns, attributes) are left? _____. (Dissolve got rid of a ton of attributes – as will be explained later.) Final_Sites is the solution to the problem that you have been working on since Chapter One!

____ **12.** Launch ArcMap with a Blank Map. Add as data Final_Sites. Use a bright green color. Also, from Line_Features, add Sewers as black lines. Zoom to full extent. Use the Measure tool to verify that all no part of any polygon is more than 300 meters away from the closest sewer line.

____ **13.** Add Streams as blue lines. Verify that no part of any polygon is closer than 20 meters to any stream. Zoom to the extent of the Sewers. Your map should look like Figure 7-24.

____ **14.** Compare your final map with the map you made when you worked this problem using manual means as Exercise 1-1 in the first chapter of the book. Give yourself a grade, based on 100, evaluating how well you did at solving this problem by hand. _____. If you didn't do very well, don't be too unhappy about it. If problems like this were easy to do well by hand, society wouldn't be spending billions of dollars on computer-based GIS.

____ **15.** Make a layout of the results of this exercise, using your own judgment as to what to include and how to present the various layers. Print your map, preferably in color. Close ArcMap. Close ArcCatalog.

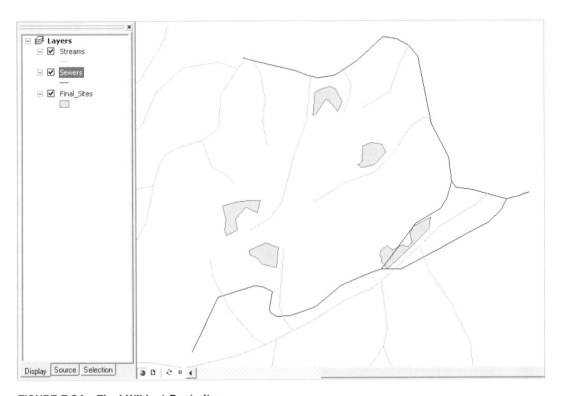

FIGURE 7-24   Final Wildcat Boat sites

Solving a Revised Wildcat Boat Problem

As often happens when some time has gone by since a problem was posed (in this case while you were busy learning GIS), the client suggests additional requirements:

——— **1.** The cost of each site should be available.

——— **2.** Unless the area is more than 7000 square meters the site eccentricity ratio (the perimeter divided by the square root of the area) must be no greater than 4.6, so that the site will not be too narrow or strangely shaped.[5]

Requirement #1 should be easy to take care of. You recall that the Landcover feature class had a field called COST_HA which stood for cost per hectare.[6] So a simple computation (like dividing the area by 10,000 and multiplying by the cost per hectare) should give us the cost of each site. The difficulty, which you can verify by looking at the table of Final_Sites, is that you no longer have the COST_HA field. It went away when you did the Dissolve. The last time you saw COST_HA was in the table for Sites_A. So you need to find a way to preserve the COST_HA field but still dissolve boundaries between smaller, adjacent polygons that meet our size criterion when they are merged. This calls for a closer look at the Dissolve tool.

Understanding Dissolve

If you use Dissolve to make feature class Y from feature class X, and you don't specify any Dissolve Fields, then no lines that separated polygons in X will show up in Y. Further, the attribute table of Y will contain only the fields OBJECTID, Shape, Shape_Length, and Shape_Area.

However, suppose that you specify a Dissolve Field in the Dissolve tool window. Then, a boundary line between each pair of adjacent polygons of X is erased in Y if, and only if, the two attribute values for that field are the same for each polygon. In any event, the specified field is retained in the attribute table of Y.

For example, suppose that X has an attribute (field) titled Owner. Assume there are two polygons (one might be growing wheat and the other corn) and they are adjacent—sharing a common border. Assume now that the Dissolve tool, specifying a Dissolve Field of Owner, is run on X to produce Y. If the Owner attribute value for each is the same (say 'Brown'), then the boundary line will not appear in Y and the attribute table will reflect a single record, with Owner 'Brown.' However, if 'Brown' owns one polygon and 'Smith' owns the other, then the boundary line between them will be present in Y and there will be a record in the attribute table for each polygon.

——— **1.** Start ArcCatalog. Look at the geography and the table of Union_Lc_Se_So_St. Use the tool Dissolve (Data Management) to bring up the Dissolve window. For input use Union_Lc_Se_So_

---

[5] This ratio (perimeter to square root of area) provides a measure of how "eccentric" a site is. A circle, which has the least perimeter for enclosing a given amount of area, has a site eccentricity ratio of about 3.5. A square's ratio is 4.0. A rectangle that is three times as wide as it is tall has a value of approximately 4.6. (Site eccentricity is not the same as the mathematical eccentricity of conic sections.)

[6] Recall, a hectare (abbreviation HA) is 10,000 square meters. For example, a square that is 100 meters on each side is a hectare. A hectare is very nearly 2.5 acres.

St. For output use Dissolve_All. In Dissolve_Fields leave all fields unchecked. Uncheck Create multipart features. Click OK. Examine the geography of Dissolve_All. Note that all boundary lines are gone. All the attributes, except the basic ones that come with any feature class, are gone.

2. Again use Dissolve (management) to bring up the Dissolve window. For input use Union_Lc_Se_So_St. For output use Dissolve_None. In Dissolve_Fields check all fields. Uncheck the box that produces multipart polygons. Click OK. Examine the geography and table of Dissolve_None. It should be virtually identical with Union_Lc_Se_So_St, with no boundary lines gone and all the attributes the same.

3. Now use Dissolve (management) to bring up the Dissolve window a third time. For input use Union_Lc_Se_So_St. For output use Dissolve_Suit. Uncheck the box that produces multipart polygons. In Dissolve_ Fields check SUIT. Click OK. Examine the geography of Dissolve_Suit using Identify. If two adjacent areas had the same suitability (say they were both 2), the boundary line between them is erased. If the suitabilities were different, the boundary line was retained. Examine the table of Dissolve_Suit. Note that the attributes are mostly gone, but the attribute field SUIT remains.

So, with polygon input, Dissolve does several things at once: If you check a field, then that field is preserved in the attribute table on the output. If two adjacent areas have different values in that field, the boundary line between them is preserved. If two adjacent areas have the same values in that field, the boundary line between them is erased.

4. Look at Sites_A. Use Identify on Sites_A to verify that the COST_HA is the same for the areas on both sides of the boundary lines that split the major polygons.

5. Housekeeping time: Use ArcCatalog to delete Sites_C. You can also rid yourself of any feature classes whose names begin with Dissolve. Close ArcCatalog.

Making New Sites that Include the COST_HA Field

6. Start ArcMap. Show ArcToolbox. Bring up the Dissolve (management) tool. Remake Sites_C from Sites_A, using Dissolve, but this time put a check in the COST_HA box, to preserve this field in the Sites_C table. Be sure to make singlepart polygons. Verify that the COST_HA field exists in the attribute table. Use Analysis Tools > Extract > Select to make Sites_D by applying the "greater than or equal to 4000" criterion to the Shape_Area of Sites_C. When computation stops, turn off all layers in the T/C except Sites_D Look at the table of Sites_D, and verify again that you do indeed have the COST_HA field. Also verify that the boundaries you wanted to lose are gone. (There is the one little worry that two adjacent polygons might have different land cover values, and that by running the Dissolve tool we might have left some troublesome boundaries. Here is where visual inspection gets into the process. Examine Sites_D to assure yourself that no connected polygons exist.)

7. Find the Options button on the attributes table of the Sites_D window. Add a field named SITE_COST to the Sites_D table. Its type should be Float, because its values will be calculated with an expression involving the floating-point numbers of Shape_Area. If any of the operands of an expression are floating point, then the results must also be floating point.

8. Right-click the SITE_COST column head. Select Field Calculator. Ignore the warning. In the Field Calculator window, part of the calculation (SITE_COST =) is already done for you. You simply need to supply the rest. Since our Shape_Area is in square meters and our cost is in hectares, calculate SITE_COST as

Shape_Area / 10000 * COST_HA

9. Run statistics on SITE_COST. To the nearest dollar, what is the cost of the average site? _____

Considering the Site Eccentricity Criterion

We have one more selection to perform. Recall that the site eccentricity may not be more than 4.6 unless the total area of the polygon is greater than 7000 square meters.

10. Add another field to the table of Sites_D. Name it Eccentricity. Since the number will contain decimal places and we are not sure of its range, specify the Type as Float. Click OK.

11. Right-click the Eccentricity column head; pick Field Calculator. Ignore the warning. Write the expression you need to use in the following blank: _____. Put the expression in the Field Calculator window, then check it against Figure 7-25. When you have it right, Click OK.

The Attributes of Sites_D window will have eccentricities ranging from just over 4 to about 6 and a half.

12. Clear any selected records or polygons. Use Analysis Tools > Extract > Select to apply the criterion. Write the SQL Expression that will let you use Sites_D to make a feature class called Final_NEW_Sites, that satisfies this additional requirement. _____ _____. Wait to look at Figure 7-26 until you have written your expression. Select the proper polygons with the Select tool. How many are there? _____

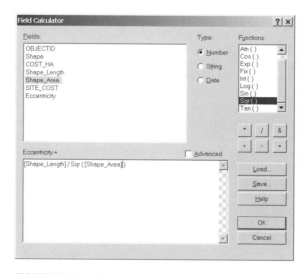

FIGURE 7-25   Wildcat eccentricity calculation

FIGURE 7-26 Size and Eccentricity selection

___ **13.** Display Final_NEW_Sites, showing the sites in bright yellow. Provide graphic data from the other data sources to show the relative location of the sites. Show the following:

❑ The sewers in black, width 3

❑ The streams in dark blue, width 2

❑ The outline of the sewers buffer emphasized in red, width 3

❑ The streams buffer in light blue

❑ The polygons of land cover, with land cover code 300, emphasized and others in subdued colors (except make code 500 dark blue, since it is water). See Figure 7-27.

❑ Save the map with whatever name you want. Now, using the features above, except for land cover, display

❑ The polygons of soils, with suitabilities 2 and 3 emphasized and the others in subdued colors. Save this map as well.

___ **15.** Zoom in on the group of sites. With the Identify tool, examine the resulting map to assure yourself that the criteria have been met.

___ **16.** Make a layout of the map. Print the layout, using a color printer if possible. Close any open ArcGIS programs.

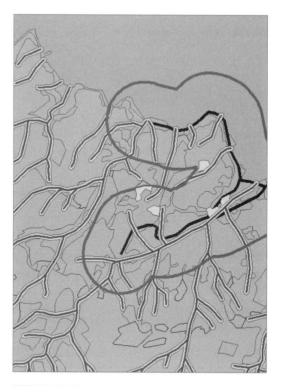

FIGURE 7-27

Making a Model of the Wildcat Boat Solution

—— **1.** Start ArcCatalog. Make a folder named Wildcat_Boat_Model in ___IGIS-Arc_*YourInitials*.

Make a folder connection to Wildcat_Boat_Model. From

[___] IGIS-Arc_AUX\Wildcat_Boat_Data

copy the Wildcat_Boat.mdb into

___IGIS-Arc_*YourInitials*\Wildcat_Boat_Model.

—— **2.** Make a toolbox in the folder Wildcat_Boat_Model. Highlight the newly created toolbox in the Catalog Tree and choose File > New > Model. A Model window opens. Close the Model window.[7] By the Toolbox icon in the Catalog Tree, change the name Model to Wildcat_Boat_Model. Right-click the new name. In the menu that appears, click Edit. The Wildcat_Boat_Model window appears.

—— **3.** *Construct a solution to the Wildcat_Boat problem,* placing the intermediate data and feature class Final_Sites in

---

[7] If you need to bring up a model window later, you can expose the Model under the Toolbox, right-click it, and select Edit from the drop-down menu.

___IGIS-Arc_*YourInitials*\Wildcat_Boat_Model\Wildcat_Boat.mdb\Area_Features.

Refer back to Exercises 7-7 and 7-8 for input numbers and field names to use. Also use the following tables for hints for building the model. Run the model after every tool so that if something is going wrong you can catch it immediately. If you go back to ArcCatalog and refresh it you will know that the right feature classes are appearing. You can, of course, reduce the model window to see ArcCatalog.

Important note: It is easy to put model results into the wrong folder. Look carefully at every output specification. Make sure the output is going into Wildcat_Boat_Model, not Wildcat_Boat_Data (which you used previously).

	FROM	WITH		MAKE
___	Streams	Multiple Ring Buffe	→	Streams_mpp
___	Streams_mpp	Multipart to Singlepart		Streams_buf
___	Sewers	Multiple Ring Buffer		Sewers_mpp
___	Sewers_mpp	Multipart to Singlepart		Sewers_buf
___	Landcover +			
___	Streams_buf			
___	Soils +			
___	Sewers_buf +	Union		Union_mpp
___	Union_mpp	Multipart to Singlepart		Union_Lc_Se_So_St

Run the model at this point so that the extraction will have the proper data available.

___	Union_Lc_Se_So_St	Extract > Select (expression)	Sites_A
___	Sites_A	Dissolve (COST_HA)(singlepart)	Sites_C
___	Sites_C	Extract > Select (4000)	Sites_D
___	Sites_D	Fields > Add Field[8] (Site_Cost)	Sites_D (2)
___	Sites_D (2)	Fields > Calculate Field[9]	Sites_D (3)
___	Sites_D (3)	Fields > Add Field (Eccentricity)	Sites_D (4)
___	Sites_D (4)	Fields > Calculate Field	Sites_D (5)
___	Sites_D (5)	Extract > Select	Final_Sites

Print the Model. Check Area Features. Close any ArcGIS software that Is open.

---

[8] Since the model runs automatically, you can't use the Add Field in the Options window of the attribute table. So look for the Add Field (Data Management) tool in ArcToolbox.

[9] Since the model runs automatically you can't get the Calculate Field window by right-clicking a column head. So look for the Calculate Field tool in ArcToolbox.

Checking, Updating, and Organizing Your Fast Facts File

The Fast Facts File that you are developing should contain references to items in the following checklist. The checklist represents the abilities to use the software you should have upon completing Chapter 7.

❏ In buffers, inside and outside are distinguished by

❏ To load the Buffer Wizard

❏ To bring more than one geodataset into ArcMap simultaneously

❏ The buffer wizard indicates areas inside the buffer by BufferDist value of

❏ When you use Calculate to change cells in records and some records are selected

❏ The Multiple Ring Buffer tool allows the user to

❏ When one polygon data set is overlaid with another, the attribute table of the resulting data set

❏ The difference between Union and Intersect is

❏ When you overlay two polygon geodata sets, the number of both _____ and _____ is likely to increase in the polygon attribute table.

❏ To form a new data set from attributes of an existing data set use

❏ SQL means

❏ When a point or line feature class is overlaid the overlay feature is usually a

❏ Identity and Intersect are different in that

❏ To join a table to a geodatabase feature class

❏ To convert multipart features to single-part features

❏ Building a model of a set of GIS operations has several benefits:

❏ To build a model in a toolbox

❏ Models can be exported to the _____ scripting language

❏ Dissolve makes new feature classes by

CHAPTER **8**

Spatial Analysis Based on Raster Data Processing

A *Really* Different Processing Paradigm[1]

You have briefly met rasters before in this book. Most images are stored in raster format. The COLE_DEM of Chapter 1 stored elevations in floating-point raster format. And you saw rasters that stored landcover in integer raster format. What is new here is that you will do GIS analysis using raster representation.

To be candid, it would take another book the size of this one to cover the immense subject of raster processing. Up until a few years ago, it was almost completely divorced from GIS vector processing. Esri has done a good job in taking down the barriers between the two. In terms of GIS data model development, raster processing (or grid processing, or cell processing, which are other terms for it) preceded vector processing. It was, at one time, about the only way to get large amounts of geographic data into the memories of the then-small computers. (100 Kilobytes of storage was a lot in 1960.) In the following years vector processing, with its greater capacity for precision, tended to supersede raster processing in terms of popularity for solving GIS problems. It is now understood that raster processing has some properties that make it better for some applications than vector processing. Specifically, analysis is why GIS analysts use it now—instead of just its advantages in speed of computation, and, sometimes, conserving computer memory. Since you now have considerable

---

[1]As you have probably inferred by now, in GIS, "paradigm" is used to describe a method of storage of geographic data. In the larger view it is a set of concepts and practices that constitutes a way of viewing reality, usually for a particular discipline.

OVERVIEW

IN WHICH you move into an entirely different realm of GIS analysis and geoprocessing: solving problems with a method of data representation—called raster, grid, or cell-based—that is dramatically different from what you are used to.

Raster is faster but vector is corrector. Berry, JK, 1995

background in the basics of GIS, and since this is the only chapter that is devoted to raster processing, the descriptions of the basics of raster processing will be succinct.

The major package in ArcGIS that processes rasters is called Spatial Analyst.[2] ArcToolbox's Spatial Analyst Tools have about 20 toolsets with a total of many more than 100 tools. Included as Toolboxes are the following:

- ❏ Conditional
- ❏ Density
- ❏ Distance
- ❏ Extraction
- ❏ Generalization
- ❏ Groundwater
- ❏ Hydrology
- ❏ Interpolation
- ❏ Surface Analysis
- ❏ Map Algebra

- ❏ Math
- ❏ Multivariate
- ❏ Neighborhood
- ❏ Overlay
- ❏ Raster Creation
- ❏ Reclass
- ❏ Solar Radiation
- ❏ Surface
- ❏ Zonal

In the Map Algebra[3] toolbox is a Raster Calculator which lets you calculate new rasters, making use of hundreds of functions, commands, and operators that are part of the Esri Spatial Analyst software.

Within these toolboxes are a large number of tools. It is true that there is considerable overlap in these software tools. They represent an evolution of raster processing techniques developed over decades. But even considering these overlaps, one cannot deny the conclusion: Spatial Analyst is *big*!

You will use Spatial Analyst get a good look at the software as you solve problems dealing with proximity, cost of moving across an area, siting a picnic park, doing hydrological and watershed analysis, and make surfaces. Also you will solve the Wildcat Boat problem using Spatial analyst.

Facts about Rasters

Each raster data set must have a name, just as a shapefile or geodatabase feature class must.

Rasters can be, and should be, lodged in a geodatabase. A raster data set may have multiple bands. If so, Spatial Analyst operates on band 1.

Rasters can be either geographic data or image data. We mostly discussed the structure of image data in Chapter 4. Here we will concentrate on geographic data.

---

[2]The package that dealt with raster processing in ArcInfo versions before ArcGIS was called GRID. Most GRID commands have been superseded by Spatial Analyst commands.

[3]The idea of "Map Algebra" was developed by Dana Tomlin and presented, in 1990, in a book entitled Geographic Information Systems and Cartographic Modeling.

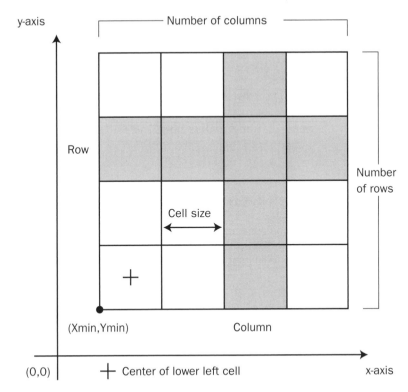

FIGURE 8-1   Schematic of a GRID or raster

A raster should encompass a study site. The raster representing the study area is called a Cartesian matrix, usually with rows parallel to the x-axis of a coordinate system. Columns are parallel to the y-axis.

A geographic raster is (usually) a rectangular matrix of (usually) square cells. Each cell is a member of a row and a member of a column. Almost always all the cells of a given raster are the same size. See Figure 8-1.

Raster cells may be almost any size. Choosing the cell size that is best for both representing the features of the Earth and for analysis is a science and an art. It is not unusual to have raster composed of hundreds of millions of small cells. If you choose a cell size that is half of what you were previously considering, the new raster will have four times as many cells. This may or may not mean four times the storage requirement (depending on the compression technique used) and four times the processing time.

Coordinate Space

A raster's coordinate system may be in real-world coordinates or "image space." For those in geographic space, cells are referenced primarily by an (x,y) location in map coordinate space, not by row and column numbers, as previous versions of grid processing were. The location referenced for a given cell is the center of that cell. For the overall raster, the geographic reference is usually the center of the lower-left (southwest) cell. Cells are square in map coordinates.

Each cell represents a specific location on the surface of the Earth. While each cell of a raster is square, the surface area of the Earth the raster represents may not be. Such real-world areas should be almost square, however. If the northing span of a raster is too great, the representation suffers. For example, take a UTM zone whose "base" is at latitude 40 degrees and "top" is at latitude 48 degrees. Suppose the cells are 100 meters on a side. Let's say a cell at the base would represent 10,000 square meters. A cell at the top of the zone, admittedly more than 500 miles away, would represent only about 8,800 square meters!

Rasters may be described as integer or floating-point. All the cells in an integer raster contain integers or, if not, are designated NoData. All the cells in a floating-point raster contain floating-point numbers, or, if not, are designated NoData.

Rasters with Integer Cell Values

Rasters with integer cell values are used to store what is variously called discrete data or categorical data. Such data sets describe a phenomenon or object that exists at the location of that cell, referenced by the cell Value.

Each cell in an integer raster belongs to a *zone*. The zone number—the Value—must be an integer—positive, negative, or zero. Cells in a given raster that have the same value all belong to the same zone. So a zone consists of all the cells with a given value. Each cell that has a Value belongs to some zone or other, even if the zone is composed of only one cell. Cells in a zone may be adjacent, but may well not be. See Figure 8-2.

An integer raster has associated with it a Value Attribute Table (VAT). [4] The table may have many fields but the two primary fields are Value and Count.

There is a record in the VAT for each zone. The raster shown in Figure 8-2, consists of eight zones. The zone numbers are 21, 22, 23, 24, 25, 26, 27, and 29. Obviously, sets of cells in the same zone may be connected or not.

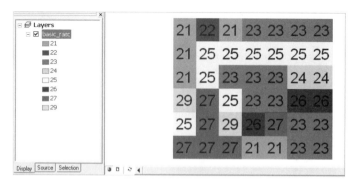

FIGURE 8-2   Basic_Raster with integer values indicating categories

---

[4]And some floating-point rasters have VAT's as well, as you will see.

ObjectID	Value	Count
0	21	6
1	22	1
2	23	13
3	24	2
4	25	9
5	26	3
6	27	6
7	29	2

Record: 1 Show: All Selected Records (0 out of 8 Selected.)

FIGURE 8-3 Basic_Raster Value Attribute Table (VAT)

The raster shown Figure 8-2, which we will call Basic_Raster, would have a VAT as shown in Figure 8-3. It indicates, for example, that Zone 21 consists of six cells. Figure 8-2 shows you that the cells with value 21 are not all connected.

There is a term for sets of cells that have the same value (i.e., are members of the same zone) and that *are* connected. That term is "region."[5] Regions are subsets of zones: They are connected cells of the same value. With a command called RegionGroup, the cells in each region can be identified. Regions can be defined in two different ways. I will define a region of Type IV as one that consists of all the cells that have the same value and are connected by sharing a side; thus, each cell may have four (IV) neighbors. A region of Type VIII consists of cells that all have the same value and are connected by sharing a side *or touching at a corner*; thus, each cell may have eight (VIII) neighbors. (There are only these two types.)

Figure 8-4 shows the regions of type IV. Note from 8-2 that Zone 23 has 13 cells. If RegionGroup is applied to Basic_Raster, and Type IV regions are asked for, the result looks like Figure 8-4 with attribute table shown in Figure 8-5. Note that the 13 cells of Zone 23 are divided into three groups; the region numbers are 4, 6, and 16. The Link number tells the cell Value in the original raster.

1	2	3	4	4	4	4
1	5	5	5	5	5	5
1	5	6	6	6	7	7
8	9	10	6	6	11	11
12	9	13	14	15	16	16
9	9	9	17	17	16	16

FIGURE 8-4 Type IV Regions of Basic_Raster (neighbors sharing sides but not corners)

[5]If you needed proof that raster models and vector models grew up independently, you have only to look at the definitions of region. For vector coverages, a region is defined as "a coverage feature class that can represent a single area feature as more than one polygon." Those polygons could be, and frequently are, disconnected.

Chapter 8

ObjectID	Value	Count	Link
0	1	3	21
1	2	1	22
2	3	1	21
3	4	4	23
4	5	7	25
5	6	5	23
6	7	2	24
7	8	1	29
8	9	5	27
9	10	1	25
10	11	2	26
11	12	1	25
12	13	1	29
13	14	1	26
14	15	1	27
15	16	4	23
16	17	2	21

Record: 14 4 | 1 | ▶ ▶I Show: All Selected Records (0 out of 17 Selected.) Options ▾

FIGURE 8-5   Attribute table of the type IV regions of the sample raster

If RegionGroup is applied to Basic_Raster, and Type VIII regions are asked for, the result looks like Figure 8-6. Note that the 13 cells of Zone 23 are divided into only two groups; the region numbers are 4 and 6. (Where there were 17 Type IV region groups before there are now only 15 Type VIII region groups. Can you see which other region group was affected by including corner neighbors?)

When rasters are composed of integers, those integers generally represent categories. They are quite frequently just nominal data. It is useful if the categories are present in the table as well, so they can be used in selection. For example, suppose the Basic_Raster numbers referred to zoning. Perhaps the categories might be as follows:

21 Heavy Industrial

22 Light Industrial

23 Commercial

24 Parkland

25 Agricultural

26 Multi-Family Housing

27 Single Family Detached

29 Single Family Attached

Then a table might be created that looks like Figure 8-7.

As mentioned, cells may contain NoData. How NoData cells are handled during analysis operations depends on the particular analysis tool and/or decisions made by the user. Sometimes NoData is ignored. Sometimes NoData prevents the operation from occurring. There is not such a thing as a "NoData zone." NoData is not represented in the attribute table, but when the raster is displayed you can indicate how you want NoData symbolized.

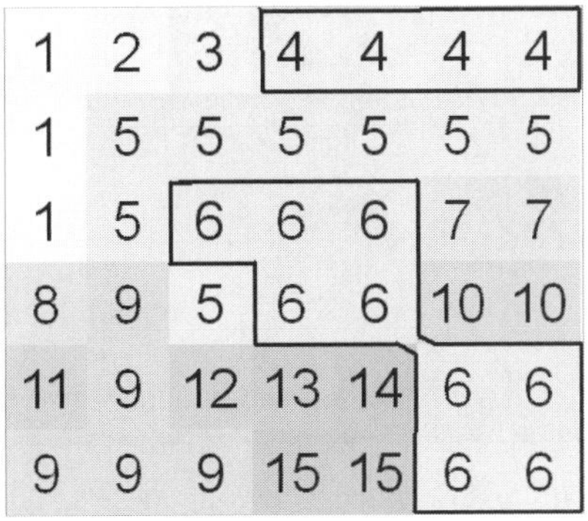

FIGURE 8-6 Type VIII Regions of Basic_Raster (neighbors sharing sides and corners)

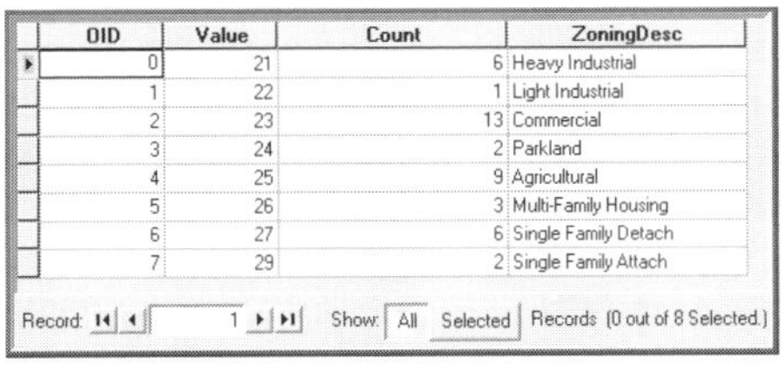

OID	Value	Count	ZoningDesc
0	21	6	Heavy Industrial
1	22	1	Light Industrial
2	23	13	Commercial
3	24	2	Parkland
4	25	9	Agricultural
5	26	3	Multi-Family Housing
6	27	6	Single Family Detach
7	29	2	Single Family Attach

Record: 1 Show: All Selected Records (0 out of 8 Selected.)

FIGURE 8-7 A raster attribute table enhanced with an additional column

Rasters with Floating-Point Values

Rasters with floating-point cell values are used to store what is called continuous data. Such data may also be referred to as nondiscrete data, field data, or surface data. The numbers are stored in a format that allows many significant digits and great range. A given cell may well have a different value in it than any other cell in the raster. The concept of a zone does not apply, since most of the zones would probably consist of a single cell and, therefore, usually, no value attribute table. The values in a floating-point raster usually represent magnitude, elevation, distance, or relationships of cells to other nearby cells in that or some other raster. A major use for floating-point rasters is to represent surfaces, such as elevation of the Earth's surface, or surfaces derived from point data such as rainfall over an area. The word "continuous" is something of a misnomer. There are abrupt changes in Values at the boundaries of cells. The amount of change at boundaries can be reduced by using more and more cells—that is, higher resolution—but, at the conceptual level, the lumpiness of the surface remains. See Chapter 6 for a discussion of continuous and discrete phenomena.

What Is Raster Storage and Processing Good For?

To start with, rasters can be used to solve the same sorts of problems that vectors can. As computers have become more powerful (bigger memories, higher speeds, lower costs, improved compression techniques), rasters are rivaling their vector cousins in solving problems like the Wildcat Boat facility

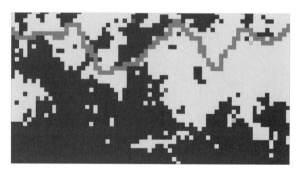

FIGURE 8-8   Raster representation creates jagged lines and polygons

siting problem. Raster methods may suffer, when tightly zoomed in during display, from what is inelegantly called the "jaggies" (see Figure 8-8), where polygonal and linear features are shown in stair-step fashion. This actually may not be a bad thing in many cases, since the precise lines separating polygons usually imply a level of accuracy that does not exist. In human-made phenomena (like parcel boundaries or building footprints), there are such precise lines, but they may not be represented in exactly the right place. In natural phenomena (e.g., the edge of a forest), there is usually little chance that the precise one-dimensional line accurately represents an actual boundary. Here the jagged boundary is a nice reminder that we don't know precisely where the boundary is—at least partially because there isn't a precise boundary.

A main objection to using raster methods on the kinds of problems we have used vector methods to solve is that the minimum units for line and point representation all have to be the same size. So, for purposes of overlaying, a tree and a highway usually wind up being the same width. But, as has often been stated in this text, GIS only serves up an approximation of the real world, and the proof of its utility is in whether it promotes comprehension of the real world, not whether it mimics the real world.

Another problem is that one size does not fit all. You probably wouldn't want to store moose habitat at the same cell size as soil type. Adjustments can be made during analysis—generally toward using the largest cell size—but again there are compromises.

Another use of rasters—an important use, mentioned before: A raster can pretend it is a surface. A surface is a mathematical construct such that, for every value of x and y on a horizontal Cartesian plane, there is a single z value that indicates a displacement from that plane. Any change in x and/or y, no matter how tiny, may result in a change in the z value. In pretending to be a surface, a raster cell clumps a whole bunch (an infinite number) of x-y points together and gives them a single z value. The surface is therefore lumpy. Thus, like much of GIS, it provides an approximation to the hypothetical (mathematical) world, which itself is endeavoring to provide an approximation of the real world.

Rasters can be effectively used in simulations. Given a raster of a forest fire covering a certain area, you could plug in other rasters that indicated wind speed and direction, slope, concentrations of fuel, and so on. From this you could make predictions about where the fire would go next. The fact that each cell relates geographically to its neighbors in only one of two ways (side by side, or diagonally) provides tremendous modeling and computation advantages.

Another simulation use is to represent, over time, the effects or concentrations of matter as it emanates from a source such as a pollution point. Further, floating-point rasters can be used to represent costs of moving or transporting objects across a surface.

Rasters and Features

As you work through this chapter, you will get a feel for the differences between vector and raster representation. Here I simply want to let you know that you can convert from one data model to the other—albeit with loss of information each time.

Let's consider operations in which the inputs are feature classes and the outputs are rasters: the simple conversion from feature data set to raster data set. To do these conversions, you can use the tools in ArcToolbox.

If you convert a point feature class to a raster, each cell that has a point within its boundary takes on the value of that point. If more than one point exists within the boundaries of a given cell, the software randomly picks one of them to represent in the cell. In terms of attributes, the VAT value is taken from the feature attribute table field that is specified by the user. If that field is an integer number, it becomes the VALUE attribute in the VAT. If the user-specified field is a floating-point number that number becomes Value in the VAT if there aren't too many of them. If the user-specified field contains text, then an additional field is added to the VAT, containing the field name, and the Value field becomes a randomly assigned integer.

If you convert a line feature class to a raster, each cell that contains any part of any line takes on the value of the line, according to the field specified by the user. The rules for what goes in the VAT are the same as for points, discussed previously. So a line in a feature class becomes a set of connected cells. Each adjacent pair is connected by either a side or a corner.

If you convert a polygon feature class to a raster, each cell that contains any part of a given polygon takes on the value of that polygon, according to the field specified by the user. The rules for what goes in the VAT are the same as for points, discussed previously. Therefore, a feature-class polygon is represented by a (usually connected) set of cells.

Rasters: Input, Computation, and Output

It would be less than candid to suggest that using spatial analyst for analysis is anything but complex. A large number of ideas, approaches, techniques, and data types are in play here. Let's return to the most fundamental level of computer operation. A computer reads bits, stirs bits, and writes bits. That is, there is input, computation, and output.

For any operation in SA, the inputs can be

❏   One or more rasters

❏   One or more feature layers

❏   Parameters specified at the time of the operation

❏   Settings provided by a spatial analyst environment

For any operation in spatial analyst the outputs can be

❏   A raster

❏   A feature layer

❏   Statistics

You have a good appreciation for feature layers by this time. You are learning about rasters—what a raster is and some of its characteristics. Parameters and statistics have been part of many of the operations you have applied to vector data sets. Environment settings will be discussed in the Step-by-Step section of this chapter.

Where Raster Processing Shines: Cost Incurred Traveling over a Distance

Calculating costs (or cost-weighted distance) has applications in many fields. If you were siting a business that needed to be accessible, you could use the ArcMap Spatial Analyst Cost Distance tool to measure distance to important points in a road network. If you were doing animal migration studies, you could use it to measure migration routes. You could also use cost distance calculations to determine sites for new roads, pipelines, or transmission lines. In fact, this approach has been used in Alaska as part of a project to determine the impact of building pipelines. You will work a problem of this type in the Step-by-Step section of this chapter.

It is important to mention another ArcMap tool designed specifically for moving across distances to points connected by linear features, such as roads or railroads. The ArcMap Network Analyst allows you to consider additional factors, including one-way streets and the costs of turning at intersections. Network Analyst is very powerful, and if your problem or application is specific to street or road networks, you should become familiar with this extension. Chapter 9 contains an exercise using Network Analyst.

Proximity Calculation with Rasters

GIS is meant to help human beings in understanding both natural and human-generated phenomena. A fundamental concept in both arenas is the idea of closeness. Raster analysis can play a key role. We will focus on the following:

❏   The importance of distance in geographic analysis and synthesis

❏   How to use ArcGIS Spatial Analyst to find the distance from anywhere to places of interest

❏   The relationship between distance and cost

❏   How to calculate the cost effects of other factors besides distance

❏   How to find the shortest or least expensive path from anywhere to places of interest

❏   How to determine which is the closest place of interest and/or the place of interest that can be reached with the least cost

Human Activity, Cost, and Distance

Life forms on our planet may be divided into plants and animals. A major difference between them (although there are exceptions) is that plants do not move (much) and animals do. In fact, for human beings in particular, moving themselves and their artifacts, such as clothing, weapons, and stereo equipment, is a major activity.

Frequently, a major concern about a given human activity is its cost—cost in terms of energy, time, money, fuel, suffering, or other parameters. Moving from one location to another is a frequent human activity and one whose cost we often want to minimize. We usually move because we want to be in a new location or leave an old one for some reason. (There are exceptions to this: a walk in the woods for soul healing, a jog for exercise.) The subject of this discussion is using GIS to

❏ Reduce the cost of going from one place to another

❏ Lower the cost of transporting artifacts

❏ Minimize the cost of building linear structures, such as roads, bridges, and tunnels

The costs of moving from one place to another can be complex to calculate. In cases where the distance is far, the trips are repeated frequently, or the cost is critical (as in the cost in time of getting an ambulance to the scene of an accident), it may well be worth doing the computations. In this discussion, you will learn how to deal quickly and simply with the complexity of performing the "cost of moving" calculation.

What are the factors that determine the cost of moving things, including ourselves? The one that perhaps comes to mind most quickly is "distance." It costs more to go from Punxsutawney, Pennsylvania to Red lands, California than to go from your living room to your bedroom. Or from your home to your office. We begin with a consideration of distance—distance of the simplest sort: straight-line distance (on a Cartesian plane) from any point to a given point.

Euclidean Distances on the Raster

Distance is measured between two points. The two points used when measuring distance on a raster (or grid) are the centers of two specific cells. That is, in the case of the raster, the measurements are made from cell center to cell center. See Figure 8-9.

If the cell size is 50, the distance shown would be 200, spanning all or parts of five cells.

For another example, if you had a raster that had cells of size 10 and 1000 columns, the longest east-west distance measurement would be 9990.

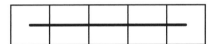

**FIGURE 8-9   Distance is measured
from cell center to cell center**

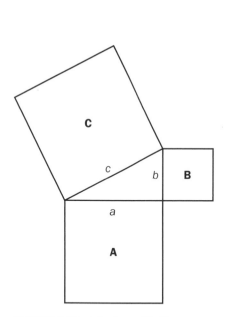

FIGURE 8-10   The law of Pythagoras (the Pythagorean theorem)

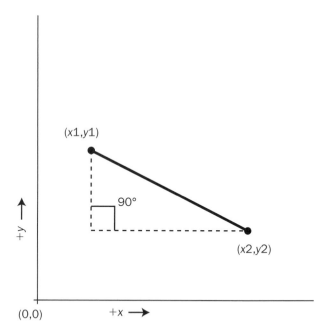

FIGURE 8-11   The straight-line distance between two points

The most fundamental "distance raster" is one in which there is a single "source" cell and all other cells indicate the distance from that cell.

Measurements are made from cell center to cell center, regardless of intervening cells or the angle of the line connecting the two cells of interest. The distance is calculated by the law of Pythagoras, who, despite having lived 200 years before Euclid, determined that the hypotenuse of a right triangle is the square root of the sums of the squares of the other two sides.

As illustrated in Figure 8-10, the area of the square *on* the hypotenuse is the sum of the areas of the squares *on* the other two sides.

When you put the law of Pythagoras (also known as the Pythagorean theorem) together with René Descartes[6] great invention you can calculate the distance between two points whose coordinates are (x_1, y_1) and (x_2, y_2) as:

$$(x_1 - x_2)^2 + (y_1 - y_2)^2 = c^2$$

where c is the straight-line distance between the points. See Figure 8-11.

---

[6]Descartes invented the x-y coordinate system on the 2-D plane in the 1600s.

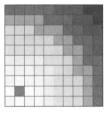

FIGURE 8-12

Euclidean Distance and the Spatial Analyst

Proximity, the closeness of one point in space to another, is most easily expressed by the Euclidean straight-line distance between the two points. The term "Euclidean" comes from the geometry that was first formally developed by Euclid around 300 BCE.

ArcGIS Spatial Analyst allows you to automatically make a raster in which each cell in the raster contains the straight-line distance from itself to the closest cell of a set of source cells in another raster that represents the same geographic space. (Actually, because this distance is represented on the two-dimensional raster surface—differences in elevation being excluded—we could refer to the values in the cells as "Cartesian distance" after the mathematician Descartes.

Figure 8-12 shows a raster whose cells are composed of Euclidean distances from each cell to the dark cell in the lower left. The lighter the color, the smaller the distance.

Proving Pythagoras Right

If you're like most people, particularly those who are primarily spatially minded, you (vaguely) know the law of Pythagoras, but probably not why it is true. So, for those who are interested, here is a proof—one that will appeal to those interested in graphical, rather than strictly algebraic, matters. (If you just want to take my word for the validity of the law, you may skip the remainder of this discussion.)

Draw a square on a piece of paper. Draw a smaller square within the first, rotated so that its corners touch the sides of the first square. Call the line segment running south from the northwest corner of the big square to the point where the smaller square touches it "a." Call the segment running north from the southwest corner of the big square to the touching point "b." Label all of the line segments around the big square with "a" or "b," as appropriate. Label each of the sides of the smaller square "c." See Figure 8-13.

Now compute the areas. The area of each of the four triangles is ½*ab*. So the total area of all four triangles is 2*ab*. The area of the smaller square is, of course, *c* squared. What is the area of the larger square? The length of each side is *a* + *b*, so the area of the square is (*a* + *b*) times (*a* + *b*). When you multiply it out, that's:

$$a^2 + 2ab + b^2$$

Note now, by looking at the figure, that the area of the larger square is equal to the area of the smaller square plus the areas of the triangles, so we have

$$a^2 + 2ab + b^2 = c^2 + 2ab$$

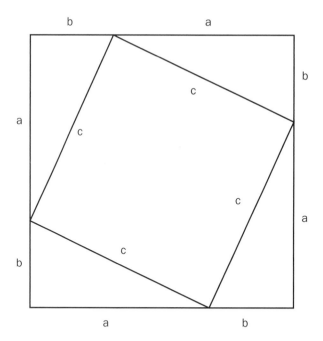

FIGURE 8-13   Re-proving the Pythagorean theorem

Subtracting 2*ab* from each side of the equation gives the Law of Pythagoras:

$$a^2 + b^2 = c^2$$

Finding the Closest of Multiple Source Cells

The raster of source cells that serves as input to Euclidean distance may consist of several single cells, clumps of cells (contiguous cells), a linear structure of several cells, or any combination of these. For each cell in the output raster, the straight-line distance to the closest source cell is calculated. This happens automatically. You simply apply the Euclidean distance tool and the newly generated raster contains the closest distances in all the other cells.

Many distance problems involve several source cells instead of a single one. Perhaps you are flying along in your private plane, and because of an ominous new noise coming from the engine, you develop a strong interest in knowing the distance to the closest of several airports in the area. The airports might be recorded as source cells. If you had time to fire up your portable GIS, connected to your GPS, no matter which cell you were in, the distance to the closest airport would be shown.

Another example: Suppose that you have a neighborhood of houses that need to be connected to a power line. The power line is represented as a line of source cells; houses are built on other cells. You want to find the shortest distance from each house to the power line, so you can calculate the minimum length of wire required. Also, because of electrical requirements, there is a maximum distance the wire can be strung, so you will also need to identify the cells that are more than that distance away.

Problems of this sort may be solved with the Euclidean Distance tool, using as input a source raster with multiple source cells.

Excluding Distances beyond a Certain Threshold

If the problem warrants it, you may limit the search for the closest source cell so that it does not exceed a given value. To do this, you select an option that says "if you have to look further than this to get to a source cell, just forget it." Such a number may be called a limit, a cap, a cutoff value, or a threshold.

Other Factors That Influence Cost

We began this discussion on proximity saying our primary concern was the cost associated with moving (moving being defined in a very general sense) from one place to another. We also said that distance between points on the Cartesian plane was a major factor in the cost, and we have spent much of the preceding discussion discussing how to obtain that distance.

Frequently, however, distance is not the only, or even the principal, cost of moving from one place to another. An extreme example is of a person living in New York City who wants to visit a second person who lives nearby. Let's conjecture: the two people live in 33rd-floor apartments, toward the middle of parallel long-block streets, and their apartments share a common back wall. They actually live only a few feet from one another. But to travel from one apartment to the other requires two elevator rides, encounters with doormen or buzzer systems, and a not-inconsiderable walk (or cab ride). Clearly, the calculation of simple Euclidean distance is not the only tool we need to determine cost.

As a second example, suppose that you want to take a hike. In between your starting and ending locations lie both swamp and dense forest. By taking a longer route, you can walk through pasture. The time and difficulty of the route you take will be a function of both the distance and the difficulty of making your way through various environments.

We can generalize these problems by creating a "cost surface." We develop this cost surface so that each raster cell contains a number that indicates the cost of going through a unit of distance in that cell. If it is three times as hard to go through the forest as it is to go through the pasture, we might assign values of 1 in the pasture cells and 3 in the forest cells. If a certain place in the swamp poses a threat of people being eaten by alligators, we might assign a very high cost (say 10,000) to traversing the distances in the swamp cells.

A third example: While off-road vehicles are becoming increasingly popular, it is still less expensive to travel by automobile from one city to another by using paved roads. The cost of traveling a straight line in this case, in terms of speed, danger, and possible encounters with law enforcement officials, is prohibitive.

Consider Figure 8-14. The time to go from A to B is greatly reduced by going around the volcanic mountain rather than over it. While the shortest path between points A and B is a straight line, it is not the fastest path, because the cost for crossing over the mountain is much greater than going around it.

In short, Euclidean distance has the limitation of assuming that the cost of crossing distances in any cell on the way to a source cell is the same as the cost of crossing distances in any other cell. You will learn how to indicate that some cells are more expensive to cross than others. You will also learn how to find the most inexpensive route (the least-cost path) even though it may not be the shortest.

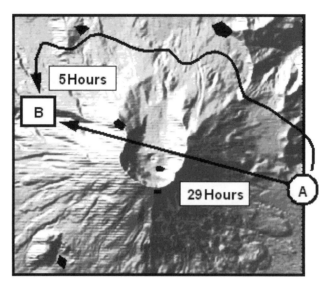

FIGURE 8-14   A straight line, while the shortest, is not always the fastest

The Cost Distance Mechanism

Spatial Analyst provides the Cost Distance tool for calculating the least cost of getting from any cell to a source cell. Two input rasters are involved: the source raster and a cost surface raster, each of whose cells indicate a price of traversing each unit of distance in that cell.

The overall cost of the path from a given cell to the closest source cell is calculated on the basis of the least-cost path between the two cells. Calculating the total cost involves summing up the costs of crossing all the individual cells along the least-cost path. The cost of crossing an individual cell is found by multiplying the distance across that cell by the value found in the geographically equivalent cell in the cost-surface raster. Thus, the accumulated cost found in the output raster is the sum of the products formed by (a) the distance across each cell and (b) the value (the "price per unit distance for going through") of each cell.

The Cost Distance Calculation

If you want to generate a raster similar to that made by Euclidean distance, but one that will place in each cell the least cost of traveling from that cell to the source cell, you can use the cost distance calculation. Cost distance operates with inputs of a source raster and a cost-raster to produce a cost-distance raster. See Figure 8-15.

The source raster you already know about. The cost raster has darker colors for the cells that it is more expensive to traverse. The cost distance raster that is formed shows the cost to travel between a given cell and the source cell. Here the darker color indicates a greater cost in traveling between the source

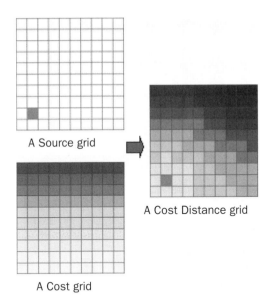

A Source grid

A Cost Distance grid

A Cost grid

FIGURE 8-15   Making a cost distance raster

cell (dark) and every other cell.[7] Notice particularly the northwest and southeast corners of the resulting raster. Even though they are the same distance from the source cell, it is clear that it is more expensive to travel from the northwest than from the southeast. The reason is that southern cells are cheaper to cross than northern ones.

When you apply the Cost Distance tool, the resulting raster shows up in ArcMap (asuming it is running) as a raster that you provide a name for.

Path Calculation in Euclidean Distance and Cost Distance

It is worth mentioning that the straight-line distance from a given cell produced by Euclidean distance compared with the least-cost path generated by cost distance are quite different; they are calculated in completely different ways. Euclidean distance calculates the direct, straight-line distance from each cell center to the closest source cell center, regardless of how this path slices through intervening cells. The cost distance path, however, must pass through cell centers, frequently generating a sequence of line segments that change direction at the center of each cell.

Therefore, even when crossing a cost surface where each cell cost is 1, so that a straight line would be the least-cost path, cost distance will usually produce a slightly longer path—because the path proceeds from cell center to adjacent cell center. This sort of "connect the dots" behavior will not result in a much longer path than the straight-line distance, but observing it illustrates the difference in the ways the two output

---

[7] A calculation is made for every cell in the raster.

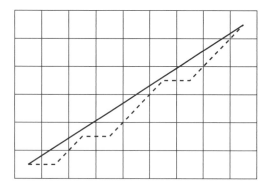

FIGURE 8-16   The cost distance path goes from cell center to cell center.

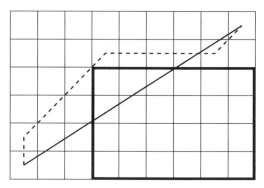

FIGURE 8-17   A cost distance path tries to avoid areas of high cost.

rasters are calculated. In cost distance, the path proceeds from one cell center to the center of one of the eight adjacent cells, always with the aim of minimizing the cost of getting to the source cell. In other words, except for a few special cases (e.g., due north, straight southwest), the path generated by cost distance will not be a straight line, but will zigzag so as to take in the cell centers along the way. See Figure 8-16.

The Euclidean distance direction is shown by a solid line. A possible path generated by cost distance is shown by a dotted line, under conditions of a uniform cost raster. If the cost is not uniform, then the cost distance path will be different. For example, look at Figure 8-17.

The EucDistance direction and path are represented by a solid line. A possible path generated by CostDistance is represented by a dotted line, under the condition that high costs exist in the cells within the outlined box.

Understanding How Total Costs Are Calculated

Calculating cost distance is somewhat involved. Take as an example a cell, size 10, that costs $1.15 per unit distance to cross. What is the cost of traveling from the center of such a cell directly to its southern edge? See Figure 8-18. The length of this segment is 5 (half the total distance of 10 across the cell). The cost associated with the cell is 1.15 per unit of distance. The cost of this segment, then, is the product of the distance and the cost per unit distance: 5 * 1.15 = 5.75.

When the path goes from cell corner to corner, the distance is longer than if the path goes from the side of a cell to the opposite side. How much longer? Multiply the side-to-side distance by the square root of 2, which is approximately 1.41421356237, which we will round to 1.414. See Figure 8-19.
For example, the cost to go diagonally across a 10 unit cell that had a toll of $1.15 would be

10 * 1.414 * 1.15 = 16.261

Each possible path from the source cell to each other cell is either calculated or considered in some way. That can be a lot of paths. This type of calculation may take place several times for each cell, there can be hundreds of thousands of cells, and the computer may make many attempts for each cell to find the least-cost path.

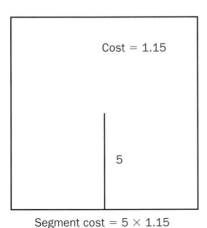

Segment cost = 5 × 1.15

FIGURE 8-18   Cost of traversing a cell from its center to a side

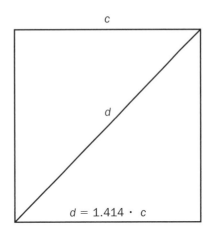

FIGURE 8-19   Cost of traversing a cell from a corner to the opposite corner

Getting More Information: Paths and Allocations

You have learned that you can calculate the shortest distance between two places by using the EucDistance request, and you can find the least-cost path with the CostDistance request, but there are still a couple of major questions: Which source cell is closest to (or least expensive to reach from) each cell in the raster, and what is the direction or path from each cell? That is, how would you identify the closest source cell and how to get there?

To answer these questions, first for straight-line distance, you can use two additional parameters in the Euclidean Distance tool.

Direction and Allocation Rasters for Euclidean Distance

The Euclidean Distance tool allows you to specify two new rasters, which we will call a direction raster (DirGrid) and an allocation raster (AlloGrid). DirGrid's cells each contain the direction to the nearest source cell, in degrees clockwise, based on due north as 360 degrees (or 0 degrees). AlloGrid's cells tell you the value of the nearest source cell. (These rasters will be created for you, with the names you specify.)

Direction and Allocation Rasters for Cost Distance

The cost distance calculation is similar to the Euclidean distance calculation in that both can be used to generate direction and allocation rasters. Both calculations allow caps to be placed on the maximum values in the primary rasters they generate. Cost distance differs from Euclidean distance in three major ways:

❏   It uses a distance-weighting raster.

❏   It differs in the way in which the direction raster points along the path.

❏   It calculates the least-cost path rather than the shortest path.

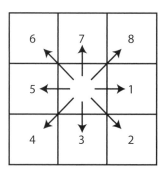

FIGURE 8-20   Direction code for the next cell in the path

Because the path generated by cost distance goes from cell center to cell center, each cell in the direction raster contains information that indicates to which of the eight neighbors the path goes next. This information is coded according to Figure 8-20.

Or, in words:

1 means the next cell in the path is to the east (90 degrees).

2 means the next cell in the path is to the southeast (135 degrees).

3 means the next cell in the path is to the south (180 degrees).

4 means the next cell in the path is to the southwest (225 degrees).

5 means the next cell in the path is to the west (270 degrees).

6 means the next cell in the path is to the northwest (315 degrees).

7 means the next cell in the path is to the north (360 degrees).

8 means the next cell in the path is to the northeast (45 degrees).

A Major Application of Raster Processing: Hydrology

Because of the regular nature of the areal unit in raster processing, a number of applications, such as a simulation of a forest fire or an oil spill, are made much simpler. Hydrologic modeling is one such application, so we look at it in some detail, both for its own sake and because it illustrates the power of spatial analyst. An important proviso: If you want to do serious hydrologic investigations, you should use packages designed for the purpose, including those provided as separate software by Esri.

You may not have thought a lot about how water flows over the surface of the Earth, or how to simulate it with a computer, so let's explore the following:

❏   The fundamental elements of hydrologic modeling and analysis

❏   How to determine the direction of water flow from cells in a study area

- ❏ The importance of eliminating sinks from the drainage area

- ❏ The importance of including the entire drainage area in the study area

- ❏ How to determine flow accumulation in the cells of a raster

- ❏ How to find the distance, both upstream and downstream, from a given cell

- ❏ The methods used to delineate and assign order numbers to segments

- ❏ How to generate watersheds for streams and points

We enter into a discussion of these topics in the text that follows. Then, in the Step-by-Step section of this chapter, you simulate a hydrological system.

Basic Surface Hydrology

First a caveat: Hydrologic analysis is a complex subject. The concepts and tools presented to you here are, in themselves, not sufficient to undertake hydrologic analysis or modeling. Real-world situations frequently do not conform to the assumptions and conditions that underlie the examples presented here. However, the concepts discussed here will help you understand the basic principles of surface hydrologic analysis.

Surface hydrologic analysis (as opposed to underground hydrologic or groundwater analysis) seeks to describe the behavior of water as it moves over the surface of the Earth. Most simply, this type of analysis includes the following:

- ❏ Obtaining a mathematically correct representation of the surface of the area to be analyzed, considering the elevation of the surface at a given point to be the value of a raster cell at that point

- ❏ Determining the direction water would flow from each cell on the surface

- ❏ Determining to which adjacent cell water would flow when each cell is doused with a given amount of water

- ❏ Finding those cells that get considerable flow accumulation and delineating them as creeks, streams, and rivers, either persistently or when flooding occurs

- ❏ Developing a network of these creeks, streams, and rivers; determining a hierarchy of them; and classifying them as to volume, relative to their upstream tributaries

- ❏ Determining the areas (watersheds) that feed into given creeks, streams, and rivers and determining the outlets (pour points) of these watersheds

- ❏ Determining into which watershed and which water entities a given quantity of liquid (such as a polluting spill) might flow

What we will discuss here are the basic hydrologic tools available in Spatial Analyst. As mentioned there are also Hydrologic Modeling tools available in other Esri products.

In Spatial Analyst, most hydrologic analysis is done by generating new rasters. This operation is usually accomplished by entering formulas in the Raster Calculator or by specific tools in ArcToolbox.

Basic Surface Hydrology Concepts

The concepts are described in terms of the tools you can use to examine surface hydrology.

❏ The FlowDirection calculation determines the direction of flow from each cell of a surface raster. The raster generated by FlowDirection must be well behaved. The sort of analysis I am describing specifically excludes land areas that contain lakes or ponds. The assumption is that all the water placed on the raster will ultimately exit the raster at one or more low points on its edge.

❏ Assuming that the study area involved does not contain lakes or ponds, one of the ways the raster can be ill behaved is to contain a cell that is lower than its surrounding neighbors; such a cell is called a sink. Sinks distort the analysis; to find them, you will use the Sink calculation. (Editing rasters with sinks is beyond the scope of this lesson.)

❏ Another requirement of the raster is that the cells of primary interest—for example, the mouth of a river near a town that might flood—must include all the "uphill" cells. That is, all the cells that constitute the drainage basin for the cells of interest must be considered. The FlowAccumulation calculation may be configured to compute the amount of water that flows into each cell from all of the cells that are uphill from it.

❏ Stream networks are characterized by small creeks flowing into larger ones, these flowing into small streams, and so on. It is useful to speak of the "order," or relative size, of such water entities. The smallest creeks are labeled order 1. Larger entities have larger integer numbers. The StreamOrder calculation handles the process of assigning order numbers to streams. Both of the two principle methods for numbering streams (Strahler and Shreve) are available.

❏ The Mississippi River has a watershed consisting of all the land that supplies water to it. The smallest creek also has a watershed that consists of all the land that supplies water to it. The creek's watershed may be contained in the Mississippi's watershed, so the delineation of watersheds (or drainage basins, catchment areas, and contributing areas, as they are also called) is not trivial, either in concept or calculation. The WaterShed calculation assigns cells to such areas.

In addition to these calculations, an important operation that precedes surface hydrology analysis is the generation of a surface raster that gives the elevation at every cell. There are several ways to do this, as previously discussed.

Calculating Flow Direction

The primary data source for hydrologic operations in ArcGIS is a raster of flow direction. This raster is formed by the FlowDirection calculation based on a surface of elevation; for our discussion here, we will call the resulting raster "DirOfFlow." Each cell in the DirOfFlow raster contains an integer number; these numbers are powers of 2: 1, 2, 4, 8, 16, 32, 64, and 128. (Just why these numbers were chosen, rather than 1, 2, 3, etc., has a historical and computer component, which will be discussed.) Each number indicates a direction, as shown by Figure 8-21.

The idea is, simply, that the precipitation that falls, or otherwise appears, on a given cell flows immediately to a single adjacent cell. To which of the eight adjacent cells? The one indicated by the number and the arrow in Figure 8-21, which points in the direction of the steepest descending slope.

For example, consider the simple raster shown in Figure 8-22. The numbers in the cells indicate elevation.

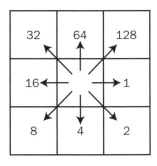

FIGURE 8-21 Direction code for the flow of water

92	94	96	98	100
91	93	95	97	99

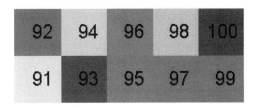

FIGURE 8-22 Elevation values in a tiny raster

4	8	8	8	8
8	16	16	16	16

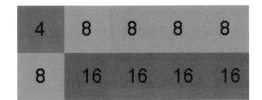

FIGURE 8-23 Flow direction following steepest slope

32	64	64	64	128
8	4	4	4	2

FIGURE 8-24 Flow allowed directly off edge cells of the raster

The range in altitude is from 100 to 91, sloping gradually from east to west and a bit from north to south. When the Flow Direction calculation is applied to this raster, the resulting raster looks like Figure 8-23.

Note that water flows from each cell to the nearest neighbor cell so that the water flows down the steepest slope, except from the cell with lowest elevation in the southwest, where it flows off the raster.

Normally, cells along the edge of the raster are treated as any other cells in the raster, except that if none of the five adjacent edge cells have lower elevation than the edge cell under consideration, the flow will be directly off the side of the raster. But there is an option (force all edge cells to flow outward), which shoves water off the raster, even if edge cells have lower elevation neighbors. Thus, the raster in Figure 8-24 would be generated instead.

The Ultimate Destination of Water Is off the Raster Area

The lowest point on the raster must be on an edge. This requirement is not as stringent as it sounds. Basically, you want to deal with land that has no ponds or lakes. You want a network of valleys that will hold only linear bodies of water, at least one of which will flow off the edge. As already indicated, the ArcGIS hydrologic tools presented here do not work with lakes. They are strictly for stream networks. Lakes, which would constitute sinks, are not allowed.

It is worth remarking on the rather strange choice of numbers used to indicate flow direction. You've learned that water flows from any given cell to one of the eight adjacent cells. In the previous exercise on proximity, the directions were indicated simply by the integers 1 through 8. Why, then, are we dealing with numbers such as 32 and 64?

One conjecture is that, in the early days of hydrologic analysis, which correspond to the early days of computers, central processing unit speeds were slow and storage space in memory was at a premium.

It was efficient to indicate direction with a single bit (a 1 or 0) in a position in a computer byte. Those positions correspond to columns in the base 2 number system. Those columns are designated 1, 2, 4, 8, and so on. Eight bits in a byte; eight neighbors for each cell. It may be that the precedent set in the early days endures in the hydrologic modeling field today.

Flow Accumulation: Drainage Delineation and Rainfall Volume

Once you have a raster that indicates flow direction, a number of other interesting and useful calculations are possible. In particular, you can determine the locations of all the linear bodies of water, and you can determine, from slope and elevation, those areas where water may accumulate during times of intense precipitation. This is accomplished with the ArcToolbox Flow Accumulation tool.

Basically, the value in each cell in the resulting raster contains the sum of the amount of water that has fallen on all the raster cells upstream from it. The intent is to simulate the flow, or potential flow, of water to form creeks, streams, and rivers. If each cell is presumed to have one unit of water (say, an inch in depth) to contribute—under a condition of "uniform rainfall," you can think of the number in a given cell as the number of cells upstream from that cell. To illustrate, examine the elevation surface in Figure 8-25. Note that the low points are in the middle of the south edge (elevation 1) and the west edge (elevation 3). All around the rest of the raster the elevations are 9 or somewhat less.

From this, you can produce a raster showing the direction of flow, using the ArcToolbox Flow Direction tool. Some arrows have been scattered on the raster to show flow direction. See Figure 8-26.

Now, applying the Flow Accumulation tool to the flow direction raster produces a raster that shows, for each cell, the water that accumulates due to adding up the accumulations from the cells "above" it. Figure 8-27 depicts some of these accumulation values.

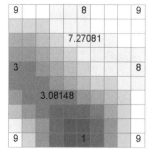

FIGURE 8-25   Elevations of a raster

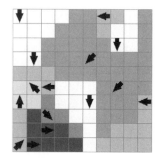

FIGURE 8-26   Some directions of flow, given the elevations of the raster of Figure 8-25

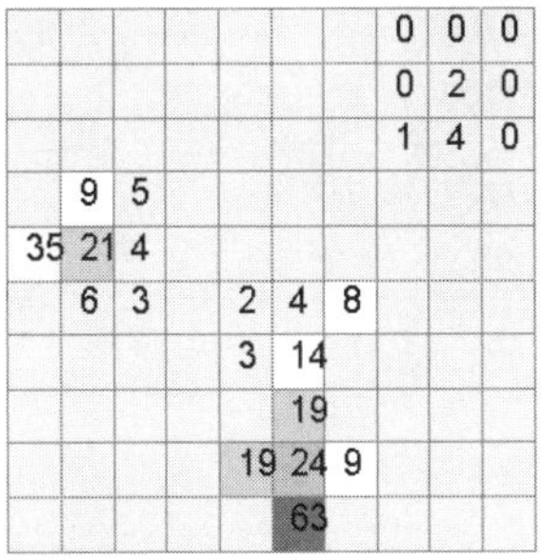

FIGURE 8-27
Accumulations of water,
based on elevations and
flow directions of the two
previous figures

The largest accumulation is in the south, which had the lowest elevation. Another point of considerable accumulation is in the middle of the western side. If you look at the flow raster and the accumulation raster, you can get an idea of where and why the stream channels developed.

Note that some cells have the value 0, indicating that no cells are uphill of them. According to the model the Value of a cell is what flows into it from another cell; the rain that falls on a cell does not affect its Value. Note also that most of the cells accumulate very little water, but some accumulate a great deal of it—just as you might expect, since most of the land around us is uncovered by water, but there are numerous creeks and streams. Finally, if you add up the values of the southern and western pour points (63 and 35), you get 98. This makes sense, since of the 100 cells total, 98 of them are above the two pour points.

Nonuniform Rainfall

The raster in the figure was developed without an input weight raster. It was assumed, therefore, that each, received the same amount of rain and was presumed to receive the same amount of water. You can change that by specifying a number for each cell in the study area. Consider an input weight raster that looks like Figure 8-28. This weight raster represents a gradation in rainfall, which was heaviest in the north.

If you consider that this was a rainfall event, and that the values in the raster cells, which vary from almost nothing to more than 1.5, constitute inches of rainfall, you can see that much more rain fell in the north than in the south across the study area. The total amount of rainfall is approximately the same as in the previous example, but in that case the rainfall was distributed uniformly.

Now you can apply the Flow Accumulation calculation with this weight raster. The results, shown in Figure 8-29, indicate that considerably more water volume showed up at the western pour point than before, because the rain was lighter in the south. In fact, with the weight raster, about as much water flows west as south. Recall that with no weight raster, almost twice as much flowed south as west.

This raster is the result of applying the Flow Accumulation calculation using the previous weight raster.

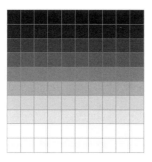

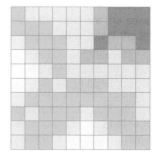

FIGURE 8-28   A weight raster: more rainfall in the northern part of the raster

FIGURE 8-29   Flow accumulation taking into account the weight raster

You can see from this example that hydrologic modeling can be a complex operation with many variables and parameters.

Calculating the Length of a Potential Linear Water Body

The length of a potential creek or stream is a useful thing to know when modeling. You can apply the Flow Length calculation to the direction of flow raster to show either the length of the channel of flowing water from each cell upstream or downstream. Upstream flow length for a given cell is the distance, totaled from cell to cell, from the given cell to the origin of the longest path of water (the top of its basin) coming into that cell. Downstream flow length from a given cell is the distance from that cell to the pour point for the water passing through the given cell. See Figure 8-30, where darker shades indicate larger flow length.

The upstream flow length looks like Figure 8-31, where darker shades indicate longer flow lengths.

The Input Weight Raster in the Flow Length tool operates in precisely the same way as the weight raster does (impedance, cost surface) in our previous discussion of proximity. It multiplies the length through

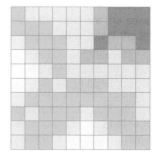

FIGURE 8-30   Downstream flow length

FIGURE 8-31   Upstream flow length

each given cell by the value in the geographically equivalent cell in the weight raster. The weight raster provides the cost or impedance for water to flow through each cell. Thus, you could simulate the fact that water flowing through forested land takes longer to cover a given distance than water flowing over rock.

You can use the output of the Flow Length tool to find the length of the longest flow path in a given basin. This is one of the values needed to calculate a more sophisticated hydrologic quantity, "time of concentration" for a basin. You can use flow length rasters to create distance-area diagrams of hypothetical rainfall/runoff events, using the optional weight raster as an impediment to downslope movement.

Assigning Identities to Streams

The most basic hydrologic unit (outside of the individual cell) is the stream segment. Generally, streams segments (also called links) run between intersections in the linear network. A segment consists of all the cells between the junctions of two or more streams or between junctions and the pour points. (The cell that is the junction is considered to belong to one of the streams.) Figure 8-32 illustrates this numbering. The integer number is only a nominal, identifying value.

On a raster, ArcGIS places the same unique identifying number in all the cells of a given stream segment. In the discussion of the Flow Direction and Flow Accumulation calculations, every cell was considered a contributor to the creeks, streams, and rivers that developed ("Into each cell some rain must fall."). But you do not want to define all the cells in the study area as part of the water network. Instead, the software will delineate specific stream channels, running from intersection to intersection. In other words, all of the study area contributes to the total amount of water to be dealt with, but only a small part of the study area carries most of that water. That area is known variously as the water network or the stream channels. This area is defined by including only those cells with flow accumulations greater than a chosen value; that value is called the cell threshold. Figure 8-33, an illustration of how stream segments are numbered,

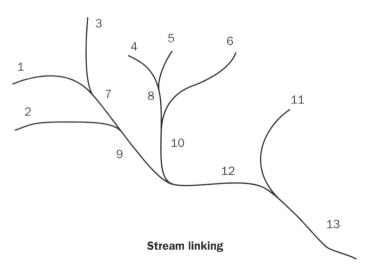

Stream linking

FIGURE 8-32   Numbered stream links

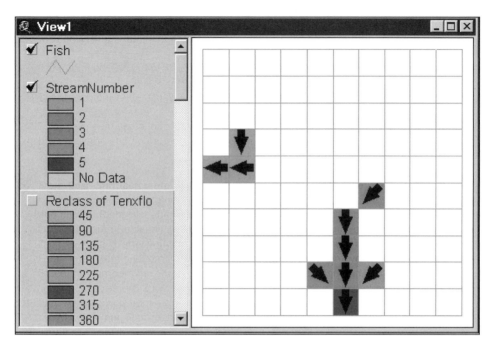

FIGURE 8-33 Stream Channels

shows five stream links. Each cell has a flow accumulation of greater than 7.0. These cells are considered to make up streams. Each stream segment is uniquely numbered, as shown by the color coding.

Vector vs. Raster Representation

The two preceding figures show the difficulties involved in representing the virtually infinite, three-dimensional environment in the memory of a computer, necessarily using only the most fundamental discrete symbols: 0s and 1s. In vector mode, a stream is represented by one-dimensional lines; the lines have no width, only length. If quantities like flow, width, or velocity are to be included, they must be part of the attribute table.

In raster mode, a stream is represented by a sequence of adjacent cells. These cells are two-dimensional—they cover area. The area each cell covers, in basic hydrologic analysis, is the same, whether a mountain creek or a major river is being represented. Again, the geographic representation is only an approximation; even information about quantities such as width must be carried along separately.

This confluence of vector representation and raster representation in storing and displaying information about streams illustrates the challenges of using a computer to represent natural phenomena. Next, an attempt is made to represent the relative "size" of streams and stream channels.

Assigning Orders to Stream Links

You can attach an order number (integer value) to each stream segment or link. Generally, streams with lower numerical values have a size that can carry smaller volumes of water, but this is not always the case, as you will see.

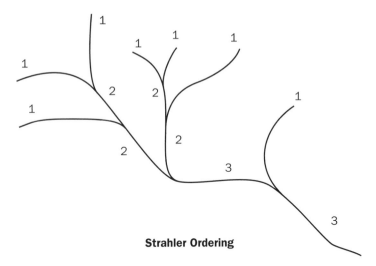

Strahler Ordering

FIGURE 8-34   The Strahler method of stream ordering

Two ways of determining stream order number have been devised (one by A. N. Strahler in 1957 and the other by R. L. Shreve in 1966), as discussed below. In both methods, the smallest originating streams are numbered 1, from the source of the stream continuing up to the first intersection.

In the Strahler method, when (any number of) streams of the *same order* merge at a point, the downhill stream takes on an order number that is the original stream order number plus 1. For example, if a stream of order 3 merges with another stream of order 3, the resulting stream is order 4.

In any other case of stream merging, the order number of the downhill stream retains the order number of the larger uphill stream. So, if a stream of order 3 is joined by a stream of order 2, the resulting stream is still of order 3. Figure 8-34 illustrates the Strahler method.

When two streams merge according to the Shreve method, the order numbers of the uphill streams are added together to produce the order value of the downhill stream. So, when merged, two streams of order 3 produce a stream of order 6. An order 3 stream joined by an order 2 stream produces an order 5 stream. Figure 8-35 illustrates Shreve ordering.

To generate stream orders in ArcGIS, you use the Stream Order tool, which allows you to choose either the Strahler or Shreve method.

Watersheds and Pour Points

A basin is an area that drains water and other substances carried by water to a common outlet as concentrated drainage. Other common terms for a basin are watershed, catchment, and contributing area. The contributing area is normally defined as the total area contributing water flow to a given outlet, also called a pour point.

A delineation of these areas is the output of the Basin tool calculation. The geographic line between two basins is referred to as a basin boundary or drainage divide. Such a line, as you might imagine, runs along ridge tops and other lines of relatively higher elevation.

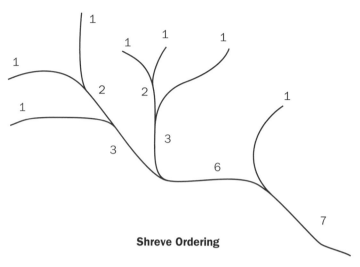

Shreve Ordering

FIGURE 8-35 The Shreve method of stream ordering

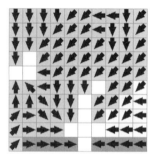

FIGURE 8-36 The five watersheds correspond to the five stream links

An outlet, or pour point, is the point at which water flows out of an area. It is the lowest point along the boundary of the basin. The cells in the source raster are used as pour points above which the contributing area is determined. Source cells may be features such as dams or stream gauges for which you want to determine characteristics of the contributing area.

In Figure 8-36, the five watersheds correspond to five stream links. The arrows indicate the direction of flow from each cell; the cell shade indicates to which watershed the cell belongs.

Another ArcGIS tool that calculates areas that contribute water to a given point is the ArcToolbox Watershed tool. The distinction between them will be made clear when you work the exercises.

Spatial Analysis Based on Raster Data Processing

—— Open your Fast Facts File.

—— Open the Color Figures file.

Exercise 8-1 (Project)

Basic Raster Principles and Operations

—— **1.** Start ArcCatalog. Make a folder named Raster_Experiments in your

___IGIS-Arc_*YourInitials*

folder. Make a folder connection to that folder. Now make a File Geodatabase named Ras_Expr.gdb inside that folder. Also make sure that the Spatial Analyst extension is turned on (You will also have to turn it on in ArcMap later.)

First, let's create a tiny, trivial, basic raster to work with. There are several ways to create a raster. We could create a point shapefile, as we did in the project where we mapped the computer terminals, and then convert that to a raster. We will do that in a future project. Another method would be to use the ASCII to Raster tool (in ArcToolbox > Conversion Tools > To Raster). This needs a text file, which we will create. (When I say "we" will use a certain tool you have, of course, noted that I make the suggestions and you do the work.)

—— **2.** Using a text processor, make a text file as shown in the lines of characters that follow. Call the text file Forty_Two_Cells.txt and save it in the Raster_Experiments folder.

NCOLS 7

NROWS 6

XLLCENTER 4150

YLLCENTER 1150

CELLSIZE 100

```
NODATA_VALUE  -32768

21 22 21 23 23 23 23

21 25 25 25 25 25 25

21 25 23 23 23 24 24

29 27 25 23 23 26 26

25 27 29 26 27 23 23

27 27 27 21 21 23 23
```

Minimize, but do not close, Forty_Two_Cells.txt.

We haven't defined a coordinate system for our raster, so a standard Cartesian system will be assumed. In the text file, the meanings of NCOLS and NROWS are obvious. XLLCENTER is the x-coordinate of the center of the cell in the lower-left corner. YLLCENTER is similar. CELLSIZE is the length of a cell side. The NODATA_VALUE can be set to whatever you want. Obviously, it would be a real mistake to set it as a number that might actually be found in your data. Here, it is set to be able to hold the smallest number that can be represented by a 2-byte (16-bit) integer. (For reference: that number is -32,768.) The numbers following the first six lines are the values the cells are to take on, in "row-major order"—meaning all the values on the first row, followed by all the values on the second row, and so on. The lines in the text file do not have to correspond to the rows in the raster.

(By the way: Only the order of the numbers prescribing the values of the cells is important. That is, the file

```
NCOLS 7

NROWS 6

XLLCENTER 4150

YLLCENTER 1150

CELLSIZE 100

NODATA_VALUE  -32768

21 22

21 23 23 23 23 21 25 25 25 25 25 25 21 25 23

23 23 24 24

29 27 25 23 23 26 26 25 27

29 26 27 23 23 27 27 27 21 21 23 23
```

would work just as well.)

_____ **3.** Start ArcMap with a new blank map. In ArcToolbox locate the ASCII to Raster (Conversion) tool and start it. For the Input ASCII file, browse to Raster_Experiments and select Forty_Two_Cells.txt.

_____ **4.** Click the Output Raster text box. Use Browse to navigate to Raster_Experiments > Ras_Expr. gdb, and type Basic_Raster for the name.)

_____ **5.** The Output data type should be INTEGER. OK the ASCII To Raster window. Watch the lower right of the window to see the progress. When the tool has finished, Basic_Raster will be added to the map.

_____ **6.** Use the mouse cursor to determine the extent of the raster. (If the cursor slips off into the Pan mode, as it seems to sometimes, click Select Elements.) Round your readings off to the nearest 100.

Y_Maximum _____.

X Minimum _____.   X Maximum _____.

Y Minimum _____.

Considering the specification coordinates of the center of the lower-left cell (x=4150 and y=1150) and the cell size (100), does it appear that the raster is in the correct position? _____ Move the cursor to the approximate center of the lower-left cell to verify your conclusion.

_____ **7.** Look at the Table Of Contents (T/C). What is the smallest value in the raster? ____. The largest? _____. Use the Identify tool to check out some cell values. Bring back the textfile window: Forty_Two_Cells.txt. Verify that the values you find in the map are in the appropriate position relative to their locations in the text file that created them—for example, the two cells showing 24 are in the third row, cells 6 and 7. Close Identify. Close the text file.

_____ **8.** Bring up the Layer Properties window of Basic_Raster. Under Symbology > Show, pick Stretched. Examine the window, then click Apply and OK.

This "stretches" values over a color ramp. The default is black to white, so you get a grayscale representation of the values—low (black) to high (white).

_____ **9.** Open the attribute table of Basic_Raster and put it where you can see both it and the map. Using the table, pick a record and highlight it. Verify that the Count field in the selected record in the table represents the number of cells highlighted. How many cells have value 23? _____. Deselect all records and close the attribute table.

It would be nice if each cell could be labeled with its value. That capability is not a part of ArcMap per se, but there is available an unsupported tool, called CellTool, that does the job. It is provided on the DVD that accompanies this book. You can add the CellTool with the following step. (This process assumes that you have the right to add software to your computer. If you don't have that right, contact your instructor or computer system administrator.)

_____ **10.** Using the CD-ROM that came with the book, find the CellTool folder with the operating system of the computer. Navigate to esriCellTool.dll. Copy the DLL to your personal folder ___IGIS-Arc_*YourInitials*. In ArcMap click Customize on the Main menu. Pick Customize Mode to bring up a Customize window. Under the Toolbars tab click Add From File. Navigate to the DLL in your personal folder. Click it. Click Open. Windows may ask if you want to allow the program to make changes to your computer. Say Yes. Windows may ask again. Say Yes again. An Added Objects window with six listed objects will appear. OK this window. Shortly, CellTool will appear, highlighted, among the toolbars. Close the Customize window.

From this point on, CellTool will be available just as any other toolbar through Customize > Toolbars.[1] Add the CellTool toolbar. Looking at ToolTips (or the Status bar), record in the spaces provided the functions of the five icons on the Cell Tool toolbar. See Figure 8-37.

_____

_____

_____

_____

_____

FIGURE 8-37

____ **11.** If you click the Value icon and then an interior cell, you will see the values of that cell and the eight around it. (If you get an error, just OK it and try again.) If you drag a box that includes a number of cells, you will see the values of all of them. Experiment. Click the Clear Graphics icon. Experiment some more. Clear Graphics.

Important note: The Cell tool works only on the raster at the top of the T/C. Should you fail to clear graphics, and then create a new raster, the graphics from the previous raster will remain—showing wrong results for the new raster.

The Raster Calculator—Integer Rasters

____ **12.** Use Search to find the tool Raster Calculator (Spatial Analyst). Write here the path to it. You will be using it a lot.

_____

(Now would also be a good time to make sure the Spatial Analyst extension is turned on: Customize > Extensions > Spatial Analyst.) The Raster Calculator lets us do fundamental operations on rasters. To use it, you put a Map Algebra expression into the calculation box. This expression can be very involved, but in the next step you will place only the name of the raster you have been working with, with the goal of creating a duplicate of that raster with a different name.

____ **13.** Start the Raster Calculator. (You may have to expand the window or make it full screen so you can see it all.) Make a second raster that is identical to the first simply by double-clicking Basic_Raster so that it is placed in Map Algebra expression box. Browse and put the output in

_____

[1]CellTool can be a bit cantankerous. You may have to quit ArcMap. Restart it, add the CellTool toolbar, then add Basic Raster. Even so, the first time you click on a cell, looking for Value, you may get an "Error in CellValueExtent" message. Just click some more.

Ras_Expr.gdb, calling it Bas_Ras2. Click OK and, shortly, Bas_Ras2 will be added to the map when the Raster Calculator is finished. Verify that it is identical to Basic_Raster (despite the fact that different colors are used for the cell values).

Arithmetic Calculation

One of the simplest such operations involving rasters is adding them. When two rasters that represent the same geographic space and have the same cell sizes are added together, each cell of the resulting raster contains the sums of the cells in equivalent locations. In the next step, you will explore the Raster Calculator a bit more and then add the two rasters you have made.

___ **14.** Start the Raster Calculator again. Click Tool Help and look at the Illustration. You can come back to this if you have questions regarding the operation of the Raster Calculator. Close the help window.

___ **15.** You will sum the two rasters to produce a third raster. The expression to do this is

"Basic_Raster" + "Bas_Ras_2"

You can do this in much the way you made expressions earlier: double-click the particular raster you want and single-click the operators or digits. Place this expression in the middle box. Browse and put the output in Ras_Expr.gdb, calling it Bas_Ras_Sum. See Figure 8-38. Click OK. The new raster will be added to the map. Use Identify, the T/C, the CellTool, and the Bas_Ras_ Sum attribute table to convince yourself that each cell contains the correct value.[2]

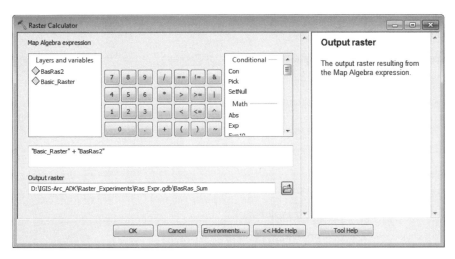

FIGURE 8-38

___

[2] You may have noticed that the raster names aren't particularly consistent when it comes to capitalization of letters. ArcMap may refer to Bas_Ras_Sum, while ArcCatalog may refer to the same raster as bas_ras_sum, Think of the ArcGIS software as an old house that got added onto. You are currently in an older part of the house where capitalization is sort of ignored. Just be content that you don't have to look at everything in all capital letters. (Also, newer versions of the software may hang onto your capitalization; if so, appreciate it.)

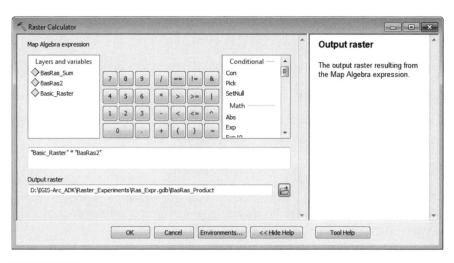

FIGURE 8-39

____ **16.** Use the Raster Calculator again–this time to multiply (use * as the multiplication operator) Basic_Raster by Bas_Ras_2. Call the result Bas_Ras_Product—making sure, as you will for the rest of this exercise, that you put it in Ras_Expr.gdb. See Figure 8-39. Use Identify, the T/C, the CellTool, and the Bas_Ras_Product attribute table to convince yourself that cells contain the product of the individual values in the two input rasters.

Boolean Operations

Rasters can be produced that indicate "truth values" or Boolean[3] values, in cells. A "1" stands for TRUE. A "0" stands for FALSE. Suppose you were interested in knowing which cells in Basic_Raster were greater than or equal to 25.

____ **17.** Calculate Basic_Raster >= 25. "Basic_Raster >= 25" is a statement that is either true or false, correct? The result, name it True_or_False, will have a 1 (for TRUE) in those cells that were 25 or more. A zero will appear in the other cells, as you can see from the T/C, the Identify tool, and the attribute table. (The CellTool may indicate ND for NoData; ignore it.)

The Boolean operators are as follows:

❏   AND (shown in the expression as &),

❏   OR (the inclusive OR, shown in the expression as |),

❏   XOR (the exclusive OR, shown in the expression as ^), and

❏   NOT (shown in the expression as ~).

---

[3]You might wonder why they aren't called Boolian values. It's because the person who developed the concept of Boolean algebra was George Boole. Fortunately, no one wanted to take the responsibility of changing his name, just to conform to standards of English.

The Comparison operators are as follows:

- ❏ Greater than (shown in the expression as >),

- ❏ Less than (shown in the expression as <),

- ❏ Greater than or equal to (shown in the expression as >=),

- ❏ Less than or equal to (shown in the expression as <=),

- ❏ Equal to (shown in the expression as ==),

- ❏ Not equal (shown in the expression as !=),

The Boolean operators may be used in general expressions as well. Zero values are assumed to be FALSE. Nonzero values are considered TRUE.

___ **18.** Subtract 25 from Basic_Raster to make the raster Bas_Ras_minus_25. Look at the attribute table of the result. From the attribute table, how many values of zero are there? _____. If you now use the Boolean operator not (~), you can make all the zeros turn into 1s, and the nonzero values into 0s.

The expression to do this is ~"Bas_Ras_minus_25". Using this expression make a raster named T_F_Swap. Check the raster T_F_Swap to verify that the cells containing zeros of Bas_Ras_minus_25 have become 1s and the nonzero cells have become 0s.

How many values of one are there? _____. Where are the zeros, compared to Bas_Ras_minus_25? _____.

Floating-Point Rasters

If the following math stuff leaves you scratching your head, don't worry about it. The point is that Spatial Analyst can do a lot of esoteric operations on rasters. Just go with the flow, even if the details escape you.

___ **19.** Bring up a new map without saving changes to the former one. Find and start the Create Random Raster (Spatial Analyst) tool in ArcToolbox. For the Output raster, browse to the folder Raster_Experiments > Ras_Expr.gdb and call the new raster Ran_Ras. Save. Leave the Seed value for the pseudorandom number generator blank.[4] For the output cell size, use 100. For the Output extent, use the four numbers you wrote down above (x & y minima, x & y maxima) for Basic_Raster. OK the window. The raster Ran_Ras will be added to the map.

___ **20.** Ran_Ras consists of 42 cells, whose values are approximately uniformly distributed between zero and one. For Symbology, pick a Color Ramp that goes from light to not too dark. Use the

---

[4]You can look on the Web for "pseudorandom number generator" to see several hundred thousand Web pages. The seed is the number that starts the process off. If you don't specify the seed, it is taken from the computer's clock, which counts the number of seconds that have elapsed since the January 1, 1970. Computers cannot actually generate true random numbers, as you would (for all we know) if you repeatedly threw a die. Computers are completely deterministic, so with a given seed and a given program you will always get the same sequence of random digits. Hence the word "pseudo."

Identify tool to examine four values (to three significant digits). _____, _____, _____, _____. Use the CellTool to light up the whole raster with values.

____ 21. Ran_Ras contains numbers in the range zero to one. Suppose that instead you wanted to have integer numbers between 1 and 10 put in the cells. You could use the Int function (make into an integer) in the Raster Calculator on Ran_Ras. You will find Int in the box to the right of the Boolean buttons. Place this expression in the Raster Calculator:

Int(("Ran_Ras" * 10) +1)

Put the output, Ran_Ras_1_to_10, in Ras_Expr.gdb. This produces an Integer raster, so it has an attribute table. How many values of "1" did you get? _____. How many of "7"? _____. Based on the digits from 1 to 10 being evenly distributed over 42 cells, about how many of each digit would you expect in the raster? _____. Check it out with the attribute table.

A second floating-point raster demonstration: The cosine, say, x, of an angle between zero and a right angle is a number between 1 and 0. The arc cosine function finds the angle whose cosine is x. So we could use our Ran_Ras raster (which you recall is bounded by zero and one) to make a raster of random angles, using the Acos function.

____ 22. Start the Raster Calculator. Find the Acos Trigonometric function in the list of the Raster Calculator.

____ 23. Create the expression

Acos("Ran_Ras")

Name the output raster Random_Angles and put it . . . you know where by this time. The floating-point raster is added to the map. Check out the results. You may have expected to see angles between 0 and 90. But this function, which you will discover if you check the Help file, produces resulting angles expressed in radians, not degrees. There are 2*pi radians in the 360 degrees of a full circle. So one radian is about 57.296 degrees. Thus you get values bounded by about zero and pi/2 (about 1.57).

____ 24. To convert the Angles raster to degrees (let's say integer degrees), use the Raster Calculator one more time:

Int("Random_Angles" * 57.296)

Make the new raster called Random_Angles_Degrees. Now check out the values. You are back to an Integer raster, so you have an attribute table. What's the smallest? _____ The largest? _____

You now have a grasp (or at least an inkling) of how both integer rasters and floating-point rasters operate. You could play indefinitely with the immense number of functions and rasters. Better that you go on to Project 8-2, where you solve the Wildcat Boat problem with rasters.

Solving the Original Wildcat Boat Problem with Rasters

The raster data model has some unique strengths. It is particularly suited to modeling the environment. Its advantages are that each area (cell) is homogeneous and is the same size and shape. Further, each area has four nearest neighbors and four next-nearest neighbors. As computers become more powerful—much larger memories and much greater speeds—raster processing is becoming a tool that started with cell sizes measured in miles to those measured in feet (and not many of those).
Raster data modeling trades geographic specificity for

❑   Speed

❑   Storage economy

❑   Simplicity

❑   Large group of tools

I want to introduce raster processing using the problem we solved in the last chapter regarding the location of the Wildcat Boat facility. This particular problem is probably better solved with vector processing, but solving it using raster processing techniques is very instructive.
We'll start by setting up a folder for your work and then converting the relevant geodatabase feature classes to rasters.

_____ **1.** Start ArcCatalog. Find the folder named

[___] IGIS-Arc\Spatial_Analyst_Data

Highlight it and choose Edit > Copy. Find

___ IGIS-Arc_*YourInitials*

and highlight it. Choose Edit > Paste. Make a folder connection to

___ IGIS-Arc_*YourInitials*

\Spatial_Analyst_Data\WC_Boat_SA.

Expand WC_Boad_SA completely to see the five feature classes that you are familiar with.

Setting the General and Raster Environment

When you are using vector feature datasets the bounding rectangle of the data pretty much takes care of itself, once you set the coordinate system. Not so with raster data. The raster is a rectangle, and when you combine rasters, you want them to represent the same area. So, we will set the output extent—in this case to the extent and projection of the vector feature class Soils.

_____ **2.** Start ArcMap with a Blank Map. Dismiss ArcToolbox, if it is present. Add as data

___ IGIS-Arc_*YourInitials*\Spatial_Analyst_Data\

WC_Boat_SA\Wildcat_Boat_SA.mdb\Area_Features\Soils

Choose Geoprocessing > Environments > Workspace. Make both the Current Workspace and the Scratch Workspace

___ IGIS-Arc_*YourInitials*

____ **3.** For the Output Coordinates, browse to

___ IGIS-Arc_*YourInitials*\Spatial_Analyst_Data\

WC_Boat_SA\Wildcat_Boat_SA.mdb\Area_Features\Soils

The Output Coordinate System should be set to "As Specified Below". For the coordinate system browse to Projected Coordinate Systems > UTM > WGS 1984 > Northern Hemisphere > UTM_Zone_18. Scroll down to Processing Extent and again browse to Soils. The four bounding values should appear. Write them here (to the nearest one-tenth meter).

Top _____   Bottom _____

Left _____   Right _____

____ **4.** Scroll down further and click Raster Analysis. Click the Cell Size drop-down menu and pick "As Specified Below". Type in 2.0. OK the Environment Settings window.

____ **5.** Save the map, which you will call Preserve_Environments.mxd, in

___IGIS-Arc_*YourInitials*\Spatial_Analyst_Data\WC_Boat_SA

(If you quit ArcMap without saving, the Environment settings may go away. Saving a map ensures the preservation of its Environment settings.)

____ **6.** Quit ArcMap. Restart ArcMap, opening Preserve_Environments.mxd from Existing Maps > Recent. Recheck the Environment Settings. If they seem alright cancel the Environment Settings window.

Converting Features to Rasters

The next task will be to convert the vector-based feature classes to raster-based geodatasets.

____ **7.** Using List By Source, verify that the T/C shows Soils as data from

___IGIS-Arc_*YourInitials*\

\Spatial_Analyst_Data\WC_Boat_SA

\Wildcat_Boat_SA.mdb\Area_Features\Soils

8. In ArcToolbox go to Conversion Tools > To Raster > Feature to Raster and run the tool. Select Soils from the drop-down menu. For Field, pick SUIT. If the Output cell size is not 10, make it 10. For the Output raster, browse to

___IGIS-Arc_*YourInitials*\Spatial_Analyst_Data\WC_Boat_SA\Wildcat_Boat_SA.mdb

For the name type Soils_rr10m (rr being our designation for raster resolution; 10m meaning 10 meters). Click Save. Check the Feature to Raster window and click OK if all the boxes say what they should. The status bar will show the progress and a popup window will tell you when the process is done.

Creating a raster in this way will put a raster dataset in the folder you specified and also add a layer to your map. By clicking List By Source in the T/C you can note that the raster dataset goes into the personal geodatabase Wildcat_Boat_SA.mdb.

9. Go back to List By Drawing Order and, if Soils isn't already the topmost layer, make it so. The raster layer Soils_rr10m should be just below it. Make the Soils layer Hollow or No Color, so you can see both the raster, and also the polygon boundaries. Zoom in on the northernmost point of land in the northwest quadrant of the map. See Figure 8-40.

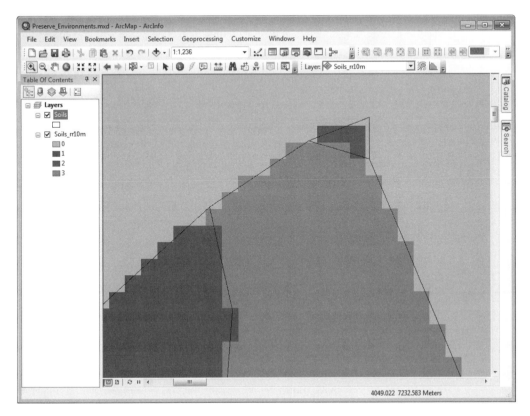

FIGURE 8-40

From this you can get a pretty good idea of the rule for assigning a value to a cell: The cell gets the value of whatever polygon exists at the center of the cell.

_____ **10.** Zoom in some more and measure the length of a side of a cell. _____ meters.

_____ **11.** Double-click the Soils_rr10m name in the T/C and choose Source in the Layer Properties window. How many columns and rows are there? _____. _____. Therefore, how many cells? _____. What is the Cellsize (X,Y)? _____, _____. What is the Uncompressed Size of the raster? _____KB. What is the Format? _____. Scroll down through the list to get an idea of the information you can get about a raster dataset. Click OK.

_____ **12.** Repeat the creation of a Soils raster dataset with the same parameters, but this time use a cell size of 5 meters. Call the dataset Soils_rr5m.

_____ **13.** From Layer Properties: How many columns and rows are there in Soils_rr5m? _____. _____. What is the Cellsize (X,Y)? _____, _____. What is the Uncompressed Size of the raster? _____KB. Notice that this is about four times as large as Soils_rr10m. Twice as many rows and twice as many columns implies four times as many cells. How many cells in Soils_rr5m? _____ Click OK.

_____ **14.** Compare Soils_rr5m with the polygon feature class by flipping Soils_rr5m off and on. Note how this change to five meters in resolution results in an improvement in precision. You can see that the edges of the squares of the raster more nearly conform to the polygon boundaries. So, the area covered by the raster is more nearly congruent to the polygons than at the 10-meter resolution. To see this clearly, flip between the 5-meter layer and the 10 meter layer at different levels of zoom.

Rasters seem to work best in ArcGIS versions 10.0 and 10.1 when they are inside a personal or file geodatabase. In earlier versions of the software, rasters resided in their own folders. In order to show yourself an interesting fact about data compression you will make two additional rasters: Soils_rr10m_X and Soils_rr5m_X. What's different about these rasters is that you will place them in

_____IGIS-Arc_*YourInitials*\Spatial_Analyst_Data\WC_Boat_SA

rather than in a geodatabase. We do this so we can look at their compressed sizes—something we cannot do when they are cloaked inside a geodatabase.

_____ **15.** Make the two rasters described above (one of 10 meter cell size, the other of 5 meter cell size) being careful to put them directly in WC_Boat_SA. Minimize ArcMap and, using Windows Explorer, navigate to

_____IGIS-Arc_*YourInitials*\Spatial_Analyst_Data\WC_Boat_SA.

Right-click the folder Soils_rr10m_X and, with Properties > General, determine the Size of the contents of the *folder*. It is _____ kilobytes (KB). Now determine the Size of Soils_rr5m_X. _____KB. Close the window.

What is interesting about the comparison of the sizes of these two rasters is that, while the 5-meter resolution raster contains four times as many cells as the 10-meter resolution one, the dataset is not four times as large (which would be 300 percent larger). In fact, it is only about 100 percent larger. So, it is not the case that doubling the resolution quadruples the compressed dataset size—far from it! This happy occurrence is the result of the data compression techniques discussed in Chapter 4, which take advantage of the fact that, with smaller cell sizes the cell next door to a given cell likely has the same value.

_____ **16.** Restore ArcMap (The title should still be Preserve_Environments.mxd) and remove Soils_rr10m_X and Soils_rr5m_X from the map, since they have illustrated their point. Since none of these datasets we have been working with is very big, and they didn't take very long to create, let's increase the resolution—say, to 2 meters. Make Soils_rr2m, being careful to put it inside the mdb geodatabase.

_____ **17.** Add the feature class Landuse[5] from

___IGIS-Arc_YourInitials\Spatial_Analyst_Data

\WC_Boat_SA\Wildcat_Boat_SA.mdb\Area_Features)

to the map. Using 2-meter resolution, make a raster dataset of Landuse, called Landuse_rr2m inside the geodatabase. Use LU-CODE as the field to make the "value variable" of the raster. Zoom to the extent of Landuse_rr2m. Turn off all layers except the Landuse pair. Zoom to the extent of the Landuse Layer. Make the Landuse Symbol No Color or Hollow. By zooming in, look at how closely the cells approximate the Landuse boundaries.

_____ **18.** In ArcMap open the attribute table of Landuse_rr2m. How many cells of value 100 are there? _____. How many square meters does each cell represent? _____. So, how much area, in square meters, is occupied by the land use with LU_CODE equal to 100? _____.

_____ **19.** Just as a check, open the attribute table of the feature class Landuse. Click Table Options. Using Select By Attributes, select those records with LU_CODE = 100. Show only selected records. Run statistics on Shape_Area. What is the sum to the nearest square meter? _____. Compare this number with the area you calculated in the previous step. Feel encouraged by the reasonably close agreement between the two areas. Clear the selected features and records.

The results of the calculations regarding area should be confirming, even though they are not exactly equal. To say it for the umpteenth time: All GIS representation, whether vector or raster, is an approximation of reality.

_____ **20.** Save the map as WC_Boat_Area_Map.mxd in WC_Boat_SA. Dismiss ArcMap.

Next, you will convert (create rasters for) the feature classes Roads, Streams, and Sewers. To do this properly, we should first verify the Environment variables, so that we can be sure the extents of all the datasets will be the same.

---

[5]We called this dataset Landcover when we worked with these datasets before. The terms land use and land cover are technically not the same, but are often used interchangeably.

Creating Rasters with Linear Features

We will create rasters for the line feature classes—Sewers, Streams, and Roads—in the same way as above, with ArcToolbox. Actually, we don't have to load the feature class into ArcMap. Also, we can easily set or check environment variables, important for determining the extent of a raster.

As mentioned, the extent of a vector dataset is pretty much determined by the data. However, raster datasets are rectangles, and if you convert a vector dataset to raster, it is important to set the extents to the same as other datasets of the study area. Since the soils feature class covers the entire study area (and the sewers dataset, for example, does not), we earlier set the extent of rasters that we were to create to that of the more extensive soils dataset. Below, we check just to make sure.

____ 21. Start ArcMap with WC_Boat_Area_Map.mxd (Hints: Start ArcMap and find the map under Existing Maps > Recent. Or, with the operating system, go to the folder WC_Boat_SA and double-click the map entry.) Open ArcToolbox. You will use the Polyline to Raster tool. Where is the tool in the ArcToolbox tree?

_____ > _____ > _____.

Double-click the tool. In the Polyline to Raster window that opens, you find a text box for Input Features. Browse to

___IGIS-Arc_*YourInitials*\Spatial_Analyst_Data

\WC_Boat_SA\Wildcat_Boat_SA.mdb\Line_Features\Sewers

and click Add. For Field, choose DIAMETER. (You may recall that the two diameters were 45 and 60.) Browse to make the Output raster Sewers_rr2m in

___IGIS-Arc_*YourInitials*\Spatial_Analyst_Data\WC_Boat_SA mdb\

For the Output Cell Size use 2.0. Do not OK the window yet.

____ 22. Press the Environments button. Check that the Environment Settings are as you specified earlier. If not, reset them, as described earlier. OK the Environment Settings window. OK the Feature to Raster window. The raster Sewer_rr2m will show up in the T/C and the map.

____ 23. Add the Sewers feature class, and make it a bright line of width 1. Make sure every other layer is turned off except the two that relate to sewers. Zoom to the extent of the feature layer Sewers. Look at it on top of Sewers_rr2m. Zoom way in on a place where the sewer lines intersect. Note how a raster dataset represents lines. Use its attribute table to determine how many cells it took to represent the 45-inch pipe. _____. The 60-inch pipe. _____.

____ 24. Using the same approach as previously, make a raster of Streams, called Streams_rr2m. For Field (which will become Value in the attribute table), pick STRM_CODE and use 2.0 again for Output Cell Size. From the attribute table of Streams_rr2m, what are the two values that were put in from STRM_CODE? _____, _____. (The Stream Code is an indicator of the size of the stream.)

____ 25. Use the same approach a third time to make Roads_rr2m, using RD_CODE.

Look at the attribute tables of each of the three raster datasets. What are the numbers you find for Value and Count?

	Values	Counts
Roads_rr2m	_____	_____
Sewers_rr2	_____	_____
Streams_rr2m	_____	_____

Buffering with Spatial Analyst (Maybe)

Recall that the site for the Wildcat_Boat facility has to be within 300 meters of a sewer line, and the site must be at least 20 meters away from any stream. Both the sewer lines and the streams are represented in raster by square cells that are 2 meters on a side, connected either at a side or a corner. To make a raster-style buffer representation of these linear entries, we need to fatten up the zones. There is a Spatial Analyst tool, Expand, which is reputed to be the raster equivalent of the vector procedure Buffer. You can find it at ArcToolbox > Spatial Analyst Tools > Generalization.

The problem with Expand is that it adds on the number of cells regardless of whether they touch side to side or corner to corner. If they touch side to side, and you have specified a cell size of 2 meters, you get 2 meters, on each side. If they touch corner to corner, however, as they would when buffering a diagonal line, the increment is about 2.828 meters (the length of the diagonal of the square which is twice the square root of 2). So, the moral of this story is that Expand is not a good equivalent for making a buffer. You can expand the width of Streams_rr2m by 10 cells to verify this fact. You'll find that when a line runs horizontally or vertically, the distance from the center of the stream to the edge of the raster zone is about 21 meters. When the line is diagonal, the "buffer" is more like 29 meters. Please see Figure 8-41, noting the difference in "buffer" width between the diagonal line and the vertical line.

FIGURE 8-41

Buffering—Plan B

ArcGIS provides you the ability to slide easily from one data model to another. Let's take advantage of the fact that you created vector buffers of both sewers and streams when you solved the Wildcat Boat problem in the last chapter.

_____ **26.** Start ArcMap with a Blank Map. Add Streams_buf from

___IGIS-Arc_YourInitials_\Wildcat_Boat_Data\Wildcat_Boat.mdb\Area_Features

to the map. Look at the Attribute Table. You should see 12 records. In each record the value of the field NoBuild will be 20, indicating the number of meters adjacent to the stream where building is not allowed. This number will become zone value in the attribute table of the raster you are about to build.

_____ **27.** Using the tool Polygon_to_Raster

(where is it? _____),

start to make a new raster named Streams_buf_rr2m from the Streams_buf feature class. The location you want to put the raster in is

___IGIS-Arc__YourInitials_\Spatial_Analyst_Data\WC_Boat_SA\Wildcat_Boat_SA.mdb.

Make the cell size 2.0 and the Value field NoBuild. Click OK.

_____ **28.** The raster will be added to the map. Check out the raster's attribute table. Notice the Value field. It will indicate "20". What is count? _____. Zoom way in on the northwest stream. There is a raster beneath the Streams_buf layer. Make the symbol for Streams_buf Hollow or No Color to see that the raster conforms (mostly) to Streams_buf. Enhance this image by using the Symbol Selector of Streams_buf to make the outline of Streams_buf red of width 1. See Figure 8-42.

_____ **29.** Using the same procedure as in the two steps above, add Sewers_buf. For the Field value, use Build. The Cell size should be 2.0. Call the raster Sewers_rr2m.

_____ **30.** The raster will be added to the map. Check out its attribute table. What is the value and count? _____, _____. From Wildcat_Boat_SA add to the map the rasters that you previously made: Landuse_rr2m and Soils_rr2m. Rearrange the T/C, though you won't be able to see everything you can make things somewhat better.

Reclassifying the Data

When we solved the Wildcat Boat problem with vector-based GIS, we combined landuse, soils, the sewer buffer, and the stream buffer. After some manipulation we will now do the same with the rasters. We have

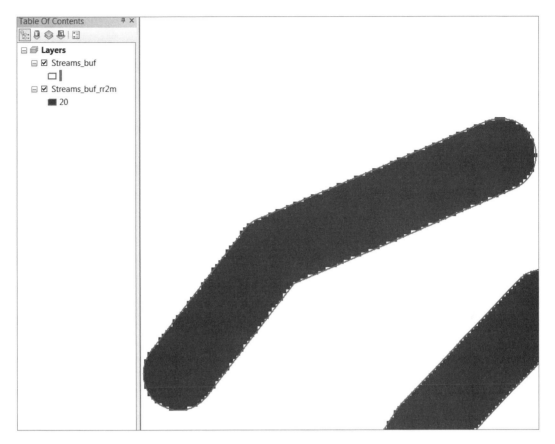

FIGURE 8-42

Landuse_rr2m

Soils_rr2m

Sewer_buf_rr2m

Stream_buf_rr2m

The strategy will be to reclassify the values of each of these rasters. We'll make the value "one" if the zone meets the requirement, and zero or NODATA otherwise.

For Landuse_rr2m, this means changing all the values to "0," except for landuse type code "300," which we change to "1."

For Soils_rr2m, we will change values of "0" and "1" (both of which indicate unsuitability) to "0." We will change the values "2" and "3" to "1."

For Sewer_buf_rr2m, we will replace the numbers of meters within which building is allowed (300) with "1," leaving NODATA alone.

For Stream_buf_rr2m, the matter is a bit different. The values of 20 we want to replace with NODATA, while we give the current NODATA values a "1."

Once we have done this, we merely add the four rasters together, so that the sum at each location is a 0, 1, 2, 3, or 4. From the way we constructed the reclassification, a cell must have a 4 in it to be eligible to be a part of the areas that would allow the Wildcat Boat facility.

_____ **31.** Using Search, find and run the Reclassify (Spatial Analyst) tool. For the Input raster, browse to Landuse_rr2m in Wildcat_Boat_SA.mdb (or select it from the drop-down menu, which is easier). Set the Reclass field to Value. Press the Unique button, so all the values of Landuse are shown. The Reclassify window shows columns titled Old Values (100, 200, and so on) and suggested New Values (0, 1, 2, and so on). Click the New Value opposite 100 and make sure it is 0. For 200, 400, 500, and 700 also supply a 0. These are areas not allowed for building. For the value opposite 300, put in a 1. For the name of the Output raster, browse to

___IGIS-Arc_*YourInitials*\Spatial_Analyst_Data\WC_Boat_SA\Wild_Cat_Boat_SA.mdb

and put in the name rcl_Landu. See Figure 8-43. Click Save. Double check that the tool hasn't changed your classifications (as it has a habit of sometimes doing). If everything is correct click OK. Once the raster is added to the map you can see the areas that meet the landuse requirement in one color (the T/C indicates 1), and those that don't in another color.

_____ **32.** Reclassify Soils_rr2m, calling the resulting raster rcl_Soils. Again, you may note that some of the categories of soil suitabilities are shown as ranges—such as 2–3. You want each category to have a unique value, so punch that button. Change the New values opposite Old values of 0 and 1 (which indicate unsuitability) to 0, and the values of 2 and 3 to 1. Click OK.

_____ **33.** Start to reclassify Sewers_buf_rr2m. You want _____ for 300 and zero for NoData. Make the name rcl_Sewers_buf, and click OK.

_____ **34.** Finally, reclassify Streams_buf_rr2m, so that 20 becomes 0, and NoData becomes 1. Call the result rcl_Streams_buf. Click OK.

FIGURE 8-43

_____ **35.** Use the Identify tool and the attribute tables to look at each of the four reclassified datasets: rcl_Soils, rcl_Landu, rcl_Sewers_buf, and rcl_Streams_buf. Make sure that the Values are 0 or 1, and that they indicate the correct places.

Adding the Rasters with the Raster Calculator

Now you may perform addition on the four rasters whose names begin with rcl. You will use the Raster Calculator.

_____ **36.** Start the Raster Calculator. By double-clicking the raster names and single-clicking the plus (+) operator, create the expression that adds together the rasters

rcl_soils,

rcl_landu,

rcl_Sewers_buf, and

rcl_Streams_buf

shown in Figure 8-44. Click Evaluate. Call the raster SUM_LuSoStSw and save it in the Wildcat_Boat_SA personal geodatabase. It will appear in the T/C and on the map. What values are produced? _____, _____, _____, _____. What value indicates those areas suitable in all respects? _____ Check it out with the Identify tool and the attribute table. In the T/C, change the colors of zones 1, 2, and 3 to light pink. Make zone 4 a dark red. The map should look a lot like the first solution to the Wildcat Boat problem that you developed in Chapter 7, but without the 2000 square meter criterion.

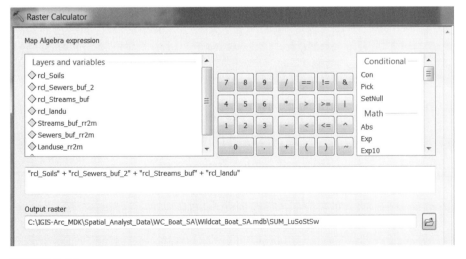

FIGURE 8-44

Converting Zones to Regions to Find Individual Sites

You noticed from the attribute table that only four numbers appear in the Value column. This means that all the cells of the areas we are interested in (indicated by Value = 4) reside together in a single zone. A *zone* consists of cells that all have the same value. So, the areas that are candidates for the Wildcat Boat facility are all connected in a way—they are part of a single zone. You may recall that a raster *region* consists of one or more sets of disconnected cells that are in the same zone. That is, for a given zone, regions are clumps of cells that are completely disconnected from other cells in the zone. So it is the regions of zone 4 that we want as distinct entities that will form the individual sites we need. We can form these using the Region Group tool in ArcToolbox.

_____ **37.** Locate the Region Group tool in ArcToolbox. Where is it?

_____ _____ _____ _____

_____ **38.** While you are in the Spatial Analyst Tools toolbox, count the number of sub-toolboxes in the Spatial Analyst toolbox. _____. Expand a few of the sub-toolboxes and note the number of tools. This is to say that this is a *big* chunk of software, all by itself.

Recall that the term "connected cells" can have two meanings. Cells can be connected by their sides, in which case each cell has up to four neighbors. Or cells can be considered connected if either their sides or corners touch, in which case each cell has up to eight neighbors.

The tool lets you add a LINK field to the attribute table of the new raster. The link value is the old zone value. You will need that, since the candidate areas are all of zone 4 and you need to be able to distinguish them in the resulting raster attribute table.

_____ **40.** For number of neighbors to use, select EIGHT. Leave Add Link Field To Output checked. Click OK.

_____ **41.** RG_WC_Boat will be added to the map. It consists of dozens of distinct zones, formed by all the regions of all the zones in SUM_LuSoStSw. Observe the map. Open the attribute table of RG_WC_BOAT. Notice, by looking at Count, that some of these zones are quite small; remember that each cell contains only 4 square meters. Notice that the Link value in each record is 1, 2, 3, or 4. Sort the table by LINK value, descending. Select the records with LINK value 4. Observe the map.

As a last step, to isolate the sites, you will use the ArcToolbox command Extract by Attributes. You will want the Link value to be 4. You will also want the areas to be at least 2000 square meters. How many cells would 2000 square meters be? _____

_____ **42.** Clear any selections you may have made and close the attribute table. Locate the ArcToolbox command Extract by Attributes. Where is it?

_____._____._____

_____ **43.** Start the tool. The Input raster should be RG_WC_Boat. Browse to make the Output raster FinalSites_SA in Wildcat_Boat_SA.mdb . For the Where clause, press the SQL (Structured Query Language) button and construct the following expression:

"LINK" = 4 AND "Count" >= 500

and click OK in Query Builder window.

FIGURE 8-45

___ **44.** Verify that the Extract By Attributes window looks like Figure 8-45 and click OK.

Turn off all layers except FinalSites_SA. Examine the sites and compare them with those sites you found using the vector approach in Chapter 7. How many sites are there? _____ Verify with the attribute table. Then add as data the Streams feature class from

___IGIS-Arc_*YourInitials*\Spatial_Analyst_Data

\WC_Boat_SA\Wildcat_Boat_SA.mdb\Line_Features

Making the color blue with width 1.

Also add as data the Sewers_buf polygon feature class from

___IGIS-Arc_*YourInitials*\Spatial_Analyst_Data

\WC_Boat_SA\Wildcat_Boat_SA.mdb\Area_Features

making its color hollow and its outline width 2.

Save the map as Final_Sites_SA.mxd in the folder WC_Boat_SA.

___ **45.** As you wish, change the symbology—perhaps even the content of the map—and switch to Layout View. Add other map components such as title, legend, and so on, as you wish. Print the layout.

Solving a Wildcat Boat Problem with Different Requirements

Raster processing, by its nature, allows a lot of flexibility. Suppose that the Wildcat Boat facility requirements could be varied somewhat. Perhaps the 300-meter rule might be waived if the builders are willing to put in a septic tank. This is not as desirable a solution, but it is a possibility. Suppose also that agricultural land is deemed as okay, and forested land is no longer absolutely forbidden, but not encouraged. As for soils, you would like to favor the very suitable soils (type 3) over the moderately suitable soils (type 2). The streams requirement is still firm, however.

When you solved the Wildcat Boat problem with vector data, the study area was partitioned into two categories: where building could take place and where it couldn't. What you could do easily now, with raster processing, is to display areas of greater or lesser degrees of suitability. You do this by going back to reclassification. Keep in mind that you are going to combine the values of the zones. In the last exercise you simply added the four rasters together (each Value variable being either 0 or 1) and took only those zones with value 4 as the sum that was acceptable. But consider the flexibility you have with raster processing and reclassification. Through reclassification you could make the most desirable areas have large sums, while the prohibited areas have small or even negative sums.

For example, you could reclassify the sewers—making the value 10 for areas within the 300 meters, and 5 for areas outside that distance, since it's now permissible to build outside the defined area, but not as desirable.

Regarding Landuse, you could make brushland (code 300) and agricultural land (code 200) have value 10—because they are now both acceptable—but forested land (code 400), which is acceptable but less desirable, have a value of only 7. For other, prohibited Landuse codes—100 (urban), 500 (water), and 700 (barren)—you could reclassify those areas with a negative number (say, −100), so the zone would show a negative value after raster addition takes place.

Regarding Soils, you could use 6 for moderately suitable (code 2) and 10 for very suitable (code 3). What about the prohibition regarding building on unsuitable soil? Again you could reclassify those areas with a large negative number so the zone would show a negative value after raster addition takes place.

What about the streams requirement? You could reclassify the NOBUILD area as negative 100, while the permissible areas could have value 10.

If you added these four reclassified rasters, you could have zones with value 40 for the most suitable, and smaller values for less suitable areas. For example, a zone depicting an area outside the 300-meter sewer zone (reclassified as 5), on forested land (reclassified as 7) that was only moderately suitable for building (reclassified as 6) would be 28. (You would know that any area that came out negative was prohibited.)

___ **31.** A summary of the reclassifications and suggested names for the new rasters are provided below. Put the new rasters in

___IGIS-Arc_*YourInitials*\Spatial_Analyst_Data

\WC_Boat_SA\Wild_Cat_Boat_SA.mdb

Reclassify Sewers_buf_rr2m to make rcl_version2_Sewers_buf

OLD VALUES		NEW VALUES
300	→	10
NoData	→	5

Reclassify Landuse_rr2m to make rcl_version2_Landuse.

100	→	−100
200	→	10
300	→	10
400	→	7
500	→	−100
700	→	−100
NoData	→	NoData

Reclassify Soils_rr2m to make rcl_version2_Soils.

0	→	−100
1	→	−100
2	→	6
3	→	10
NoData	→	NoData

Reclassify Streams_buf_rr2m to make rcl_version2_Streams_buf.

20	→	−100
NoData	→	10

2. Add these four rasters together, as before, to make Final_Sites_version2, and examine the results. Open the attribute table and sort descending according to Value. The most desirable sites should appear at the top. The map looks very different from the ones you made previously, which showed only a few permissible sites. Select the records with Value 40 to see the best sites, under these new criteria. Then look at the Value 37 sites. (Of course, you would have to run Region Group to get the individual sites.)

As a further demonstration of flexibility, you could make a version3 of the final sites in which the land use issue was twice as important as the soils or sewers issue, so you could weight that variable. Your function might look like this:

rcl_version2_Soils +

(2 * rcl_version2_Landuse) +

rcl_version2_Sewers_buf +

rcl_version2_Streams_buf

You would, of course, have to rethink the maximum and other values to form conclusions as to what was ultimately desirable and what wasn't.

_____ **3.** (Optional) Investigate creating a map in which the land use variable is twice as important as the other variables. Or choose some other combination of rasters to produce a map that shows varying degrees of suitability. Or go on to Exercise 8-4 where you will learn about making rasters that represent surfaces that are based on values as specific points.

Exercise 8-4 (Demonstration)

Making Surfaces with IDW, Spline, Trend, Nearest Neighbor, and Kriging

I will now introduce a very big subject with a very small exercise. The subject is basically this: Given a finite set of points with known values, say, "z" values (e.g., altitudes, pressures, pollution levels), what values could reasonably be assigned to the remaining (infinite number) of points that are in the same area? That is, how would one interpolate between the known points?

This is an important area, because all we can really measure with a high degree of precision are conditions at a point. Think about altitude, temperature . . . . We can be very sure what the situation is at a given point (in space and time). We have to infer what conditions exist in the vicinity of that point. So what follows is a quick and very cursory look at the tools in ArcGIS that let you do that. An entire course could be built around the techniques these tools represent and the statistics involved. The Help files are some help, but to make real use of these tools, you need to do a lot of reading, or engage a statistician.

(You have, of course, seen one approach to developing a continuous surface from a set of points: TIN. What is different about these surfaces you are about to explore is that they have no sharp line breaks—that is, they are differentiable, in the calculus use of the word—and they are represented by continuous mathematical functions.)

_____ **1.** Start ArcCatalog. In ___IGIS-Arc_*YourInitials* make a folder named Experiment_with_ Interpolation. Make a folder connection. (This might also be a good time to clear up previous folder connections that you have accumulated but no longer need.) Within this folder make a file geodatabase named Surfaces.gdb.

_____ **2.** From ___IGIS-Arc_*YourInitials*\Trivial_GIS_Datasets, copy over, to the Experiment_with_ Interpolation folder, three shapefiles:

Known_Altitudes.shp

Known_Populations.shp

Only_points.shp

_____ **3.** Start ArcMap with a Blank map and add as data Known_Altitudes.shp.

_____ **4.** *Consider the shapefile Known_Altitudes.shp:* Make the map units and the display units Meters. Look at the attribute table. Each number is the altitude in hundreds of feet of the particular point. What is the range of altitudes: Lowest? _____. Highest? _____. Label the map using Altitude as the label field; use a bright green color with size 16. Very roughly, on average, how far from each point is its nearest neighbor? _____ meters. How would you characterize the lay of the land if the numbers represented hundreds of feet of altitude. High areas, low areas?

_____.

Close the attribute table if it is open.

The question now is, interpolating from these known altitudes that are at known positions, what are the altitudes at other nearby points we might randomly select. There are a number of tools that use different statistical techniques that can make very good, educated guesses at the answers to this question. No technique answers the question with great accuracy—which may be why there are so many approaches.

____ **5.** Display ArcToolbox. Choose Spatial Analyst Tools > Interpolation. Disregarding the Topo to Raster tools (read about them if you want), list all tools in your Fast Facts File.

Done? Yes___. No ___.

____ **6.** We'll start by using the Spline tool. Pounce on Spline. Read the Help panel. Click Show Help > Tool Help. Click Learn more about how spline works. Read the Summary and Usage sections. (You are welcome to read more, but, without statistical and calculus knowledge you are likely to just go: "huh?") Close the Help window. Click the down arrow on Input Point Features and choose Known_Altitudes. For the Z value field, use Altitude. Attempt to put the Output raster directly in the file folder

___IGIS-Arc_*YourInitials*\Experiment_with_Interpolation

and call it simply Spline_surface. You will probably get an error. Click on the red "X". You will be told that the name you used is longer than 13 characters. Note down that this is the maximum length of a raster name that goes directly into a folder. Now attempt to put Spline_surface into the file geodatabase Surfaces.gdb. That should work. Make the output cell size 5.0. Leave the rest of the text boxes as they are—after reflecting that REGULARIZED, Weight, and Number Of Points are options that would affect the surface you are about to create. (Again, all the tools, with their parameters, are intended to make guesses at the values between the points. Those guesses are affected by the options chosen by the user.) Click OK.

____ **7.** The raster Spline_surface has now been added to the map. If necessary, move the point feature class to the top of the T/C, so you can see the point locations and the associated elevations. From the T/C, what are the values related to the range of altitudes that the generated surface exhibits (to the nearest tenth)? Lowest _____. Highest _____. Differences between displayed altitudes? _____. What classification method would you say was being used to divide the results into classes? _____ (Hint: use subtraction.) Contrast what you see with your impressions of what the points implied about the surface before you saw the raster.

____ **8.** *Examine the Properties of the raster:* Turn on MapTips for the raster (under the Display tab). Move the Select Elements cursor around the map. (Actually, these rasters don't look like rasters initially (no jags) because ArcMap displays them smoothed out, but they are rasters, as you can see from the Properties > Source. If you select the Display tab and pick "Resample during display using: Nearest Neighbor" you can see the traditional raster appearance.)

____ 9. Start the Identify tool (specifying Spline_surface) and check some cell (Pixel) values. Check close to points and between points. Recall from the Help file that the generated surface is created to pass through all the points. Also notice that you get a "Class Value." This number comes from the T/C. What is the smallest class number? _____ The largest? _____Turn off Map Tips.

____ 10. *For all the steps that follow, use a cell size of 5.0:* Put each result into the Surfaces.gdb file geodatabase. Always move the point feature class to the top of the T/C, above all the rasters if necessary to see it.

____ 11. Create a surface called IDW_surface using the IDW (Inverse Distance Weighted) tool, after reading the first three paragraphs of the help file. By flipping IDW_surface off and on, you can compare it with the results of Spline_surface. You see a fair amount of difference. Eyeball a location between two of the known points and get its altitude with Spline_surface. _____. Then, pick approximately the same point with the IDW surface. _____. You see why I use words like "guess" and "reasonable estimate" to describe the outputs of these tools.

____ 12. Create a surface, called Ntrl_Ngbrs_surface, with the Natural Neighbors tool. Read the Usage Tips. Explore. What is one obvious difference between Ntrl_Ngbrs_surface and the other two presentations? (Hint: Look at the high points with the Identify tool set to Ntrl_ Ngbrs_surface.) _____.

____ 13. Create a surface with the Trend tool. Call it Trend1_surface. Your immediate impression may be that something went wrong, because you just get stripes! What has happened here is that the process fitted a plane, a flat surface, through the space occupied by the set of points. The angle of the plane was set so that the sum of the squared distances between the plane and the points was minimized. If you identify cells along a given stripe, you'll see what I mean. You will be able to see the plane better if you change the Symbology to Stretched.

____ 14. Create another surface with the Trend tool, calling it Trend2_surface: The difference? You'll put in 2 for the Polynomial Order instead of the default value of 1. This allows for a surface whose defining equation can contain squared terms, rather than just linear ones. Again go to Stretched Symbology to view it. Still, the surface makes no pretense of going through the points.

____ 15. Make Trend3_surface, with the value 3 for the polynomial order. The surface is generated by a cubic equation. You could also make a Trend4 surface. If you try to make Trend5 you will be told that there aren't enough points (for a polynomial of the 5th order).

____ 16. Make a surface with the Kriging tool[6] (Call it Surface_from_Kriging), after looking through the Help, including Learn More About How Kriging Works. You will also notice that you have lots of options in the Kriging tool window—Ordinary vs. Universal, Spherical vs. Circular vs. Exponential vs. Gaussian vs. Linear—and so on. Choose the defaults. Also, at the bottom of the window you will find a field called Output variance of prediction raster. Call it Kriging_variance and put it into the geodatabase as well. Explore Surface_from_Kriging with the Identify tool. Then look at Kriging variance. This is a layer which, with some arithmetic manipulation described in Tool Help, will give you an idea of the quality of the interpolation at each cell on the map.

---

[6]Comes from the name of Danie Krige, the co-developer of the method, along with Georges Matheron.

The lesson of this section is that, as you start your career in GIS (after all, you are almost done with this book), when you need to create a surface, you should seek help from people or documents that will let you know the most appropriate tool to use and with what parameters to use it.

Points and Density

While we are on the subject of making surfaces and areas out of points, let's consider the Density tool. This tool's procedure takes the values that are associated with points and, with a raster, spreads them out over the landscape around the points. For example, suppose that you have data that places all the population of some lightly populated counties in a state at the points designated as courthouses. Obviously, everyone in a county doesn't live at the courthouse. We will spread them out over a rectangular area (to make checking the results easier).

____ **17.** Start a new map. Add as data Known_Populations.shp.[7] Look at its attribute table, then close it. Label each point with the Population value, in green, size 12. Make the map units and the display units Kilometers. Approximately how far apart are the courthouses (population centers)? _____ kilometers. What we will do is spread the population at each courthouse over an area of several hundred square kilometers, just to illustrate how the Density tool works.

____ **18.** Find the Point Density tool (Spatial Analyst > Density) in ArcToolbox. Start the tool. The Input Point Features would be Known_Populations. The Population field is just called Population. Make the Output raster Density_test, in

___IGIS-Arc_*YourInitials*\Experiment_with_Interpolation\Surfaces.gdb

using an output cell size of 1.0 kilometer on a side. For the neighborhood (populated area of the county), pick Rectangle. For the Neighborhood Settings, make a square from Map Units (not Cell Units) of 40 kilometers on a side. How many square kilometers (and therefore, cells) will there be in each neighborhood? _____. OK.

____ **19.** Make sure Known_Populations is at the top of the T/C. Examine the Density_test raster with the Identify tool. Zoom in on a square which has a population of 3200. When you click in the square you are clicking on a cell that is one kilometer on a side. So how many people are there per square kilometer? _____ (If you didn't get 1600 for the number of square kilometers in each neighborhood, go back and rethink the previous step.) Click some other neighborhoods, including one with a population of 1600 (which should have one person per square kilometer).

Thiessen, Dirichlet, Voronoi (and, of course, Decartes)

While we are on the subject of points and their surrounding areas, let's look at the creation of a set of polygons, each created from a single point Each polygon has the following property: Every interior point of the polygon is closer to its generating point than to the generating point of any other polygon. This idea

---

[7]This is basically the same sort of file as Known_Altitudes.shp, but we are not planning on making a surface, although one could consider a surface of population—with the z being, say, the number of people per acre.

has "been invented" by at least three different people (and was used informally by Descartes), giving rise to Thiessen polygons, Dirichlet domains, and Voronoi cells. The polygons tessellate a portion of the Cartesian plane. Unlike the procedures used above, the value at the point has no bearing on the size, shape, or extent of the area around it. Only the position of the point is important.

Making Thiessen polygons is a little more involved than making the other rasters we have used, because they expand the extent of the raster area. There are ways to correct this, but we will use a "workaround" to accomplish the same thing. This will (a) make the polygons, and (b) acquaint you with the idea of workarounds which is a handy frame of mind to be in when dealing with today's very complex and not particularly robust software.

We will first use "Only Points" to make a shapefile. Then we will convert that shapefile into a raster.

_____ **20.** In ArcMap start a new blank map. Add as data Only_Points.shp. Label the points layer with the FID field. Open the attribute table. (You will find a field there named Population. If you try to delete it you will be told NO and given the reason that comes down to Tables must have at least three fields. Whatever.) With Search, find the Create Thiessen Polygons tool. Where is it? _____

Start the tool. For Input Features click the drop-down menu arrow, and you should see the shapefile Only_Points. Click it. For the Output feature type browse to the Experiment_with_ Interpolation folder in the Look in box and, in the Name box type T_D_V.shp – noting that you are making a shapefile, not a raster, and that it is in the folder that is one level up from the Surfaces file geodatabase. Accept ONLY_FID as the output field. OK. Observe the lower right of screen for the progress and completion of the execution of the tool. The shapefile will be added to the map.

_____ **21.** Make sure that Only_Points is at the top of the T/C. Label the T_D_V layer with its Input_FID field with text size 10 points, color red. Click the Full Extent button and examine the shapefile display.

_____ **22.** Right-click the display and, in Data Frame Properties, make the Map and Display units both kilometers. Use the Measure tool to convince yourself that the lines separating the polygons lie halfway between pairs of adjacent, generating points, by measuring from the first point to the second point, clicking, and then measuring back to the line. The segment distance should be about one-third the total distance.

_____ **23.** *Use Polygon to Raster to convert the shapefile to a raster.* Where is the tool? _____ Start the tool. Convert the shapefile T_D_V to the raster Thiessen_Dirichlet_Voronoi with the following parameters: Value field should be Input_FID. Cell assignment type should be MAXIMUM AREA. Set the Cellsize to 5.0. Make sure the Thiessen_Dirichlet_Voronoi raster goes into the Surfaces.gdb file geodatabase.

_____ **24.** Make sure the point feature class is at the top of the T/C. Turn off the T_D_V shapefile, leaving the Thiessen_Dirichlet_Voronoi raster on. Go to full extent. Use the Identify tool on the raster to verify that the raster zone number (its Input_FID) is the same as the point value. Bearing in mind that the cell size is 5 kilometers on a side, how large is the area designated as zone 9? _____ square kilometers.

_____ **25.** Close ArcMap.

Rasters: Distance and Proximity

Making a Raster Showing Straight-Line Distances to a Single Place

In the following steps, you will see how ArcMap Spatial Analyst makes a raster of Euclidean (straight-line) distances **from** each cell **to** a single "source" cell.

___ **1.** Start a new blank map. From Customize > Extensions make sure the Spatial Analyst box is checked.

___ **2.** Click the Add Data button. In the Add Data dialog box, navigate to the shapefile Square_grid .shp (it's in

   ___IGIS-Arc_*YourInitials*\Spatial_Analyst_Data\Proximity_Data_SA)

   and add it to the map. Make the lines of the shapefile bright red. Save the map as Proximity_ startup.mxd, in

   ___IGIS-Arc_*YourInitials*\Spatial_Analyst_Data\Proximity_Data_SA

Note that "Square_Grid" appears as a 10 by 10 matrix of transparent squares. Each square is 10 units (say the units are kilometers) on a side, so the square represents a space 100 square kilometers. Square_Grid will serve as a backdrop for the next several raster datasets. You will work in your folder Spatial_Analyst_Data\Proximity_Data_SA and with the file geodatabase Prox-Exp.gdb (named for Proximity Experiments).

___ **3.** Bring up the ArcCatalog tree. Navigate to

   ___IGIS-Arc_*YourInitials*\Spatial_Analyst_Data\Proximity_Data_SA\Prox-Exp.gdb

   file geodatabase, and drag the raster named Onecell onto the map. Onecell is a raster consisting entirely of No Data values, except for a single cell with a value of 999. Use Identify to verify this. Open, examine, and close the attribute table for Onecell. In the T/C, change the color of OneCell to solid black. See Figure 8-46.

___ **4.** From the T/C, bring up the Properties of Onecell. Under Source, notice that the cell size is 10 by 10 and there are 10 rows and 10 columns in the raster. Scroll down to look at the rest of the information, though it's pretty uninteresting, given that the entire raster consists of only one cell that isn't NoData. Under Symbology, check to see that Unique Values highlighted. Cancel the Layer Properties window.

By using Spatial Analyst, you will be able to make a raster whose cells contain the distances from the center of each cell in the raster to the center of the source cell (Onecell).

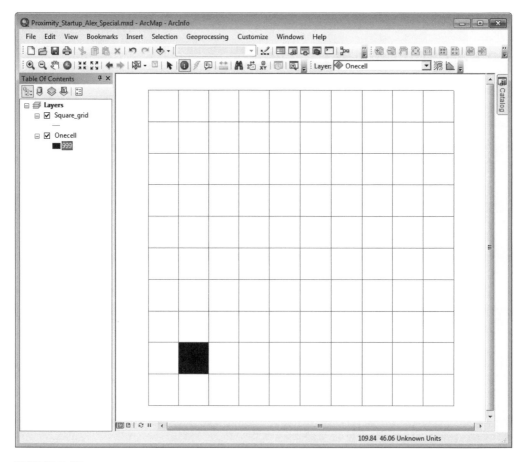

FIGURE 8-46

___ **5.** ***Prepare to calculate the Euclidean distance:*** Start the Euclidean Distance tool which you find in ArcToolbox > Spatial Analyst Tools > Distance. For Input raster, browse to

    ___IGIS-Arc_*YourInitials*\Spatial_Analyst_Data\Proximity_Data_SA\Prox_Exp.gdb\Onecell.

(Alternatively, click the arrow for the drop-down menu and pick Onecell. From now on in this exercise you will be working in the Prox_Exp.gdb in the Proximity_Data_SA folder, so I will only tell you the name of the dataset.) For the Output raster, browse to the correct gdb and type the name MapDist1. The Output Cell Size should say 10. Press OK. View full extent.

___ **6.** Spatial Analyst will calculate a raster with the distances and add that dataset to the map. Change the order of entries in the T/C, so that from top to bottom they are ordered: Square_ Grid, Onecell, MapDist1. See Figure 8-47.

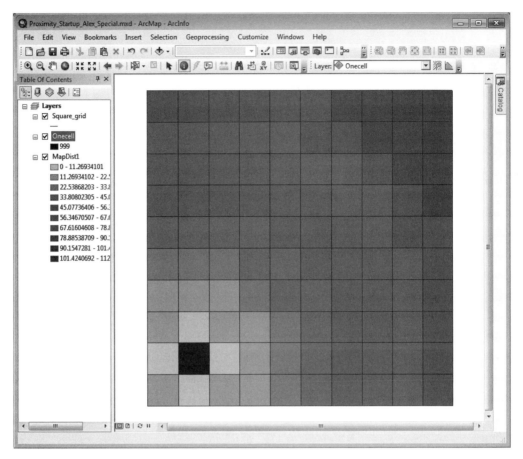

FIGURE 8-47

——— **7.** *Explore the* **MapDist1** *raster with the Identify tool:* If you wish, change the Symbology of MapDist1 from gray scale to something more colorful. Click the Identify tool and make MapDist1 the "Identity from" dataset. Run the mouse over the raster and note that the cursor position (both X and Y between zero and 100) is given by the numbers on the Status bar of the ArcMap window. Now click one of the cells. The cell's Pixel value, which is the distance from its center to the center of the source cell, is displayed. Set Map Tips on (MapDist1 Layer Properties > Display > Show Map Tips) to look at that cell. Using the Identify tool or Map Tips, query several more cells.

——— **8.** *Understand the raster values by checking some other cells:* Each cell "Pixel value" represents the straight-line distance from the center of that raster cell to the center of the source cell. (Obviously, the value of the cell on the new raster that corresponds geographically to the source cell on the Onecell raster has a value of zero.)

Note that if you go three cells to the right of Onecell, the distance is 30. If you go from that cell up four cells (that is, 40 units), the (diagonal) distance to the source is 50. That is reassuring, because Pythagoras tells us that the sum of the squares of the two sides of a right triangle equals the square of the hypotenuse (900 plus 1600 equals 2500, whose square root is 50).

_____ **9.** *Verify distances using the Measure tool:* Look at the Measure tool icon. You may be disappointed to find that it is "grayed out." This disappointment is caused because we haven't defined units for the raster. You know how to solve this problem: Data Frame Properties, and so on. Use kilometers for both Map and Display units. Once Measure is operating, click a cell and use the Identify tool to obtain its value. Now click the Measure tool, fixed up to display in kilometers, and verify that the distance from the center of your chosen cell to Onecell is approximately the same as the value displayed in the Identify Results window.

_____ **10.** *Determine the distance from a source cell to a diagonal neighbor:* Use the Identify tool to check the distance from the source cell to the cell in the lower-left corner of the raster. What is it? _____. Use the measure tool to determine the distance from the center of the lower left cell to the center of OneCell. _____.

The distance should be 14.142136, which is approximately 10 times the square root of 2. ($10^2 + 10^2 = 200$; the square root of 200 is approximately 14.142136.)

_____ **11.** In Customize > Toolbars find the Spatial Analyst toolbar and turn it on. In the drop-down menu select MapDist1, then click the Histogram button at the right of the toolbar. (Enlarge that window if necessary.) Looking at the T/C (for the "colors" of the vertical bars) and using the Identify tool on the histogram (right-click on vertical bars in the bar graph), how many cells are about 30 units away from Onecell? _____. Dismiss the Histogram, the Measure window, and the Identify Results window.

To this basic understanding of Euclidean distance in rasters, we can add some capabilities. The new features that you will work with are: (a) more than one source cell in the raster and (b) a cap, or cutoff value, so that no value greater than the cap will be recorded in any cell.

Examining Many Source Cells and the Capping Distance

_____ **12.** In the T/C turn off Onecell and MapDist1. Add the raster Manycells to the map. Notice the presence of several cells containing data, while the rest of the raster shows No Data. There is still a single source cell in the southwestern portion of the raster, but there is also a line of source cells in the northeast and a clump (cluster) of source cells in the northwest. What are the values of the three zones? _____, _____, _____. Change the colors of all the source cells to black.

You will generate a raster containing distances to the nearest source cells, provided that the distance is no further than 30 kilometers. You can limit the search for the closest source cell so that it does not exceed a given value, thereby excluding cells in the output raster that are more than a certain distance away from any source cell. (Those cells will contain NO DATA.) You are telling the software "if you have to look further than this to get to a source cell, just forget it." Such a number may be called a limit, a cap, a cutoff value, or a threshold.

_____ **13.** Use the same Spatial Analyst tool (Euclidean Distance) as you did in the previous steps to create the new distance raster from Manycells. Browse to make its name DistManycells. In the text box for Maximum distance, type 30. The Output cell size should be 10. Click OK.

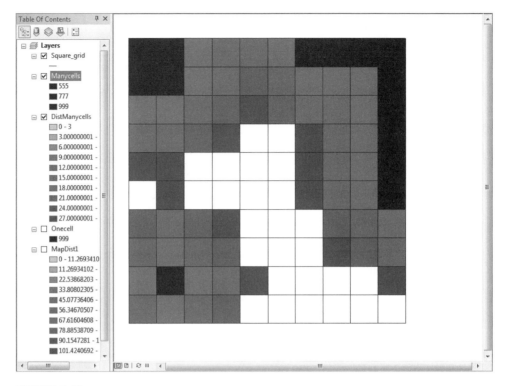

FIGURE 8-48

The raster of source cells that serves as input to this tool may consist of several single cells, clumps of cells (contiguous cells), a linear structure of several cells, or any combination of these. For each cell in the output raster, the straight-line distance to the **closest** source cell is calculated. This happens automatically. You simply apply the straight-line distance tool, as you did in the previous steps, and the newly generated raster contains the closest distances.

___ **14.** Arrange the entries in the T/C so that from top to bottom they are ordered Square_Grid, Manycells, then DistManycells. In the Properties > Symbology window of DistManycells Display NoData as white. See Figure 8-48.

___ **15.** Examine the distance raster, making sure that DistManycells is the active dataset in the Identify tool. Verify that some distances shown by cells in the new dataset are indeed associated with the closest source cells in Manycells. The cutoff value of 30 kilometers for the distance resulted in some NODATA cells in the new dataset. Use the Measure tool to verify that the NODATA cells, shown in white, are more than 30 units away from any source cells. Note that 30 is the largest value in the legend of DistManycells. Use the Spatial Analyst toolbar to make a histogram of DistManyCells. How many cells are about 20 kilometers, away from a source cell. Count them on the map as well. Using CellToool to help. Close the Histogram window.

Developing a Raster with Cost Distance

The following steps will illustrate the principle of calculating cost that is based on both distance and a cost surface. First, a quick review question:

____ **16.** What would be the cost of the path going horizontally through a cell (cell size 10, cell cost 0.85)? Answer: 8.5. How about the cost of moving through that same cell, going diagonally? _____.

____ **17.** Turn off all visible entries except Square_Grid. In the steps that follow, keep Square_Grid at the top of the T/C to delineate the cells of the displayed rasters.

The Tolls dataset is an artificial cost surface raster dataset. The surface is divided into rows of cells; all the cells in a given row have the same value.

____ **18.** Add the cost surface raster, Tolls,[8] from

___IGIS-Arc_*YourInitials*\Spatial_Analyst_Data\Proximity_Data_SA\Prox_Exp.gdb

In Layer Properties, under the Symbology tab, click Stretched under Show. Right-click the color ramp and turn off Graphic View. Make the color ramp of Tolls "Purple Bright". Press the Labeling button. Type in "9" for Number of Intervals and press Generate. Press OK. Press OK in Layer Properties. Look at the T/C for the values of the rows. See Figure 8-49 for the image.

Imagine that the center of each cell contains a toll booth. Depending on which row the cell is in, the toll booth charge varies from 25 cents to $1.15 to pass through **one unit of distance** of the cell. In the Tolls raster, all the cells in the bottom row have values of 0.25; the top row cell values are 1.15. So, to cross a single cell in the top row in an east-west direction would cost $11.50 (1.15 dollars times 10 distance units). (Since the cost is based on each unit of distance, it costs more to cross the cell in a diagonal direction because the distance is longer.) Use the Identify tool to examine the values of cells in the rows.

____ **19.** *Examine the source raster:* In the T/C move Onecell to just under the square grid and turn it on. Recall that Onecell was the source raster you originally worked with that contained a single source cell. Turn off the tolls raster.

____ **20.** *Determine cost weighted distance using the Cost Distance tool:* Where is it? _____ _____ Start the tool. The Input raster should be Onecell. The Input Cost Raster should be Tolls. Call the Output Raster TCostOnecell (the total cost of the path from the center of each cell to the center of the source cell in the raster Onecell). Click OK. When the new raster, TCostOnecell, arrives in the T/C, doubleclick the name to bring up its Layer Properties. Click Symbology. Find the Invert box and click it on. Click Apply. Click OK.

____ **21.** Arrange the top entries (Square_Grid, Onecell, TCostOneCell) and look at the new legend. Each cell contains the cost of moving across the cost surface raster from that cell to the source cell. You can get a general idea of the relative costs by looking at the "colors" of the raster.

---

[8]If you are using version 10.0 and you see only a gray square on the map, do the following: Right click Tolls, the Properties > Symbology > Show: Stretched. Then Stretch Type: Standard Deviations. Say Yes to Do you want to compute statistics. OK. You should now see gray stripes for Tolls.

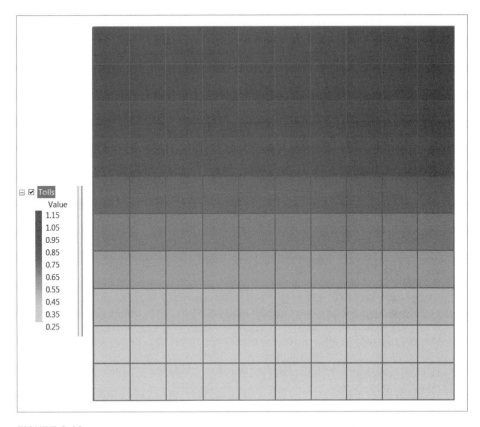

FIGURE 8-49

____ 22. Using the Identify tool, with the "Identify from" field being TCostOnecell, examine the total cost of traversing the cells due north of the source cell (pick the third cell directly above the source cell). What is that cost? ____. Now look at the total cost of traversing the same distance but crossing from the east of the source cell (pick the third cell directly to the east of the source cell). What is it? ____. It is obviously cheaper to come from the east (cost is $10.50) than from the north ($15). The overall range of costs is from 0 (from the source cell to the source cell) to approximately $82. Again use the Identify tool to display the values of the Toll cells that are traversed. To come from the east, one need only traverse cells with costs of 0.35 per kilometer or less. To come from the north, the cell costs are 1.15, 1.05, 0.95, and so on. The most expensive trip is from the northeasternmost cell—it is furthest away and in an area where the cost per cell is highest.

____ 23. ***Check out the contouring capability of the Spatial Analyst:*** In the Spatial Analyst toolbar use the drop-down menu to select TCostOneCell. Click the Create Contour button. Then click various places on the raster. The lines that appear are estimates of constant cost. That is, from any point on a given contour line, it should cost about as much to go to Onecell as from any other point on that line. Of course, this won't be exact, since the cost jumps incrementally each time you cross a cell boundary, but you see the idea. Use MapTips on TCostOneCell to check out the cells that are crossed by a contour line. Pick another line and check those cells.

____ 24. Erase the contour lines by pressing the Select Elements button, dragging boxes around the contour lines, and pressing the Delete key or Delete button on the Standard toolbar.

Creating Direction and Allocation Rasters

You learned that you can calculate the shortest distance between two places using the Euclidean Distance request and you can find the least-cost path with the Cost Distance request, but there may still be a couple of major questions: (1) Which source cell is closest to (or least expensive to reach from) each given cell in the raster, and (2) what is the direction or path from each given cell?

Going back to the Euclidean (straight line on the map) calculation, you will be able to make not only a raster showing the distance to a source cell as you did previously, but you will make two additional rasters. One of these will show, in each cell, approximately, the direction one must travel to get to a source cell. The second will show, in each cell, the number of the closest source cell. You will use the two parameters in the Euclidean distance request that were left unchecked in previous examples.

_____ **25.** In ArcMap click File > Open. Open Proximity_startup.mxd.[9] Add the layer Twocells to the map.

_____ **26.** Find and start the Euclidean Allocation tool. The Input raster should be Twocells. Source field: Value. The output cell size should be 10. Max Distance should be 50. Browse to the Prox_Exp. gdb and name the Output raster Eucl_Allocation_Twocells. Make the Output distance raster Eucl_Distance_Twocells and the Direction raster Eucl_Direction_Twocells_dir. Click OK.

_____ **27.** Arrange the layers in the T/C this way:

Square_Grid

Twocells

Eucl_Allocation_Twocells

Eucl_Distance_Twocells

Eucl_Direction_Twocells

Turn on these rasters; turn any other layers off. Change the colors of the two cells to bright red, so you can identify them when they are surrounded by other cells.

_____ **28.** Using the colors and legend of Eucl_Allocation_Twocells and the legend of Twocells, notice that cells of the same zone are allocated to (clustered around) the nearest source cell, and (using Identify) that the allocated cells take on the value of that source cell. Cells greater than 50 distance units away from a source are set to NoData. Note that there are several of these.

_____ **29.** Turn off Eucl_Allocation_Twocells so that Eucl_Distance_Twocells is visible. Using the Symbology Properties, set the Layer Properties Show box to Stretched. Examine Eucl_Distance_Twocells with the Identify tool. Note the distances of several cells. How far is it from the cell (45,35) to the nearest source cell? _____. Turn off Eucl_Distance_Twocells so that Eucl_Direction_Twocells appears. Change the Symbology Color Ramp (classified, with 18 classes) of Eucl_Direction_Twocells to Yellow to Green to Dark Blue. (Hint: Find it by right clicking on the Color Ramp Bar.) Notice that the T/C breaks the raster into (pretty much) 45-degree categories based

---

[9]If you don't have Proximity_startup.mxd for some reason you can re-make it by just adding Square_grid to Blank map.

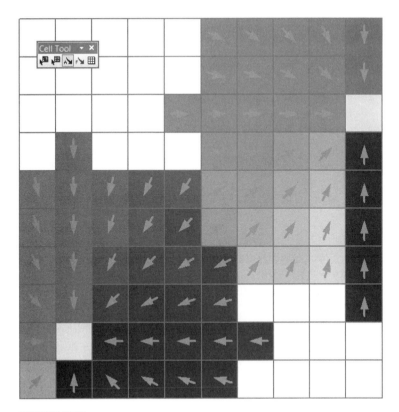

FIGURE 8-50

on the cardinal directions. Verify (by looking at Pixel value) that the direction in each cell appears to be the compass direction you would travel from that cell to reach the nearest source cell. The direction 360 is used to indicate north; the value 0 is indicates those cells that correspond to the source cells. Use Identify to check out the direction from cell (45,35) to its closest source cell. _____. Figure 8-50 has arrows drawn to show the value of several cells. Figure 8-51 shows the values in the cells. Use the CellTool toolbar to create the arrows and value numbers. (You will have to move the direction raster to the top of the T/C to make CellTool work.)[10]

Using Cost Distance to Make Direction and Allocation Rasters

Solve the following problem in a new map. Basically, you are repeating the previous exercise, but using cost distance instead of Euclidean distance. The distances are calculated differently. In the Euclidean case they are straight line distances; using cost distance the distances are calculated along a path which goes from cell center to cell center.

[10]CellTool did not work in ArcGIS Version 10.0 for some of those who tested the exercises. See the Color Figures File for the resulting image should look like.

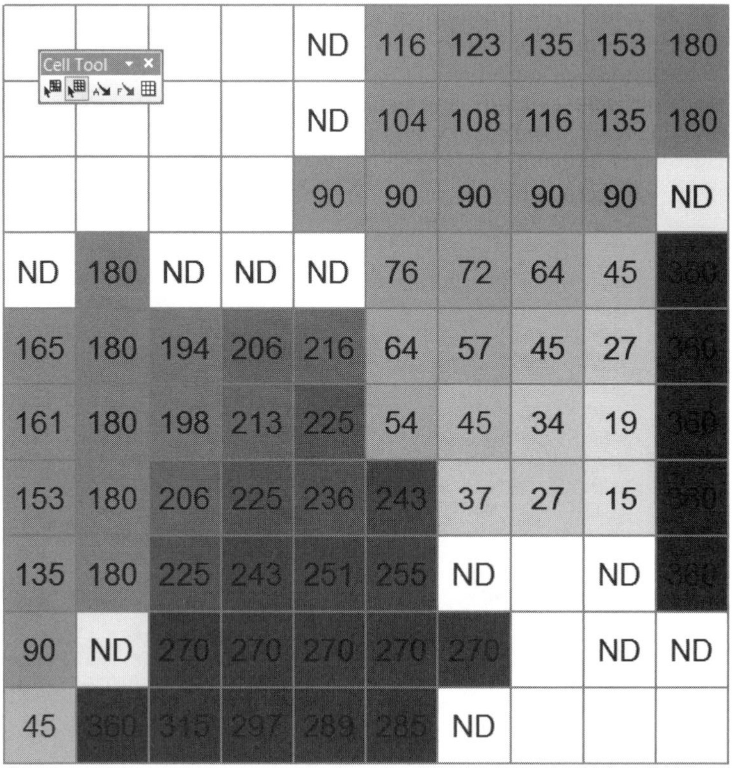

					ND	116	123	135	153	180
	Cell Tool ▾ ✕				ND	104	108	116	135	180
				90	90	90	90	90	90	ND
ND	180	ND	ND	ND	76	72	64	45	360	
165	180	194	206	216	64	57	45	27	360	
161	180	198	213	225	54	45	34	19	360	
153	180	206	225	236	243	37	27	15	360	
135	180	225	243	251	255	ND		ND	360	
90	ND	270	270	270	270	270		ND	ND	
45	360	315	297	289	285	ND				

FIGURE 8-51

____ **30.** *Open* Proximity_startup.mxd. Add TryThis as a source raster and Tolls as a cost raster. Make the color ramp of Tolls "Purple Bright" treating its T/C labeling as you when you added it last time.

____ **31.** Find and start the Cost Allocation tool. TryThis is the Input raster. Value should be in the Source field. Tolls is the Input cost raster. Browse to Prox_Exp.gdb and name the Output allocation raster CostAllo. Make the Maximum distance 30. Call Output Distance raster CostDist. Call the Output backlink raster CostDir. Double-check that you are using Prox_Exp.gdb for all output rasters. Click OK.

____ **32.** The new rasters should appear in the T/C. Collapse the legends in the T/C. Click on List By Drawing Order, and arrange the entries in the T/C in the following order: Square_Grid, TryThis, CostAllo, CostDir, CostDist, Tolls. Turn off all except Square_Grid and TryThis. Examine TryThis with Identify, and the T/C (which you may have to re-expand), so you know what is where.

____ **33.** Start the Identify tool and set it so that it shows the values of CostDist, even though that layer is turned off. Position the cursor in the cell whose center is at location approximately (85, 15). The cost (Pixel value) from that cell should be about 21. Examine other cells in the accumulated cost raster. What is the cost from (5, 55)? _____.

____ **34.** In the T/C, turn on and expand the allocation raster CostAllo. Examine CostAllo with Identify. Verify that the cell at (85, 15) is allocated to (generates the least-cost path to) a source cell whose value is 50. Determine where the cells of value 50 are. To what zone is the cell at (65, 45) allocated? _____. (Recall: A *zone* consists of cells that all have the same value.)

___ **35.** ***Compare the allocation raster (CostAllo) with the source raster (TryThis):*** Display TryThis with a grayscale ramp by choosing Properties > Symbology > Classified. Set the number of classes to 8. Repeat the process for raster CostAllo. Turn both entries on and everything else (except Square_Grid) off. Now, by toggling CostAllo off and on, you can see the cells in CostAllo that are clustered around the source cells in TryThis. Since the two entries have the same grayscale scheme, CostAllo seems to "grow out of" TryThis. However, because of the cost surface Tolls, there are more cells clustered toward the southwest. This is the effect of the weighted allocation operation. Turn off CostAllo.

___ **36.** Turn CostDir on and expand its legend. Explore the values of some of its cells with Identify. The values tell toward which of the eight nearest neighbors the path to the desired cell runs. The code is as follows:

Code	Direction	Degrees
1	East	90
2	Southeast	135
3	South	180
4	Southwest	225
5	West	270
6	Northwest	315
7	North	360
8	Northeast	45

As you can see, the code is not particularly intuitive.

___ **37.** ***Reclassify the direction code to degrees:*** Find and start the Spatial Analyst > Reclass > Reclassify tool. For the Input raster, use CostDir. The Reclass field should be Value. Click the Unique button to show all values. Call the new raster CostDir_Deg. Type in the new values from the list that follows. The new values are in degrees, based on the compass direction north as 360. Not all the old values listed will be represented in the Reclassify window (there is no 2); just type values for those that are.

Old Value	New Value
0	NoData
1	90
3	180
4	225
5	270
6	315
7	360
8	45
No Data	NoData

Click OK to generate the new dataset called CostDir_Deg. It should be at the top of the T/C under Square_Grid. Turn all other layers off. Use CellTool to fill up the raster with arrows. Verify that to head toward a the source cell at (15,15) from cell (85,15), you would first go to cell (75,5). That is, cell (85,15) will have the value 225, indicating a direction of southwest. Trace the rest of the path from cell (85,15). Notice how it takes a southern route and then turns northwest again, despite the fact that the shortest distance (the straight-line distance) is due west. Why is this?

_____.

Calculating a Least-Cost Path from "A" to "B"

In the following steps, you will find the least-cost path between a starting cell and a destination cell. You have probably noticed that you could figure out the least-cost path by investigating the direction raster. If so, you will have certainly noticed that doing so was a total nuisance. In the next steps you will generate a polyline featureclass that will show you the least-cost path from a given, specified cell to a "source" cell.

The software can handle multiple cells in both the starting raster and the destination raster, but this exercise will demonstrate the request using only one cell in each. If it has not become clear by this point, I should state the obvious: The cost (and/or the path) of going from A to B, as calculated by the raster manipulations you have been directing, is the same as the cost (and/or the path) of going from B to A. There are certainly instances in which this assumption is not warranted. One example would be if A is at the bottom of a mountain and B is at the top. Another example is driving in a city that has one-way streets. But the tool can be useful anyway.

The goal of this exercise is to use Shortest Path to create a *polyline* that shows the least-cost path between the single cell in a raster called Single_cell and the cell in the raster Onecell. You could, of course, determine this path by exploring the destination raster, but instead we will generates a raster based on least cost, then create a polyline with the path delineated.

_____ **38.** Find and run the Cost Allocation tool. Make Onecell the Input raster. The Source field should be Value. Tolls is the Input cost raster. Name the Output allocation raster Onecell_allo. Call the Output distance raster" Onecell_dis. Call the "Output direction raster" Onecell_dir.

_____ **39.** Find and run the Cost Path tool. You want the least cost path from Single_cell so that becomes the Input raster. The "Input cost distance raster" is Onecell_Dis. The "Input cost backlink raster" is Onecell_Dir. Call the output raster Cost_Path_from_Single_Cell. Click OK. This should put Cost_Path_from_Single_cell at the top of the T/C and turned on. Rearrange the layers in the T/C: Put Onecell and Single_cell just below Square_grid. Cost_Path_from_Single_Cell should be next. Turn everything else off and, to reduce clutter, collapse the legends of all the layers that are turned off.

What you are seeing now is a raster of cells that indicate the least cost path for getting from the cell at (95,75 – called Single_cell) to the cell at (15,15—called Onecell), taking into account the cost surface Tolls. It is, of course, the cost surface that is forcing the least cost path to not be a diagonal set of cells following an almost straight line, but rather a path that goes south, then southwest, and then west—because this is the cheapest way to go. We can make a vector representation of this path easily.

___ 40. Find and start the Raster to Polyline tool. The Input raster should be Cost_Path_from_Single_ Cell. Name the output Least_Cost_Polyline and plan to put it in Prox_Exp.gdb. Make sure the box "Simplify polylines" is unchecked. Click OK.

___ 41. Turn on Tolls. Examine the new layer Least_Cost_Polyline (make it black, width 3). Note the somewhat circuitous route that was calculated as the least-cost path runs from cell center to cell center. Is the path surprising in light of the Tolls cost surface raster? The polyline illustrates well that the cost surface Tolls has a major impact on the path! See Figure 8-52.[11]

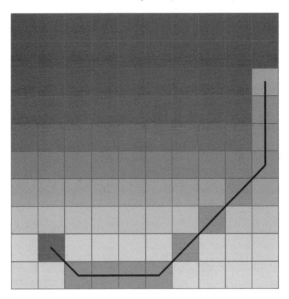

FIGURE 8-52

___ 42. Exit ArcMap.

Putting the Tools Together: Find a Site for a Regional Park

The power of GIS is most evident when you use a number of tools in concert. This is also the stage where the creativity comes in—and sometimes the frustration as well, because frequently, even if datasets are available, they are not in the form you can most easily use.

This exercise uses the Cost Path tool, which calculates and tries to minimize costs according to the "price you pay" in traveling between a source and a destination. This price has many names, among them: cost surface, friction, resistance to flow, resistance to travel, and impedance.

---

[11]What would have happened if the Simplify polylines box were checked? You might try it to see. It makes a slight difference in the length of the polyline in this case. In real cases, where you are dealing with millions of tiny cells rather than one hundred large ones, the difference would probably be negligible.

Suppose that the county of San Bernardino, California, would like to build a regional park somewhere near the city of Redlands. One of the criteria for the park location is that it must be accessible to those who might drive some distance using freeways or interstate highways.

For this exercise, accessibility is defined very specifically: The less time it takes to drive from any freeway off-ramp, the more accessible the park is deemed to be. The source raster, therefore, will consist of cells containing off-ramps.

Your approach will be to create a cost surface, expressed in minutes required to travel. You will derive travel cost weights from speed limits on the nonfreeway roads in the area. The cost surface, or weighting raster, will be the inverse of the speed; that is, instead of, say, miles per hour you want hours per mile. To get hours per mile you divide 1.0 by the speed. The raster will be based on the time it will take a car to traverse each given cell in the study area.

Here are the steps:

❏ In ArcMap, add the shapefiles STUDY_AREA and ROADS.

❏ Join a table that augments the ROADS attribute table with road types and speed limits.

❏ Select the roads records with road type Offramp.

❏ Make a raster called Offramps of these selected roads.

❏ Select all the other road type records.

❏ Make a raster called Roads_raster of the selected records.

❏ Reclass Roads_raster with appropriate speed costs to make a raster called Speed.

❏ Calculate a Cost raster, Mn_per_Ft, as an inverse function of Speed.

❏ Use Offramps and Mn_per_Ft to calculate Drive_Time, the driving minutes to each cell.

Setting Things Up

___ **1.** With ArcCatalog make a file geodatabase named Redlands.gdb in

___IGIS-Arc_*YourInitials*\Spatial_Analyst_Data\Proximity_Data_SA\Park_Data

Start ArcMap. To a blank map add the data files ROADS.shp and STUDY_AREA.shp from the Park_Data folder. Make sure that ROADS is at the top of the T/C. Save the resulting map as Site_Park.mxd in the Park_Data folder. All rasters you make should go into this file geodatabase.

___ **2.** Examine the Site_Park T/C entries: The study area is a simple, single polygon representing the Redlands, California, area. ROADS is a line file. Each single feature in ROADS is an amalgamation of all the roads of a given type in the study area.

___ **3.** Change the mapping units and the display units of the data frame to Feet.

4. Set Environment Variables: Geoprocessing > Environments. Leave the coordinate system undefined. Make the Processing Extent Same as Layer "STUDY_AREA.shp." Under Raster Analysis Settings, set the cell size to 150 (under As Specified Below). Click OK. Resave the mapfile Site_Park.mxd, in case you have to come back to this point.

Now that setup is complete and you begin thinking about the problem, a couple of issues might come to mind.

(a) How are you going to identify the off ramps?

(b) How are you going to exclude cells that have no roads, and cannot be traversed? Recall that your previous experience has been with rasters in which any cell could be traversed.

As to (a), you will be able to select the ramps based on values in the Roads table, which indicate the road types. For (b), once you have identified cells with road segments, you can set their impedance based on speed limits, which will also be available in the Roads table. For the cells without roads, you can set a high impedance, or cost, so that it would be prohibitive to drive through very many of them.

5. Open the table containing attributes of Roads. Some of the data massaging has already been done for you. The roads have been clipped to the study area, and the number of roads has been reduced to 12 sizable networks (from 29,326 road segments in the valley). The 12 road networks were derived by merging similar road types based on their Census Feature Classification Codes (CFCCs). (Note: Roads.shp is a feature-based layer, not a raster dataset; it will be in the Park_Data folder, not the Redlands geodatabase.)

The assumption has been made that all roads of the same CFCC will have the same speed limit. (The speed limit data is in a table that you will later join to the Attributes of Roads table).

6. *Add the CFCC_Speeds.dbf table to the map:* Open it. Notice the different road classifications and the speed limits associated with each.

7. *Join the two tables:* Right-click ROADS in the T/C. Choose Joins and Relates. Click Join. In the Join Data window, you want to use the "Join attributes from a table" selection. The field that the join will be based on is CFCC in both layers (that is, window items 1 and 3). The table to join (item 2) is CFCC.dbf. Click OK. The data contained in CFCC.dbf is added to the Attributes of Roads table. Close CFCC.dbf and open the now-enhanced Attributes of Roads table and scroll or widen the window so that you can see all its fields.

The Roads table identifies the Off_ramps with a code (A63). You want to make a raster containing source cells that represent road segments with the A63 designation; you want to assign No Data to all other cells. You will select the Off_ramps and then make a raster dataset of the selected features.

8. *Select the Off_Ramps:* With the Attributes of Roads table active, select the record containing Off_ramps. Arrange things so that you can see the Off_ramps selected in the map, highlighted in cyan.

9. *Create the source raster:* Find and start the Feature To Raster tool. The Input Features should be ROADS. Make the Field ROADS.CFCC. (The field chosen here is only informational; the integer value of the one record in the raster will be 1.) If necessary, set the Cell Size by typing 150 (recall that the units are feet). To create the name, browse to

___IGIS-Arc_*YourInitials*\Spatial_Analyst_Data

\Proximity_Data_Spatial Analyst\Park_Data\Redlands.gdb

and type as the name Off_ramps. Click OK. The tool will convert the selected features (offramps) to raster.

_____ **10.** If, after processing is complete, Off_ramps does not show up in the T/C, add it from the file geodatabase Redlands. Click the Display tab in the T/C. Make the T/C order, from the top, ROADS, Off_ramps, and Study_Area. Have Off_ramps on, ROADS and STUDY_AREA off. Zoom in on a region of colored cells, so you can see the individual pixels. Measure the cell size. _____ feet. Zoom back to full extent.

_____ **11.** Turn on the ROADS Layer. By panning and zooming, you should assure yourself that the selected Roads (Off_ramps) did, in fact, create cells in the Off_ramps raster.

_____ **12.** Open the Off_ramps attribute table. How many cells are there representing the Off_ramps? _____. All the other cells are set to NoData. (There is no record that gives the number of NoData cells.)

Preparing to Create a Cost Surface

You have just made the raster containing source cells. The next step is to create a cost surface raster that represents "resistance to travel" by assigning travel costs to every cell in the study area. In those cells where there are roads, you will base the travel cost on speed limits. In areas where there are no roads, you will simply assign a high cost based on a very low speed limit (5 miles per hour), primarily to keep cars on the roads. It would be reasonable to ask, "Why not just assign a super-large impedance to off-road cells?" You won't do this, because you want to allow every cell in the study area the possibility of access. If you assign an impossibly high impedance value to cells that don't have roads in them, you exclude those cells from consideration for the park because they could not be accessed unless a road ran through them. Presumably, if a group of cells is to become a park, a low-speed access road would be built. You do, however, want to completely exclude freeways to force travel on local roads and to prevent a path from crossing the freeways. In the following steps, you will select highways, county roads, and local roads to make a raster based on their speed limit values. You will then reclassify the nonroad cells to give them a speed limit of 5 mph.

_____ **13.** *Select the proper roads and create the Speeds raster:* The Speed limits data will be a raster representation of the selected roads; the values in the cells are the speed limits. Make the attributes of ROADS table visible. Clear any selections. Now select the Interstate, US Highway, County Road, Local Road, and Not A Public Road records. Do not include the Offramp records. Prepare to convert these features to a raster by starting the Feature_To_Raster tool. For the Features To Raster window input features text box navigate to ROADS.shp (or select it from the drop-down menu). For the Field, pick CFCC.SPEEDLIMIT and type 150 as the cell size. Browse to Redlands.gdb and name the Output raster SpeedLimits. Click Save. Click OK. Once the SpeedLimits raster is built and added to the map, move it to just below ROADS in the T/C. Pan and zoom to check it out with the Identify tool. Notice from the map that the cells between the roads were assigned the No Data value and that cells occur where there are roads. Zoom to an off-ramp. Click the Off_ramps layer off and on to assure yourself that those cells are not included in SpeedLimits. In the attributes of Roads table, clear the selections.

Let's now make a raster in which the speed in each cell is the same as the speed limit in that cell, except for Interstate cells (which we will make low because we don't want our path to the park to try to cross an interstate). The No Data cells we will set to 5 mph. To do all this, we will use the Reclassify tool.

———— **14.** ***Make a raster with speeds in the cells:*** Find and start the Reclassify(Spatial Analyst) tool. The Input raster should be SpeedLimits. Press the Unique button to be sure that you get all the different values of the cells. The Old Values are 1 (for private roads), 30, 45, 50, 65, and No Data. The suggested new values are 0, 1, 2, 3, 4, and No Data. You do not want these suggested values. Mostly, you want the new values to be the same as the old values, except for the No Data value, which you want to be 5, and the Interstate value (65), which we'll make 2. Type in the new values: 1, 30, 45, 50, 2, and 5. Browse and call the new Output raster Speeds (in Redlands.gdb, of course). Click Save, then OK.

———— **15.** ***Using Identify, examine the cells containing the speeds by panning and zooming:*** Use the Identify tool (operating on the Speeds raster) to verify that the data cells that were formerly NoData cells have a value of 5 and that other cells have appropriate values. Turn off the Speed_Limit, Off_Ramps, and the Study_Area entries, if they are on. Move the Roads feature class to the top of the T/C and make sure it is on. Turn on the Speeds raster. Next, zoom in on an area that includes part of the interstate where the ramps are. Notice how the speed limit cells follow along with the local roads. Notice also that all the cells between the roads now have the 5 mph value. Zoom in some more to see the relationship between the feature dataset Roads and the raster dataset Speeds. When you are finished looking, choose Full Extent from the map menu.

———— **16.** ***Examine the Properties of the Speeds raster:*** In the T/C right-click the Speeds layer name and click Properties > Source. How many columns rows are there? _____. Rows? _____. What is the uncompressed size of the raster? _____. Dismiss the Layer Properties window.

Building a Cost Surface

Now you have the raw material (those speeds at which cars going to the park may travel) for building a cost surface. What numbers do you put in the cost surface cells? You cannot put miles per hour (mph), because then higher speeds would imply greater resistance to flow—the opposite of what you want. You could use the reciprocal, hours per mile. To get this reciprocal, you simply divide "1" by miles per hour, which gives you hours per mile. For example, 50 mph is 0.02 hours per mile; 25 mph is 0.04 hours per mile. Since 0.04 is a larger amount than 0.02, and that larger amount represents a slower speed, 0.04 is what you want to see in your cost raster.

But there is another consideration: It is important to express the cell costs in terms of the units of the raster and the problem statement. You are interested in drive times in minutes. And the raster is measured in feet. The numbers you have been given to work with are miles and hours, but what you want, strange as it may seem, is minutes per foot! How do you convert hours per mile to minutes per foot? Multiply hours per mile by a conversion factor: 0.0113636, which is the number of minutes it takes to travel a foot at the rate of one hour per mile.[12]

---

[12]One hour per mile is 60 minutes per 5280 feet. 60/5280 equals 0.0113636—your conversion factor. So, for example, 50 miles per hour is 1/50 hours per mile. 1/50 * 0.0113636 5 = 0.000227272 minutes per foot. If a cell is 150 feet across, say, from east to west, it therefore takes 150 * 0.00022727 or 0.034 minutes to cross it at 50 mph. (Or you can just take my word for it.)

_____ **17.** ***Create the travel cost surface:*** Find and start the Raster Calculator. Enter the expression[13] that will convert mph into minutes per foot for every cell in the raster:

1.0 / "Speeds" * 0.0113636

Call the output raster Cost_Surface, heading it into Redlands.gdb.

This may seem like a lot of trouble just to divide the speeds into 1.0 and multiply by a constant, but the arithmetic operations take place over the entire study area, composed of more than 200,000 cells (~500 * ~400). You get a lot of computation for your expression. Click OK.

_____ **18.** ***From Cost_Surface create a Minutes_per_Foot raster and make it a unique value raster:*** If necessary, start ArcCatalog. Navigate to Proximity_Data_SA and refresh it. Find Redlands.gdb and click it. Right-click the name Cost_Surface and choose Copy from the drop-down menu. Click the name Redlands.gdb, right-click, and select Paste. This will create a raster named Cost_Surface_1 after you click OK. Right click the new name and Rename it to Minutes_per_Foot. In ArcMap, add Minutes_per_Foot to the map and Remove Cost_Surface from the T/C. Since the numbers have fractional parts, the software assumed you wanted to see them as a "continuous raster." But we know that there are only a small number of unique values because they were derived from a few number of speeds. So let's change the symbology: Right-click the Minutes_per_Foot raster name, choose Properties > Symbology > Unique Values. Ignore the warning. Click OK.

You now have the ingredients necessary to run the Cost Distance tool. The source raster is called Off_ramps; the cost surface is Minutes_per_Foot.

_____ **19.** Find and start the Cost Distance tool. For the Input raster browse to Off_ramps in Redlands .gdb. The Cost raster is, of course, Minutes_per_Foot. For the Output raster browse to Redlands.gdb and put in the name Drive_Time. The next field says Maximum distance, but that is misnamed in this case, since it is minutes we are producing. Put in 60 for the maximum number of minutes. Click OK.

Depending on the speed of your computer, this may take a while to calculate. Why does it take so long? Realize that about one-fifth of a million cells must have values calculated for them. You will know that Cost Distance is finished when the new raster is added to the map.

_____ **20.** Turn all entries off except Drive_Time and Roads. Keep the Roads dataset above Drive_Time in the T/C. Examine the Drive_Time layer. Explore carefully, at different levels of zoom, the Drive Time raster. While many areas are within a 10-minute drive of an off ramp, others are an hour away, and still others show NoData because of the 60 minute cap we put on Drive_Time.

Of course, drive time is only one factor to consider in siting the park. Perhaps the areas closest to the Off_ramps are the most developed and therefore least suitable for a park.

---

[13]You may have noticed that sometimes an expression uses square brackets ([xxx]) around an argument for an expression, and sometimes double quotation marks ("*xxx*") are used. The syntax for the expression differs depending on the data source. For example, if you're querying a file geodatabase, or shapefiles, you enclose field names in double quoatation marks. If you're querying personal geodatabases, you enclose fields in square brackets:

Improving the Understandability of the Map

You can get a better idea of travel times by redoing the legend of Drive Time. The current legend simply breaks the drive times up into equal intervals, resulting in unwieldy ranges. A better set of classes can be defined, and a better legend can be created.

____ 21. Through Drive_Time Properties, bring up Symbology. In the Show box, select Classified. Specify 10 classes, then press Classify. In the Classification window, pick Manual for the Classification Method and make the break values 3, 5, 10, 15, 20, 25, 30, 40, 50, and 60. Click OK. Back in Layer Properties pick a color ramp you like. In the Label column, type in the top value of each range: 3, 5, and so on. Click Apply, then OK. Look again at the T/C entry for Drive_Time. Better?

The new legend uses more appropriate and readable divisions of driving time.

____ 22. Close all legends except ROADS, Off_ramps, and Drive_Time. Clear any selected features. Make the symbol for Off_ramps and ROADS solid black. From the top to the T/C, order the layers: ROADS, Off_ramps, and Drive_Time. Turn them on and everything else off. Explore the resulting map, using Identify on the Drive_Time raster. Click in a number of cells—both at full extent and zoomed in.

It should be apparent that the mathematical massaging you did earlier (calculating minutes per foot) to make the ultimate results show up as minutes was worthwhile.

____ 23. Use File > Save As to save the project with a new name: Time_to_Park_Sites_*YourInitials*.mxd. Exit ArcMap.

Exercise 8-7 (Project)

Watershed Analysis

To review from the Overview and prepare you for the next exercise, let's recap the various steps in determining a model of surface water flow that you encountered in the Overview of this Chapter. You might want to quickly review what was said there.

It starts with elevation information, since water (unless extremely provoked) runs downhill. In this exercise the elevation layer is called Elevpts. From Elevpts we will create ElevSurface. From ElevSurface we will create FlowDir, which will define, for each cell in the raster, the adjacent cell into which the water will flow. We can also create a raster that shows the maximum downhill slope for each cell, called ElevDrop.

From FlowDir we can determine three other rasters:

❑ UniqueBasins—A basin is an area where the elevations and slopes are such that the ultimate drainage all winds up in the same place

❑ FlowAccu—A measure of how much water arrives in each cell, given that every cell initially gets one unit of water

❑ AnySinks—Showing any areas inside the study area where water could not drain to an adjacent cell

From FlowAcc we can determine StrmChannels by saying that if a cell has more than a prescribed amount of water in it, it is part of a stream. StrmChannels lets us produce two new rasters:

❑ StrmOrder_StM—In which each stream is assigned an integer value that is an indication of its size, in terms of the volume of water in it. The Strahler Method (hence the StM in the name) is used.

❑ StrmID—In which each stream is assigned an integer value that is unique within the study area.

Finally, to show the utility of this information, we will consider that stream pollution has been discovered, indicated by a raster called POLLUTION_PTS, from which we will determine SuspectAreas.

In summary, in this exercise you will create an elevation surface from a large number of points, generate a flow direction raster, and then check it for sinks. You will also create a drainage network, assign stream order to linear water bodies, and identify watershed basins.

Important note: The four input data files for this exercise (which are the shapefiles elevpts, fishnet72, and the rasters ElevSurface2 and Pollution_Pts) will be found in

___IGIS-Arc_*YourInitials*\Spatial_Analyst_Data\Hydrology_Data_SA

However, you will put all the rasters you create in a file geodatabase Hydro.gdb.

_____ **1.** Start ArcMap. Make sure the 3D Analyst Extension is turned on and the 3D Analyst toolbar is shown. Using the ArcCatalog sidebar, make a file geodatabase named Hydro.gdb in

___IGIS-Arc_*YourInitials*\Spatial_Analyst_Data\Hydrology_Data_SA

Refresh the Spatial_Analyst_Data folder (to be sure Hydro.gdb will show up later when you need it),

From the folder

___IGIS-Arc_*YourInitials*\Spatial_Analyst_Data\Hydrology_Data_SA

add elevpts.shp to the map. Examine its attribute table. SPOT is the elevation. How many points are there? _____. What is the lowest? _____. Highest? _____. How many points are there? _____

_____ **2.** *Use the elevation point dataset to make a surface raster of elevations:* Find the Spline tool (in ArcToolbox > 3D Analyst Tools > Raster Interpolation). Start the Spline tool. The Input points should be elevpts. The Z value field should be SPOT. Accept the default cell size. In the Output raster specification, browse to the Hydro.gdb file database, in the Hydrology_Data_SA, folder and type ElevSurface for the name. Accept the other defaults. Click OK. ElevSurface should be added to the map.

Examining the Surface with Various Spatial Analyst and 3D Tools

_____ **3.** On the 3D Analyst toolbar click the Create Contour icon. Then click a cell near the center of the map. You will see a line that represents all the locations you could visit without changing your elevation from the selected cell. If you tap Delete on your keyboard that closed line will go away. If you click elsewhere on the map you will get another contour line. Make several more contour lines.

Now click the icon of Create Steepest Path on the 3D Analyst toolbar. Again click a cell. Note that a line runs from that point downhill to a point of lower elevation. When you are through being experimenting, use the Select Elements cursor to draw a large box around the entire map. Press the Delete key to remove the contour lines and steepest path lines.

____ **4.** Find and run the Contour (Spatial Analyst) tool. From the drop-down menu of Input raster, select ElevSurface. Make the contour interval 100, but use a base contour of some even hundred-number (ends in 00) under the altitude minimum, say, 1500. For output features browse to

___IGIS-Arc_*YourInitials*\Spatial_Analyst_Data\Hydrology_Data_SA

and name the feature ElevContours.shp. Click OK. Examine the contour lines added to the map and the conformance with the surface ElevSurface. Label the contour lines with the field CONTOUR. Zoom in on the hill top in the center of the map. Use Identify on two cells of the surface (ElevSurface) somewhere on opposite sides of a contour line labeled 3500. What are the values given by the raster? _____, _____.

____ **5.** Find the tool Slope (Spatial Analyst). Make the input surface ElevSurface. Make the output measurement PERCENT_RISE. Note that the output will be a raster—call it ElevSlope—and put it into Hydro.gdb. Click OK. The Slope raster will cover up the ElevSurface raster. Notice that the slope is greatest (like about 60 percent—100 percent is 45 degrees) where the contour lines are closest together.

____ **6.** Find the tool Aspect (Spatial Analyst). Make the input surface ElevSurface and the output ElevAspect. Again, a raster is placed on the map that covers up the previous ones. As you know, aspect is the direction the slope faces. Zoom to full extent. What you see is something of a hodgepodge—a riot of colors. Examine it further in the next step.

____ **7.** On the Spatial Analyst toolbar, set the drop-down menu to ElevAspect. Click the Histogram button. Comparing the colors of the histogram with those of the T/C for Aspect of ElevSurface, in what direction would you say most of the slopes faced? _____.

____ **8.** Find and run the Hillshade (Spatial Analyst) tool. Make the input surface ElevSurface and call the Output ElevHillshade. (You are putting these created rasters in the Hydro.gdb file database, aren't you? Sorry to nag.) Click on the Azimuth and Altitude fields to understand what is meant by these terms. Accept the default setings of 315 and 45. Click OK. Turn off all layers except ElevSurface and ElevHillshade. Go to full extent. By flipping the Hillshade layer off and on, you can get an idea of what the area would look like with sun in the northwest, 45 degrees above the horizon, and, from ElevSurface, why it would look that way.

____ **9.** *Generate a flow direction raster:* Find and start the Flow Direction tool. Use ElevSurface as the Input and FlowDir as the Output flow direction raster. Set the option that does not to force water off the edge of the raster. (Click the Output Drop Raster text box and read about it. Create such a raster, called ElevDrop.) Click OK.

____ **10.** FlowDir and ElevDrop will both be added to the map. From the T/C you can determine the maximum and minimum slopes based on ElevDrop. What are they? _____%, _____%. Remove ElevDrop from the T/C.

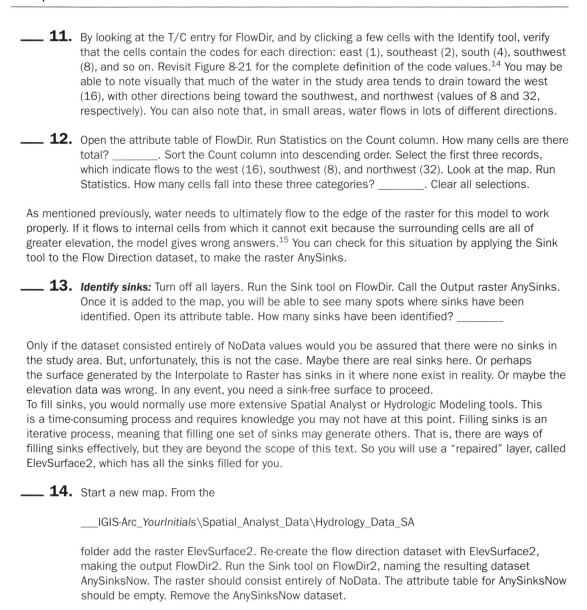

___ 11. By looking at the T/C entry for FlowDir, and by clicking a few cells with the Identify tool, verify that the cells contain the codes for each direction: east (1), southeast (2), south (4), southwest (8), and so on. Revisit Figure 8-21 for the complete definition of the code values.[14] You may be able to note visually that much of the water in the study area tends to drain toward the west (16), with other directions being toward the southwest, and northwest (values of 8 and 32, respectively). You can also note that, in small areas, water flows in lots of different directions.

___ 12. Open the attribute table of FlowDir. Run Statistics on the Count column. How many cells are there total? _____. Sort the Count column into descending order. Select the first three records, which indicate flows to the west (16), southwest (8), and northwest (32). Look at the map. Run Statistics. How many cells fall into these three categories? _____. Clear all selections.

As mentioned previously, water needs to ultimately flow to the edge of the raster for this model to work properly. If it flows to internal cells from which it cannot exit because the surrounding cells are all of greater elevation, the model gives wrong answers.[15] You can check for this situation by applying the Sink tool to the Flow Direction dataset, to make the raster AnySinks.

___ 13. *Identify sinks:* Turn off all layers. Run the Sink tool on FlowDir. Call the Output raster AnySinks. Once it is added to the map, you will be able to see many spots where sinks have been identified. Open its attribute table. How many sinks have been identified? _____

Only if the dataset consisted entirely of NoData values would you be assured that there were no sinks in the study area. But, unfortunately, this is not the case. Maybe there are real sinks here. Or perhaps the surface generated by the Interpolate to Raster has sinks in it where none exist in reality. Or maybe the elevation data was wrong. In any event, you need a sink-free surface to proceed.
To fill sinks, you would normally use more extensive Spatial Analyst or Hydrologic Modeling tools. This is a time-consuming process and requires knowledge you may not have at this point. Filling sinks is an iterative process, meaning that filling one set of sinks may generate others. That is, there are ways of filling sinks effectively, but they are beyond the scope of this text. So you will use a "repaired" layer, called ElevSurface2, which has all the sinks filled for you.

___ 14. Start a new map. From the

___IGIS-Arc_*YourInitials*\Spatial_Analyst_Data\Hydrology_Data_SA

folder add the raster ElevSurface2. Re-create the flow direction dataset with ElevSurface2, making the output FlowDir2. Run the Sink tool on FlowDir2, naming the resulting dataset AnySinksNow. The raster should consist entirely of NoData. The attribute table for AnySinksNow should be empty. Remove the AnySinksNow dataset.

---

[14]I feel like apologizing for the assortment of codes you have to cope with to determine direction. You have degrees (starting at both north and east), Euclidean direction (1, 2, 3 … 8), and now you have, powers of two (1, 2, 4, 8, …, 128). However, I didn't write the software. I'm just the person communicating the convoluted message. If you needed proof that computer software just develops in unfortunate ways that become "standards" you have it here. But it's not just computers. Think about the absurd layout of letters on a "standard" keyboard. Or key layout on phone vs. calculator.
[15]One other problem in modeling the environment rears its head. In the real world, there are such things as sink holes, which connect the surface water with the under surface water (ground water). Here, we choose to ignore this complication.

Recall that the Flow Accumulation tool computes the amount of water that flows into each cell from *all* the upstream cells. In the absence of a weight raster, the request assumes that one unit of water will come from each cell and will flow into one adjacent cell. The one unit of water for a given cell does not show up in the accumulation of water for that given cell. Therefore, some cells that have no water draining into them will have an accumulation value of zero. As you saw earlier, some cells at the bottom of the drainage can have very large values. The output raster of accumulation can be used as a measure of runoff for hypothetical rain fall events. The optional weight raster could be a surface of rainfall values interpolated from weather station measurements and modified with a model to adjust for loss from evapotranspiration and soil absorption. With a weight raster, the Flow Accumulation tool would return an estimate of actual runoff.

____ **15.** *Calculate flow accumulation:* Use the Flow Accumulation tool on the FlowDir2 layer. Name the new raster FlowAccu. Do not use an Input weight raster. From the T/C, what is the highest value of flow accumulation? _____ Compare that with the number of cells in the study area, recalling that each cell contributes one unit of water. Notice how the map delineates the accumulations of water, and therefore the streams, and perhaps rivers.

Determining the Stream Channels

You can use the Flow Accumulation raster to identify a drainage network. We will identify those cells that have high accumulated flow values. The cells that contain the most water are the stream channels. We can create a drainage network of any detail by choosing those cells that have more than a certain minimum threshold value. In this case, say that a stream exists at a given cell if the flow accumulation in that cell is equal to or greater than 140—the equivalent of 1 unit of rain over at least 140 other cells that drain to that cell. Let's say that a cell either is or is not in the drainage network, so we will make a raster consisting of binary values: 0 or 1. We can do this with the Raster Calculator.

____ **16.** In the Raster Calculator, build and evaluate this expression:

"FlowAccu" >= 140

Call the resulting raster StrmChannels. Turn off all layers except StrmChannels. The cells that constitute the stream channels have values of 1; the rest of the cells have values of 0.

Calculating Stream Order

The volume of water flowing through a stream is a function of many things, including the stream's width (number of cells) and its depth. For that and other reasons, the number of "width cells" that depict a stream is not a good indicator of its size or volume. As previously discussed, one way to get an idea of stream size is to assign a stream order number, which indicates the relative volume of water in a stream segment. In this step, you will use the Strahler ordering.

____ **17.** Use the Stream Order tool on the StrmChannels layer (the input stream raster) and the FlowDir2 layer (the input flow direction raster). Make this dataset active with the name to StrmOrder_StM. Use the STRAHLER method. How many cells are in the channels of 1st order streams? _____. How many in 4th order streams? _____

The Strahler method is more conservative than the Shreve method—that is, the numbers tend to be smaller. In Shreve, every time one stream joins another, the order number goes up. Not so with Strahler. Only when two streams of the same magnitude join does the order number increase, and then only to identify the downstream by a number greater by 1 than those above. Notice that there are only four categories of this Strahler layer, even though you had a large number of stream segments. With Shreve, the largest order is almost 40, as you can prove to yourself if you choose to by making StrmOrder_ShM.

Numbering Each Stream Individually

____ **18.** To individually designate each stream segment, use the Stream Link tool. This tool operates on the StrmChannels dataset, and it uses FlowDir2 as its flow direction raster. Call the new dataset StrmIDs, leave it on, and turn all other entries off. A large of individual stream segments have been produced. How many? _____. Zoom in on a portion of StrmIDs and use the Identify tool to examine a few assigned numbers—found in the (Pixel) Value field. Click the Full Extent button to see the complete map. Open the attribute table of StrmIDs. Select the one with Value 189. How many cells make up that stream? _____. Check it out on the map with Zoom In. Clear selections. Close the table.

Identifying Basins

Basins can be defined simply by the flow directions of water that falls on the study area. As you know, the flow directions are defined by the elevations and slopes.

____ **19.** Start the Basin tool. Note that the only input is FlowDir2. Call the output UniqueBasins. How many are there? _____. Zoom to full extent.

____ **20.** Using the Raster to Polyline tool, convert the raster dataset StrmChannels to the feature class shapefile Strms.shp. Browse so you can put it in the Hydrology_Data_SA folder. Display Strms.shp with bright yellow. Turn all layers off except Strms.shp and Unique_Basins. It should be pretty clear how the basins are delineated.

____ **21.** Set the Unique Basins layer at 85 percent transparency. (Wait! Wait! Don't tell me! It's in my Fast Facts File.[16]) In the T/C put Strms.shp at the top, UniqueBasins next, and ElevSurface2 third. Turn those three on and every other layer off so you can get a picture of why the basins are the way they are, in terms of elevation, and why the streams form as they do.

Finding Pollution Culprits

Suppose there are monitoring stations on streams, looking for dangerous substances in the water. If a station reports a problem we would like to know the location of the source of the pollution. Since each of the streams has associated with it a watershed we may narrow the search, using the WaterShed tool, which finds all the upstream cells that flow down to a given point—the point where pollution is discovered.

---

[16]Under Display in Layer Properties.

Specifically, the WaterShed request finds the up-gradient cells of a specific set of cells; those cells may or may not be cells in stream segments. The POLLUTION_PTS dataset (points at which pollution has been found) indicates pollution in streams at two points.

____ **22.** From

> ___IGIS-Arc_*YourInititalsHere*\Spatial_Analyst_Data\Hydrology_Data_SA

> add as data the raster pollution_pts. Turn off all other layers. In the pollution_pts attribute table, select both records so that you have a better chance of spotting the points, which are single pixels, on the map. If that doesn't work, turn the symbols of the two cells of the pollution_pts raster (550 and 675) bright red. (If that fails use the Raster to Point tool (in Conversion Tools) to make a points shapefile of the two raster cells, and make the points large enough so you can't miss them.)

____ **23.** Use the Watershed tool to determine those areas that drain to the points. The Input flow direction raster is FlowDir2. The "pour point data" comes from pollution_pts. Make the Pour Point field Value. Call the result SuspectAreas, and make sure it goes into Hydro.gdb. Move pollution_pts above SuspectAreas in the T/C. Open the attribute table of SuspectAreas. How many cells might be involved in the search for the location that is polluting 550? _____. 675? _____. This tool could help trace sources of pollution.

____ **24.** Using Raster to Polygon (in Conversion Tools) make a polygon shapefile of UniqueBasins. Call it Basins_Outline.shp and put it into the Hydrogology_Data_SA folder. Make its color Hollow and the width of its outline 3 picas, in black. Turn off all layers except Basins_Outline and FlowDir2. With Basins_Outline at the top of the T/C, note that all flows are either away from boundaries or parallel to boundaries of the basins (look again at Figure 8-21).

The distinction between basins and watersheds is a subtle one. Also, they are not completely distinct. In terms of ArcMap, basins are determined only by flow direction, which is determined by the elevation surface, That is, topography is the primary ingredient in basin determination. The water shed calculation is based on the aggregation of cells that feed a particular point, usually on a stream.

____ **25.** Save the map as Boundaries & Streams.mxd in the folder Hydrology_Data_SA. Make a layout and print a map that consists of: Basins_Outline (with dark outlines of hollow polygon basins) StrmOrderStM (with a symbology that goes from Cyan-Light to Blue-Dark). Note how streams, being confined to basins, become larger as they move toward the points at which they finally flow off the edge of the map.

____ **26.** Close ArcMap.

Exercise 8-8 (Review)

Checking, Updating, and Organizing Your Fast Facts File

The Fast Facts File that you are developing should contain references to items in the following checklist. The checklist represents the abilities to use the software you should have upon completing Chapter 8.

Chapter 8

❏ ___ To create a raster or raster from a text file

❏ ___ Row-major order means

❏ ___ The Symbology "stretch" option

❏ ___ The two fields in a VAT for a bare bones raster are

❏ ___ A zone is

❏ ___ A region is

❏ ___ The Cell tool

❏ ___ Rasters are of two types:

❏ ___ To do arithmetic computation with rasters use the

❏ ___ The Boolean operators are

❏ ___ When using the Raster Calculator, you must put blanks around

❏ ___ To keep environment settings

❏ ___ To convert features to rasters

❏ ___ When the cell size is halved, the storage requirement does not necessarily go up by a factor of four because

❏ ___ EXPAND does not work well as a buffer because

❏ ___ Reclassification

❏ ___ To convert zones to regions

❏ ___ SQL is

❏ ___ Extract by Attributes allows the user

❏ ___ The ways of creating a surface from a set of points are

❏ ___ The Density tool

❏ ___ Thiessen polygons

❏ ___ To create rasters for elevation, slope, aspect, and hillshade

❏ ___ Straight-line distances on a raster are calculated

❏ ___ Cost of moving across a raster is calculated

❏ ___ Distances on a raster are measured from cell center to

❏ ___ The threshold distance is

❏ ___ A Direction raster is

❏ ___ An Allocation raster is

❏ ___ A Cost Surface raster is

❏ ___ CFCC means

❏ ___ To join two tables

❏ ___ Flow direction is the basis for much of hydrologic investigations; it means

❏ ___ Two ways of delineating stream order are

Other Dimensions, Other Tools, Other Solutions

OVERVIEW

IN WHICH we examine the third spatial dimension in GIS; time and GIS; address geocoding; network analysis; and linear referencing.

Two Different Third Dimensions: The Temporal and the Vertical Spatial

So far in our GIS work, although it pains me to say it, we have had all of the disadvantages and none of the advantages of the fact that the world (and everything in it) resides in four dimensions. The disadvantages have come about partly because (1) the Earth is approximately spherical—requiring all that projection complication in moving from three spatial dimensions to zero, one, or two dimensions—(2) three-dimensional stuff is just harder to deal with, so we have contented ourselves with "flatland" in which nothing is quite right, and (3) it's hard enough just to get a dataset right at a given moment or period in time—never mind historical or anticipatory data sets.

But now we take on, separately, the third spatial dimension and the time dimension.

I present the Overview and the Step-by-Step sections together for each topic. When we say 3-D GIS, we usually mean three spatial dimensions. The title of the next section is meant to convey that idea. However, in the section after that, which is "3-D: 2-D (Spatial) Plus 1-D (Temporal)," we will take up looking at what happens to two-dimensional data over time.

The Third Spatial Dimension

3-D: 2-D (Spatial) Plus 1-D (Spatial)

This Overview will be short—because

❏ ArcGIS 3D Analyst Extension is so rich in capabilities that covering it in detail in an introductory course is out of the question.

❏ Many of the capabilities of ArcScene and ArcGlobe, the constituents of 3D Analyst, fall into the "a picture is worth a thousand words" category of explanation.

The size of 3D Analyst Extension is testified to by the fact that Esri has a manual[1] on the subject, and John Wiley and Sons has a 200-page text on 3-D modeling using the extension.[2]

Throughout the text I have hinted at the possibilities of three-dimensional GIS. Before you start into the exercises, let's look at a list of things you can do with this 3-D visualization and analysis extension. One caveat: There is a lot of overlap between the capabilities of Spatial Analyst and 3D Analyst. For example, both deal with surfaces. Spatial Analyst is more concerned with analysis and 3D Analyst emphasizes display, but these two extensions share some tools and work as a team.

ArcScene

With ArcScene you can:

❏ Create surfaces with a number of tools and techniques.

❏ Analyze surfaces in a variety of ways.

❏ Drape raster images over surfaces.

❏ Drape vector features over surfaces.

---

[1] Using ArcGIS 3D Analyst Tutorial—available through Esri
[2] *Data in Three Dimensions: A Guide to ArcGIS 3D Analyst* by Heather Kennedy. Full disclosure: Heather Kennedy is the author's daughter.

- ❏ View three-dimensional surfaces in perspective from multiple observer points.
- ❏ View scenes differently by changing the shading, transparency, and illumination properties of 3-D layers.
- ❏ Change the "Z" visual component of 3-D scenes (exaggerate the vertical).
- ❏ Create surface models from rasters and TINs.
- ❏ Make queries on raster values.
- ❏ Get instantaneous information about the elevation, slope, and aspect of TINs.
- ❏ Create contours of surfaces.
- ❏ Find the steepest path on a surface.
- ❏ See a surface under different levels of illumination.
- ❏ Calculate the volume between a horizontal plane and three-dimensional surface specified by a TIN.
- ❏ Determine what parts of a landscape can be seen from various vantage points.
- ❏ Draw lines of sight across landscapes.
- ❏ Create a 2-D cross-section profile graph of a 3-D surface, given a line across the surface.
- ❏ Create 3-D features from 2-D data.
- ❏ Digitize 3-D features.
- ❏ Extrude 3-D features from 2-D features.
- ❏ Simulate "flying through" a landscape, using animation capabilities

ArcGlobe

The other software package in 3D Analyst is ArcGlobe—a startlingly effective piece of software that ought to be used in every elementary school in the nation. ArcGlobe is to a manual globe what ArcMap is to a paper map. It is a software lever that gives you an intellectual advantage in looking at the world. With ArcGlobe you can

- ❏ Get marvelous views of the Earth, rotating and panning its surface
- ❏ Add large data sets which are made part of the Earth's surface, correctly placed regardless of their coordinate systems.
- ❏ Automatically spin the globe around its axis, looking from any vantage point.
- ❏ Measure distances along great circle routes.
- ❏ Rotate an image either around the center of the Earth or around the center point of the image—doing global or local navigation.
- ❏ Make animations of the images you generate.

Again, the capabilities of 3D Analyst are best experienced rather than described. You do that next.

The Third Spatial Dimension

STEP-BY-STEP

_____ Open your Fast Facts File.

_____ Open the Color Figures file.

An (Almost) New Software Package: ArcScene

ArcScene is the software package you saw briefly in Chapter 2. We will use it to look at multiple 3-D data sets simultaneously. If we want to look at an individual data set, we can use ArcCatalog, using 3D View Preview. ArcScene operates a lot like ArcMap.

_____ Before starting the exercise that follows, let's get into a three-dimensional, "large view of the world" frame of mind. Start Menu > All Programs > ArcGIS > ArcGlobe. Initialize with a Blank Globe. Make the window occupy the full screen. In Globe Layers, turn off all layers except Imagery. Under the Customize menu, click Toolbars and place a check mark beside Spin to bring up the Spin toolbar. In the Speed text box, type 2.0. Click the Spin Counter Clockwise button (that's counterclockwise when looking down on the North Pole—this is the way Earth spins). (I suggest you turn off the stars[3] – although they look kind of neat. Right-click on "Globe layers" and select Properties. Click off the Stars.) Sit back and be impressed for at least three revolutions. Stop the Earth (Stop Spin button). Click Draped layers. Add as data:

[___] IGIS-Arc\Other_Data\Countries\Countries.lyr

Press Navigate on the Tools toolbar. (Hint: Find it with the ToolTip). Restart rotation. When your country shows up stop the spin. Use the left and right mouse buttons to position the globe so that you are looking at the middle of your location in your county from about a thousand kilometers away.[4] Restart the counterclockwise spin. Let one revolution of Earth go by—observing other countries that are at the same latitude as yours. Minimize the ArcGlobe window and begin the following exercises with Step 1, knowing that, as you do, the Earth will continue to turn.

---

[3] Actually, the stars in the background are a little confusing. If we were at a point in space looking at a spinning earth the stars wouldn't be moving. Our view is more like the one we would have from a high-altitude satellite, in which we are orbiting Earth. Our satellite would be moving from east to west (which is the direction opposite of most satellites). The stars are realistic when we move the Earth with a cursor, but not when it is spinning.

[4] See the Distance value in the lower-right corner of the window.

Experimenting with 3-D

___ **1.** Preliminaries: Start ArcCatalog. Let's check the defaults: Under Customize > ArcCatalog Options > General, make sure Hide File Extensions is unchecked. Click OK. Also be sure that ToolTips is turned on: Customize > Toolbars > Customize > Options and show tooltips on toolbars and on menus. Click Close. Under Customize > Extensions, make sure a check is in the box labeled 3D Analyst. Click Close.

___ **2.** More Preliminaries: Under Customize > Toolbars, make sure 3D View Tools is turned on. Locate the toolbar and drag it into the display window (right pane). In your Fast Facts File, using ToolTips, list the names of the buttons on the 3D View Tools toolbar.

Navigate,

Zoom In/Out,

and so on.

Done? Yes ___. No ___.

Double-click the title bar of 3D View Tools to put it back among the other toolbars.

___ **3.** *Final Preliminaries:* In ___IGIS-Arc_*YourInitials* make a folder named 3-D_Data. For use later, make a File Geodatabase named 3-D.gdb

within the 3-D_Data folder. Make a folder connection to 3-D_Data. From [___] IGIS-Arc\River, use ArcCatalog to copy the following data sets into the folder 3-D_Data.

❏ Boat_SP83.shp

❏ cole_dem

❏ cole_tin

❏ cole_DRG.TIF

❏ COLE_DOQ64.JPG

ArcScene

___ **4.** Start ArcScene, with a Blank map, from ArcCatalog, using one of the buttons you identified previously. Make ArcScene full screen. Locate the ArcScene Tools toolbar, and move it onto the display area. List the buttons on the Tools toolbar in your Fast Facts File.

Done? Yes ___. No ___.

Notice that while some of the buttons are the same as with ArcCatalog, some are different. Use boldface type to indicate the ones that are different. Put the toolbar back with the others by double-clicking its header.

What's 3-D and What's Not

The dataset you have worked with that truly includes the vertical dimension as *spatial data* is the TIN. All other feature classes, even if they have the possibility of the third dimension, are in flatland.

___ **5.** From ___IGIS-Arc_*YourInitials*\3-D_Data add as data, to ArcScene:

❏ Boat_SP83.shp (symbolize it with a red dot, size 4)[5]

❏ cole_dem

❏ cole_DRG.TIF

❏ COLE_DOQ64.JPG

❏ cole_tin

Push the Full Extent button. A lot of visual information will appear, although it is mostly covered up by the TIN. You are viewing it from a point southwest of the image and well above it. This is what is considered full extent in ArcScene. We will explore these data sets using the Navigate tool in ArcScene. The Navigate tool is a bit like a Swiss Army knife—lots of different tools in one artifact. As such, it is both useful and dangerous.

___ **6.** Click the Navigate tool on the Tools toolbar. Move its cursor into the display pane and play with it. If you drag left and right with the *left* mouse button, you can rotate the image around its center. If you drag up and down with the *left* mouse button, you can tilt the scene. If you drag (up and down) with the *right* mouse button, you can zoom the image. If the center mouse button is a wheel, you can use that to zoom the image. If you drag with the center mouse button or depressible wheel (or the left and right mouse buttons, held down together), you can pan the image. Experiment. Throughout all this, your good friend is the Full Extent button. Use it liberally.

___ **7.** Using the Zoom In/Out tool on the Tools toolbar, zoom in on the GPS track and the area where it cuts through the landscape made by the Kentucky River. View the scene from the southeast. See Figure 9-1.

___ **8.** *Use the rotating and tilting capabilities of the Navigate Tool:* If you rotate the scene so that you are looking at it from the southeast (along the GPS track) and from elevation zero, you will see two sorts of images. On the zero elevation level most of the data sets have merged into pretty much a single line. "Floating" above this line is the TIN, with its true 3-D representation.

We are not stuck with 2-D representation of these other data sets. First, we will give the DEM three-dimensional representation.

---

[5] You may come across some differences between ArcMap and ArcScene. So far, for one, the color palette is in a different order. For another the T/C is labeled Table of Contents, rather than Table Of Contents. Different programmers; different time frame.

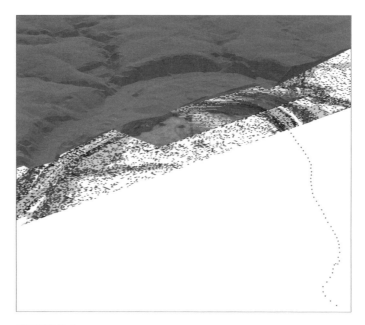

FIGURE 9-1

___ **9.** In the T/C, turn off all the images except the Digital Elevation Model (DEM) and Boat_SP83. Open the attribute table of the DEM. Note the Value field. Each record in that field indicates the elevation of the posts—the number of which is indicated by the Count field. Close the attribute table.

The DEM, while it contains elevation information as an attribute was represented as flatly as everything else (except the TIN). We can alter that. First, let's display it with more color and more differentiation between the heights.

___ **10.** Bring up the Properties window of cole_dem. Choose Symbology. Set the window to Show: Classified. Make 25 classes. Right-click the bar of the Color Ramp. Uncheck Graphic View. From the drop-down menu, choose Elevation #2. Click Apply, then OK. Take a quick look at the Value symbols in the Table of Contents. Move cole_dem around with the navigation tool to look at it.

Things are still flat—but more colorful and obvious. Now we add the vertical dimension. We actually have two sources of elevation information: the TIN, of course, and the elevation attribute (VALUE) contained in the DEM's table.

___ **11.** Bring up the Properties window of cole_dem again. Note that there is a Base Heights tab. Select it. In the "Elevation from surfaces" area choose the radio button "Floating on a custom surface". Choose the DEM itself as the surface to use. Slide the Layer Properties window away from the display, so you can see the effect as you press Apply. Close the Layer Properties window. Then concentrate on the river bend area, using pan, zoom, and navigation. One image you can see is shown in Figure 9-2.

FIGURE 9-2

_____ **12.** Just to reassure yourself that you have only changed the way cole_dem is displayed, and not the original data, add cole_dem again as data, and examine the situation with the Navigation Tool. What is going on here is that every pixel of the original cole_dem in the display has been given a height in accordance with the Value associated with that pixel. The original data set remains the same. (There is a way to permanently give a data set height values; you'll learn about this later.)

_____ **13.** Right-click this last cole_dem data set and remove it. Since you have gone to some trouble to make the first cole_dem—changing symbology and base height—make it a layer file (right-click on the name in the T/C, then select Save As Layer File) named cole_dem_layer.lyr **in the 3-D_ Data** folder. Add cole_dem_layer.lyr to the map.

Now that you have discovered the base heights, you can add an elevation component to the digital raster graphics image of the Coletown topographic map: cole_drg.

_____ **14.** Turn off cole_dem. Turn on cole_drg. Zoom to full extent. Then zoom to the river bend. Set things up so that you are looking at the bend in the river from the southeast, somewhat above. Bring up cole_drg properties and select the Base Heights tab. In the Elevation from surfaces area, choose the radio button "Floating on a custom surface". Last time you chose DEM as the surface to use, but pick the TIN this time. Click Apply, then OK. You should see the DRG "draped" over the elevations of the area. See Figure 9-3.

_____ **15.** Turn the Coletown DEM layer file back on. Zoom up on the southeast quadrant. Notice that you have a mess on your hands. In some places you see the DEM. In other places you see the DRG. Flip each on and off to see what is happening.

With 2-D images you will recall that the drawing order was controlled by the Table of Contents. With 3-D images this really doesn't work, but you can control the order in another way. Suppose that if both the DEM and the DRG are on, you want the DEM to prevail visually.

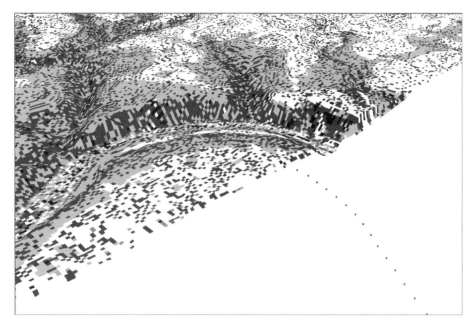

FIGURE 9-3

_____ **16.** Display the properties of the DRG. Select the Rendering tab. Under Effects, read the statement that suggests you have at least some control over the drawing order. Change the priority to seven, which lowers the priority, since "1" is the highest. Click Apply. Click OK.

_____ **17.** Make the display priority for the DEM two. Now as you navigate the scene you still will see places where the DRG will poke up through the DEM, but the DEM should predominate. Set the drawing priority of the TIN to 4 and turn it on.

_____ **18.** Navigate to a horizontal view of the river bend from the east. Note how the GPS track is decidedly underneath the other drawn layers. Set the base height of the GPS track (Boat_SP83) to the same as the DEM.

_____ **19.** Finally, set the base heights of the DOQ to that of the DEM. Turn off all layers except the GPS track and the DOQ. Notice how the DOQ is draped over the landscape. You can see the boat's path through the valley. Save the map as 3-D_Map_1.sxd in the 3-D Spatial folder.

Viewing 3-D Data with Animation

_____ **20.** Call for a new, Blank Scene file. Add as data COLE_TIN. From

[____]IGIS-Arc\Elevation_Data\Animation_Data

add Boat_SP83_elevated.shp.

In the T/C, double-click COLE_TIN to bring up the Properties window. Click the Symbology tab, and then click the Add button beneath the Show box. You will see an Add Renderer window. Add "Face elevation with graduated color ramp" and then press the Dismiss button in the Add renderer window. In the Layer Properties Show box, make sure that the Elevation symbology is at the top, and uncheck any other symbology type that may have shown up.

We will be performing several complex operations on the data. Navigating and other operations have to be applied to every feature and every pixel, which takes CPU horsepower and time. In general, you want only to display in the T/C the data sets in ArcScene that you need.

_____ **21.**   Boat_SP83_elevated.shp is the GPS track with which you are familiar, but converted to a polyline ZM (3d) shapefile. Zoom to the layer, and note that there is a high altitude portion (2000 feet actually) and then it descends rather sharply to an altitude of 640 feet, which puts it 100 feet above the river (which you may or may not remember has a normal pool elevation of 540 feet). Make the symbol for the elevated GPS track: a red line, size 2.

To set the background color, right-click Scene layers at the top of the T/C. Click Scene Properties. Under the General tab, choose a deep blue as the background color. Click Apply, then OK.

_____ **22.**   Zoom to Full Extent. Pan, navigate, then Zoom to COLE_TIN, so that the scene looks somewhat like Figure 9-4. Pretend that this is the beginning point of a helicopter flight that will take you through the gorge.

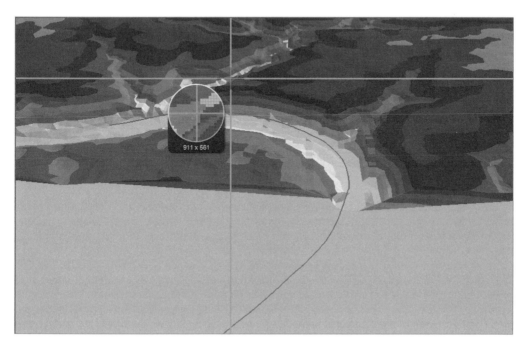

FIGURE 9-4

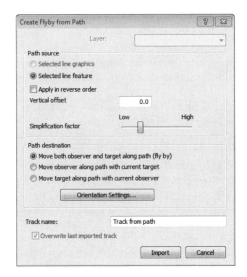

FIGURE 9-5

_____ **23.** In the T/C, right-click Boat_SP83_elevated, choose Selection, then Select All. (You will be creating a Flyby from Path, and the animation manager needs a selected line feature to perform that operation.)

_____ **24.** Select Customize > Toolbars > Animation and turn the Animation toolbar on. On the Animation toolbar drop-down menu, click Create Flyby from Path. Make sure that the dialog settings look like those in Figure 9-5. Click Import.

The idea here is that you are importing the process that creates the animation effect into the scene. This is a pretty new idea for you. Previously, maps and scenes were basically static entities that you could manipulate (and save for later), but here you have **_put a process into a scene._**

_____ **25.** On the Animation toolbar, click the rightmost button to open the Animation Controls. Since you imported the Flyby from Path animation, the "Play" button on the Animation Controls should be black, which means that there's animation ready to be played, once we set up some parameters.

Let's set things up so that the animation will last 30 seconds. Click Options in the Animation Controls window and set "By duration" to 30.0 seconds. Also, click off Restore state after playing. Click Options again to reduce the size of the Animation Controls window. One other thing: Turn off the Boat_SP83_ elevated line. Otherwise, it will just visually get in the way of the flight.

Ready? Click the Play button. Your view should be transported to the beginning of the polyline path. You will then fly toward the river gorge, suddenly dive down a few hundred feet (whoa!), level off, and fly steadily at 100 feet above the river. When the animation ends your helicopter will hover above the river, but below the top of the gorge. You can play the animation again just by clicking the play button. Experiment with different durations, and with the Boat_SP83_elevated line turned on. See Figure 9-6. Play the animation in reverse.

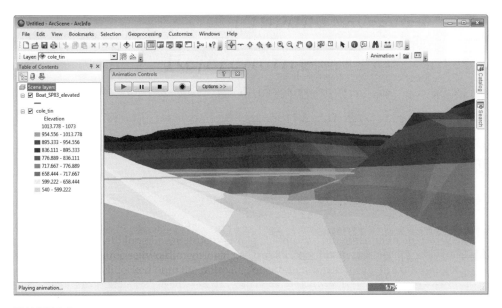

FIGURE 9-6

_____ **26.** Add cole_doq64.jpg to the scene. In its Properties menu set its Base Height to float on the custom surface of cole_tin. Turn off all layers except the DOQ. Replay the animation.

_____ **27.** Once you have a sequence you are happy with, save the Animation File (via a menu item on the Animation toolbar drop-down menu) as Flight_through Palisades.asa in your

IGIS-Arc_*YourInitials*\3-D_Data

folder. Also save the overall ArcScene file as Flight_through Palisades.sxd.

_____ **28.** Also, from the drop-down menu on the Animation toolbar, take a look at the Animation Manager, which gives you specifics about each keyframe created along the polyline path. Examine. There is a lot of detailed information there. How many keyframes made up this animation? _____. Click Close. When you're finished exploring, close ArcScene without saving changes.

_____ **29.** Using the operating system of your computer, navigate to Flight_through_Palisades.sxd and Open it. Bring up the Animation Controls and Click Play. The lesson here is that the **_animation is stored_** with the SXD file. Click on Full Extent, and save the Scene.

Making a TIN and Other 3-D Representations of Elevation

Suppose that some buried ruins have been discovered on a mountain top in a European country. An archeologist comes to you wanting a graphic representation of the site that might also serve as a repository for information about artifacts that are found at the site. We'll call the site Mount Paolo; it is small (about 80 meters by 60 meters). A surveyor has generated the elevations of a number of points, which are identified with UTM coordinates; they are supplied on an Excel spreadsheet. We want a map of the site that we can analyze and identify the locations of found artifacts. A perfect GIS problem!

Here is the plan: Assume that you save the spreadsheet in comma-separated variable (CSV) format. In the folder where the spreadsheet exists, you right-click on the name, select Open With, and use WordPad to see the data. You examine the various columns of data and create, as the first line of the file, a single header record that identifies the columns.

You then save the data as a text file named Mt_P.txt in [__]IGIS-Arc\3D_Data\The_Dig

Since you previously converted a spreadsheet to a text file earlier in the text (Exercise 6-4 on the census) we will skip that step and you will start with an already created textfile.

____ **30.** With the operating system, copy the folder

        [__]IGIS-Arc\Elevation_Data to

        ___ IGIS-Arc_*YourInitials*.

____ **31.** Inside Elevation_Data folder you will find the folder The_Dig. Inside that you will find Mt_P. txt. Right-click on the name and open it with WordPad. Examine it. The "columns" you are interested in are Northing, Easting, and Elev_MSL. None of the rest of the data will be used. Close the file.

____ **32.** Start ArcCatalog. Navigate to

        IGIS-Arc_*YourInitials*\Elevation_Data\The_Dig

        Right-click > New > File Geodatabase. Name the new geodatabase Mount_Paolo.gdb. Within the geodatabase make a Feature Dataset named Dig_Zone and click Next. Select the Projected Coordinate System:

        UTM > WGS 1984 > Northern Hemisphere\

        WGS 1984 UTM Zone 30N

        Click Next.

____ **33.** Under Vertical Coordinate Systems select World > WGS 1984. Click Next. Accept the default tolerances and click Finish. Refresh the Catalog Tree and exit ArcCatalog.

____ **34.** Start ArcMap and add as data Mt_P.txt from

        IGIS-Arc_*YourInitials*\Elevation_Data\The_Dig

        Open the table from the T/C to be sure it arrived intact. Close the table. Right-click the Mt_P.txt entry in the T/C and select Display XY Data.

____ **35.** In the Display XY Data window use the drop-down menus to make the X Field Easting and the Y Field Northing. Press Edit and make the XY Coordinate System the UTM Zone 30 North system you used earlier. Click OK. Read and OK the warning. Notice that Mt_P.txt Events has been added to the table of contents. Open this table and inspect it, then close it.

____ **36.** Right-click on the Mt_Ptxt Events entry and select Data > Export Data. For Output feature class click the browse folder. Navigate to

       ___ IGIS-Arc_*YourInitials*\Elevation_Data\The_Dig\Mount_Paolo.gdb\Dig_Zone

This should appear in the "Look in" field in the Saving Data window. Type Site in the Name field. Make sure that you pick File and Personal Geodatabase feature classes from the Save as type drop-down menu. Click Save. In the Export Data window click OK. Add the exported data (Site) to the map if it doesn't happen automatically.

____ **37.** Use Search to find Create TIN (3D Analyst). Where is the tool?
_____ Start the tool. (In version 10.0 this tool creates only an empty TIN; you will use Edit TIN to supply the data. In 10.1 the behavior is different.)

____ **38.** For the Output TIN window field browse to

IGIS-Arc_*YourInitials*\Elevation_Data\The_Dig

For the Name type Site_TIN. Click Save. Provide the correct spatial reference (UTM zone 30 north). Click OK. For Input Feature Class use Site. Click on the box below Height Field, and pick Elev_MSL from the drop-down menu. Click OK. When the tool finishes, you should see Site_TIN displayed as the map. Remove the Mt_Ptxt Events entry from the Table of Contents.

____ **39.** Use the Identify tool to look at both the points of Site and the triangles of Site_TIN. Pick a triangle, zoom in on it, and write down the elevations of its three vertices.

_____   _____   _____

Now write down the elevation, slope and aspect of a point near the middle of the triangle.

_____   _____   _____

Creating the TIN from a table of points is a bit convoluted, but the surface is a great way to represent elevation.

Creating DEM files with Kriging

Assume now that you show the results to the archeologist. She is suitably impressed but she is really used to contour lines and would like to see an elevation map in that form. Not a problem.

There are many ways to obtain three-dimensional surfaces from a discrete set of zero-dimensional points. None is exactly right (recall, there is no "exactly" with GIS points, lines, and surfaces, nor is there any "exactly" in any real-world coordinate set), but we can get "close." We will use the mathematical method called Kriging that you met before to create a digital elevation model (DEM, composed of square elevation posts). We will then use the DEM to make contour lines. If you are faced with a real-world problem of this sort, you may not want to do it this way (or you may), but the main idea here is for you to learn by being exposed to several methods.

_____ **40.** Start a new map in ArcMap, without saving the changes. From

___ IGIS-Arc_*YourInitials*\Elevation_Data\The_Dig

add as data the table Mt_P.txt. Open the table to be sure you have a 180-record table, with headings of Point, Northing, Easting, X, Y, H, and Elev_MSL. Close the table.

_____ **41.** Using Search find the Kriging (3D Analyst) tool in ArcToolbox. Where is the tool? _____ _____ Double-click the tool name to start the tool.

_____ **42.** In the Kriging window, for Input point features, browse to

___ IGIS-Arc_*YourInitials*\Elevation_Data\The_Dig

and click OK. (You may be able just to pick up Mt_P.txt Events from the drop-down menu.) Otherwise, navigate to

___ IGIS-Arc_*YourInitials*\Elevation_Data\The_Dig

to get the events table. For the Z value field use Elev_MSL. Accept the Output surface raster name—something like Kriging_Site1. The raster should go into

___ IGIS-Arc_*YourInitials*\Elevation_Data\The_Dig\Mount_Paolo.gdb.

While we could change a lot of the parameters on the Kriging window, we would probably need a statistician to tell us what the various options are. We will accept the default values, except for the cell size. Since we are dealing with a small area we can afford to make the cell size tiny. We'll use one-tenth of a meter.

_____ **43.** For Output cell size type 0.1. Click OK. Wait. When Kriging_Site1 is added to the map, click Close.

_____ **44.** Examine the raster. Use the Identify tool to look at some elevations. Zoom in to one of the higher areas of the map. You will notice the lack of "stair steps" that usually accompany rasters or grids. This is because the Layer Properties (Display tab) is set to show continuous data. To see the actual DEM cells, you can "Resample during display using" Nearest Neighbor (for discrete data) and zooming in. Measure the side of a DEM cell in meters. _____ If you want to see a table of the DEM, however, you are out of luck. Kriging creates a "continuous" surface, which has no attribute table because of the large number of possible values.

Add Site_TIN to the map. Experiment by looking at the elevations of each feature class. Click on the Display tab. Make sure the TIN is the first entry in the Table of Contents. In the Properties window (Display tab) of the Kriging site show MapTips. Then use the Identify tool on the TIN and compare the identified value with the map tip value.

Creating a Map of Contour Lines

Contours showing elevation can be created from a DEM surface or a TIN surface.

_____ **45.** Locate the tool Contour (3D Analyst), which makes contour lines from a DEM. Run the tool. The Input raster is the DEM – Kriging_Site1. Call the "Output polyline features" Dig_Zone_Contours. Make the Contour interval a quarter of a meter: 0.25 meters. Click OK. The contours will be added to the map. Make the contours bright red lines of width 1. Use the Identify tool to look at individual contours, paying attention to the Contour field. Again, you can compare the contour height with the MapTip generated by the DEM. You can also open the attribute table of Dig_Zone_Contours.

_____ **46.** Locate the Surface Contour tool, which makes contour lines from a TIN. Repeat the step above, but use the TIN (or Terrain) as the input data source. Make the name Site_TIN_Contours. Make the contour lines bright blue. Turn off everything except the two contour line feature classes. Which set of lines do you like better? _____ Why? _____ Save the map under any name you choose. _____

You have now seen three distinctly different ways of representing a surface—DEM, TIN, and contour lines. Each has its advantages and disadvantages. Each was, directly or indirectly, created from an initial set of measured points. The ways in which you created these representations are not the only ones; there are many others. In a real project, to get a representation with the accuracy you require, you would need to do some research on the mathematical and statistical procedures you should use.

Two-and-a-Half Dimensions (2.5-D): Calculating Volumes

There is sometimes a tension, if not confusion, within a given description regarding which dimension is being addressed by a term in geometry. Is a square the four lines that meet at vertices of 90 degrees, or is a square the surface formed by those four lines? Does a cube consist of (1) 12 lines in 3-D space, (2) the surfaces that those lines imply, or (3) the solid defined by those planes? That is, is the cube the wire frame of the lines, the surfaces of the planes, or the volume contained within?

When we say that a road network is represented by one-dimensional elements, we have to note that the network wanders all over the two-dimensional plane. When we talk about ArcGIS doing 3-D, we are usually talking about surfaces that are 2-D at any point, but overall occupy 3-D space. What we haven't addressed are volumes in three dimensions—that is, true three-dimensional objects. For example, a seam of coal underground is a true three-dimensional object. You can think of it as a volume that is defined by a multitude of planes, in the same way you think of a polygon as an area defined by a multitude of line segments. Frankly, ArcGIS doesn't do much with defining volumes, but it does have some capabilities.[6] The volume under (or over) a TIN can be calculated relative to some horizontal base plane. This takes place in the 3-D world, but the result might be described as 2.5-D, since one plane is the trivial, horizontal one. What follows is an example of calculation in two and a half dimensions.

_____

[6] There are computer programs that operate on three-dimensional objects. Mechanical engineering and other fields make use of these "solids modeling" programs.

Calculating a Volume with ArcGIS

Between two and three miles west of the river bend that you have looked at so often is a little hill peak that we would like to build a house on. We have to do some grading, because of the rugged terrain. We have surveyed the area and determined that if we create a flat surface at the 1020 contour elevation, we will have enough space to build and have a garden, a lawn, and maybe a tennis court. How many cubic feet of dirt would we have to deal with to flatten it out? We will use as data cole_TIN and our surveyor's contour lines to determine the answer.

___ **47.** Start ArcCatalog. From

___ IGIS-Arc_*YourInitials*\Trivial_GIS_Datasets

copy Four_contour_lines.shp to

___ IGIS-Arc_*YourInitials*\3-D_Data.

After looking at the geography of the lines, use Identify to determine the elevation, given by the field name CONTOUR, of the longest, and, in elevation, lowest line. _____. What is the elevation of the highest line? _____. Measuring up from the 1020 foot elevation to the top, what difference in altitude are we dealing with? _____ feet.

___ **48.** Start ArcMap with a Blank map. Make sure the 3D Analyst Extension is turned on. Make sure the 3D Analyst toolbar is turned on. From ___IGIS-Arc_*YourInitials*\3-D_Data, add as data cole_TIN followed by Four_contour_lines.shp. Make the symbol for the contour lines bright red, size 2. The location of interest will show up as a red dot on the TIN. Find it and zoom in on the area of interest. See Figure 9-7.

FIGURE 9-7

___ **49.** Explore Four_contour_lines.shp. What is the maximum distance across, indicated by the outermost, and lowest, line? _____ to the nearest foot.

Here is the plan (the workflow) that you will use for determining the volume of earth that lies under the TIN down to a plane at the 1020 foot level:

❑ Work entirely in the 3-D_Data folder. Using ArcCatalog, you will make a copy of Four_contour_lines.shp in the 3-D_Data folder. Rename the copy One_contour_line.shp.

❑ Edit One_contour_line.shp to remove the three innermost contour lines.

❑ Make a polygon that is bounded by the lines of One_contour_line line at the 1020 foot level.

❑ Add a field to the polygon that specifies that it has an altitude of 1020 feet.

❑ Use a tool named Polygon Volume to calculate the volume between the plane and the TIN above it. The volume will be added to the table of the polygon.

___ **50.** In ArcCatalog, make a copy of Four_contour_lines.shp to 3-D_Data. Rename the copy One_contour_line.shp. Add One_contour_line.shp to the map. Remove Four_contour_lines from the map.

___ **51.** Turn on the Editor toolbar. Start the Editor – editing One_contour_line.shp. Select and delete each of the three inner contour lines. Save edits, stop editing, and dismiss the editor toolbar.

___ **52.** Using the Feature to Polygon tool (where is it? _____) make a **polygon** named Mtn_Retreat.shp from the shapefile One_contour_line. When the tool finishes Mtn_Retreat.shp will be added to the map. Save the map as Mtn_Retreat.mxd in the 3-D_Data folder.

___ **53.** Open the attribute table of Mtn_Retreat.shp. Add a floating-point field named Altitude to the table, using Precision 6 and Scale 0. Start editing and put 1020 in the Altitude field of the single record of the polygon. Save edits; stop editing; close the table.

___ **54.** Use the Polygon Volume tool (where is it? _____) . What would you use for the Input Surface? _____. What would be the Input Feature Class? _____. What is the necessary Height Field? You want ABOVE for the Reference Plane. Accept the defaults for Volume Field and Surface Area Field. Look at Figure 9-8. Run the tool.

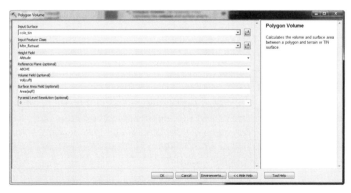

FIGURE 9-8

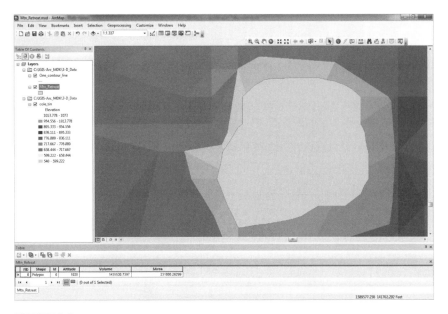

FIGURE 9-9

Open the attribute table of Mtn_Retreat.shp. Note that two new fields have been added to the table. What is the Volume of material that must be considered? _____ to the nearest cubic foot. (Put commas in, even though the software didn't.) Sarea indicates the number of square feet obtained by summing up the areas of all the triangles above the 1020 plane. Look at Figure 9-9.

Other Neat Stuff You Can Do with 3D Analyst: Viewshed and Hillshade

Suppose that you are interested in knowing what part of the Coletown landscape you can see from the river. ArcMap can calculate a "viewshed" that will graphically portray that information. Basically, you can select a set of points, and the software will show you all the surface area that can be seen from one or more of them. In our case, we will use the points from the GPS track to look at areas on cole_dem.

___ **55.** Add Boat_SP83 from ___IGIS-Arc\River. Make the size 2 and the color bright red. Find Viewshed (3D Analyst) (Where is it? _____.) Make the Input surface cole_dem. The observer points should be Boat_SP83. Make the result a raster called From_the_Boat and place it in the File Geodatabase that you made earlier: 3-D.gdb. The raster will be added to the map. Click OK. Wait. The percentage of the calculation that is complete may show up at the bottom of the screen as a blue bar. Realize that ArcMap is calculating a ray from each point in the GPS track to each 100 by 100 foot square in the entire Coletown DEM. If your computer is slow you probably have time to hand-calculate how many such pixels there are in the 6.7 mile by 8.6 mile TIN. Or you could just go get some coffee.

___ **56.** The Viewshed of From_the_Boat will appear in the Table of Contents. Change the Not Visible area to hollow or no color. Change the Visible area to a blue-green. The colored squares that

FIGURE 9-10

appear represent the cells of the DEM that can be seen from the positions on the river by people on the boat at the points where GPS readings were taken. This isn't much use until we add some additional graphical information. Here you may experiment. Start by adding the cole_doq. In the T/C, drag it below the viewshed layer and zoom to the layer. Unfortunately, the viewshed blots out the details of the imagery. Let's try various levels of transparency in the viewshed layer (Layer Properties > Display). Maybe 80% to start with. Notice that some of the water plant is visible but most is hidden from the GPS points on the river. See Figure 9-10. Now experiment with different colors and transparencies of From_the_Boat.

___ **57.** Turn off the viewshed and cole_doq. Add cole_dem to the map. Add cole_drg to the map. Zoom to the GPS track. Pan the map to the east so that it covers the area shown in Figure 9-11. Find the Hillshade (3D Analyst) tool. Where is it? _____. Start the tool. For the input surface choose cole_dem. Note that the default has the light source coming from 315 degrees (northwest) and at an angle above the horizon of 45 degrees. Call the resulting raster Hillshade_1 and put it into 3-D.gdb. Leave everything the same. Click OK.

Once this process completes, a Hillshade of cole_dem should appear at the top of the Table of Contents and a grayscale raster will appear. Note that everything is pretty bright except the southeast-facing slopes of the river gorge.

___ **58.** Rerun Hillshade (being careful to specify cole_dem), lowering the sun altitude to 30 degrees above the horizon and shifting the sun so it comes from the southeast (135 degrees). Call the resulting raster Hillshade_2. Note the reduced amount of light everywhere and the deeper shadows. Pick a color scheme that shows the sunniest spots as yellow, using the invert box in the Symbology window. Notice how the steep, southeast facing slopes of the river canyon are lit up with yellow. Look at the DEM, the DRG, and the output of Hillshade_2, adjusting the transparencies so the image looks something like Figure 9-11.

FIGURE 9-11

___ 59. Rerun Hillshade, making Hillshade_3, putting the sun overhead at 90 degrees. There are no shadows, of course, but the image reflects different amounts of illumination on steep slopes. Dismiss ArcMap.

A Closer Look at ArcGlobe and Adding Data to It

___ 60. While you have been doing the preceding steps (presuming that you have been working straight through the steps—which may not be the case), ArcGlobe has been running in the background, spinning the Earth. Restore the ArcGlobe window or restart the program. Stop any spinning. Zoom to Full Extent. Move the globe around, so you can see the United States. Zoom so that it pretty much fills the window.

___ 61. Using the same procedure as with ArcMap, add the following data sets, one at a time, to ArcGlobe.[7] After each one, use Zoom To Layer. Plan to wait for the computations to take place. (Display them as 3D vectors.) (Show layers at all distances.)

[___] IGIS-Arc\Other_Data\

KY_County_Boundaries_Geo.shp

---

[7] This is one place where you don't have to worry much about the warning regarding different coordinate systems, for two reasons: (1) one of ArcGlobe's features is that it converts data on the fly to the globe's coordinate system, and (2) with the large environment we are dealing with here, small problems in accuracy won't usually be relevant.

(This is a geographic coordinate version of the Kentucky Boundaries coverage that you used previously, when its projection was the Kentucky state plane north zone.)(Change the symbol color to light green. Display at all distances.)(Then zoom to this layer.)

[___] IGIS-Arc\River\Lexington.mdb\Roads
(Change the symbol color to black.) (Then zoom to this layer.)

[___] IGIS-Arc\River\cole_DEM
(Use this layer as an image source)

[___] IGIS-Arc\River\cole_drg.tif

[___] IGIS-Arc\River\COLE_DOQ64.JPG

[___] IGIS-Arc\River\Boat_SP83.shp
(Display symbol in point units. Use a red hexagon symbol, size 5.)

____ **62.** Experiment with the Zoom control (right mouse button or mouse wheel) and zooming to various layers. Notice that the image remains centered, allowing you to keep an area of interest in the middle of the map, even though you can't see it. Look at the image at *about* the following kilometer distances: 5, 11 (altitude of a jet airliner), 20, 50, 100, 370 (altitude of the International Space Station), 3000, 20,200 (altitude of the GPS satellites), and 35,786 (geosynchronous (e.g., TV)) satellite distance.

____ **63.** Save the 3000-kilometers-away map as

_____ IGIS-Arc_*YourInitials*\3-D_Data\Globe_Map.3dd

____ **64.** Using the operating system, determine the size of Globe_Map.3dd. _____. This small size should tell you that ArcGlobe functions like ArcMap: The map file only serves to point to the data sets and how they are drawn. If you send someone a 3DD file without sending the underlying data, you might as well not send anything.

What's interesting and useful here, other than the startling graphics and the ability to zoom from a space ship view or moon view of Earth, is that you can easily add your own data to the globe—something not yet possible with other Earth view programs that you find on the Internet.

____ **65.** Start a new ArcGlobe session with a Blank Globe. Turn off the Elevation layers if they are on. Right click on "Globe layers"Add as data

[___] IGIS-Arc\River\Lexington.mdb\Roads
(Display features as 3D vectors—although they really aren't 3-D; use red lines, width 3)

[___] IGIS-Arc\Other_Data\Crystal_Lake_GPS.shp (use a red square, size 4)

[___] IGIS-Arc\Image_Data\Pilgrim.tif

____ **66.** Zoom to the image layer. Order the Layers so that you see a GPS track that outlines Crystal Lake, Michigan, and goes through the town of Frankfort, on the western coast of Lake Michigan. Zoom to the GPS layer. Zoom out so you can determine the general location.

____ **67.** Zoom out so that you are looking from about 500 kilometers away. Use the Pan tool to position the globe so you can see both the GPS track in the north and Lexington Roads in the south.

____ **68.** *Use the Measure tool to determine the approximate distance from Lexington, KY to Frankfort, MI.* Zoom to the approximate center of the Lexington roads using Zoom To Layer. Start the measure tool. Make the measurement units miles. Zoom to the GPS layer. Double-click the approximate center of Frankfort, MI. Answer: _____ miles.

____ **69.** Click on the Navigate icon. Find the Navigation Mode button. Read the explanation on the Status bar or the ToolTip.

Until you change the sense of this button (or the software does, as it seems to sometimes without really telling you), the Navigate tool operates in "globe navigation mode" signified by a blue background of the icon. When you move the cursor, the globe revolves around the center of the Earth. When you go into "surface navigation mode," (the icon has a white background) the center of rotation of the map becomes whatever point you specify by clicking on the globe.

____ **70.** Click the Navigation Mode button so that its background no longer appears blue. Experiment with the Navigate cursor. You can actually look at the Earth from the inside. With Zoom, move out from Earth some 5000 kilometers. Use Navigate again and note the effect: The center of rotation stays locked at the point on the surface where it was before.

____ **72** Add another GPS track as data to the map:

[___] IGIS-Arc\Other_Data\Mystery_Location _1.shp (use a red square, size 18). Where is it? _____. What is it?[8] _____.

____ **73.** Add as data

[___] IGIS-Arc\Other_Data\Countries\Countries.lyr

Also add another GPS track as data to the map:

[___] IGIS-Arc\Other_Data\Mystery_Location _2.shp (use a red circle, size 4).

Zoom out. Use the Identify tool on the Countries layer. Which ones are represented? _____.

____ **74.** Dismiss ArcGlobe.

Making a Terrain

Because of the very large data sets involved, making a Terrain is somewhat involved. Here is a summary of the steps that you will use.

_____
[8] Hint: Hot. Dangerous.

❏   In the 3-D_Data folder in your own workspace you will navigate to Terrain_Data

❏   Examine TXT and XYZ files of LIDAR points

❏   Look at some ancillary files

❏   Make a geodatabase and feature dataset to hold the terrain.

❏   Import the ancillary files

❏   Determine the average point spacing of the LIDAR data

❏   Convert the XYZ data to multipoints and place them into the feature dataset

❏   Determine pyramid levels and make the Terrain feature class

❏   Build the Terrain feature class

❏   And finally you will examine the Terrain

Note: Another way to make a Terrain is to use what is called the New Terrain Wizard. You can find it by right-clicking the feature dataset name in the geodatabase (in our case you would right-click Zone in Example.gdb), click New, then Terrain to bring up a New Terrain window. [9]

In Chapter 2 you looked at a Terrain feature class in the area of a bend in the Kentucky River. In the following steps, you will actually produce a Terrain from basic LIDAR data and some ancillary data. One thing about LIDAR data is that the dataset is usually quite large. It is not unusual for a LIDAR file to contain tens of millions of points. The standard, public format of LIDAR data is the LAS file. It is a binary file that stores not only point coordinates (x, y, and z) but also intensity values. Because we are primarily interested in making a feature class that just shows elevation, this LAS file has been converted to a simple text file consisting of three values per line, separated by commas. An initial line was added, consisting of "Easting", "Northing", and "Elevation" so that ArcGIS would recognize the file as a table.

___   **75.**   Using the operating system navigate to

___ IGIS-Arc_*YourInitials*\Elevation_Data\Terrain_Data\LIDAR_Points.txt

and open that file with WordPad or Notepad. Note that you see "columns" of three numbers—the columns being headed by text identifiers. Note also that it is a file of many lines—more than a third of a million actually. Make a copy of that file by clicking on it, using Ctrl-C to put a copy on the clipboard, navigate up the hierarchy to the folder Terrain_Data, and pasting the copy using Ctrl-V. Rename the copy LIDAR_Points.xyz. Close the file. Why make a duplicate file? We do this because, while ArcCatalog will display a properly formatted .txt file it won't display a .xyz file, and we need a .xyz file to feed to ArcToolbox later when we make the terrain. [10]

---

[9] It's a little off the point to say this, but since, if you are learning this subject you are probably a bright, creative person who will sometimes in your professional career have the responsibility of naming things. My suggestion: Don't call anything "New". It gets embarrassing later. I live in a city that has a New Circle Road that is coming up on being half a century old. When I taught at the University of North Carolina my office was in the New East Building (built around 1860. It could be contrasted with the Old East Building (built in 1793).

[10] ArcCatalog can be set to display the existence of files with an xyz extension. Customize > ArcCatalog Options > File Types > New Type. In the File Type window, put in xyz; in the Description of type put Input for Terrain. OK. Apply. OK. (ArcCatalog will not display the contents but will show you that the file exists.)

Easting	Northing	Elevation
1309535.41	759774.71	962.24
1309528.69	759789.84	966.96
1309508.39	759835.15	973.52
1309501.96	759830.65	973.99
1309508.91	759815.16	972.71
1309522.72	759784.24	968.05
1309529.36	759769.24	963.52
1309536.06	759754.11	958.58
1309542.78	759738.94	953.95
1309549.55	759723.7	950.05
1309556.43	759708.24	946.59
1309563.46	759692.56	945.49
1309570.48	759676.92	944.01
1309577.45	759661.32	942

FIGURE 9-12

____ **76.** Start ArcCatalog. Navigate to

IGIS-Arc_*YourInitials*\Elevation_Data\Terrain_Data.

There you will find LIDAR_Points.txt, water_body.shp, and break_lines.shp. (The break_lines shapefile serves to limit the triangulation that is going take place. The feature class water_body is there primarily to prevent the embarrassment that would be created by slight errors in the elevation, creating differences in elevation when there were none, or when the differences were cause by instantaneous altitude differences caused by ripples or waves. We will use these later in the Exercise. For now just look at the two shapefiles—both the geographics and tables—to get an idea of the area.)

Select the table LIDAR_points.txt. When ArcCatalog finds a comma-delimited file with text headings, it sees it as a table it can work with. Get to the end of the file, which will look like Figure 9-12. How many records are there? _____. (Hint: click the "Move to end of table" button (in the same line as "Record") and wait.)

____ **77.** *Make a File Geodatabase that will contain the Terrain:* Navigate to

IGIS-Arc_*YourInitials*\Elevation_Data\Terrain_Data

Right-click the folder Terrain_Data, and select New from the drop-down menu. Make a new File Geodatabase and name it Example.gdb.

Right-click Example.gdb and make a new Feature Dataset named Zone. You have to assign a coordinate system to the feature data set. The feature classes water_body.shp and break_lines.shp have the proper coordinate system (including projection, datum, and units), so you can just import the coordinate system from one of them. Highlight Projected Coordinate Systems, find the Add Coordinate System drop-down menu (a world shaped icon), and press Import. Browse to

IGIS-Arc_*YourInitials*\Elevation_Data\Terrain_Data\water_body.shp

and press Add. The coordinate system name will be included under Favorites. Click Next. Expand Vertical Coordinate Systems, expand North America and pick NAVD 1988. Click Next. Click Finish to complete making the empty feature data set.

Terrains in ArcGIS, although they are feature classes, don't really exist in the way that other feature classes exist. Terrains resemble TINS but because of the massive number of data points ArcGIS creates terrains on the fly, depending on the spatial extent that is being viewed. A terrain sort of half-exists; the underlying points are stored, of course, but the location and degree of triangulation that takes place in order to use or display the resulting TIN is variable, depending on what is needed. When you are using larger scales (zoomed in), only a limited portion of the underlying points need be used to create the image—which saves drawing time. At smaller scales (zoomed out, so you would expect to see less detail), only a subset of the points in the area is used to make the triangulation. What you see when you view a terrain is an image that is calculated "on the fly." The person who builds the terrain specifies parameters that determine levels of detail used. Since the terrain is a dynamic entity the feature classes that are used in producing the terrain (in our case, the points and shapefiles) have to be accessible to ArcGIS. The software requires that these data must reside, in a new form, within a single feature dataset.

_____ **78.** *Import the needed shapefiles into the feature dataset:* In ArcCatalog, right-click Zone, pick Import and then Feature class (multiple). In the resulting window browse to Terrain_Data and, in the Input Features window, using the Ctrl key and clicking, pick both water_body.shp and break_lines.shp. Click Add. Make sure the Output Geodatabase reads

IGIS-Arc_*YourInitials*\Elevation_Data\Terrain_Data\Example.gdb\Zone

if it doesn't already. Click OK. When processing is completed (watch the Status bar), expand Zone to check that the features of the two shapefiles have been duplicated inside the feature dataset as feature classes. They aren't shapefiles anymore, but they contain the same information as the originals.

As you determined earlier, LIDAR_Points.txt consists of hundreds of thousands of individual geographic locations. While we could represent each of these as a point feature, it is more sensible to group them into multipoints. This preserves the geographic location of each but allows the software to produce many single records, each of which subsume many positions.

_____ **79.** *Make a multipoint feature class based on LIDAR data:* For this you need ArcToolbox. Since ArcCatalog is already open use the toolboxes there. In ArcToolbox expand 3D Analyst Tools. Expand Conversion, followed by From File. The main tool you will use in a moment will ask you for the average point spacing, so we will get that information first. There is a tool you can use to obtain this number. Double-click on Point File Information. Browse for the input file

IGIS-Arc_*YourInitials*\Elevation_Data\Terrain_Data\LIDAR_Points.xyz

Click Open. Accept the default Output Feature Class name. The File Format must be XYZ. Click OK. In the Catalog Tree highlight

LIDAR_Points_PointFileInform.shp.

The Geography of that shapefile is completely unimpressive. But the Table is more interesting. It consists of a single record. What is the number of points indicated _____. What is the average spacing (to the nearest tenth of a foot) between the points _____. What is the elevation of the lowest point? _____. The highest. _____.

In that same "From File" toolbox/tool in ArcToolbox, double-click ASCII 3D to Feature Class to start that tool. You want to Browse for Files. Again the file you want is:

IGIS-Arc_*YourInitials*\Elevation_Data\Terrain_Data\LIDAR_Points.xyz

The Input file format should, of course, be XYZ. To set the Output Feature Class navigate to

IGIS-Arc_*YourInitials*\Elevation_Data\Terrain_Data\Example.gdb\Zone

and type LIDAR_multipoints in the Name field. Click Save. The Output Feature Class Type should be MULTIPOINT. For the Average Point Spacing use either the value you wrote down above or 16. Click OK. Wait while processing finishes. At this point you have converted the points represented in the ASCII .xyz file to a multipoint feature class.

___ **80.** Using ArcCatalog navigate to LIDAR_multipoints (in Zone) and Preview the Geography. It will look like a Rorschach inkblot test[11] because of the high point density (indicated by black) and the fact that the original LIDAR file had those points which hit water masked out (appearing white).

Zoom in on an area of the map—perhaps where a strip of white meets the black. You will see multitudinous, irregularly spaced points. Look now at the table associated with the multipoint file. How many records are there in the file? _____

___ **81.** Obviously, this is quite a reduction from the number of records there were when each point had its own record. What is the maximum number of points in a multipoint? _____

___ **82.** Now it is finally time to make the Terrain feature class. It will exist inside the feature dataset Zone, so right-click that name and select New > Terrain. Make the name My_Terrain. The feature classes that will participate in the terrain are LIDAR_multipoints, break_lines, and water_body, so place checks beside each. Again supply the average distance between points, and click Next.

___ **83.** Indicate how each feature class will participate in the terrain. There are lots of choices and, if you are building a terrain on your own you will have to do some research to determine which to use. In this case, make this window contain the information of Figure 9-13 – using drop-down menus by clicking, as it turns out, the one item you need to change. Click Next. Select the pyramid type as Z Tolerance. Click Next.

Here you come to the issue of building pyramids. Pyramids, as you may recall, is the term used to describe files of data that are based on extensive datasets, but that are thinned down so as to make for faster drawing. This is a convenience for some datasets. But for the massive terrain datasets it is virtually a requirement. Further, a number of pyramid levels can be specified for a given terrain, depending on the scale at which the terrain is being viewed.

---

[11] The Rorschach inkblot text was widely used, controversially, for differentiating psychotic from nonpsychotic thinking. Make a Web search for more information if you are interested.

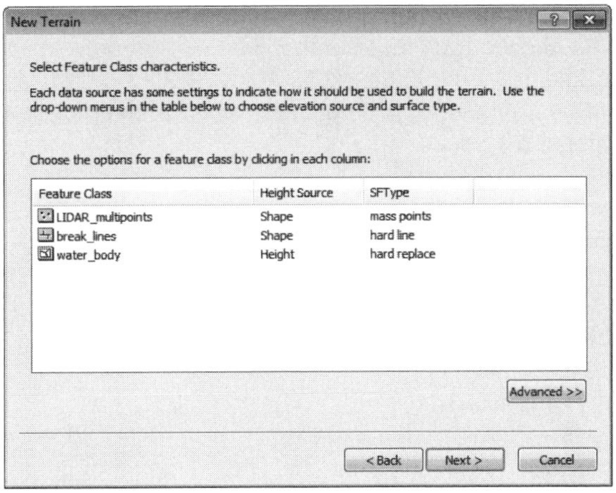

FIGURE 9-13

Specifying terrain pyramids can seem a bit confusing but need not be, once you understand the concept. First off, if you view a terrain in detail—at a large scale (that is, 1 divided by a small number, e.g. 1:500)—you get all the elevation information that can be wrung out of the data. Pyramids come into play as you decrease the scale, zooming out, viewing larger areas with less detail. The idea is that you can accept information from fewer data points and only look at elevation differences if they are greater than a given threshold. So, perhaps, when you look at the image at a scale of 1:5000, you need only those points that result in an elevation difference of 10 feet. If you use a scale of 1:20000, you might be content with elevation differences of 40 feet. The area we are looking at fills the computer screen, very approximately, at a scale of about 1:24000, which is the scale used by USGS 7.5 minute topological quadrangles. To set the pyramid threshold levels and elevation differences do the following.

_____ **84.** Study Figure 9-14. It implies that if the image is zoomed in to a scale of greater than 1:2500 (e.g., 1:800), you will get the full TIN detail. From 1:2500 to 1:5000 you will see less detail—

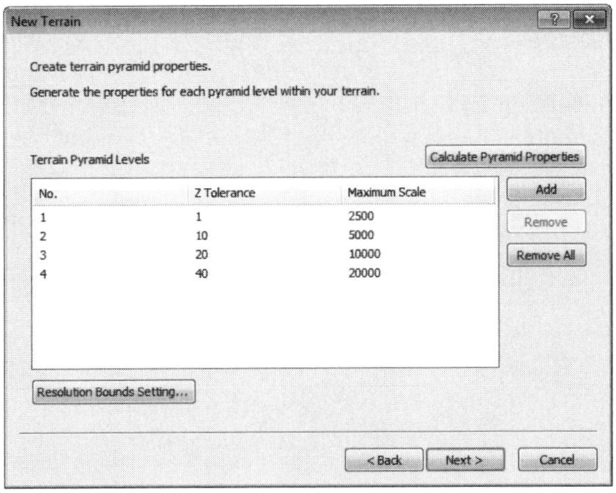

FIGURE 9-14

with elevations differences greater than or equal to 5 feet shown, and elevation differences of less than that not shown. Even less detail will be evident between 1:5000 and 1:10000. The same holds for differences between 1:10000 and 1:20000. The least detail, with only elevation differences of 40 feet or greater, will be shown when zoomed out beyond 20000. In all, how many levels of detail are specified here? _____

___ **85.** In the New Terrain window that lets you specify pyramid properties, press Add four times. Change the numbers so that the window looks like Figure 9-14. Click Next.

___ **86.** Review the window of terrain settings. If they are not correct, go back and fix things up. If they are, press Finish. You will be informed that the Terrain has been created but not built. Build it, and be prepared to wait a bit.

___ **87.** Once the terrain (My_Terrain) is built, use ArcCatalog to look at it with different zoom levels, starting with Full Extent. Note the difference in the amount of detail that is apparent, which is illustrated by being able to see more triangles as the scale increases.

___ **88.** Start ArcMap with a Blank map. Dismiss ArcCatalog. Add My_Terrain. What does the scale textbox in the standard tool bar say? _____ Type 5001 into that box and press Enter. Observe the result. Now type 4999 into the scale box (which gives you almost precisely the same scale), but carefully watch what happens to the map immediately after your press enter. What you are seeing is the software recalculating the TIN to the specifications you gave earlier. That is, you see differences of 5 feet, whereas before you saw 10 foot differences.

___ **89.** Investigate the differences using scales of 1:2501 and 1:2499. Any scale greater than 1:2499 will give you all the detail the data set is capable of, which is a lot!

___ **90.** Zoom in more and use the Identify tool on a triangle. What is the elevation at the centroid of the triangle? _____. What is the slope? _____. What is the aspect? _____.

___ **91.** Because of the shading of the TIN that ArcGIS provides you can probably estimate the aspect of a triangle (i.e., the compass direction it faces). Pick one whose direction seems obvious. What do you guess it to be? _____ degrees (north being 360). What is the aspect as indicated by the Identify tool? _____ What are the elevations of the triangle with the cursor placed close to each of the three apexes. _____, _____, and _____. Close ArcMap.

As you can see, building an Esri Terrain is not a trivial matter, but LIDAR and the Terrain bring us closer to the (ultimately unrealizable) goal of a perfect model of the Earth's surface.

The Time Dimension

3-D: 2-D (Spatial) Plus 1-D (Temporal)

In this section, we look again at three-dimensional GIS—but the third dimension is time rather than a spatial dimension.

Time is usually considered an enemy of GIS. Time, among other things, ensures that a GIS is always out of date in some respect or other. The world around us changes constantly—because of both natural effects and human action. The bits in a computer's memory persist in their states unless changed. Time marches on, gradually making our data obsolete or forcing us to update continually. Such updates are time-consuming and expensive. Imagine, for example, that a major feature of your database consisted of ortho photos, obtained from aircraft. How often do you re-fly the area so the photos reflect new development or changes to the environment? And at what cost?

Another issue with time and GIS is that very little attempt is usually made to preserve states of the database with an eye to analyzing changes later. If a piece of property is subdivided, or a road built, the affected databases may well be updated, but the time at which the update happens is usually lost, as far as any easy access by those who might want to compare "what is" with "what was."

Again we face the issue of continuous versus discrete. The movement of time is a continuous phenomenon. Incremental changes happen within the unfolding of time. Are the changes large enough to warrant modification of the database? Consider a house in an historic district. The paint on the house deteriorates slowly. When is the record of the appearance of the house changed from excellent to good to fair to poor? So, Question 1 is how and how often do you update your database. Question 2 is what does it take to trigger a change. Suppose the house is repainted a different color. Suppose that an extension is added. Suppose the house is demolished.

These are merely the complications time poses for our two-dimensional GIS. Looking to the future, can GIS be harnessed to allow analysis of changes to the environment over time? Certainly, old maps can be scanned in and interpreted. Such data can be compared with more recently installed GIS data. Those interested in using GIS for history should read Anne Kelly Knowles' book *Past Time, Past Place* (Esri Press, 2002).

But consider: Old maps have dates on them. Suppose that you are interested in making comparisons between conditions now and those 10 years ago in a venue that has been using, and continually updating, a GIS database. How would you know what changed when?

Since the beginning of this book you have grappled, at least theoretically, with representing the continuous world in the discrete memory of a computer. Now another continuum, time, is tossed in. Your reaction, and, for the most part, the reaction of the "GIS industry," has been to ignore the topic. After all, it's hard enough to get things right in two dimensions. But this issue, along with the problem of information quality control in general, is one that will have to be faced. For an older but still valid theoretical treatment of the temporal issues related to GIS read Gail Langran's book *Time in Geographic Information Systems* (Taylor & Francis, 1993).

The Time Dimension

Let's run through a quick example of how files of the same variety of data, taken at different times, might be compared. Just as an example of one way of doing the comparison, we'll make use of the topology tool available to us in geodatabase feature data sets. We'll use the buffer tool as well. For a second example, we use the ArcMap Time Slider to look at the history of road repaving over several years.

Suppose that we are interested in knowing what new streets were built in a city in the United States over approximately a decade. We have available a feature data set derived from the 1994 TIGER/Line files from the 1990 census. We also have more current files: roads from the 2000 census. Suppose that we place both files in a personal geodatabase's feature data set. Then, we could make a display showing both files. There would, of course, be a great amount of overlap. We can identify that overlap using topology. If we could then delete the streets that are (almost) coincident, we would be able to see the new streets that were added. You do that in the following project.

Exercise 9-2 (Project)

Looking at Infrastructure Changes Occurring over Time

____ **1.** Start ArcCatalog. Copy the folder

[___] IGIS-Arc\Historical_Data

to your

___ IGIS-Arc_*YourInitials* folder.

____ **2.** Create a folder connection to this new Historical_Data folder.

____ **3.** Inside the Historical_Data folder you will find the Geodatabase Roads.mdb. It contains the feature data set Lexington_SP_North_83_Feet. Within that are the feature classes Lex_Roads_1994 and Lex_Roads_2002. Make sure the Catalog Tree is showing (Windows > Catalog Tree).

____ **4.** Highlight the feature dataset name Lexington_SP_North_83_Feet. Using Ctrl-C and Ctrl-V, make a copy of it in Roads.mdb. Make a second copy, also in Roads.mdb. Rename the copies Lex_

Backup_A and Lex_Backup_B. You will be using topology operations on the Roads database and anytime you involve topology operations, you need backups of your original data, because those operations can move objects. Two backups may be overdoing it, but it is unusual to have too many backups. More frequently, one does not have enough.

___ **5.** Preview the Geography of each feature class: Lex_Roads_1994 and Lex_Roads_2002. You can note, as you flip back and forth, that a fair number of roads were added in the intervening eight years.

___ **6.** Look at the table of each feature class. How many road segments are represented in the 1994 roads file?[12] _____. In the 2002 file? _____.

___ **7.** Look at the Description of one of the feature classes. What is the Geographic Coordinate system name? _____. What are the Units? _____.

We are going to attempt to identify those road segments that are congruent, so we can ignore them and be left with only those segments that are not part of both datasets. You will recall that it is inadvisable to ask if two numbers that are computed by a machine are equal. Applying the same principle, we will ask if two segments are almost congruent—that is, their vertices are almost identical. We will do this by using topology. As you will recall, a topology may exist on the feature classes that are within a feature dataset.

___ **8.** Highlight the feature dataset Lexington_SP_North_83_Feet. Choose File > New > Topology > (read the window) > Next. For the name, accept the default. But do not accept the default for the cluster tolerance. Read the description of the cluster tolerance. Let's assume that two road segments are coincident if all their vertices fall within 50.0 survey feet of each other. Type that in. Click Next.

___ **9.** In the New Topology window, select both feature classes. Click Next.

It would probably be a safe bet that the 2002 TIGER files are more accurate than the 1994 files. So we should set things up so that it will be the 1994 lines that will move in case they are not coincident with the 2002 lines.

___ **10.** Set the rank for Lex_Roads_1994 to 5. Click Next. Add the rule "Must Not Overlap **With**".[13] In the Add Rule window use Lex_Roads_1994 for one feature class and Lex_Roads_2002 for the other. Toggle the Show Errors button a couple of times and observe the Rule Description. Red indicates the sort of error that the topology rule is looking for. This operation is only for your information; it has no effect on the functioning of the software. Click OK, then Next. Check the Summary and click Finish.

___ **11.** After the new topology has been created, select Yes to validate it. Be prepared to wait for a bit. Think of all the road segments the computer has to look at, and maybe reposition. It takes time.

___ **12.** Lexington_SP_North_83_Feet_Topology should appear in the Catalog Tree, along with the Roads feature classes. If you click on it with the Preview tab pressed, you will see a lot of pink,

---

[12] These files were modified somewhat; records in which the FEATURE NAMES were blank were deleted.
[13] Note the rule name. Four words!

representing the road segments that were coincident. Those that weren't coincident are not shown. Check the drop-down menu in the Preview text box at the bottom of the display. You will notice that there is no table associated with the topology.

____ **13.** Launch ArcMap with a Blank Map. Dismiss the ArcCatalog program but bring up the Catalog sidebar (Windows > Catalog) in ArcMap. Drag Lex_Roads_2002 and Lex_Roads_1994 from

IGIS-Arc_*YourInitials*\Historical_Data\Roads.mdb\Lexington_SP_North_83_Feet

onto the map.

____ **14.** In the T/C make the 1994 roads black, width 1. Make the 2002 roads bright green, width 1. Arrange it so that the 1994 roads are drawn last—that is, place its entry at the top of the T/C. You can see that most development occurred along the southern and eastern edges of the city. Running "topology" on the data sets has snapped some features of the data sets together, where appropriate. Add the

Lexington_SP_North_83_Feet_Topology

to the map. (Don't add the associated feature classes—they are already there.) Now individual road segments can be identified, so the new ones can be picked out. Generally, whatever is green is new. Not so good is this: If you zoom in on some areas, you will notice that roads that are probably meant to be the same still show up in both data sets. For example, Label Features for both Lex_Roads layers. Then use Selection > Select By Attributes. For Layer pick Lex_Roads_2002. Noting the single quotation marks, put in the query

[FENAME] = 'Elmendorf'

Click Apply. Click OK. Now use Selection > Zoom to Selected Features. Clear all selections. What you will see is two segments of Elmendorf that go their separate ways for a bit and then re-converge. The topology operation did not see them as almost coincident, although all the other roads around them were flagged, in pink, as errors.

Frankly, older TIGER/Line data left a lot to be desired, so there are segments such that at least one vertex in a feature was more than 50 feet away from vertices in the same feature in the other data set.

____ **15.** Let's further improve the presentation. It would be nice if we could use the topology data set to erase the 2002 roads that are coincident with the 1994 roads. Bring the Topology toolbar and the Editor toolbar into view. In the Table of Contents, make sure Lex_Roads_2002 is on, as well as the topology. Turn off Lex_Roads_1994.

____ **16.** Choose Selection. Select By Attributes. Select those roads in Lex_Roads_2002 with FENAME equal to 'Strawberry', putting Strawberry in single quotation marks. Zoom to the selection and then clear selected features. Using Layer Properties, make sure the features in Lex_Roads_2002 are labeled with FENAME. Using Placement Properties, in the Duplicate Labels section, select Place one label per feature part. OK. Apply. OK. Note that portions of Strawberry in Lex_Roads_2002 are flagged, by the pink of the topology, as overlapping Lex_Roads_1994.

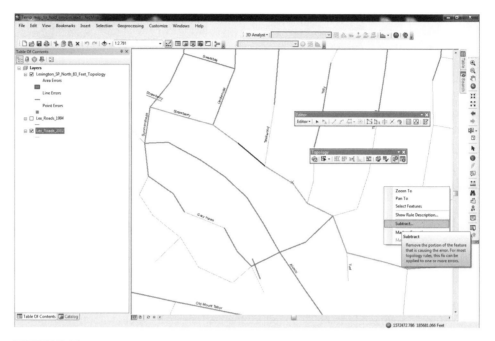

FIGURE 9-15

_____ **17.** *Erase parts of Strawberry:* Bring up the Editor toolbar and the Topology toolbar. Start Editing and then make the Fix Topology Error tool active on the Topology toolbar. Click on a portion of Strawberry that is overlapped by the pink topology line. It will turn black. See Figure 9-15. Right-click, then click Subtract. In the Subtract window, pick Lex_Roads_2002. Click OK. Both the feature and the topology disappear. Erase a few more road segments in the area, by "subtracting" the segments that are overlaid with the "topology error" lines. Click the 1994 roads on and off to demonstrate to yourself that the roads are really gone from the 2002 roads. (Leave the 1994 Roads turned off.) So, we have a technique for erasing roads that existed in both datasets — which would ultimately leave us with our goal: those roads built since 1994. However, this would be a long process for the thousands of segments in the city. How could we do it differently?

_____ **18.** One possibility for erasing all the Lex_Roads_2002 identified by topology would seem to be to select a large number of them at once. Drag a large box that encompasses several of the features identified by the topology. They will turn black. Now right-click. Unfortunately, the Subtract option doesn't exist when you right-click. If you check the Help files, you will discover that multiple subtracts are not implemented. Turn off the labels of Lex_Roads_2002. Stop editing without saving edits.

We certainly don't want to go through Lex_Roads_2002 subtracting segment by segment. Let's try another approach: Buffering the Lex_Roads_1994 (which have been snapped to the Lex_Roads_2002 where both exist) and then erasing, with an analysis tool, those Lex_Roads_2002 that lie within the buffer.

_____ **19.** With the Multiple Ring Buffer tool in ArcToolbox, start to buffer Lex_Roads_1994 by 50 – well, let's go to 150 feet, since the topology of 50 feet evidently missed some roads. Call the output Buffer_1994, and make sure it goes into

IGIS-Arc_*YourInitials*\Historical_Data\Roads.mdb\Lexington_SP_North_83_Feet.

The rest of the Multiple Ring Buffer window is OK as is, provided that you indicate the Buffer Unit should be Feet. Run the tool. Turn off the Topology layer. Explore the results.

_____ **20.** Use ArcToolbox > Analysis Tools > Overlay > Erase to erase those features in Lex_Roads_2002 (the Input Features) that lie within Buffer_1994 (the Erase Features), calling the result No_Coincident_Roads (and placing it in Lexington_SP_North_83_Feet). Click OK.

_____ **21.** No_Coincident_Roads appears in the Table of Contents. Make it bright red. Turn off all other layers. What appears should be the remaining roads. Zoom up on various areas to inspect. Flip Lex_Roads_2002 on and off to see the context.

_____ **22.** While our results are hardly perfect, you can see that we have managed to use two historical data sets to provide new information. How many segments are represented in No_Coincident_Roads? _____. What is the approximate total length of road segments (in feet, use Shape_Length) that were added? _____. How many miles is that? _____.

Sliding through Time—Seeing Changes in Features at Intervals

_____ **23.** Bring up a new map in ArcMap. Add as data Lex_Roads_2002 from

IGIS-Arc_*YourInitials*\Historical_Data\Roads.mdb\Lexington_SP_North_83_Feet

_____ **24.** Right-click the Lex_Roads_2002 name in the T/C and open its attribute table. Find the field Repaved_Year. Run Statistics on that column.[14] What is the earliest year? _____. The latest year? _____.

_____ **25.** Using Select By Attributes, display the roads that were repaved in 1995. How many were there? _____. How about 1999? _____. 2000? _____. Clear selections. Close the table.

_____ **26.** Open the layer Properties window of Lex_Roads_2002. Click the Time tab. Enable time features in the layer. Pick the Time Field: Repaved_Year. Apply. OK.

_____ **27.** Click the Time Slider button on the Tools toolbar. Click the far left icon until it **_doesn't_** say Time is disabled on this map. Click the Options button on the Time Slider window and set the

---

[14] None of the data sets in this book should be used for anything other than learning how the ArcGIS software works. The datasets have in some case been modified to make some pedagogical point or other. The data on road repaving is particularly bogus, bearing no relationship to reality.

Time step interval to 1 year. Click the Playback tab. Press the radio button that lets you specify a duration in seconds. Put in 30. OK.

_____ **28.** Press the Play button and watch the map for half a minute. You will see the repaving history of the roads. When the process finishes you can manipulate the slider bar to see any particular year. Close ArcMap.

This data set is particularly course grained (yearly jumps), but you can see the potential here for time stamped data (like hourly runs of emergency vehicles). Or you can see how the tool might be used to display any subsets of features, based on a field of sequentially increasing numbers. Time is just the most obvious one.

In the preceding two Exercises we dealt with using GIS in three-dimensional settings: First we worked with 3-D in which the third dimension was the spatial. The we considered an entirely different situation in which we considered changes to 2-D by including the time dimension). Now, in the next three Exercises, we will consider what might be called 1-D GIS, where we will work networks—roads, pipes, wires, rivers, and such—and how locations along them are recorded. While these networks exist in 4-D (what doesn't?), we'll look specifically at data structures that are made of linear, one-dimensional elements.

Address Geocoding

OVERVIEW

A Second Fundamental Way of Defining Location

As discussed in detail earlier, a location on the surface of the Earth may be referenced in a spherical coordinates system by a ray, emanating from the center of the Earth, defined by an angle of latitude measured from the equator and an angle of longitude, measured from the prime meridian. This method is "perfect" in that no error is introduced by the system itself; our accuracy is limited only by our ability to measure. Latitude and longitude descriptions may be applied anywhere on Earth. (I also discussed earlier how these two angles may be converted into myriad other pairs of coordinates through processes of projection.)

A large part of the Earth that is most interesting to us consists of human-made infrastructure. Almost always, such infrastructure includes streets and buildings. Further, the most usual way to describe a location in such areas is with a street address. Such an address is a text string: a sequence of letters, numbers, and other characters.

Outside of all addresses being text strings, addresses appear in a wide variety of formats. Some examples:[15]

26376 Alpine Lane, Twin Peaks, CA 92391, USA

76-20 34th Avenue, Jackson Heights, Queens, NY 11372, USA

305 W 100 S, Salt Lake City, UT 84119, USA

N84W 16301 W Donald Ave, St Charles, IL 60175, USA

Rua Aurora 735 01209001 Sao Paula, Brazil

Wendenstrasse 403, 20537 Hamburg, Germany

Such addresses are usually quite precise—locating a structure to within several meters, if you have a good map of the area. By themselves, however, they tell us nothing about the location with respect to most other locations, except perhaps for other structures located on the same street. For purposes of mapping and analysis, we often want to turn addresses such as these into latitude-longitude coordinates, or other coordinate pairs.

---

[15] Courtesy of *Geocoding in ArcGIS,* published by Esri.

Making use of addresses requires either personal or institutional knowledge of the location, or requires a map. To use a map, one must first find the right map—not necessarily an easy task and not one guaranteed of success if the address is in a sparsely populated area.

Suppose that we have this problem: A municipality wants to determine the spatial distribution of burglaries. It has records of the addresses of the crimes. One way to do this would be to put a map on a wall and mark each instance according to the address. Person A might look at a police report and call out the street address. Person B would look at an alphabetical list of streets and announce the row-column zone that the street was in (e.g., C-6). Person C could then locate the street on the map and make a guess at the position of the structure along the street by knowing the house number. In other words, this is not a particularly simple operation. What would be required to perform the equivalent process with a GIS?

First, we would need some sort of base map showing streets that could be referenced by spatial coordinates. Next, we would need an extensive database that related locations of structures to the coordinate system of the base map. Finally, we need some automated way to parse[16] addresses, such as those exemplified in the preceding list, so that they may be matched against the database. Such a parser must be sophisticated enough to derive the "true" address from any of a large number of text strings that might represent an address in varying forms (with variation in blanks, commas, hyphens, letter case, and so on).

TIGER/Line Files

Determining geographic coordinates from addresses is a tall order. Let's begin by restricting ourselves to the United States and the forms of address used there. You have already met TIGER/Line files, developed by the Bureau of the Census. If the terms behind the acronym[17] TIGER leave you confused, consider the predecessor description of such files: DIME, standing for Dual Independent Map Encoding. The idea behind "dual" is this: Each record in the file contains two geographic references. One is a range of address; the other is latitude-longitude coordinates. So this sort of file ties together (a) locations specified by text strings with (b) locations that are specified by geographic coordinates.

Each record in the TIGER file specifies a geographic line (called a chain). For streets in urban and sub urban areas, the chain usually represents a single block. At the ends of chains are intersections. See Figure 9-16.

Each record also contains the street name (Ninth, Jenkins, etc) and street type (Lane, Drive, etc.) together with any prefixes and suffixes (NW, E, and so on). The referenced line has a beginning point and an ending point. Each such point is defined by a latitude and longitude pair, in the NAD 1983 geographic (latitude, longitude) coordinate system. Further, at the beginning of each line, two structure numbers (street address numbers) are specified—one for the structure on the left and one for the structure on the right. The same is true for the ending point. Finally, if the line curves, its path is prescribed by vertices—familiar to you from your previous work with linear, vector representations in shapefiles, coverages, and geodatabase feature classes.

So, you can see that the TIGER database has all the ingredients that are logically necessary to convert an address into coordinates that will be "spatially close" to the actual coordinates of the structure at the address. A record in TIGER also contains lots of other information, useful in taking the census and

---

[16] Parse: To breakdown a sequence of letters or numbers into meaningful parts, based on a set of rules.
[17] Topologically Integrated Geographic Encoding and Referencing

FIGURE 9-16

making use of census results. It is worth mentioning that the TIGER file is a "flat" file. This means that there is no hierarchy—such as putting the records of a given state together and then requiring the user to know which state file a record came from. Each record contains all the information necessary to locate the place in the geographic space of the United States to which the record refers.

What follows is what the primary record in the TIGER line files looks like.

The primary TIGER record (called a Type 1 record) is 228 characters long. Broken up into six logical pieces, here is a typical record. Each of the six pieces is explained in the list that follows, beginning with "Record type."

<><><><><><><><><><><><><><><><><>

10021 31922071 AN Martin Luther King Blvd A41

301 313 300 342

01004050840508 2121067067

9122491224 46027460270003 0003 310 307

- 84490912138047734

- 84489656138048778

<><><><><><><><><><><><><><><><><>

Record type: 1

File version: 0021

Record ID: 31922071

Single side: Blank, meaning data exists on both sides of the chain

Source code: A (see www.census.gov/geo/www/tiger for explanation)

Feature direction prefix: N, meaning north

Feature name: Martin Luther King

Feature type: Blvd (Boulevard)

Feature direction suffix: none

Census Feature Classification Code (CFCC): A41, meaning local street

Address number at FROM node, LEFT side: 301

Address number at TO node, LEFT side: 313

Address number at FROM node, RIGHT side: 300

Address number at TO node, RIGHT side: 342

Impute flags (four digits): 0100 (see www.census.gov/geo/www/tiger for explanation)

Zip code on LEFT: 40508

Zip code on RIGHT: 40508

State code on LEFT: 21 (Kentucky)

State code on RIGHT: 21 (ditto)

County code on LEFT: Fayette

County code on RIGHT: (ditto)

County subdivision on LEFT: 91224

County subdivision on LEFT: 91224

Codes such as census tract and block numbers that provide keys to census data: 46027460270003 0003 310 307 (see www.census.gov to use this information)

Longitude of the FROM node: -84490912 (read as 84.090912 West)

Latitude of the FROM node: 138047734 (read as 38.047734 North)

Longitude of the TO node: -84489656 (read as 84.489656 West)

Latitude of the TO node: 138048778 (read as 38.048778 North)

Precision of the Geographic Coordinates in TIGER Files

Consider that the 90-degree angle from the equator to the North Pole (latitude) is about 10,000 kilometers. So:

A degree of latitude is about 111 kilometers.

A tenth of a degree of latitude is 11.1 kilometers.

A hundredth is 1.11 kilometers or 1110 meters.

A thousandth is 111 meters.

A ten-thousandth is 11.1 meters.

A hundred-thousandth is 1.11 meters.

A millionth is 0.111 meters.

The precision of the recorded number (e.g., 38.047734), where the last digit is in the "millionths" position, might suggest that the position is known to about the nearest tenth of a meter (i.e., 0.111 meters). Here is where your knowledge of the difference between accuracy and precision comes in. TIGER data is rarely good to 10 centimeters accuracy. Ten meters might be more like it. Sometimes the last two digits are zeros in the files, which implies an accuracy of the nearest ten-thousandth of a degree, or about 10 meters—roughly 33 feet.

Address Locators

The second piece required for converting a set of addresses to a set of latitude-longitude coordinates is a mechanism for parsing the addresses so that queries to a TIGER-style database[18] will be recognized by the database. With ArcGIS this takes the form of creating an "address locator." You kick the process off in ArcCatalog by finding Address Locators in the Catalog Tree, expanding it, and double-clicking on Create New Address Locator. This opens a window that allows you to pick an address locator style. The idea here is to provide information about the general format of addresses in the region of interest. You have more than three dozen different styles to pick from. Your choice depends on the structure of your reference data and what the form of address in the region of interest looks like.

Once the choice of address locator is made, you will encounter a dialog box that will let you specify a number of parameters that will allow the form of your input addresses to be matched with the form of the locations in the reference database. If you are only typing one or a few addresses into the address locator, you can monitor whether or not you were successful. If you have a large file of addresses to find, however, the problem gets a little stickier. Some addresses may pass muster and be represented on the map; others may be rejected. Still others may find close, but not exact, matches. These latter ones are assigned a score as to the closeness of the match, based on a number of factors. Successfully converting a large file of addresses requires both some tweaking on your part and something coming close to artificial intelligence on the part of the software. In the Step-by-Step section, you will go through the process of converting addresses.

---

[18] Many agencies and companies have taken the basic TIGER files and enhanced them.

Address Geocoding

Exercise 9-3 (Project)

Experimenting with Addresses and Coordinates

____ **1.** Using the operating system of your computer, use Notepad to open the text file

[____] IGIS-Arc\Address_Geocoding

\Clinton_County_OH_Type1_1994_TIGER_Records.txt

Clinton_County_OH_Type1_1994_TIGER_Records.txt is a raw TIGER file. Notice it is composed of only letters and digits—228 of them in each record. I show you this so you can understand that the basic TIGER files are simply character files. They contain geographic information, but to plot this information, you must convert the files into GIS files (e.g., geodatabase feature classes, shapefiles).

____ **2.** Locate the sixth record from the top of the file. Verify that it describes the 4000 block of state highway 729. Verify that the longitude of the FROM node of that block is 83.613958 WEST. What is the latitude of the TO node of that block? _____.

____ **3.** Looking at that same record you can discover that the zip code on the left of the highway is the same as the zip code on the right. In what zip code does this block of highway lie? _____.

____ **4.** Start ArcCatalog. Copy the folder

[____] IGIS-Arc\Address_Geocoding

into

____ IGIS-Arc_YourInitials

Make a folder connection to the new folder.

___ **5.** Start ArcMap. Add as data

_____ IGIS-Arc_*YourInitials*\Address_Geocoding

\Lexington_Roads.mdb\Lex_Roads_DD83\Lex_Roads_2002

You will see the road network of Lexington, Kentucky. This dataset was derived from the totally character-based TIGER files of the 2000 census. The geographic component, present in the TIGER files as only character-based information giving longitude and latitude, has been converted to (geo)graphic coordinates so it can be plotted. Some of the other character-based information remains.[19]

___ **6.** Make sure the display units Decimal Degrees. To one decimal place, what is the longitude of Lexington? _____. The latitude? _____. Now make the display units Feet.

___ **7.** Open the attribute table. How many records are there? _____. (Hint: click on the icon that gives you the last record.)

___ **8.** Look at the field names. Here you see the usual sorts of items: Feature Identifier (FID), the FROM and TO nodes, and the length (meaningless, because it is in decimal degree representation, which has no consistent scale). You also see the TIGER/Line ID (TLID), which is a unique identifier for the road segment—unique nationwide, actually.

Recall that we want to use this data set to find a map-able geographical point by specifying the address of the building at that point, For this, we need the street name (here, in the field FENAME); its type, such as avenue, road, and so on (here, in the field FETYPE); direction prefixes and suffixes (FEDIRP and FEDIRS); and the numerical address ranges for each segment (i.e., block), which we will examine in detail shortly.

___ **9.** About how many unique feature types (e.g., Rd, Ave, Blvd) are there in this feature data set? _____. Try to identify what each of the abbreviations means. (Hint: Use Selection > Select By Attributes. Construct the beginnings of an expression

[FETYPE] =

and then ask for Unique Values.)

Finding the Geographic Position of an Address "Manually"

To further understand address ranges, let's look at a particular block.

___ **10.** Using Select By Attributes (from Selection on the Main menu), locate all street segments whose FENAME is Chinoe (shin-o-ee). Zoom to selected features. Now, using the Method "Add To Current Selection" find those streets with FENAME of Cooper OR FENAME of Cochran. Again zoom to

---

[19] If you are interested in geographically augmented street data for your city, it is available free from the Esri Web site. It comes in zipped-up shapefile format, so getting to it requires a few steps, but keep in mind the word "free." (These steps were detailed in Chapter 6, Exercise 6-4 (Combining Demographic and Geographic Data) where you retrieved TIGER data for Knoxville, Tennessee.) By the way, you can also unzip the files for free with Microsoft Windows.

FIGURE 9-17

selected features. The map should look like Figure 9-17 (and you really need the color file to see this one). In the attribute table, show only selected records. How many records (and hence segments of Chinoe, Cochran, and Cooper are there? _____. Assuming that you have already zoomed up on the area of the selections, label the roads on the map with FENAME. If you click the box to the left of a given record it turns yellow, as does the segment on the map. Inspect the records. Use the Select Feature tool on the Tools toolbar to select the section of Chinoe that runs between Cooper and Cochran. What is its OBJECTID? _____. Dismiss what's left of the table.

The block of Chinoe we are interested in is the one between Cochran and Cooper.

____ **11.** Use the Identify tool to examine the block of interest. What is the From Address on the right-hand side? _____. The From Address on the left-hand side? _____. The To Address on the right-hand side? _____. The To Address on the left-hand side. _____. In which direction of Chinoe are the house numbers increasing? ___ North or ___ South. The designation of the left or right side of a road is determined by moving forward along that road in the direction of increasing structure numbers. So, on which side of the road would you expect to find the house with number 525? _____. About what percent of the way along the block, going south from the intersection of Chinoe and Cochran, would you expect to find this house, assuming even spacing of the houses? _____.

____ **12.** Make sure that the Draw toolbar is visible. At the right end there is a drop-down menu that allows you to change the marker color. Change the color to red. Near the left end is a drop-down menu that lets you select the sort of graphics to be drawn (e.g., Rectangle, Polygon, etc.). Pick Marker. Place the cursor crosshairs over the spot on Chinoe where you think 525 is, and click. Press Esc to unselect the marker.

___ **13.** Save the map file in

     ___ IGIS-Arc_*YourInitials*\Address_Geocoding

with the name 525_Chinoe_Road. Dismiss ArcMap.

Making an Address Locator

___ **14.** Back in ArcCatalog use Search to find Create Address Locator. Where is it? _____. Start the tool. In the window you will be requested to specify an Address Locator Style. Browse to choose: US Addresses – Dual Ranges (meaning that addresses on both sides of a street will be considered). (If a red X appears in the Field Map: click it, read it, then ignore it.) Browse to find and Add the Reference Data:

     ___ IGIS-Arc_*YourInitials*\Address_Geocoding

\Lexington_Roads.mdb\Lex_Roads_DD83\Lex_Roads_2002.

Click the entry under Role. Make sure the drop-down menu says Primary Table. Look at the Field Map. This lists the fields that should be in the table. As you could tell by inspection, they are. Browse for Output Address Locator field. Put the folder Address Geocoding (in your *YourInitials* folder) in the Look in field. Make the name Lexington_Streets. Save. Double-check the path. Click OK.

Wait while Create Address Locator scrolls by in the Status bar. And wait some more. (The tool is dealing with a big table.) When the tool is done and you see the pop-up window, look at the contents of the Address_Geocoding file folder. You should see Lexington_Streets. If you click on it, the status bar should tell you that it is a Locator. Right-click on the name and pick Properties. Examine. Dismiss.

Finding the Geographic Position of an Address "Automatically"

___ **15.** Restart ArcMap. Open the map 525_Chinoe_Road. With the Drawing toolbar, change the color of future markers you might draw to black. Click the binoculars on the Tools toolbar to bring up a Find window. Move the Find window out of the way so that you can see Chinoe Road. Select the Locations tab. For "Choose a locator" browse to the one you just made in the Catalog Tree under Address Locators: Lexington Streets. Double-click. For Full Address type

     525 Chinoe Road

and press Find. In case you missed it, click Find again (and again) while looking at the map. The location 525 Chinoe Road will be drawn to your attention. Click Options in the Find window. Under Offset Options put in 30 Feet, since that is the distance required in this neighborhood from the street centerline. Click OK.

____ **16.** In the lower pane of the find window right-click 525 Chinoe Rd, 40502 to bring up a context window. Click Flash. Bring up the context menu again and Add Point. A new marker symbol (black) should appear, near the one you placed earlier. Zoom way up so that the block of interest fills the screen. Repeatedly bring up the context menu, clicking Zoom To, Create Bookmark (check it out under Bookmarks on the Main menu), Add Labeled Point, and Add Callout. Look at how these results compare to the red marker you set when you found the position manually. Dismiss the Find window.

____ **17.** If you have an Internet connection, click the drop-down menu arrow next to the Add Data and pick Add Basemap. Select Imagery. Ignore any warnings. You will (shortly) see an orthophoto that covers the area, so you can actually see the houses. The centerline of Chinoe Road comes reasonably close to the TIGER representation of the road. About how many feet different near the house? _____ feet.

____ **18.** In the Data Frame Properties window, make sure the display units are in Feet. Measure the distance from the street shown by the TIGER line to the black marker. It should be about 30 feet. Measure the distance from the red marker symbol to the black one. _____. If it is more than a couple of hundred feet, try to determine why.

In this exercise, we looked at the software's ability to find a single address. This is the sort of thing that web programs (e.g., ArcGIS StreetMap, Mapquest) and GPS mobile devices (e.g., TomTom) do. Because you used the TIGER files to do the process "manually," you should understand now how such programs work and that addresses are found by interpolation within a given block—not by having every address's geographic location stored. Quit ArcMap.

TIGER Files and ZIP Codes

____ **19.** Restart Restart ArcMap with the 525_Chinoe_Road map. Zoom to full extent. "X" off the Draw toolbar. Turn off Labels Features. Using Selection > Select By Attributes, find those records in Lex_Roads_2002 in which the ZIPL is not equal (use <>) to the ZIPR. Dismiss the selection window, so you can see the map, with the selected segments. Zoom up on different parts of the map.

What I had planned on showing you, by using this selection, was a clever way to delineate zip code areas. After all, when a street segment has one zip code on one side and another on the other, the street is obviously a dividing line between the two zip code areas. While you can sort of get an idea of the zip code areas, really what I have shown you instead is how wretched the dataset is. Clearly, the zip code data in this set of TIGER records leaves a lot to be desired. If you want, you may peruse the fields to see what sorts of problems give rise to the unfortunate map, but the main lesson is that there are data sets out there, from reputable sources like the U.S. Bureau of the Census, that are shot through with errors. User beware!

____ **20.** To see the actual zip codes of Lexington, add the shapefile

____ IGIS-Arc_*YourInitials*\Address_Geocoding\ZIP_Codes_in_405.shp to the map.[20] Increase the outline width of the polygon shapefile to 4.0, make its outline color red and its Fill Color Hollow or No Color. Compare the zip code boundaries with the selected element, which, with perfect data, would be coincident. Using Identify, with "ZIP_Codes_in_405" click on some polygons. In what section of the city is ZIP code 40511? _____. Perhaps label the features with ZIP. After your exploration, save the map as Lexington_ZIP_Codes. Dismiss ArcMap.

More to Know—More Information Available

There is much, much more to know about address geocoding, but you have the basic idea. Addresses in text form can be used to create geographic coordinate points. The next step, which we won't undertake, would be to convert files of many addresses. To refer back to the example I used in the Overview—suppose that you have a file of thousands of addresses on crime reports. You might like to see where, on a map, these events occurred. You can submit this file to the software and get a map of the events.

Problems occur when an address yields no geographic point, or when an address presents you with two or more possible points. The address geocoding software is sophisticated. For example, when there are multiple possibilities, the software provides you with a "score" that tells you how good a match the input address is with what's in the database.

To learn more, read *Geocoding in ArcGIS*, which exists as a paper manual available from Esri, or as a PDF (Portable Document Format) file.[21] The filename is Geocoding_in_ArcGIS. You may find it in a folder named Esri_Library\ArcGIS_Desktop.

---

[20] ZIP codes data sets for the entire United States are available in the data sets that came with your ArcGIS suite. Look for a folder named ESRIDATA\census.
[21] Software for displaying such files is Acrobat Reader, from Adobe Systems (www.adobe.com).

Analysis of Networks

In ArcGIS Version 9.1, Esri introduced the Network Analyst in ArcGIS Desktop. The capability to do network analysis has been present in ArcInfo for years, but it had to be accessed through typed commands, rather than by point-and-click activities.

The sorts of networks that Network Analyst deals with are primarily composed of roads. You might think of network analysis as sort of a simulation of a vehicle, confined to streets, whose driver wants to get from A to B by the shortest route or in the least time—staying on the roads, of course. Or network analysis can develop service areas, considering the network, such as assigning voters to the nearest polling place.

To analyze a network, one needs a network to analyze—and this is no mean feat. First, you have to have all the sort of data that is, very roughly, supplied by TIGER-like files. Then the actual lengths, in linear units—not degrees of latitude and longitude—of segments must be developed. The connectivity of segments must be ensured. Since time is usually as important as distance, the time to traverse each segment needs to be produced. While traffic signals and turns from one segment to the next aren't reflected in distance calculations, they very much affect time. So tables, for each intersection, that describe averages of how long it takes to turn right, turn left, or go straight are needed. Also, the 99 percent of you who have driven a car know that the time it takes to drive a particular block can depend on what time of day it is, whether there is construction ahead, and whether there is an accident. In other words, a truly effective network needs to be sensitive to the time of day, and, better yet, sensitive to traffic conditions. The need for "dynamism" of this sort is well beyond our scope here. What follows in the next section are the basics of network analysis, given that the network is already built.

Analysis of Networks

Experimenting with Routes and Allocations

_____ **1.** Start ArcCatalog. Copy the folder Network_Analyst_Data from [___] IGIS-Arc to your

____ IGIS-Arc_*YourInitials* folder

Make a connection to the Network_Analyst_Data folder.

_____ **2.** Start ArcMap with a Blank Map. Choose Customize > Extensions and make sure that there is a check mark by Network Analyst.

_____ **3.** Display the Network Analyst toolbar. On the Network Analyst toolbar, find the Network Analyst window button. Click to display the pane.

What you will see in the following steps is made possible by data and a network dataset that were constructed with considerable effort. We will not look at the construction effort nor the building of the network database, called streets_nd.ND. For further information, take the Network Analyst tutorial that comes from Esri. The data for this exercise was derived from that tutorial dataset.

_____ **4.** Add the dataset streets_nd.ND from

____ IGIS-Arc_*YourInitials*\Network_Analyst_Data\Route_&_Allocate\Network.

You will be asked if you want to also add all participating feature classes. You do. Make the Streets line symbol black, size 1.

_____ **5.** In the T/C, turn off all layers except "Streets". You will see the streets of the San Francisco area. On the Network Analyst toolbar, note that the network dataset is streets_nd.

_____ **6.** In the T/C make Streets the only selectable layer.

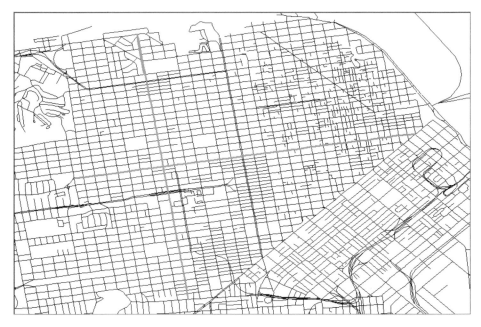

FIGURE 9-18

---

7. Open the Streets attribute table. Make it occupy the full screen. How many street segments are there? _____. Notice that the field names suggest that this dataset was derived from TIGER/Line files. Notice also that several fields have been added. You will see Full_Name, street length in meters, FT_Minutes (some sort of average time for traversing the street, moving from the from node (F) to the to node(T)), and TF_Minutes (same sort of thing, but traversing the opposite direction).[22] Notice also the field Oneway, which contains either FT (meaning traffic may proceed from the FNode to the TNode) or TF. Minimize or dismiss the attribute table.

8. Choose Selection > Select By Attributes. Make the layer Streets. In order to find the intersection of Pine and Fillmore Streets, create **_exactly_** the following query, being aware that every character is case-sensitive:

"Full_Name" = 'Fillmore St' OR "Full_Name" = 'Pine St'

and press Verify. If it works, press Apply. Now choose Selection > Zoom To Selected Features.

9. Obviously, from all the selections you see, there are Pine and Fillmore streets scattered all over the Bay Area. You are interested in the two that intersect, so zoom in on those. See Figure 9-18. Then, zoom in tightly on the intersection itself.

---

[22] These two numbers might be used to indicate one-way streets (by making the "wrong way" a very large number), but the one-way issue is handled in a different way: by the field Oneway.

____ **10.** Use the Identify tool with the Streets layer to find the block of Pine Street immediately west of the intersection. What is the address range for the buildings on the right side of this block? _____, _____. (If you want, bring up the Basemap Imagery of the area.)

In the next few steps, you are going to build a route that traverses some San Francisco streets. To do so, you will create "stops". Perhaps you have volunteered to drive a vehicle for the Meals-on-Wheels organization and need directions to the various locations.

____ **11.** On the Network Analyst toolbar drop-down menu, click New Route. The Table of Contents will fill up with lots of features associated with the about-to-be-created route. The Network Analyst window will show that, so far, there are no stops, no routes, and no barriers.

____ **12.** On the Network Analyst toolbar, find the Create Network Location Tool button and click. Using the mouse cursor, which now contains a flag, place the crosshairs midway down that western block of Pine Street, and click. This will put Stop 1 in position. Choose Selection > Clear Selected Features. (It is probably a good idea to turn off any basemap imagery you have added. The imagery will only slow things down and increase clutter.)

To make Stop 2, you will use the Find capability of the Address Locator named SanFranStreets.

____ **13.** Bring up the Find window by clicking the binoculars icon on the Tools toolbar. In the Find window, click Locations. Browse in the "Choose a locator" field to find SanFranStreets in

_____ IGIS-Arc_*YourInitials*\Network_Analyst_Data\Route_&_Allocate

Select it and press Add. For Street or Intersection type ***exactly***

Cabrillo St & 5th Ave

(The & is the symbol used to specify an intersection.) Press Find.

____ **14.** Two possibilities appear (use the horizontal slide bar): the one we want (score 100) and one on 45th Avenue—obviously not of interest. Right-click on the first candidate and pick Zoom To. Right-click on the first candidate and pick Flash. You probably didn't see the flash because the Find window was in the way. Move it and activate Flash again. Use Identify on the connecting streets to assure yourself that this is the right intersection.

____ **15.** Right-click the candidate again and pick Add as Network Analysis Object. A disk with a "2" in it should appear at the intersection.

____ **16.** In the Networks Analyst window right-click on Stops (2) and select Zoom To Layer, so you can see the extent of the map that contains both stops.

____ **17.** Using the same procedure as in the steps above (don't forget the &), place Stops 3, 4, 5 and 6 at

❏ Teddy Ave and Delta St

❏ Clement St and 18th Ave

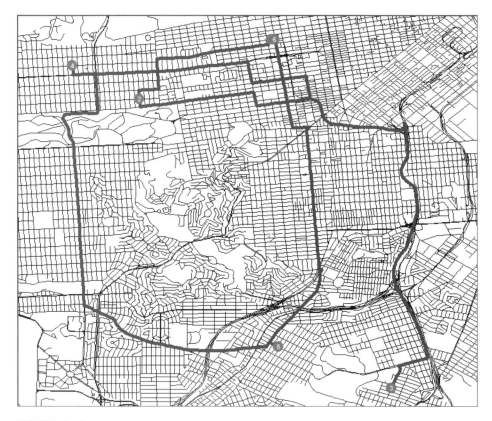

FIGURE 9-19

❑   245 Brazil Ave

❑   Pine St and Fillmore St (near the starting point)

(To see each stop on the map click the location under Stops. The selected stop turns cyan on the map. If you got them out of order you can put them back in order by dragging them in the list. The order is important.)

____ **18.** Dismiss the Find window. Zoom to the Stops (6) layer extent.

Now that the stops are established, you can have Network Analyst create a route.

____ **19.** Click the Solve (Run the current analysis) button on the Network Analyst toolbar. Right-click on Routes (1). The route will be shown. Click the button that will give you the Directions window. How long is the trip? _____ miles. How many minutes will it take? _____. Peruse the Directions window. Click on the word Map next to a direction to see a detailed map of the described turn. In the T/C, add Label Features to Streets to understand the directions. Examine the map. Dismiss the Directions window.

___ **20.** Turn off the labels (to make drawing faster). Zoom to the extent of the route. Expand Stops, if necessary, in the Network Analyst window. Click each graphic pick (a point that you indicated with the cursor), intersection, and address to select it and note its position on the map. Notice that you would have to do a lot of extra driving to visit these stops in their present order. See Figure 9-19. You could make the trip more efficient if you switched the order of some of the stops.

___ **21.** You can reorder the stops (confine yourself to numbers 2, 3, 4, and 5 only) by dragging a stop to a different position in the list. Try changing the order to something that looks more efficient. Then click the Solve button again. Better? What is the most efficient trip you can manage? _____Miles. _____Minutes.

Not only does Network Analyst provide a very efficient algorithm for finding the shortest (or fastest) path that visits a number of stops in which the order is predefined. It also attempts to solve the famous "Traveling Salesman" problem, which minimizes trip length or time.[23] You can have Network Analyst decide on the order of the stops.

___ **22.** In the Network Analyst window, at the top right, is a button that will give you Layer Properties. Under Analysis Settings, check the box Reorder Stops To Find Optimal Route, while preserving the first and last stop. Also, you will notice that you can choose to have the impedance (the factor that Network Analyst tries to minimize in selecting a route) be set to either minutes or meters. You can also notice that you can restrict the route so that it does not go the wrong way on one-way streets—which seems like a good idea, so check that box. Click Apply, then OK.

___ **23.** Solve the network again. With the Directions window, determine if the new network analysis improved on your last effort for an efficient trip. (If not, blame the one-way streets.)

Zoom in on one of your stops. With the Select/Move Network Locations tool (on the Network Analyst toolbar), drag the stop a few blocks away. Re-solve the network. Notice how the route is changed to go through the moved point.

There is one other button on the Network Analyst toolbar that deserves your attention. Find and press the Build Entire Network Dataset button. (Answer No if asked if you want to build the network data set.) You will be told is that the network dataset has already been built. What you should realize, from this statement and this whole exercise, is that a lot of effort goes into building the database that allows a user to do all the nifty operations that you just performed.

Finding the Shortest Route to a Facility

Suppose that an automobile accident occurs and routes from the closest hospitals for ambulance service are needed quickly. Network Analyst can provide this information almost instantly.

---

[23] There is no known solution for ordering a set of points so that the path that encounters each one of them once (and then returns to the starting point) is the shortest possible, except for examining every possible path, which is computationally out of the question for a large number of points. (It is known as the Travelling Salesman Problem (TSP).) However, one can get very close to an optimal path by a variety of techniques, which is what this software does.

___ **24.** Add as data to the map the data set Hospital.shp, found in

IGIS-Arc_*YourInitials*\Network_Analyst_Data\Route_&_Allocate\Layers

Zoom to that layer. Make the symbol for a hospital green, with a size 12.

___ **25.** In the Network Analyst toolbar drop-down menu, click New Closest Facility. Note that headings for Facilities, Incidents, Routes, and three kinds of Barriers appear in the Network Analyst window.

___ **26.** Right-click Facilities (0). Choose Load Locations. In the window that appears, choose Hospital from the Load From drop-down menu. Click OK. Notice that several hospitals appear on the map with the symbol indicated in the Table of Contents under Closest Facility > Facilities > Located. These are, of course, in the same locations as the Hospital layer you added earlier. Facilities (0) should become Facilities (9).

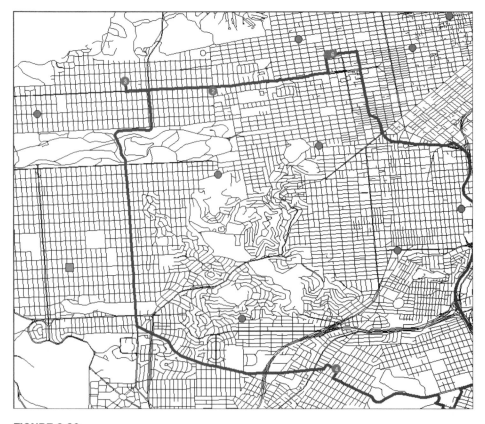

FIGURE 9-20

_____ **27.** On the Network Analyst Window make sure the drop-down menu is set to Closest Facility. Right-click the text Closest Facility and click Properties to bring up a Layer Properties window. Select the Analysis Settings tab. Make the Impedance Minutes. Under Facilities To Find, change the number to 3. Travel From should be set at Facility To Incident. Restrictions should be set to respect Oneway streets. Click Apply, then OK.

_____ **28.** Click Incidents (0). Click on the Create Network Location Tool button on the Network Analyst toolbar. Move the cursor onto the map.

Pretend that this cursor, rather than being pushed around by your mouse, is directed by an automatic signal coming from a vehicle whose airbag has just deployed. The car knows where it is because it has an onboard GPS receiver. That location is identified with a marker. Figure 9-21 shows the location with a dark, solid dot.

_____ **29.** Click on the map at approximately the location shown by the dot. Click Routes (0). Click Solve. The path from the three closest hospitals to the accident should appear. Also, the names of those hospitals will show up if you expand Routes (3).

_____ **30.** Click on one of the hospital names under Routes (3). One path will be selected. Press the Directions button. How long will it take an ambulance to reach the scene of the accident? _____. How about for the other two hospitals? _____. _____.

Allocating Territories to Facilities

Assume that you want to do an analysis of how well fire stations can respond. Perhaps a new standard is proposed that every point in the downtown area of the city should be reachable within 1.5 minutes from some fire station.

_____ **31.** In the T/C, under Layers, remove all layers except Streets, Streets_ND_Junctions, and Streets_nd. Add as data Fire_Station.shp. Make each fire station show up with a yellow circle of size 12. Choose Zoom To Layer.

_____ **32.** On the Network Analyst toolbar drop-down menu, choose New Service Area. Right-click on Facilities in the Network Analyst window and load the fire station locations. Click OK. How many are there? _____.

_____ **33.** Click Polygons (0) (under Service Area > Facilities (40)). Click the Service Area Properties text to bring up Layer Properties > Analysis Settings. The Default Breaks field specifies the extent of the service area by calculating how far, along each possible route from the fire station, a vehicle could reach in the time allotted. Set it at 1.5. Respect one-way streets. Click Apply. Under the Polygon Generation tab, read the text of the various options, then click on Generalized if necessay. (If solving the problem with generalized polygons doesn't take too long, you can come back and ask for the detailed polygons.) Click Apply, then OK.

_____ **34.** Press the Solve button and be prepared to wait. The software is computing, for each fire station, the 1.5-minute range along each possible path. That's a lot of computing.

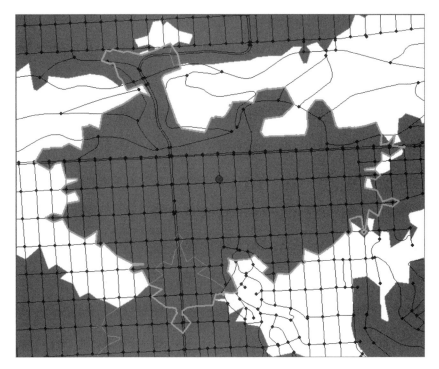

FIGURE 9-21

____ **35.** In the Network Analyst window click the + sign to expand Polygons (40). Click on Station 22 to highlight the name and the boundary of the polygon. In the Selection menu pick Zoom To Selected Features. Notice the area that is covered by the station in the 1.5-minute time allotment, as well as the area left uncovered. See Figure 9-21. After examination, close ArcMap.

Linear Referencing | OVERVIEW

Managers of artifacts with long linear structures (e.g., highways, pipelines, railroads) are generally unconcerned with the precise two-dimensional geographic coordinates of the entity they deal with. A locomotive is pretty well locked onto the railroad's one-dimensional linear structure—unlike, say, an airplane, which can operate in three dimensions. As a result, many of the coordinate systems that have grown up around linear structures are different from those we have been talking about up to now. Terms like road miles, river miles, and rail miles are in common use in those industries. The average citizen who uses the Interstate Highway System will probably be familiar with the small green signs with numbers on them that indicate the number of miles from some origin—frequently a state border.

We are in the habit of representing linear structures with lines drawn between junction points (e.g., nodes). (Of course, these structures are portrayed on a two-dimensional field, but the smallest part of each such line is one-dimensional [a vector], so we call them one-dimensional, or linear.) The difficulty with simply representing such a structure—let's take a highway, for example—from intersection to intersection is related to attributes. The requirement, so far in your studies, is that any GIS feature (whether point, line, or polygon), must be homogeneous in all its attributes. For example, all of a given cadastral polygon is owned by one entity, the taxes are paid on all of it or none of it, and so on. If different attributes apply to two different parts of the feature, then you need two features (e.g., if different people own different parts of a piece of land, then you need more polygons).

Imagine that you have a stretch of highway that is 2 miles long between its intersections A and B with other roadways. Among the attributes you want to store for this length of road are speed limit, pavement type, pavement quality, political jurisdiction, and number of lanes. The difficulty is that the speed limit changes four times and the road goes from four lanes down to two and then back to four. It crosses a county boundary. Part of it is blacktop and part concrete. Also, repairs have taken place on different segments of the road at different times. To represent this "traditionally" in ArcGIS, we would have to have a plethora of features. Every time an attribute changed (e.g., the speed limit changed from 55 to 45), a new feature would have to be declared. So, the single-line feature that represented the 2 miles between intersections might have to turn into dozens of short features. The complications created then—for example, how would you find the distance from A to B?—are considerable.

The invention that turned out to make attribute representation tractable for such linear features is called "linear referencing." The fundamental idea is that you can have several sets of attribution information accompany a single linear feature. What makes this possible is the concept described earlier in this section: The thing that fixes the point at which a change takes place is a number—a distance—that is related to the origin of the feature, in the style of, say, road miles.

To recap, linear referencing lets you store geographic information without a large number of explicit x-y coordinates. Instead, a measure (distance) along a linear feature is provided. Linear referencing is a mechanism that allows you to associate multiple sets of attributes to portions of linear features. One way to view linear referencing is "features within features." Because of this, some new terminology and tools are necessary to understand and operate linear referencing. We address those in the Step-by-Step section that follows.

Linear Referencing

The first concept to grasp is that of a "route." A route is a linear feature, probably made up of several or many other linear features. A route has a unique identifier and has a measurement system stored with its geometry.

Exercise 9-5 (Project)

Experimenting within Linear Features

_____ **1.** Start ArcCatalog. In the Catalog Tree find the folder

[____] IGIS-Arc\Linear_Referencing.

Copy it into ____ IGIS-Arc_*YourInitials*.

_____ **2.** In

____IGIS-Arc_*YourInitials*\Linear_Referencing

\Pittsburgh.mdb\PITT_Roads_Routes

preview the Geography of Just_Roads. Using the Identify tool, click on a few features. You will notice that you that you get some standard information (e.g., feature name) and also a route identifier (called ROUTE1), a beginning mile point, and an ending mile point. Look at the attribute table of Just_Roads. How many road segments are there? _____.

_____ **3.** In the Catalog Tree, click All_Routes. Switch back to the Geography display, which thins out a lot compared to Just_Roads. These are road features that have been combined and designated as "routes"—all under the name of ROUTE1. Each route is composed of sets of features from Just_Roads.

4. Again use Identify. Notice that you get completely different results from those you got before. The shape is still polyline, but the Shape_Length is much longer. When you click on a feature, the line that flashed is usually lengthy—composed of many of the features you saw before. Look at the table. How many routes are there? _____.

5. Lastly, display Some_Routes from the Catalog Tree. Some_Routes is a subset of All_Routes that we will use for demonstration purposes. How many features of Some_Routes are there? _____.

From the preceding, you can see that routes are features, but they are features with remarkable, complex, and useful characteristics, as you will see from the steps that follow.

6. Start ArcMap with a Blank Map. Add two data sets from

___IGIS-Arc*YourInitials*\Linear_Referencing:

(1) County and (2) Cities, making sure that Cities is at the top of the Table of Contents. From the geodatabase Pittsburgh.mdb feature dataset PITT_Roads_Routes, add Some_Routes. Make the line feature dark green, with a width of 2. Make County a beige color.

7. Open the Some_Routes attribute table. Find the route 30000030. Select it in the table and note its location in the northeast quadrant of the map. See Figure 9-22. How long is the route? _____ feet. Dismiss the Some Routes attribute table. Zoom to the selected feature.

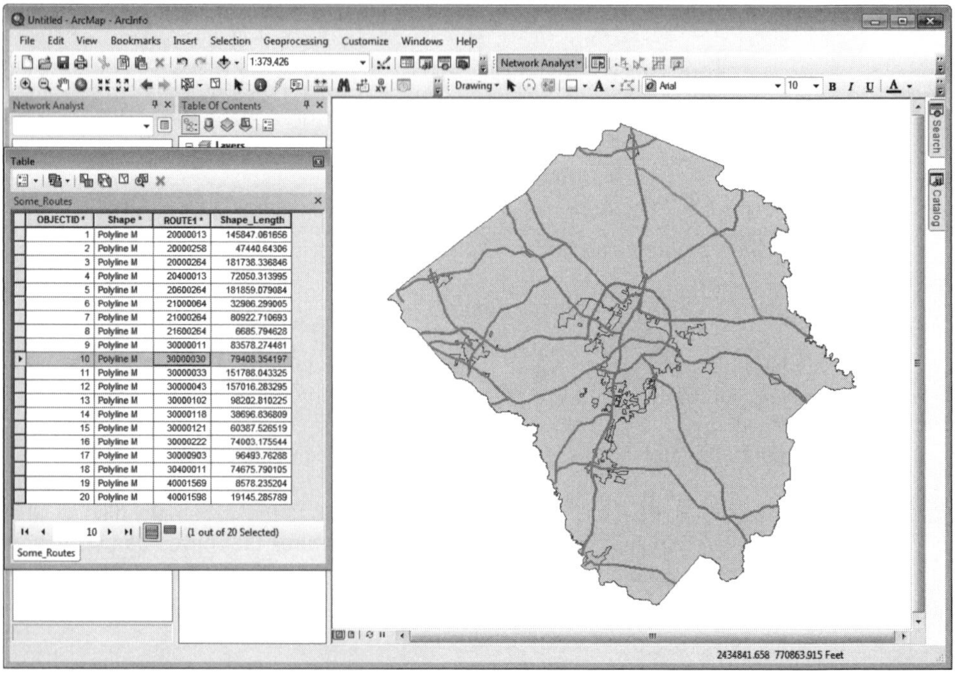

FIGURE 9-22

You have identified 30000030 in Some_Routes as a feature. Next, you will Identify it as a route, but you have to add a button to a toolbar to do it.

____ 8. Choose Customize > Add-In Manager > Customize > Commands. Find Linear Referencing in the Categories list. Drag the Command "Identify Route Locations" to the end of the Tools toolbar, where it becomes a button that will bring up a tool. Close the Customize window.

____ 9. Make the Identify Route Locations tool active, and click at a random point along the selected route. What is the maximum value of the Measure of the route? _____. What would you say the minimum value was? _____. Of how many parts does the route consist? _____. Click on different points of the route so you can determine, based on the Measure value, in which direction (northwest or southeast) the Measure increases? _____. Dismiss the Identify Route Location Results window.

Suppose that you know that a call box has malfunctioned at Measure 3.35. You want to find that point on the map.

____ 10. Click Find (the binoculars) > Linear Referencing. For Route Reference, pick Some_Routes. The Route Identifier is, as always, ROUTE1. Click Load Routes. Pick the route that you have been working with: 30000030. The Type should be Point. Put in the Location 3.35 and press Find. Information about the route should appear in the bottom panel of the window.

____ 11. Right-click the row containing 30000030. Flash the route. Flash the route location that was found. Draw the route location. Label the route location. Click the Select Elements pointer and place it over the route label; click once and pause. Drag the label around, noting that it is a call-out box. Press Delete to remove the box. Select the drawn location by dragging a box around it; delete the drawn location. Close the Find window.

____ 12. Add the table accident.dbf as data from the Linear_Referencing folder. Open the table. How many accidents are recorded? _____. With Table Options > Select By Attributes, select the records with ROUTE1 = 30000030.[24] Show the selected records. How many are there? _____. Right-click the gray box to the left of one record, click Identify, and examine the Identify box that comes up. Note that the record contains a MEASURE field that indicates the position along the route where the accident occurred. Lots of other data about the accident is also recorded. Close the Identify window.

____ 13. Move the table around so that you can see the map, and particularly Route 30000030. Notice that no points along the route where accidents occurred are shown. That's because the table is just that: a table. To see the locations of the accidents, you can use the table to make a layer. (This layer will exist only in memory and will go away once you close ArcMap, unless you save the map. Of course, you may specifically save it as a layer file by right-clicking and choosing save as LayerFile.)

____ 14. Show all records in the table. With Table Options, choose Clear Selection. Close the table. In the T/C right-click Accident. Choose Display Route Events. Set up the window so that it looks like Figure 9-23. OK the window.

---

[24] As mentioned before, sometimes a field name in a query is enclosed in square brackets and sometimes in double quotes. It depends on what database management system is in use for the layer or table.

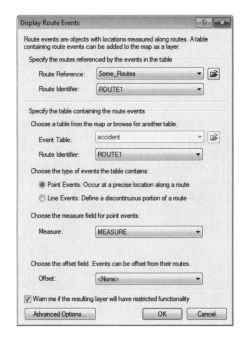

FIGURE 9-22

_____ **15.** Examine the map. Now you see the accidents from the table depicted on the map—put there by a layer called Accident Events. Click the List By Source tab in the T/C.

For reasons whose logic or history escapes the author, points or segments within a route are termed "events."

_____ **16.** Open the Accident Events attribute table. Again use Select By Attributes to select ROUTE1 to be 30000030. Show only selected records. Close the select by attributes window and resize. Resize and rearrange the table so that you can see the map. Notice that the accidents on our favorite route are highlighted. Sort the MEASURE attribute values, smallest to largest. Click the gray box to the left of the record with MEASURE value 7.23. The record should turn yellow. The point on the map should turn yellow. Display All Records. Choose Selection > Clear Selected Features. Dismiss the table.

_____ **17.** Add and open the Pavement.dbf table. How many different pavement events are there? _____. Select those in Route 30000030. How many? _____. Notice that there is a beginning mile point and an ending mile point field. Sort BEGIN_MP is ascending order. That should also put END_MP in ascending order. The segments connect but do not overlap. Display All Records. Clear all selections and close the table.

_____ **18.** Right-click Pavement in the T/C and make a layer from the Pavement.dbf table by using Display Route Events. This time, of course, the event table should be "Pavement" **and it is a line event.** The From-Measure should be BEGIN_MP and the To-Measure END_MP. Always the route identifier is ROUTE1. Click OK. Click the Source tab in the T/C.

___ **19.** Turn off Some_Routes. Make the Pavement Events layer bright red. Open its attribute table. Using Options, select those pavement events on Route 30000030, and show selected records. Reselect ("Yellow-fy") the record of the segment that runs from mile point 4 to mile point 7. Zoom in and look at it on the map. By inspecting visually, how many accidents would you say occurred on this stretch of road? (Some accidents may be extremely close together and appear as one. You may have to turn off "pavement events" to see the accident points, since the yellow selection line almost covers them up. Alternatively, you could make the accident point symbol bigger.) _____. In the Table, examine the RATING attribute, which is a value from 0 to 100, indicating the quality of the roadway. What is it? _____. What is the **range** of RATING values for all selected records? High _____. Low _____.

___ **20.** Show all records. Close the Some_Routes table if it is open. Save the map as Pittsburgh_ Routes in

   ___IGIS-Arc_*YourInitials*\Linear_Referencing.

Intersecting Route Events

Just as you could combine a set of polygons with another set of polygons to create a third set of polygons with appropriate (geo)graphic and attribute information, you can also combine the graphic and linearly referenced attributes of routes.

Suppose that someone suggests that there is a correlation between auto accidents and road conditions on Route 30000030: if the Rating is less than 75 on a section of the road the number of accidents (per mile) is greater. To test this, you want to combine the accident data with road condition data.

___ **21.** If necessary, start ArcMap with Pittsburgh_Routes, as saved in the previous step.

___ **22.** Open the attribute table of Some_Routes. What is the Shape_Length of 30000030 to one decimal place? _____ feet. Close the table. Under Selection, clear any selected features and records.

The length of a route, considered as a feature, is, of course, in the units of the coordinate system. However, as noted previously, the event measures are in miles. Just for the sake of confidence, let's compare one against the other.

___ **23.** Restore the attribute table of Pavement Events. Select the records where Route1 is 30000030. Display Selected Records.

Nowhere in the table is the length of each segment, but we can fix that, since we have the beginning and ending mile point number. With this few records (six), you could easily verify that there are no overlaps (a requirement for valid segmented data) and determine the total length in miles covered by the segments. But let's add a field to the table whose value is the segment length.

___ **24.** With Table Options, bring up the Add Field window. Call the new field Seg_Len. Make its type Float. Set the Precision (the number of digits possible) at 6. Set the Scale (the number of digits to the right of the decimal) at 2. Click OK.

_____ **25.** Right-click on the field name Seg_Len and click Field Calculator. Ignore any warning.

The segment length can be calculated by subtracting the beginning mile point from the ending mile point.

_____ **26.** Calculate Seg_Len as [END_MP] - [BEGIN_MP]. The calculation will take place only for selected records. Click OK. By looking at the table, verify that Seg_Len does indeed contain the positive difference between the beginning and ending points.

_____ **27.** Run Statistics on Seg_Len. What is the Sum? _____. Divide the length of the route in feet that you found previously by the number of feet in a mile (5280). The result should be reassuring. _____.

_____ **28.** Using Table Options > Select By Attributes (Method: Select From The Current Selection), obtain those segments of 30000030 that have RATING >75. How many miles are in these segments? _____.

_____ **29.** Since you know the total length of 30000030 >75 you can easily calculate the total of the segments that are <= 75. How many miles? _____.

The intersection of point events and line events has a lot in common with its polygonal counterpart. The graphic result will consist of those points and segments that occupy common space. The resulting attribute table will reflect both event tables. We will do an intersection operation for all route segments and all accident event points. (There are certainly more efficient ways to do this, as far as number of calculations is concerned, but this is the most straightforward.)

The process will be to intersect the accident and pavement event layers for the entire region to produce a new event table, consisting of records of pavement event segments where the segments contain accidents. This table will then be made into a (temporary) event layer and displayed. From that we will select those with ROUTE1 value 30000030.

_____ **30.** Display All Records of Pavement Events. Clear Selected Records. Close that table. Show ArcToolbox. In Linear Referencing Tools, right-click on Overlay Route Events and select Open.

_____ **31.** In the Overlay Route Events window, click the down arrow at the end of the Input Event text box. Select Pavement Events. The next four text boxes should be filled in for you. For the Overlay Event table, pick Accident Events in the same way as Pavement Events.

_____ **32.** Browse for where to put the Output Event Table to

___IGIS-Arc*YourInitials*\Linear_Referencing

and call it Accidents_and_Pavement.dbf. Accept the rest of the form. Click OK. You may have to wait a bit, since thousands of point events are being matched up with hundreds of segments.

What you have now is a table named Accidents_and_Pavement.dbf. Open it and explore it. As with most output from overlay operations there are lots of fields (from each of the contributing feature classes) and lots of records. To see the graphics, you must make a layer, using a tool in ArcToolbox.

____ **33.** In Linear Referencing Tools, right-click on Make Route Events Layer and select Open. In the Input Route Features text box, take the only choice offered: Some_Routes. The Route Identifier field is, of course, ROUTE1. The Input Event Table, from the drop-down menu, should be Accidents_and_Pavement. Again, the Route Identifier field is ROUTE1. The rest of the window should be filled in for you. Click OK.

____ **34.** Dismiss ArcToolbox. Turn off both Accident Events and Pavement Events. Open the Accidents_ and_Pavement Events table. Select the records with Route1 equal to 30000030. Show only those records. How many are there? _____.

____ **35.** Click on the gray box to the left of some record. Use Identify on the yellow dot that appears on the map. Notice the large number of fields and values in the Identify Results box. Dismiss the Identify Results window. Under Table Options, click Select By Attributes. In the window, under Method, pick Select From Current Selection. For the Select FROM expression, use RATING <= 75.

____ **36.** How many accidents occurred on the stretch of road where the rating is less than or equal to 75? _____. From your previous calculation, how many miles were involved? _____.

____ **37.** By using subtraction, determine of the number of accidents that occurred on segments with a RATING > 75. _____. From a previous step, copy the length of road involved: _____. Based on accidents per mile, what is your conclusion as to whether the condition of the road was related to the number of accidents? _____.

As with all the sections in this chapter, we have barely scratched the surface. There is the issue of calibrating routes, and, as with everything we do, editing is a major issue. To make use of linear referencing, you should refer to Esri's publication Linear Referencing in ArcGIS, which exists as a paper manual available from Esri or as a PDF file. Search the web for

Linear Referencing in ArcGIS PDF

In addition to a Quick Start Tutorial (from which the data for this section of this book are adapted), the topics covered there are Creating Route Data, Displaying and Querying Routes and Events, Editing Routes, and Creating and Editing Event Data.

What's Not Covered Here

While this is the end of the book it's hardly the end of GIS or ArcGIS capability. From looking at the extensions, you can get an idea of other capabilities of the software. Look at the possible extensions:

ArcScan

Geostatistical Analyst

Publisher

Schematics

Tracking Analyst

Your browser can tell you the capabilities of these packages.

You can also gain insight as to the depth of ArcGIS software by considering the plethora of toolbars we didn't consider:

ArcScan

COGO

Data Driven Pages

Distributed Geodatabase

Edit Vertices

Effects

Feature Cache

Feature Construction

Geocoding

Geodatabase History

Geometric Network Editing

GPS

Graphics

Image Classification

LAS Dataset

Parcel Editor

Raster Painting

Representation

Route Editing

Schematic

Schematic Editor

Schematic Network Analyst

Tablet

TIN Editing

Transform Parcels

Utility Network Analyst

Versioning

Esri also has other software packages:

Data Interoperability

Business Analyst

VBA

Workflow Manager

Defense Mapping

Nautical

Bathymetry

Aeronautical

Airports

Esri Roads and Highways

Of course, all this is in addition to the myriad menu options and tools we did not explore. It's probably safe to say that no one knows all the capabilities of this software. But is also probably safe to say that, with the GIS foundation you now have, you are well equipped to pursue these other capabilities, which are rooted in what you have already learned. Good luck. Have fun. Be productive.

Exercise 9-6 (Review)

Checking, Updating, and Organizing Your Fast Facts File

The Fast Facts File that you are developing should contain references to items in the following checklist. The checklist represents the abilities to use the software you should have gained upon completing Chapter 9.

❑ ___ The two Esri software packages that deal with 3-D spatial data are

❑ ___ To automatically spin the globe

❑ ___ To turn on 3D Analyst

❑ ___ The data set type that is naturally, truly three-dimensional is

❑ ___ Base Heights refers to

❑ ___ In 3D Analyst, drawing order is controlled by

❑ ___ Keyframes are

❑ ___ One way to produce animation is

❑ ___ To get to the animation controls

❑ ___ To calculate volumes

❑ ___ To make a TIN from line features

❑ ___ A viewshed is a scene that shows

❏ ___ To add your own data to ArcGlobe

❏ ___ To change the center of rotation in ArcGlobe from the center of the Earth to another target

❏ ___ When dealing with two data sets and topology, the one with the greater number indicating the rank is likely to move _____.

❏ ___ To erase features flagged as topology errors

❏ ___ An address locator is

❏ ___ Distances obtained directly from TIGER files are not useful because the coordinate system

❏ ___ TIGER files define location in two different ways, which are

❏ ___ To make an address locator

❏ ___ The reason for an offset in an address locator is

❏ ___ Network Analyst has capabilities that allow the user to determine

❏ ___ To use Network Analyst, one needs ArcMap version

❏ ___ Best route is defined as

❏ ___ To create a network location (called a stop)

❏ ___ You can manually reorder the stops by

❏ ___ The traveling salesman problem

❏ ___ To move a stop

❏ ___ Finding closest facilities is done by

❏ ___ Allocating geographic areas to facilities is done by

❏ ___ Routes are features that are composed of

❏ ___ To find a particular location on a particular route

❏ ___ A route event is

❏ ___ Just as polygons can contain lines, linear route events can contain

❏ ___ Intersecting route events means

❏ ___ Intersecting route events produces a table, which, in order to have its geographics displayed, must be converted to

Afterword: From Systems to Science

Michael F. Goodchild[1]

In the Preface, Michael Kennedy listed three objectives for this book, the last of which was to provide a "basis for the reader to go on to the advanced study of GIS or to the study of the newly emerging field of GIScience, which might be described as the scientific examination of the technology of GIS and the fundamental questions raised by GIS." This idea of a "science behind the systems" goes back to the early 1990s. The GIS software industry was flourishing, courses were being instituted in universities, and GIS was being adopted as an essential tool in a wide range of applications. Nevertheless several prominent academic geographers began asking awkward questions. If GIS was simply a tool, then why was so much attention being devoted to it? After all, universities didn't feel the need to teach courses in word processing or spreadsheets. GIS was clearly useful to powerful interests, but what was it doing for the poor and the marginalized? Wasn't GIS simply "nonintellectual expertise," a bag of tricks?

By the middle of the decade, a consensus began to emerge that there was indeed "more to it," that the use of GIS raised some very interesting and fundamental questions, and that there were important research issues to be resolved if the design of GIS was to be improved in the next generation. In addition to training in the latest version of GIS software, a course in GIS ought to be an education in some very basic principles that would still be true ten or twenty years into the future. The term "geographic information science" was widely adopted in the titles of books and degree programs and the names of journals and conferences, though other terms such as "geomatics" were also used, particularly outside the United States.

So what exactly are these questions, issues, and principles? Many of them have been raised or at least hinted at in the pages of this book, but here are some of the more important and challenging:

❏ The problem of uncertainty. Every GIS database attempts to represent some selected aspect of reality in the geographic world, but none succeeds exactly because reality is almost infinitely complex, but the storage capacity of a digital computer is always limited. Choices must always be made about how much detail to include, how much to generalize or approximate in the interests of creating a database at reasonable cost, and what to leave out. In short, a GIS database always leaves its users

[1]Professor Emeritus, Department of Geography, University of California, Santa Barbara, CA 93106-4060, USA. Phone +1 805 893 8049, FAX +1 805 893 3146, Email good@geog.ucsb.edu.

with some degree of uncertainty about the real world. How should that uncertainty be described and measured, and what impact does it have on the results of GIS analysis? Over the past two decades, a very valuable body of knowledge has accumulated on the answers to these questions.

❑   Representation choices. The reader has encountered several different approaches to the representation of geographic phenomena in GIS—shapefiles, coverages, geodatabases, TINs, rasters, vectors—and there are many others. Over 300 formats are recognized by software packages that specialize in conversions. Yet many types of phenomena are still difficult to capture in GIS. There are still only very rudimentary techniques for representing three-dimensional data, or time-dependent data on dynamic phenomena, or data on flows and interactions. Many GIScientists specialize in this area, developing new data models and formats that will likely form the foundation of future generations of GIS software.

❑   Public participation. GIS is complex technology, and it can take several courses at university level to master it. But if it is to be truly useful in everyday life it needs to be accessible to people who cannot afford to take the time to read books, sit through lectures, and work through complex lab exercises. A branch of GIScience called *public-participation GIS* (PPGIS) examines its use in public decision-making and in community planning. Instead of a single database, PPGIS allows each stakeholder to develop a database that expresses his or her own views. This is particularly important in decisions that involve indigenous cultures who may have entirely different ways of thinking about their local environments.

❑   Spatial data infrastructures. Geographic data is essential in many aspects of human activity, from day-to-day way-finding to disaster management and from logistics to forestry. Many of the feature classes that the reader will have encountered in this book are useful for many different purposes, and it makes sense if they can be effectively shared. The term spatial data infrastructure describes all of the arrangements society makes to produce and share geographic data. In the United States, the National Spatial Data Infrastructure was initiated under President Clinton in 1994 and has grown into a complex of committees, format standards, data warehouses, and training programs that together seek to improve access to geographic data. Similar strategies have been adopted in other countries, and research continues into better approaches to achieve these objectives and into the impediments that stand in the way.

These four are only a few of the topics that make up the research agenda of GIScience. An excellent source of more information is the Web site of the University Consortium for Geographic Information Science (www.ucgis.org), which has published perhaps the best-documented research agenda and also organizes conferences and workshops and offers scholarship programs for students. Another good source is the book *Geographic Information Systems and Science* by Longley, Goodchild, Maguire, and Rhind, also published by Wiley.

This book has covered a large amount of material, and its exercises will have brought the reader to an advanced state of expertise in today's technology. But what are the wider implications of today's technology, and where is it headed? How accurate are the results it produces, and what broader impacts will it have on society and the decisions it makes? These are all important questions, and they are the stuff of GIScience. Like all good books, this one will have succeeded if it leaves the reader with as many questions as it answers.

Index

Index

Index

Index

Index

Notes

Customer Note: If this book is accompanied by software, please read the following before opening the package.

This software contains files to help you utilize the models described in the accompanying book. By opening the package, you are agreeing to be bound by the following agreement:

This software product is protected by copyright and all rights are reserved by the author, John Wiley & Sons, Inc., or their licensors. You are licensing to use this software on a single computer. Copying the software to another medium or format for use on a single computer does not violate the U.S. Copyright Law. Copying the software for any other purpose is a violation of the U.S. Copyright Law.

This software product is sold as is without warranty of any kind, either express or implied, including but not limited to the implied warranty of merchantability and fitness for a particular purpose. Neither Wiley nor its dealers or distributors assumes any liability for any alleged or actual damages arising from the use of or the inability to use this software. (Some states do not allow the exclusion of implied warranties, so the exclusion may not apply to you.)